WOOD
Handbook

Basic Information on Wood
as a Material of Construction
with Data for Its Use in
Design and Specification

Prepared by
The Forest Products Laboratory
Forest Service
U. S. Department of Agriculture

AGRICULTURE HANDBOOK NO. 72

(Supersedes an unnumbered publication of the
same title and a later slightly revised edition)

1955

WILDSIDE PRESS

PREFACE

Forests, distinct from all their other services and benefits, supply a basic raw material—wood—which from the earliest times has furnished mankind with necessities of existence and with comforts and conveniences beyond number. To return maximum values, however, forests must not only be maintained in vigorous condition for the continuous production of timber crops but must be fully and profitably utilized.

As an aid to better and more efficient use of wood as a material of construction, this handbook was prepared by the Forest Products Laboratory, a unit of the research organization of the Forest Service, U. S. Department of Agriculture. The Laboratory, which was established in 1910, is maintained at Madison, Wis., in cooperation with the University of Wisconsin. It was the first and for several years the only institution in the world conducting general research on wood and its utilization. The vast accumulation of information that has resulted from its engineering and allied investigations of wood and wood products during the past 40 years is the chief basis for this handbook.

CONTENTS

STRUCTURE OF WOOD

BARK, WOOD, AND PITH

A cross section of a tree shows the following well-defined features in succession from the outside to the center: (1) Bark (fig. 1), which may be divided into (a) the outer, corky, dead part that varies greatly in thickness with different species and with age of trees, and (b) the thin, inner, living part; (2) wood, which in merchantable trees of

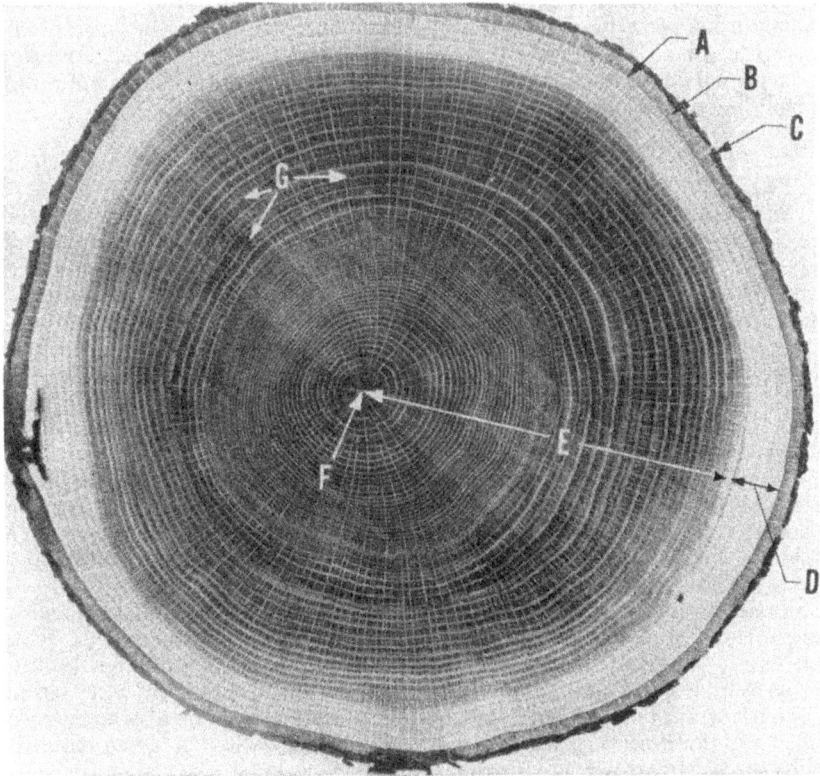

M 88620 F

FIGURE 1.—The tree trunk: *A*, Cambium layer (microscopic) is inside of inner bark and forms wood and bark cells. *B*, Inner bark is moist and soft. Carries prepared food from leaves to all growing parts of tree. *C*, Outer bark or corky layer is composed of dry dead tissue. Gives general protection against external injuries. *D*, Sapwood is the light-colored wood beneath the bark. Carries sap from roots to leaves. *E*, Heartwood (inactive) is formed by a gradual change in the sapwood. Gives the tree strength. *F*, Pith is the soft tissue about which the first wood growth takes place in the newly formed twigs. *G*, Wood rays connect the various layers from pith to bark for storage and transference of food.

1

most species is clearly differentiated into sapwood and heartwood; and (3) the pith, indicated by a small central core, darker in color, which represents primary growth formed when woody stems or branches elongate. Pith is not separated from the wood in rough lumber and timbers, but it is excluded from finished surfaces.

Most branches originate at the pith, and their bases are intergrown with the wood of the trunk as long as they are alive. These living branch bases constitute intergrown knots. After the branches die, their bases continue to be surrounded by the wood of the growing trunk. Such enclosed portions of dead branches constitute the loose or encased knots. After the dead branches drop off, the dead stubs become overgrown, and, subsequently, clear wood is formed. In a tree, the part containing intergrown knots comprises a cylinder, extending the entire length of the tree; the part containing loose knots forms a hollow cylinder, extending from the ground to the base of the green crown. Clear wood constitutes an outer cylinder covering overgrown branch ends. In second-growth trees, the clear zone and even the zone of loose knots may be absent.

GROWTH RINGS

Between the bark and the wood is a layer of thin-walled living cells, invisible without a microscope, called the cambium, in which all growth in thickness of bark and wood arises by cell division. No growth in either diameter or length takes place in wood already formed; new growth is purely the addition of new cells, not the further development of old ones. New wood cells are formed on the inside and new bark cells on the outside of the cambium. As the diameter of the woody trunk increases, the bark is pushed outward, and the outer bark layers become stretched, cracked, and ridged in patterns often characteristic of a species.

With many species in temperate climates, there is sufficient difference between the wood formed early and that formed late in a growing season to produce well-marked annual growth rings. The age of a tree at the stump or the age at any cross section of the trunk may be determined by counting these rings (fig. 2). If the growth of trees in diameter is interrupted by drought or defoliation by insects, more than one ring may be formed in the same season. In such an event, the inner rings usually do not have sharply defined boundaries and are termed false rings. Trees that have only very small crowns or that have lost most of their foliage accidentally may form only an incomplete growth layer, sometimes called a discontinuous ring, until the crown is restored.

Growth rings are most readily seen in such native ring-porous hardwood species as ash and oak, and among softwoods in the several species belonging to the yellow pine group owing to the sharp contrast between springwood and summerwood in these species. In some other species, such as water tupelo, sweetgum, and soft maple, differentiation of early and late growth is slight, and the annual growth rings are difficult to recognize. In some tropical regions, growth may be practically continuous throughout the year, and no well-defined annual rings are formed.

M 10712 F

FIGURE 2.—Cross section of a log showing annual growth rings. Each light ring is springwood. Each dark ring is summerwood.

SPRINGWOOD AND SUMMERWOOD

The inner part of the growth ring formed first in the growing season is called springwood or early wood, and the outer part formed later in the growing season, summerwood or late wood. Actual time of formation of these two parts of a ring may vary with environmental and weather conditions. Springwood is characterized by cells having relatively large cavities and thin walls. Summerwood cells have smaller cavities and thicker walls. The transition from springwood to summerwood may be gradual or abrupt, depending on the kind of wood and the growing conditions at the time it was formed. In some species, such as the maples, gums, and yellow-poplar, there is little difference in the appearance of the inner and outer parts of a growth ring.

When growth rings are prominent, as in the southern yellow pines and ring-porous hardwoods, springwood differs markedly from summerwood in physical properties. Springwood is lighter in weight, softer, and weaker than summerwood; it shrinks less across and more lengthwise along the grain of the wood. Because of the greater density of summerwood, the proportion of summerwood is sometimes used to judge the quality or strength of wood. This method is useful with such species as the southern yellow pines, Douglas-fir, and the ring-porous hardwoods, ash, hickory, and oak.

SAPWOOD AND HEARTWOOD

Sapwood contains living cells and has an active part in the life processes of the tree. It is located next to the cambium and functions in sap conduction and storage of food.

The sapwood layer may vary in thickness and in the number of growth rings contained in it. Sapwood commonly ranges from 1½ to 2 inches in radial thickness. In certain species, such as chestnut and black locust, the sapwood contains very few growth rings and sometimes does not exceed one-half inch in thickness. The maples, hickories, ashes, some of the southern yellow pines, and ponderosa pine may have sapwood 3 to 6 inches or more in thickness, especially in second-growth trees. As a rule, the more vigorously growing trees of a species have wider sapwood layers. Many second-growth trees of merchantable size consist mostly of sapwood.

Heartwood consists of inactive cells formed by changes in the living cells of the inner sapwood rings, presumably after their use for sap conduction and other life processes of the tree have largely ceased.

The cavities of heartwood also may contain deposits of various materials that frequently give much darker color to the heartwood. All heartwood, however, is not dark colored. Species in which heartwood does not darken to a great extent include the spruces (except Sitka spruce), hemlock, true firs, Port-Orford-cedar, basswood, cottonwood, and buckeye. In some species, such as the ashes, hickories, and certain oaks, the pores become plugged to a greater or lesser degree with ingrowths, known as tyloses, before the change to heartwood is completed.

Heartwood having pores tightly plugged by tyloses, as in white oak, supplies lumber suitable for tight cooperage. The infiltrations or materials deposited in the cells of heartwood usually make lumber cut from it more durable when used in exposed situations than that from sapwood. Unless treated, all sapwood is nondurable when exposed to weather.

There is no consistent difference between sapwood and heartwood in weight when dry and in strength. These properties are influenced more by growth conditions of the trees at the time the wood is formed than they are by the change from sapwood to heartwood. In some instances, as in redwood, western redcedar, and black locust, considerable amounts of infiltrated material may somewhat increase the weight of the wood and its resistance to crushing.

WOOD CELLS

Wood cells that make up the structural elements of wood are of various sizes and shapes and are quite firmly grown together. Dry wood cells may be empty or partly filled with deposits, such as gums and resins, or with tyloses. Many cells are considerably elongated and pointed at the ends; they are customarily called fibers or tracheids. The length of wood fibers is highly variable within a tree and among species. Hardwood fibers average about one twenty-fifth of an inch in length; softwood fibers range from one-eighth to one-third of an inch in length.

In addition to their fibers, hardwoods have cells of relatively large diameter known as vessels. These form the main arteries in the movement of sap. Softwoods do not contain special vessels for conducting sap longitudinally in the tree; this function is performed by the tracheids.

Both hardwoods and softwoods have cells that are oriented horizontally in the direction from the pith toward the bark. These cells conduct sap radially across the grain and are called rays or wood rays. The rays are most easily seen on quartersawed surfaces. They vary greatly in size in different species. In oaks and sycamores, the rays are conspicuous and add to the decorative features of the wood.

Wood also has other cells, known as wood parenchyma cells, that function mainly for the storage of food.

HARDWOODS AND SOFTWOODS

Native species of trees are divided into two classes—hardwoods, which have broad leaves, and softwoods or conifers, which have scalelike leaves, as the cedars, or needlelike leaves, as the pines. Hardwoods, except in the warmest regions, are deciduous and shed their leaves at the end of each growing season. Native softwoods, except cypress, tamarack, and larch, are evergreen. Softwoods are known also as conifers, because all native species of softwoods bear cones of one kind or another.

The terms "hardwood" and "softwood" have no direct application to the hardness or softness of the wood. In fact, such hardwood trees as cottonwood and aspen have softer wood than the white pines and true firs, and certain softwoods, such as longleaf pine and Douglas-fir, produce wood that is as hard as that of basswood and yellow-poplar.

CHEMICAL COMPOSITION OF WOOD

The four major components of wood are extractives, ash-forming minerals, lignin, and cellulose. The extractives are not part of the wood structure, but they contribute to the wood such properties as color, odor, taste, and resistance to decay. They include tannins, starch, coloring matter, oils, resins, fats, and waxes, and can be removed from the wood by neutral solvents, such as water, alcohol, acetone, benzene, and ether.

Ash-forming minerals, lignin, and cellulose make up the wood structure. Due to their uniform distribution throughout the wood structure, any one of these three components when isolated retains the microstructural pattern of the wood.

Cellulose, the most abundant constituent, comprises about 70 percent of wood. It may be subdivided into alpha-cellulose and hemicelluloses. Alpha-cellulose is the main basis of the useful products, such as paper, explosives, synthetic textiles, and plastics, obtained from cellulose. Further research is needed before complete and profitable utilization of the hemicelluloses can be realized.

Lignin comprises from 18 to 28 percent of the wood. It cements the structural units of wood together and thus imparts rigidity to the wood.

The ash-forming minerals, which are left as ash when the lignin and cellulose are burned, make up from 0.2 to 1.0 percent of the wood and comprise the nutrient plant-food elements of the tree.

CHARACTERISTICS OF SOME IMPORTANT COMMERCIAL WOODS GROWN IN THE UNITED STATES

The commercial forest land (table 1) of the United States includes about 622 million acres, of which about 8 percent is classed as old-growth sawtimber. Second-growth sawtimber comprises nearly 25 percent. Thus, sawtimber areas having sufficient volume for economic sawlog operation constitute slightly more than one-third of the total commercial forest area. The greater part of the remaining forest land supports immature second growth.

The principal hardwoods may be classified in about 25 groups of individual or closely related species and the softwoods in the same number of similar groups. In the case of the oaks and southern yellow pines, a considerable number of species are included in one group. A map showing distribution of the general forest types in the United States is shown in figure 3.

TABLE 1.—*Forest land area of the United States by class of forest, character of growth, and region and State*

Region and State	Total forest land	Commercial forest [1]						Reserved for parks [5]	Non-commercial forests [6]
		Total	Old-growth sawtimber [2]	Young-growth sawtimber [2]	Re-stocking [3]	Poorly stocked and denuded [4]		
	Thousand acres	*Thousand acres*	*Thousand acres*	*Thousand acres*	*Thousand acres*	*Thousand acres*	*Thousand acres*	*Thousand acres*
New England:								
Connecticut	1,907	1,900		403	1,264	233	7	
Maine	16,788	16,665	150	9,189	5,905	1,421	103	20
Massachusetts	3,310	3,297	2	930	1,842	523	12	1
New Hampshire [7]	4,848	4,682		1,808	2,248	626	25	141
Rhode Island	452	447		28	355	64	5	
Vermont	3,835	3,820	16	1,588	1,745	471		15
Total	31,140	30,811	168	13,946	13,359	3,338	152	177
Middle Atlantic:								
Delaware	442	442	2	200	195	45		
Maryland	2,742	2,722	4	848	1 495	375	18	2
New Jersey	2,348	2,329	2	423	1,494	410	17	2
New York	13,500	11,114	49	5,327	4,238	1,500	2,358	28
Pennsylvania	15,228	15,127	20	3,725	8,100	3,282	96	5
West Virginia	9,954	9,852	84	4,129	5,392	247		102
Total	44,214	41,586	161	14,652	20,914	5,859	2,489	139
Lake:								
Michigan	19,000	17,380	1,125	1,645	10,810	3,800	552	1,068
Minnesota	19,700	16,700	260	1,340	10,100	5,000	500	2,500
Wisconsin	17,000	16,265	125	1,975	9,075	5,090	86	649
Total	55,700	50,345	1,510	4,960	29,985	13,890	1,138	4,217

See footnotes at end of table.

TABLE 1.—*Forest land area of the United States by class of forest, character of growth, and region and State*—Continued

Region and State	Total forest land	Commercial forest [1]					Reserved for parks [5]	Non-commercial forests [6]
		Total	Old-growth saw-timber [2]	Young-growth saw-timber [2]	Re-stocking [3]	Poorly stocked and de-nuded [4]		
	Thousand acres	Thousand acres	Thousand acres	Thousand acres	Thousand acres	Thousand acres	Thousand acres	Thousand acres
Central:								
Illinois [7]	3,996	3,941		1,824	1,712	405	55	
Indiana	3,445	3,358		1,448	1,486	424	87	
Iowa	2,248	2,226		1,168	734	324	22	
Kansas	1,121	1,011	110	350	450	111	10	100
Kentucky	11,857	11,694	145	4,073	5,900	1,576	80	83
Missouri [7]	15,188	15,074		2,034	11,263	1,777	114	
Nebraska	1,112	987	75	246	486	180	25	100
North Dakota	621	470		110	220	140	15	136
Ohio	4,831	4,779		2,079	1,892	808	52	
Total	44,419	43,540	320	13,332	24,143	5,745	460	419
South Atlantic:								
North Carolina	18,400	17,997		9,588	6,584	1,825	332	71
South Carolina [7]	11,943	11,900		5,042	5,969	889	40	3
Virginia	14,832	14,377		7,138	6,462	777	271	184
Total	45,175	44,274		21,768	19,015	3,491	643	258
South:								
Alabama	18,878	18,800	80	10,071	5,167	3,482	61	17
Arkansas	20,036	19,928	180	12,647	5,004	2,097	16	92
Florida [7]	23,047	21,451		3,233	5,826	12,392	46	1,550
Georgia	21,432	21,107	150	11,252	5,300	4,405	227	98
Louisiana	16,196	16,169	189	9,852	2,902	3,226	10	17
Mississippi [7]	16,533	16,509		5,945	8,664	1,900	24	
Oklahoma	10,646	4,308	25	2,034	1,127	1,122	7	6,331
Tennessee	12,165	11,850		6,566	4,513	771	254	61
Texas	36,553	10,788	143	6,966	2,091	1,588	8	25,757
Total	175,486	140,910	767	68,566	40,594	30,983	653	33,923
Total East	396,134	351,466	2,926	137,224	148,010	63,306	5,535	39,133
Pacific Northwest:								
Douglas-fir subregion	29,145	26,027	9,348	3,810	7,434	5,435	1,032	2,086
Pine subregion	24,710	20,177	9,171	3,969	6,222	815	472	4,061
Total	53,855	46,204	18,519	7,779	13,656	6,250	1,504	6,147
Oregon	29,755	26,330	11,904	4,810	6,469	3,147	448	2,977
Washington	24,100	19,874	6,615	2,969	7,187	3,103	1,056	3,170
Total	53,855	46,204	18,519	7,779	13,656	6,250	1,504	6,147
California	45,515	16,405	8,400	2,497	3,451	2,057	705	28,405
North Rocky Mountain:								
Idaho	18,813	10,149	2,236	2,530	4,333	1,050	269	8,395
Montana	24,238	14,758	3,445	1,503	9,327	483	673	8,807
Total	43,051	24,907	5,681	4,033	13,660	1,533	942	17,202

See footnotes at end of table.

TABLE 1.—*Forest land area of the United States by class of forest, character of growth, and region and State*—Continued

Region and State	Total forest land	Commercial forest [1]						Reserved for parks [5]	Non-commercial forests [6]
		Total	Old-growth saw-timber [2]	Young-growth saw-timber [2]	Re-stocking[3]	Poorly stocked and denuded [4]		
	Thousand acres	*Thousand acres*	*Thousand acres*	*Thousand acres*	*Thousand acres*	*Thousand acres*	*Thousand acres*	*Thousand acres*
South Rocky Mountain:								
Arizona	19,538	2,815	1,969	600	221	25	332	16,391
Colorado	19,902	7,874	2,324	1,052	3,584	914	800	11,228
Nevada	4,720	98	31	9	58			4,622
New Mexico	20,001	3,465	1,834	360	872	399	710	15,826
South Dakota	1,979	1,765	392	388	770	215	65	149
Utah	8,494	1,530	987	251	276	16	41	6,923
Wyoming	8,878	3,102	1,244	404	1,317	47	1,869	3,997
Total	83,512	20,559	8,781	3,064	7,098	1,616	3,817	59,136
Total West	225,933	108,075	41,381	17,373	37,865	11,456	6,968	110,890
Total United States	622,067	459,541	44,307	154,597	185,875	74,762	12,503	150,023

Source: Forest Service, U. S. Department of Agriculture. Estimates as of Jan. 1, 1945, from a Reappraisal of the Forest Situation, except as noted in footnote 7.

[1] Land capable of producing timber of commercial quantity and quality and available now or prospectively for commercial use.

[2] Includes areas characterized by timber large enough for sawlogs (lumber) and in sufficient volume per acre for economic operation.

[3] Includes areas characterized by timber of cordwood size (diameter breast high 5 inches and larger) but too small for sawlogs and seedling and sapling areas on which at least 40 percent of the growing space is occupied by commercial species predominantly below poletimber size and below minimum volume per acre for sawtimber or for poletimber.

[4] Includes lands that do not qualify in any previous class.

[5] Commercially valuable land in parks, preserves, and other holdings withdrawn from timber use.

[6] Land chiefly valuable for purposes other than timber production, such as watershed protection, reducing soil erosion, and protecting wildlife.

[7] Data from the Forest Survey, 1946–49.

The following brief discussion of the principal localities of growth, characteristics, and uses of the main commercial species or groups of species will aid in selecting woods for specific purposes. More detailed information on the properties of these and other species is given in various tables throughout this handbook. The publications listed in references (*1*) to (*7*) [1] may also be helpful.

The common and botanical names given for the different species conform to the official nomenclature for trees of the U. S. Forest Service.

HARDWOODS

Most of the commercial hardwood species in the United States grow east of the Great Plains. The hardwoods of the West, which grow principally in California, Oregon, and Washington, amount to only about 6½ billion board-feet of the 315 billion total. Hardwoods

[1] Italic numbers in parentheses refer to Literature Cited, p. 35.

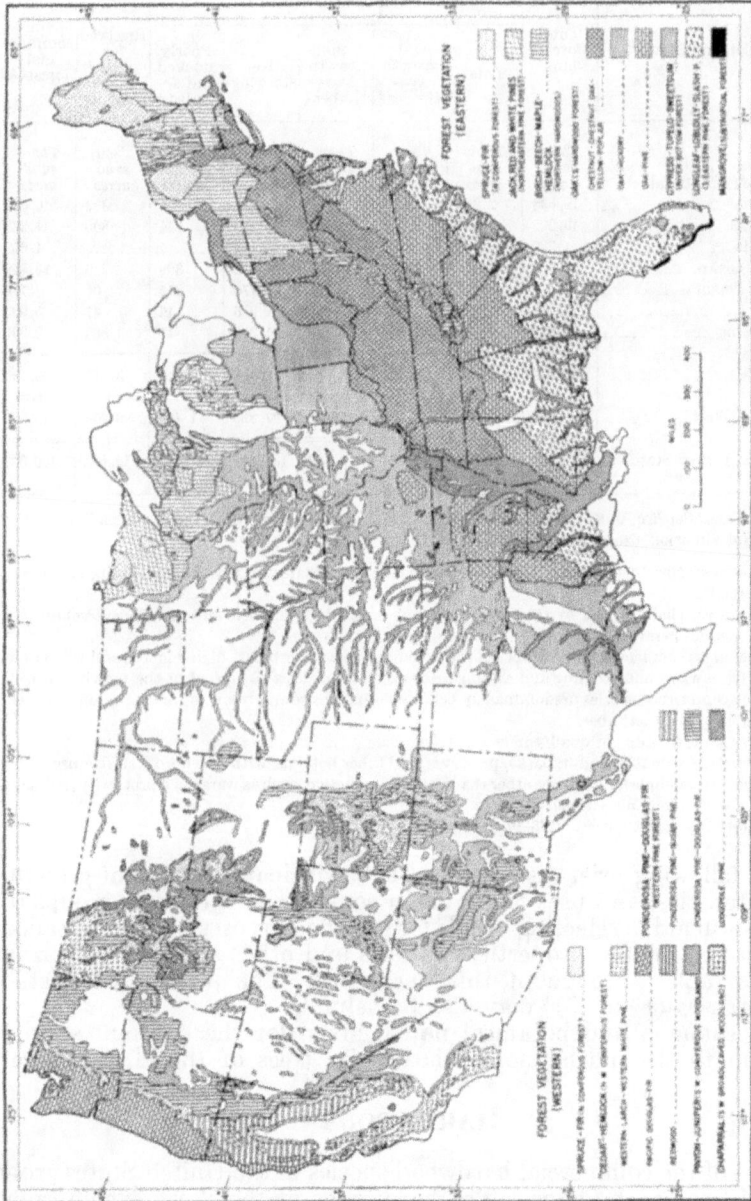

FIGURE 3.—Forest vegetation of the United States (adapted from Shantz and Zon's natural vegetation map of the United States in the Atlas of American Agriculture).

comprise less than 20 percent of the total timber resources of the United States.

The predominant hardwood species are the oaks, constituting about 7 percent of the total sawtimber volume. Other hardwoods make up 12 percent of the total.

Alder, Red

Red alder (*Alnus rubra*) grows along the Pacific coast between Alaska and California. It is used commercially along the coasts of Oregon and Washington and is the most abundant commercial hardwood species in these two States.

The wood of red alder varies from almost white to pale pinkish brown and has no visible boundary between heartwood and sapwood. It is moderately light in weight, (5) intermediate in most strength properties, but low in shock resistance. Red alder has relatively low shrinkage.

The principal use of red alder is for furniture, but it is also used for sash, doors, and millwork.

Ash

Important species of ash are white ash (*Fraxinus americana*), green ash (*F. pennsylvanica*), blue ash (*F. quadrangulata*), black ash (*F. nigra*), pumpkin ash (*F. profunda*), and Oregon ash (*F. latifolia*). The first five of these species grow in the eastern half of the United States. Oregon ash grows along the Pacific coast.

States with the greatest production of ash are Louisiana, Pennsylvania, Wisconsin, Michigan, Ohio, and Tennessee.

Commercial white ash is a group of species that consists mostly of white ash and green ash. Blue ash is also included in this group. Heartwood of commercial white ash is brown; the sapwood is light colored or nearly white. Second-growth trees have a large proportion of sapwood. Old-growth trees with little sapwood are scarce.

Second-growth commercial white ash is particularly sought because of the inherent qualities of this wood; it is heavy, strong, hard, stiff, and has high resistance to shock. Because of these qualities, such tough ash is used principally for handles, oars, vehicle parts, and sporting and athletic goods. Some handle specifications call for not less than 5 or more than 17 growth rings per inch for handles of the best grade. The addition of a weight requirement of 43 or more pounds a cubic foot at 12-percent moisture content will assure excellent material.

Oregon ash has somewhat lower strength properties than white ash, but it is used locally for the same purposes.

Black ash is important commercially in the Lake States. The wood of black ash and pumpkin ash runs considerably lighter in weight than that of commercial white ash. Ash trees growing in southern river bottoms, especially in areas that are frequently flooded for long periods, produce buttresses that contain relatively lightweight wood. Such wood is sometimes separated from tough ash when sold.

Ash wood of lighter weight, including black ash, is sold as cabinet ash, and is suitable for cooperage, furniture, and shipping containers. Some ash is cut into veneer.

Aspen

"Aspen" is a generally recognized name applied to bigtooth aspen (*Populus grandidentata*) and to quaking aspen (*P. tremuloides*). Aspen does not include balsam poplar (*P. balsamifera*) and the species of *Populus* that make up the group of cottonwoods. In lumber statistics of the U. S. Bureau of the Census, however, the term "cottonwood" includes all of the preceding species. Also, the lumber of aspens and cottonwood may be mixed in trade and sold either as poplar or cottonwood. The common term "popple" is sometimes applied to the aspens. The name "popple" or "poplar" should not be confused with yellow-poplar (*Liriodendron tulipifera*), also known in the trade as "poplar."

Aspen lumber is produced principally in the Northeastern and Lake States. There is some production in the Rocky Mountain States.

The heartwood of aspen is grayish white to light grayish brown. The sapwood is lighter colored and generally merges gradually into heartwood without being clearly marked. Aspen wood is usually straight grained with a fine, uniform texture. It is easily worked. Well-seasoned aspen lumber does not impart odor or flavor to foodstuffs.

The wood of aspen is lightweight and soft. It is low in strength, moderately stiff, moderately low in resistance to shock, and has a moderately high shrinkage.

Aspen is cut for lumber, boxes and crating, pulpwood, excelsior, matches, veneer, and miscellaneous turned articles.

Basswood

American basswood (*Tilia americana*) is the most important of the several native basswood species; next in importance is white basswood (*T. heterophylla*). Other species occur only in very small quantities. Because of the uniformity of the wood of the different species, no attempt is made to distinguish between them in lumber form. Other common names of basswood are linden, linn, and beetree.

Basswood grows in the eastern half of the United States from the Canadian provinces southward. Most basswood lumber comes from the Lake, Middle Atlantic, and Central States. In commercial usage, the term "white basswood" is used to specify white wood or sapwood of either species.

The heartwood of basswood is pale yellowish brown with occasional darker streaks. Basswood has wide, creamy-white or pale-brown sapwood that merges gradually into the heartwood. When dry, the wood is without odor or taste. It is soft and light in weight, has fine, even texture, and is straight grained and easy to work with tools. Shrinkage in width and thickness during drying is rated as large; however, basswood stays in place well and does not warp while in use.

Basswood lumber is used mainly in venetian blinds, sash and door frames, moldings, apiary supplies, woodenware, and boxes. Some basswood is cut for veneer, cooperage, excelsior, and pulpwood.

Beech, American

Only one species of beech, American beech (*Fagus grandifolia*), is native to the United States. The terms "red beech" or "red-heart beech" are applied to the darker colored heartwood and "white beech" or "white-heart beech" to the lighter colored heartwood.

Beech grows in the eastern one-third of the United States and adjacent Canadian provinces. Greatest production of beech lumber is in the Central and Middle Atlantic States.

Beech wood varies in color from nearly white sapwood to reddish-brown heartwood in some trees. Sometimes there is no clear line of demarcation between heartwood and sapwood. Sapwood may be 3 to 5 inches thick. The wood has little figure and is of close, uniform texture. It has no characteristic taste or odor.

The wood of beech is classed as heavy, hard, strong, high in resistance to shock, and highly adaptable for steam bending. Beech has large shrinkage and requires careful drying. It machines smoothly, wears well, and is rather easily treated with preservatives.

Largest amounts of beech go into flooring, furniture, handles, veneer, woodenware, containers, cooperage, and laundry appliances. When treated, it is suitable for railway ties.

Birch

The important species of birch are yellow birch (*Betula alleghaniensis*), sweet birch (*B. lenta*), and paper birch (*B. papyrifera*). Other birches of some commercial importance are river birch (*B. nigra*), gray birch (*B. populifolia*), and western paper birch (*B. papyrifera* var. *Commutata*).

Yellow birch, sweet birch, and paper birch grow principally in the Northeastern and Lake States. Yellow and sweet birch also grow along the Appalachian Mountains to northern Georgia. They are the source of most birch lumber and veneer.

Yellow birch has white sapwood and light reddish-brown heartwood. Sweet birch has light-colored sapwood and dark-brown heartwood tinged with red. Wood or yellow birch and sweet birch is heavy, hard, strong, and has good shock-resisting ability. The wood is fine and uniform in texture. Paper birch is lower in weight, softer, and lower in strength than yellow and sweet birch. Birch shrinks considerably during drying.

Yellow and sweet birch lumber and veneer go principally into the manufacture of furniture, boxes, baskets, crates, woodenware, cooperage, interior finish, and doors. Birch veneer goes into plywood used for flush doors, furniture, radio and television cabinets, aircraft, and other specialty uses. Paper birch is used for turned products, including spools, bobbins, and toys.

Cherry, Black

Black cherry (*Prunus serotina*) is sometimes known as cherry, wild black cherry, wild cherry, or choke cherry. It is the only native species of the genus *Prunus* of commercial importance for lumber production. It occurs scatteringly from southeastern Canada throughout the eastern half of the United States. Production is centered chiefly in the Middle Atlantic States.

The heartwood of black cherry varies from light to dark reddish brown and has a distinctive luster. The sapwood is narrow in old trees and nearly white. The wood has a fairly uniform texture and very satisfactory machining properties. It is moderately heavy. Black cherry is strong, stiff, moderately hard, and has high shock resistance and moderately large shrinkage. It stays in place well after seasoning.

Black cherry is used principally for furniture and for backing blocks on which electrotype plates are mounted. Other uses include burial caskets, woodenware novelties, patterns, and paneling in buildings and railway coaches. It has proved satisfactory for gunstocks.

Chestnut, American

American chesnut (*Castanea dentata*) is known also as sweet chestnut. Before chestnut was attacked by a blight, it grew in commercial quantities from New England to northern Georgia. Practically all standing chestnut has been killed by blight, and supplies come from dead timber. There are considerable quantities of standing dead chestnut in the Appalachian Mountains, which may be available for some time because of the great natural resistance to decay of its heartwood.

The heartwood of chestnut is grayish brown or brown and becomes darker with age. The sapwood is very narrow and almost white. The wood is coarse in texture, and the growth rings are made conspicuous by several rows of large, distinct pores at the beginning of each year's growth. Chestnut wood is moderately light in weight. It is moderately hard, moderately low in strength, moderately low in resistance to shock, and low in stiffness. It seasons well and is easy to work with tools.

Chestnut is used for poles, railway ties, furniture, caskets, boxes, crates, and core stock for veneer panels.

Cottonwood

Cottonwood includes several species of the genus *Populus*. Most important are eastern cottonwood (*P. deltoides* and varieties), also known as Carolina poplar and whitewood; swamp cottonwood (*P. heterophylla*), also known as cottonwood, river cottonwood, and swamp poplar; and black cottonwood (*P. trichocarpa*).

Eastern cottonwood and swamp cottonwood grow throughout the eastern half of the United States. Greatest production of lumber is in the Southern and Central States. Black cottonwood grows in the West Coast States and in western Montana, northern Idaho, and western Nevada.

The heartwood of the three cottonwoods, eastern, black, and swamp, is grayish white to light brown. The sapwood is whitish and merges gradually with the heartwood. The wood is comparatively uniform in texture, and generally straight grained. It is odorless when well seasoned.

Eastern cottonwood is moderately low in bending and compressive strength, moderately limber, moderately soft, and moderately low in ability to resist shock. Black cottonwood is slightly below eastern cottonwood in most strength properties. Both eastern and black cottonwood have moderately large shrinkage. Some cottonwood is difficult to work with tools because of fuzzy surfaces. Tension wood is largely responsible for this characteristic.

Cottonwood is used principally for lumber, veneer, pulpwood, excelsior, and fuel. The lumber and veneer go largely into boxes, crates, and baskets.

Elm

There are six species of elm in the United States: American elm (*Ulmus americana*), slippery elm (*U. rubra*), rock elm (*U. thomasii*), winged elm (*U. alata*), cedar elm (*U. crassifolia*), and September elm (*U. serotina*). American elm is also known as white elm, water elm, and gray elm; slippery elm as red elm; rock elm as cork elm or hickory elm; winged elm as wahoo; cedar elm as red elm or basket elm; and September elm as red elm.

American elm grows throughout the eastern half of the United States, except in higher elevations of the Appalachian Mountains. Slippery elm occupies about the same area, excepting the Atlantic Coastal Plain, most of Florida, and along the gulf coast. Rock elm occurs from New Hampshire to northern Tennessee and Nebraska. Winged elm grows from the Ohio Valley southward to the gulf and westward to Texas. Cedar elm extends from southern Arkansas and eastern Mississippi into Texas. September elm is most abundant in the central Mississippi Valley.

The sapwood of the elms is nearly white and the heartwood light brown, often tinged with red. The elms may be divided into two general classes, hard elm and soft elm, based on the weight and strength of the wood. Hard elm includes rock elm, winged elm, cedar elm, and September elm. American elm and slippery elm are the soft elms. Soft elm is moderately heavy, has high shock resistance, and is moderately hard and stiff. Hard elm species are somewhat heavier than soft elm. Elm has excellent bending qualities.

Production of elm lumber is chiefly in the Lake, Central, and Southern States.

Elm lumber is used principally in boxes, baskets, crates, and slack barrels; furniture; agricultural supplies and implements; caskets and burial boxes; and vehicles. For some uses the hard elms are preferred. Elm veneer is used for fruit, vegetable, and cheese boxes, baskets, and panels.

Hackberry

Hackberry (*Celtis occidentalis*) and sugarberry (*C. laevigata*) supply the lumber known in the trade as hackberry. Hackberry grows east of the Great Plains from Alabama, Georgia, Arkansas, and Oklahoma northward, except along the Canadian boundary. Sugarberry overlaps the southern part of the range of hackberry and grows throughout the Southern and South Atlantic States.

The sapwood of both species varies from pale yellow to greenish or grayish yellow. The heartwood is commonly darker. The wood resembles elm in structure.

Hackberry lumber is moderately heavy. It is moderately strong in bending, moderately weak in compression parallel to the grain, moderately hard to hard, high in shock resistance, but low in stiffness. It has moderately large to large shrinkage but keeps its shape well during seasoning.

Most hackberry is cut into lumber, with small amounts going into dimension stock and some into veneer. Most of it is used for furniture and some for containers.

Hickory, Pecan

Species of the pecan group include bitternut hickory (*Carya cordiformis*), pecan (*C. illinoensis*), water hickory (*C. aquatica*), and nutmeg hickory (*C. myristicaeformis*). Bitternut hickory grows throughout the eastern half of the United States. Pecan hickory grows from central Texas and Louisiana to Missouri and Indiana. Water hickory grows from Texas to South Carolina. Nutmeg hickory occurs principally in Texas and Louisiana.

The wood of pecan hickory resembles that of true hickory. It has white or nearly white sapwood, which is relatively wide, and somewhat darker heartwood. The wood is heavy and sometimes has very large shrinkage.

Heavy pecan hickory finds use in tool and implement handles and flooring. The lower grades are used in pallets.

Hickory, True

True hickories are found throughout most of the eastern half of the United States. The species most important commercially are shagbark (*Carya ovata*), pignut (*C. glabra*), shellbark (*C. laciniosa*), and mockernut (*C. tomentosa*).

The greatest commercial production of the true hickories is in the Middle Atlantic and Central States. The Southern and South Atlantic States produce nearly half of all hickory lumber.

The sapwood of hickory is white and usually quite thick, except in old, slowly growing trees. The heartwood is reddish. From the standpoint of strength, no distinction should be made between sapwood and heartwood having the same weight.

The wood of true hickory is very tough, heavy, hard, and strong, a combination not found in any other native commercial wood. Hickory shrinks considerably in drying.

About three-fourths of the hickory production is used for tool handles, which require high shock resistance. It is also used for ladder rungs, athletic goods, agricultural implements, dowels, gymnasium apparatus, poles, shafts, well pumps, and furniture.

A considerable quantity of lower grade hickory is not suitable, because of knottiness or other growth features and low density, for the special uses of high-quality hickory. It appears particularly useful for pallets, blocking, and similar items.

Holly, American

American holly (*Ilex opaca*) is sometimes called white holly, evergreen holly, and boxwood. The natural range of holly extends along the Atlantic coast, gulf coast, and Mississippi Valley.

Both heartwood and sapwood are white, the heartwood with an ivory cast. The wood has a uniform and compact texture; it is moderately low in strength when used as a beam or column and low in stiffness, but it is heavy and hard, and ranks high in shock resistance. It is readily penetrable to liquids and can be satisfactorily dyed. It works well, cuts smoothly, and is used principally for scientific and musical instruments, furniture inlays, and athletic goods.

Locust, Black

Black locust (*Robinia pseudoacacia*) is sometimes called yellow locust, white locust, green locust, or post locust. This species grows from Pennsylvania along the Appalachian Mountains to northern Georgia. It is also native to a small area in northwestern Arkansas. The greatest production of black locust timber is in Tennessee, Kentucky, West Virginia, and Virginia.

Locust has narrow, creamy-white sapwood. The heartwood, when freshly cut, varies from greenish yellow to dark brown. Black locust is very heavy, very hard, very high in resistance to shock, and ranks very high in strength and stiffness. It has moderately small shrinkage. The heartwood has high decay resistance.

Black locust is used extensively for round, hewed, or split mine timbers and for fence posts, poles, railroad ties, stakes, and fuel. An important product manufactured from black locust is insulator pins, a use for which the wood is well adapted because of its strength, decay resistance, and moderate shrinkage and swelling. Other uses are for rough construction, crating, ship treenails, and mine equipment.

Magnolia

Three species comprise commercial magnolia—southern magnolia (*Magnolia grandiflora*), sweetbay (*M. virginiana*), and cucumbertree (*M. acuminata*). Other names for southern magnolia are evergreen magnolia, magnolia, big laurel, bull bay, and laurel bay. Sweetbay is sometimes called swamp magnolia, or more often simply magnolia.

The natural range of sweetbay extends along the Atlantic and gulf coasts from Long Island to Texas, and that of southern magnolia from North Carolina to Texas. Cucumbertree grows from the Appalachians to the Ozarks northward to Ohio. Louisiana leads is production of magnolia lumber.

The sapwood of southern magnolia is yellowish white, and the heartwood is light to dark brown with a tinge of yellow or green. The wood, which has close, uniform texture and is generally straight grained, closely resembles yellow-poplar. It is moderately heavy, moderately low in shrinkage, moderately low in bending and compressive strength, moderately hard and stiff, and moderately high in shock resistance. Sweetbay is reported to be much like southern magnolia. The wood of cucumbertree is similar to that of yellow-poplar, and cucumbertree growing in the yellow-poplar range is not separated from that species on the market.

Magnolia lumber is used principally in the manufacture of furniture, boxes, venetian blinds, sash, doors, veneer, and millwork.

Maple

Commercial species of maple in the United States include sugar maple (*Acer saccharum*), black maple (*A. nigrum*), silver maple (*A. saccharinum*), red maple (*A. rubrum*), boxelder (*A. negundo*), and bigleaf maple (*A. macrophyllum*). Sugar maple is also known as hard maple, rock maple, sugar tree, and black maple; black maple as hard maple, black sugar maple, and sugar maple; silver maple as white maple, river maple, water maple, and swamp maple; red maple as soft maple, water maple, scarlet maple, white maple, and swamp maple; boxelder as ash-leaved maple, three-leaved maple, and cut-leaved maple; and bigleaf maple as Oregon maple.

Sugar maple grows from Maine to Minnesota, and southward to northern Georgia, Alabama, Louisiana, and Texas. Black maple occupies mainly a belt from New York through southern Michigan, southward to Kentucky, and westward through Iowa. Silver maple grows through most of eastern United States except the southern Atlantic and gulf coasts. Red maple grows east of the Great Plains and south to the Gulf of Mexico. Boxelder grows from Minnesota to Texas and eastward to the Middle Atlantic States. Bigleaf maple grows along the Pacific coast.

Maple lumber comes principally from the Middle Atlantic and Lake States, which together account for about two-thirds of the production.

The wood of sugar maple and black maple is known as hard maple; that of silver maple, red maple, and boxelder as soft maple. The sapwood of the maples is commonly white with a slight reddish-brown tinge. It is from 3 to 5 or more inches thick. Heartwood is usually light reddish brown, but sometimes is considerably darker. Hard maple has a fine, uniform texture. It is heavy, strong, stiff, hard, resistant to shock, and has large shrinkage. Sugar maple is generally straight grained. Soft maple is not so heavy as hard maple.

Maple is used principally for lumber, veneer, crossties, distillation wood, and pulpwood. A large proportion is manufactured into flooring, furniture, boxes and crates, shoe lasts, handles, woodenware, novelties, motor-vehicle parts, spools, and bobbins.

Oak (Red Oak Group)

Among the numerous species of red oaks in the United States, 10 have considerable commercial importance.

(1) Northern red oak (*Quercus rubra*), also known as eastern red oak, grows in the eastern half of the United States to the lower Mississippi Valley, Florida, and the Atlantic Coastal Plain. It is the most important lumber tree of the red oak group. (2) Scarlet oak (*Q.*

coccinea) grows in the eastern third of the United States, except the southern border States. (3) Shumard oak (*Q. shumardii*), also known as Schneck oak, Texas oak, and southern red oak, grows chiefly along the Atlantic and gulf coasts. (4) Pin oak (*Q. palustris*), also known as swamp oak, grows principally in the central Mississippi Valley. (5) Nuttall oak (*Q. nuttallii*) grows in the lower Mississippi Valley region from Missouri southward and from Alabama to Texas.

(6) Black oak (*Quercus velutina*), also known as yellow oak, grows in the eastern half of the United States to northern Florida. (7) Southern red oak (*Q. falcata*) grows from New Jersey to Missouri, Arkansas, and Texas. (8) Water oak (*Q. nigra*) grows in the South Atlantic and Gulf States from Maryland to Texas. (9) Laurel oak (*Q. laurifolia*) grows in the South Atlantic and Gulf Coastal Plains from Maryland to Louisiana. (10) Willow oak (*Q. phellos*) grows along the Atlantic and gulf coasts and the lower Mississippi Valley.

Most red oak lumber comes from the Southern States, the southern mountain regions, and the Atlantic Coastal Plain.

The sapwood is nearly white and usually 1 to 2 inches thick. The heartwood is brown with a tinge of red. Sawed lumber of red oak cannot be separated by species on the basis of the characteristics of the wood alone. Red oak lumber can be separated from white oak by the number of pores in summerwood and because, as a rule, it lacks the hairlike growth known as tyloses in the pores. The open pores of the red oaks make these species unsuitable for tight cooperage. Quartersawed lumber of the oaks is distinguished by the broad and conspicuous rays, which add to its attractiveness.

Wood of the red oaks is heavy. Rapidly grown second-growth oak is generally harder and tougher than finer textured old-growth timber. The red oaks have fairly large shrinkage in drying.

The red oaks are largely cut into lumber, railroad ties, mine timbers, fence posts, veneer, and fuelwood. Ties, mine timbers, and fence posts require preservative treatment for satisfactory service. Red oak lumber is remanufactured into flooring, furniture, general millwork, boxes and crates, agricultural implements, caskets and coffins, woodenware, and handles. It is also used in railroad cars and boats.

Oak (White Oak Group)

There are nine commercially important species of the white oak group, and all grow mainly in the eastern United States.

(1) White oak (*Quercus alba*) grows throughout the eastern half of the United States and adjacent Canada. It is the most important lumber tree of the white oak group. (2) Chestnut oak (*Q. prinus*), also known as rock chestnut oak or rock oak, grows from southern Vermont and New Hampshire, southward along the Appalachian Mountains to central Georgia and Alabama. (3) Post oak (*Q. stellata*) grows throughout the eastern half of the United States from southern New England to the Great Plains. (4) Overcup oak (*Q. lyrata*), also known as swamp white oak, grows in the Atlantic Coastal States and westward to Texas through southern Illinois and Indiana. (5) Swamp chestnut oak (*Q. michauxii*), also known as basket oak

and cow oak, grows along the Atlantic coast and westward to Texas through southern Illinois and Indiana.

(6) Bur oak (*Quercus macrocarpa*) grows mainly from New York to Montana and southward through Kentucky to Texas.   (7) Chinkapin oak (*Q. muehlenbergii*) grows from New York, southern Michigan, and southern Minnesota southward to the Gulf of Mexico, except for the Atlantic Coastal Plain.   (8) Swamp white oak (*Q. bicolor*) grows from southern Maine through the Central States to the Great Plains.   (9) Live oak (*Q. virginiana*) is limited to the Atlantic Coastal Plain, Florida, and the gulf coast.

White oak lumber comes chiefly from the South, South Atlantic, and Central States, including the southern Appalachian area.

The heartwood of the white oaks is generally grayish brown, and the sapwood, which is from 1 to 2 or more inches thick, is nearly white. The pores of the heartwood of white oaks are usually plugged with the hairlike growth known as tyloses.   These tend to make the wood inpenetrable by liquids, and for this reason most white oaks are suitable for tight cooperage.   Chestnut oak lacks tyloses in many of its pores.

The wood of white oak is heavy, averaging somewhat higher in weight than that of the red oaks.   The heartwood has moderately good decay resistance.

White oaks are used for lumber, railroad ties, cooperage, mine timbers, fence posts, veneer, fuelwood, and many other products. High-quality white oak is especially sought for tight cooperage.   Live oak is considerably heavier and stronger than the other oaks, and was formerly used extensively for ship timbers.   An important use of white oak is for planking and bent parts of ships and boats, heartwood often being specified because of its decay resistance.   It is also used for flooring, agricultural implements, railroad cars, furniture, doors, millwork, and many other items.

Sweetgum

Sweetgum (*Liquidambar styraciflua*) grows from southwestern Connecticut westward into Missouri and southward to the gulf. Lumber production is almost entirely from the Southern and South Atlantic States.

The lumber from sweetgum is usually divided into two classes— sap gum, the light-colored wood from the sapwood, and red gum, the reddish-brown heartwood.

Sweetgum has interlocked grain, a form of cross grain, and must be carefully dried.   The interlocked grain causes a ribbon stripe, however, that is desirable for inside finish and furniture.   The wood is rated as moderately heavy and hard.   It is moderately strong, moderately stiff, and moderately high in shock resistance.

Sweetgum is used principally for lumber, veneer, plywood, slack cooperage, railroad ties, fuel, and pulpwood.   The lumber goes principally into boxes and crates, furniture, radio and phonograph cabinets, interior trim, and millwork.   Sweetgum veneer and plywood are used for boxes, crates, baskets, and interior woodwork.

Sycamore, American

American sycamore (*Platanus occidentalis*) is also known as sycamore and sometimes as buttonwood, buttonball tree, and planetree. Sycamore grows from Maine to Nebraska, Texas, and northern Florida. In the production of sycamore lumber, the Central States rank first.

The heartwood of sycamore is reddish brown; sapwood is lighter in color and from 1½ to 3 inches thick. The wood has a fine texture and interlocked grain. It shrinks moderately in drying. Sycamore wood is moderately heavy, moderately hard, moderately stiff, moderately strong, and has good resistance to shock.

Sycamore is used principally for lumber, veneer, railroad ties, cooperage, fence posts, and fuel. Sycamore lumber is used for furniture, boxes (particularly small food containers), flooring, handles, and butcher's blocks. Veneer is used for fruit and vegetable baskets.

Tupelo

The tupelo group includes water tupelo (*Nyssa aquatica*), also known as tupelo gum, swamp tupelo, and gum; black tupelo (*N. sylvatica*), also known as blackgum and sour gum; swamp tupelo (*N. sylvatica* var. *biflora*), also known as swamp blackgum, blackgum, tupelo gum, and sour gum; and Ogeechee tupelo (*N. ogeche*), also known as sour tupelo, gopher plum, tupelo, and Ogeechee plum.

All except black tupelo grow principally in the southeastern United States. Black tupelo grows in the eastern United States from Maine to Texas and Missouri. About two-thirds of the production of tupelo lumber is from the Southern States.

Wood of the different tupelos is quite similar in appearance and properties. Heartwood is light brownish gray and merges gradually into the lighter colored sapwood, which is generally several inches wide. The wood has fine, uniform texture and interlocked grain. Tupelo wood is rated as moderately heavy. It is moderately strong, moderately hard and stiff, and moderately high in shock resistance. Buttresses of trees growing in swamps or flooded areas contain wood that is much lighter in weight than that from upper portions of the same trees. For some uses, as in the case of buttressed ash trees, this wood should be separated from the heavier wood to assure material of uniform strength. Because of interlocked grain, tupelo lumber requires care in drying.

Tupelo is cut principally for lumber, veneer, pulpwood, and some railroad ties and cooperage. Lumber goes into boxes, crates, baskets, and furniture.

Walnut, Black

Black walnut (*Juglans nigra*) is also known as American black walnut. Its natural range extends from Vermont to the Great Plains and southward into Louisiana and Texas. About three-quarters of the walnut timber is produced in the Central States.

The heartwood of black walnut varies from light to dark brown; the sapwood is nearly white and up to 3 inches wide in open-grown trees.

Black walnut is normally straight grained, easily worked with tools, and stays in place well. It is heavy, hard, strong, stiff, and has good resistance to shock. Black walnut wood is well suited for natural finishes.

The outstanding use of black walnut is for furniture. Other important uses are gunstocks, cabinets, and interior finish. It is used either as solid wood or as plywood.

Willow, Black

Black willow (*Salix nigra*) is the most important of the many willows that grow in the United States. It is the only one to supply lumber to the market under its own name.

Black willow is most heavily produced in the Mississippi Valley from Louisiana to southern Missouri and Illinois.

The heartwood of black willow is grayish brown or light reddish brown, frequently containing darker streaks. The sapwood is whitish to creamy yellow. The wood of black willow is uniform in texture, with somewhat interlocked grain. The wood is light in weight. It has exceedingly low strength as a beam or post and is moderately soft and moderately high in shock resistance. It has moderately large shrinkage.

Willow is cut principally into lumber. Small amounts are used for slack cooperage, veneer, excelsior, charcoal, pulpwood, artificial limbs, and fence posts. Black willow lumber is remanufactured principally into boxes, baskets, crates, caskets, and furniture. Willow lumber is suitable for roof and wall sheathing, subflooring, and studding.

Yellow-Poplar

Yellow-poplar (*Liriodendron tulipifera*) is also known as poplar, tulip poplar, tulipwood, and hickory poplar. Sapwood from yellow-poplar is sometimes called white poplar or whitewood.

Yellow-poplar grows from Connecticut and New York southward to Florida and westward to Missouri. The greatest commercial production of yellow-poplar lumber is in the South.

Yellow-poplar sapwood is white and frequently several inches thick. The heartwood is yellowish brown, sometimes streaked with purple, green, black, blue, or red. These colorations do not affect the physical properties of the wood. The wood is generally straight grained and comparatively uniform in texture. Old-growth timber is moderately light in weight and is reported as being moderately low in bending strength, moderately soft, and moderately low in shock resistance. It has moderately large shrinkage when dried from a green condition but is not difficult to season and stays in place well after seasoning.

Much of the second-growth yellow-poplar is heavier, harder, and stronger than old growth. Selected trees produce wood heavy enough for gunstocks. Lumber goes mostly into furniture, interior finish, siding, core stock for plywood, radio cabinets, and musical instruments. Boxes and crates are made from lower grade stock. Yellow-poplar plywood is used for finish, furniture, piano cases, and various other special products. Yellow-poplar is used also for pulpwood, excelsior, and slack-cooperage staves.

Lumber from the cucumbertree (*Magnolia acuminata*) sometimes may be included in shipments of yellow-poplar because of its similarity.

SOFTWOODS

Most of our softwood timber is west of the Great Plains, and nearly three-fourths of our eastern softwood stands are in the South Atlantic and Southern States. Western softwood timber stands of Washington and Oregon contain more than three-fifths of the western softwood timber, those of California about one-fifth, and another one-fifth is scattered along the mountains from Idaho and Montana southward.

Alaska-Cedar

Alaska-cedar (*Chamaecyparis nootkatensis*) grows in the Pacific coast region of North America from southeastern Alaska southward through Washington to southern Oregon. In Washington and Oregon, it is confined to the west side of the Cascade Mountains, usually above an elevation of 2,000 feet. It reaches its best development along the coast and on the nearby islands of southern Alaska and British Columbia.

The heartwood of Alaska-cedar is bright, clear yellow. The sapwood is narrow, white to yellowish, and hardly distinguishable from the heartwood. The wood is fine textured and generally straight grained. It is moderately heavy, moderately strong and stiff, moderately hard, and moderately high in resistance to shock. Alaska-cedar shrinks little in drying, stays in place well after seasoning, and the heartwood is very resistant to decay. The wood has a mild, unpleasant odor.

Alaska-cedar is used locally for interior finish, furniture, small boats, cabinetwork, and novelties.

Baldcypress

Baldcypress (*Taxodium distichum*) is commonly known as cypress, also as southern cypress, red cypress, yellow cypress, and white cypress. Commercially, the terms "tidewater red cypress," "gulf cypress," "red cypress (coast type)," and "yellow cypress (inland type)" are frequently used.

Baldcypress grows along the Atlantic Coastal Plain from Delaware to Florida, westward through the gulf coast region to the Mexican border in Texas, and up the Mississippi Valley to southern Indiana. The heaviest stands are found in the extensive swamps of the lower Mississippi Valley and in Florida. About one-half of the cypress lumber comes from the Southern States and one-fourth from the South Atlantic States.

The sapwood of baldcypress is narrow and nearly white. The color of the heartwood varies widely, ranging from light yellowish brown to dark brownish red, brown, or chocolate. The wood is moderately heavy, moderately strong, and moderately hard, and the heartwood is one of our most decay-resistant woods. Shrinkage is moderately small, but somewhat greater than that of cedar and less than that of southern yellow pine.

Frequently the wood of certain cypress trees contains pockets or localized areas that have been attacked by a fungus. Such wood is known as pecky cypress. The decay caused by this fungus is arrested when the wood is cut into lumber and dried. Pecky cypress therefore is durable and useful where appearance is not important and water-tightness is unnecessary.

Cypress is used principally for building construction, especially where resistance to decay is required. It is used for beams, posts, and other members in docks, warehouses, factories, bridges, and heavy construction.

It is well suited for siding and porch construction. It is also used for caskets, burial boxes, sash, doors, blinds, and general millwork, including interior trim and paneling. Other uses are in tanks, vats, ship and boat building, refrigerators, railroad-car construction, greenhouse construction, cooling towers, and stadium seats. It is also used for railroad ties, poles, piling, shingles, cooperage, and fence posts.

Douglas-fir

Douglas-fir (*Pseudotsuga menziesii* and var. *glauca*) is also known locally as red fir, Douglas spruce, and yellow fir.

The range of Douglas-fir extends from the Rocky Mountains to the Pacific coast and from Mexico to central British Columbia. Most of the Douglas-fir production comes from the coast States of Oregon, Washington, and California, principally Oregon, and some from Idaho and Montana.

Sapwood of Douglas-fir is narrow in old-growth trees but may be as much as 3 inches wide in second-growth trees of commercial size. Fairly young trees of moderate to rapid growth have reddish heartwood and are called red fir. Very narrow-ringed wood of old trees may be yellowish brown and is known on the market as yellow fir.

The wood of Douglas-fir varies widely in weight and strength. When lumber of high strength is needed for structural uses, density can be determined by applying the density rule. This rule uses percentage of summerwood and rate of growth as a basis.

Douglas-fir is used mostly for building and construction purposes in the form of lumber, timbers, piling, and plywood. Considerable quantities go into fuel, railroad ties, cooperage stock, mine timbers, and fencing. Douglas-fir lumber is used in the manufacture of various products, including sash, doors, general millwork, railroad-car construction, boxes, and crates. Small amounts are used for flooring, furniture, ship and boat construction, wood pipe, and tanks.

Firs, True (Eastern Species)

Two of the true firs, balsam fir (*Abies balsamea*) and Fraser fir (*A. fraseri*), grow in the East. In the United States, balsam fir grows principally in New England, New York, Pennsylvania, and the Lake States. Fraser fir grows in the Appalachian Mountains of Virginia, North Carolina, and Tennessee.

The wood of the true firs, both eastern and western species, is creamy white to pale brown. Heartwood and sapwood are generally indistinguishable. The similarity of wood structure in the true firs makes it impossible to distinguish the species by an examination of the wood alone.

Balsam fir is rated as light in weight, low in bending and compressive strength, moderately limber, soft, and low in resistance to shock.

The eastern firs are used mainly for pulpwood, although there is some lumber produced from them, especially in New England and the Lake States.

Firs, True (Western Species)

Six commercial species make up the western true firs: Subalpine fir (*Abies lasiocarpa*), California red fir (*A. magnifica*), grand fir (*A. grandis*), noble fir (*A. procera*), Pacific silver fir (*A. amabilis*), and white fir (*A. concolor*).

White fir grows from the Rocky Mountains to the Pacific coast. Subalpine fir grows at high altitudes in the Rocky Mountain region and the Cascade Mountains of Oregon and Washington. Pacific silver fir is found in Oregon and Washington. Grand fir's range is western Montana, northern Idaho, northeastern Oregon, and along the coast from Washington to northern California. Noble fir grows in the mountains of northwestern Washington, western Oregon, and northern California. California red fir is limited to the mountains of southwestern Oregon and northern and eastern California.

The western firs are light in weight, but, with the exception of grand fir, have somewhat higher strength properties than balsam fir. Shrinkage of the wood is rated from small to moderately large.

The western true firs are largely cut for lumber and marketed as white fir throughout the Western States. Lumber of the western true firs goes principally into building construction, boxes and crates, planing-mill products, sash, doors, and general millwork. In small house construction, the lumber is used for framing, subflooring, and sheathing. A considerable amount goes into boxes and crates. High-grade lumber from noble fir is used mainly for interior finish, moldings, siding, and sash and door stock. Some of the best material is suitable for aircraft construction. Other special and exacting uses of noble fir are for venetian blinds and ladder rails.

Hemlock, Eastern

Eastern hemlock (*Tsuga canadensis*) grows from New England southward along the Appalachian Mountains to northern Alabama and Georgia, and in the Lake States. Other names are Canadian hemlock and hemlock spruce.

The production of hemlock lumber is divided fairly evenly between the New England States, the Middle Atlantic States, and the Lake States. North Carolina, South Carolina, and Virginia also produce considerable amounts.

The heartwood of eastern hemlock is pale brown with a reddish hue. The sapwood is not distinctly separated from the heartwood but may be lighter in color. The wood is coarse and uneven in texture (old trees tend to have considerable shake); it is moderately light in weight, moderately hard, moderately low in strength, moderately limber, and moderately low in shock resistance.

Eastern hemlock is used principally for lumber and pulpwood. The lumber is used largely in building construction for framing, sheathing, subflooring, and roof boards, and in the manufacture of boxes and crates.

Hemlock, Western

Western hemlock (*Tsuga heterophylla*) is also known by several other names, including west coast hemlock, hemlock spruce, western hemlock spruce, western hemlock fir, Prince Albert fir, gray fir, silver fir, and Alaska pine. It grows along the Pacific coast of Oregon and Washington and in the northern Rocky Mountains, north to Canada and Alaska.

The heartwood and sapwood of western hemlock are almost white with a purplish tinge. The sapwood, which is sometimes lighter in color, is generally not more than 1 inch thick. The wood contains small, sound, black knots that are usually tight and stay in place. Dark streaks often found in the lumber and caused by hemlock bark maggots as a rule do not reduce strength.

Western hemlock is moderately light in weight and moderately low in strength. It is moderate in its hardness, stiffness, and shock resistance. It has moderately large shrinkage, about the same as Douglas-fir. Green hemlock lumber contains considerably more water than Douglas-fir, but it is comparatively easy to kiln-dry.

Western hemlock is used principally for pulpwood and lumber. The lumber goes largely into building material, such as sheathing, siding, subflooring, joists, studding, planking, and rafters. Considerable quantities are used in the manufacture of boxes, crates, and flooring, and smaller amounts for refrigerators, furniture, and ladders.

Incense-Cedar

Incense-cedar (*Libocedrus decurrens*) grows in California, southwestern Oregon, and to a small extent in Nevada. Most of the incense-cedar lumber comes from the northern half of California and the remainder from southern Oregon.

Sapwood of incense-cedar is white or cream colored, and the heartwood is light brown, often tinged with red. The wood has a fine, uniform texture and a spicy odor. Incense-cedar is light in weight, moderately low in strength, soft, low in shock resistance, and low in stiffness. It has small shrinkage and is easy to season with little checking or warping.

Incense-cedar is used principally for lumber, fence posts, and ties. Nearly all the high-grade lumber is used for pencils and venetian blinds. Some is used for chests and toys. Much of the incense-cedar lumber is more or less pecky; that is, it contains pockets or areas of disintegrated wood caused by advanced stages of localized decay in the living tree. There is no further development of peck once the lumber is seasoned. This lumber is used locally for rough construction where cheapness and decay resistance are important. Because of its resistance to decay, incense-cedar is well suited for fence posts. Other products are railroad ties, poles, and split shingles.

Larch, Western

Western larch (*Larix occidentalis*) grows in western Montana, northern Idaho, northeastern Oregon, and on the eastern slope of the Cascade Mountains in Washington. It is found at elevations of 2,000 to 7,000 feet. About two-thirds of the lumber of this species is produced in Idaho and Montana and one-third in Oregon and Washington.

The heartwood of western larch is yellowish brown and the sapwood yellowish white. The sapwood is generally not more than 1 inch thick. The wood is stiff, moderately strong and hard, moderately high in shock resistance, and moderately heavy. It has moderately large shrinkage. The wood is usually straight grained, splits easily, and is subject to ring shake. Knots are common but small and tight.

Western larch is used mainly in building construction for rough dimension, small timbers, planks, and boards and for railroad ties and mine timbers. It is used also for piling, poles, and posts. Some high-grade material is manufactured into interior finish, flooring, sash, and doors.

Pine, Eastern White

Eastern white pine (*Pinus strobus*) grows in the United States from Maine southward along the Appalachian Mountains to northern Georgia and in the Lake States. It is also known as white pine, northern white pine, Weymouth pine, and soft pine.

Lumber production of eastern white pine is confined largely to the New England States, which produce about one-half the total. About one-third comes from the Lake States and most of the remainder from the Middle Atlantic and South Atlantic States.

The heartwood of eastern white pine is light brown, often with a reddish tinge. It turns considerably darker on exposure. The wood has comparatively uniform texture, and is straight grained. It is easily kiln-dried, has small shrinkage, and ranks high in ability to stay in place. It is also easy to work and can be readily glued.

Eastern white pine is light in weight, moderately soft, moderately low in strength, and low in resistance to shock.

Practically all eastern white pine is converted into lumber, which is put to a great variety of uses. The largest proportion, which is mostly second-growth knotty lumber of the lower grades, goes into boxes. High-grade lumber goes into patterns for castings. Other important uses are sash, doors, furniture, trim, knotty finish, caskets and burial boxes, shade and map rollers, toys, and dairy and poultry supplies.

Pine, Jack

Jack pine (*Pinus banksiana*), sometimes known as scrub pine, gray pine, or black pine in the United States, grows naturally in the Lake States and in a few scattered areas in New England and northern New York. In lumber, jack pine is not separated from the other pines with which it grows, including red pine and eastern white pine.

The sapwood of jack pine is nearly white, and the heartwood is light brown to orange. The sapwood may make up one-half or more of the volume of a tree. The wood has a rather coarse texture and is somewhat resinous. It is moderately light in weight, moderately low in bending strength and compressive strength, moderately low in shock resistance, and low in stiffness. It also has moderately small shrinkage. Lumber from jack pine is generally knotty.

Jack pine is used for pulpwood, box lumber, and fuel. Less important uses include railroad ties, mine timber, slack cooperage, poles, and posts.

Pine, Lodgepole

Lodgepole pine (*Pinus contorta*), also known as knotty pine, black pine, spruce pine, and jack pine, grows in the Rocky Mountain and Pacific coast regions as far northward as Alaska. The cut of this species comes largely from the central Rocky Mountain States; other producing regions are Idaho, Montana, Oregon, and Washington.

The heartwood of lodgepole pine varies from light yellow to light yellow brown. The sapwood is yellow or nearly white. The wood is generally straight grained with narrow growth rings.

The wood is moderately light in weight, fairly easy to work, and has moderately large shrinkage. Lodgepole pine rates as moderately low in strength, moderately soft, moderately stiff, and moderately low in shock resistance.

Lodgepole pine is used for lumber, mine timbers, railroad ties, and poles. Less important uses include posts and fuel. It is being used in increasing amounts for siding, finish, and flooring.

Pine, Pitch

Pitch pine (*Pinus rigida*) grows from Maine and northern New York south along the mountains to eastern Tennessee and northern Georgia. The heartwood is brownish red and resinous; the sapwood is thick and light yellow. The wood of pitch pine is medium heavy to heavy, medium strong, medium stiff, medium hard, and medium high in shock resistance. Its shrinkage is medium small to medium large. It is used for lumber and fuel.

Pine, Pond

Pond pine (*Pinus serotina*) grows in the coast region from New Jersey to Florida. It occurs in small groups or singly mixed with other pines on low flats. The wood is heavy, coarse-grained, and resinous, with dark, orange-colored heartwood and thick, pale-yellow sapwood. At 12-percent moisture content it weighs 38 pounds per cubic foot. Shrinkage is moderately large. The wood is moderately strong, stiff, medium hard, and medium high in shock resistance. It is used for general construction, railway ties, posts, and poles.

Pine, Ponderosa

Ponderosa pine (*Pinus ponderosa*) is known also as pondosa pine, western soft pine, western pine, California white pine, bull pine, and black jack. Jeffrey pine (*P. jeffreyi*), which grows in close association with ponderosa pine in California and Oregon, is usually marketed with ponderosa pine and sold under that name.

Ponderosa pine has extensive distribution from Arizona and New Mexico to South Dakota and westward to the mountains of the Pacific Coastal States.

Major producing areas are in Oregon, Washington, California, and Nevada. Other important producing areas are in Idaho and Montana; lesser amounts come from the southern Rocky Mountain region.

Botanically, ponderosa pine belongs to the yellow pine group rather than the white pine group. A considerable proportion of the wood, however, is somewhat similar to the white pines in appearance and properties. The heartwood is light reddish brown, and the wide sapwood is nearly white to pale yellow.

The wood of the outer portions of ponderosa pine of sawtimber size is generally moderately light in weight, moderately low in strength, moderately soft, moderately stiff, and moderately low in shock resistance. It is generally straight grained and has moderately small shrinkage. It is quite uniform in texture and has little tendency to warp and twist.

Ponderosa pine is used mainly for lumber and to a lesser extent for piling, poles, posts, mine timbers, veneer, and ties. The clear wood goes into sash, doors, blinds, moldings, paneling, mantels, trim, and built-in cases and cabinets. Lower grade lumber is used for boxes and crates. Much of the lumber of intermediate or lower grades goes into sheathing, subflooring, and roof boards. Knotty ponderosa pine is used for interior finish.

Pine, Red

Red pine (*Pinus resinosa*) is frequently called Norway pine. It is occasionally known as hard pine and pitch pine. This species grows in the New England States, New York, Pennsylvania, and the Lake States. In the past, lumber from red pine has been marketed with white pine without distinction as to species.

The heartwood of red pine varies from pale red to a reddish brown. The sapwood is nearly white with a yellowish tinge, and is generally from 2 to 4 inches wide. The wood resembles the lighter weight wood of southern yellow pine. Summerwood is distinct in the growth rings.

Red pine is moderately heavy, moderately strong and stiff, moderately soft and moderately high in shock resistance. It is generally straight grained, not so uniform in texture as eastern white pine, and somewhat resinous. The wood has moderately large shrinkage but is not difficult to dry and stays in place well when seasoned.

Red pine is used principally for lumber and to a lesser extent for piling, poles, cabin logs, hewed ties, posts, and fuel. The wood is used for many of the purposes for which eastern white pine is used. It goes mostly into building construction, siding, piling, flooring, sash, doors, blinds, general millwork, and boxes and crates.

Pine, Southern Yellow

There are a number of species included in the group marketed as southern yellow pine lumber. The most important, and their growth range, are:

(1) Longleaf pine (*Pinus palustris*), which grows from eastern North Carolina southward into Florida and westward into eastern

Texas. (2) Shortleaf pine (*P. echinata*), which grows from southeastern New York and New Jersey southward to northern Florida
and westward into eastern Texas and Oklahoma. Northern limits of
growth are the Ohio Valley and southern Missouri. (3) Loblolly
pine (*P. taeda*), which grows from Maryland southward through the
Atlantic Coastal Plain and Piedmont Plateau into Florida and westward into eastern Texas. The northern limit of growth west of the
Appalachian Mountains is near the southern Tennessee border. (4)
Slash pine (*P. elliottii*), which grows in Florida and the southern parts
of South Carolina, Georgia, Alabama, Mississippi, and Louisiana
east of the Mississippi River.

Lumber from any one or from any mixture of two or more of these
species is classified as southern yellow pine by the grading standards
of the industry. These standards provide also for lumber that is
produced from trees of the longleaf and slash pine species to be classified as longleaf yellow pine if conforming to the growth-ring and summerwood requirements of such standards. The lumber that is classified
as longleaf in the domestic trade is known also as pitch pine in the
export trade.

Southern yellow pine lumber comes principally from the Southern
and South Atlantic States. States that lead in production are Georgia,
Alabama, North Carolina, and Texas.

The wood of the various southern yellow pines is quite similar in
appearance. The sapwood is yellowish white and the heartwood
reddish brown. The sapwood is usually white in second-growth
stands. Heartwood begins to form when the tree is about 20 years
old. In old, slow-growth trees, sapwood may be only 1 or 2 inches
in width.

Longleaf and slash pine are classed as heavy, strong, stiff, hard, and
moderately high in shock resistance. Shortleaf and loblolly pine are
usually somewhat lighter in weight than longleaf. All the southern
yellow pines have moderately large shrinkage but stay in place well
when properly seasoned.

In order to obtain heavy, strong wood of the southern yellow pines
for structural purposes, a density rule has been written that specifies
certain visual characteristics for structural timbers.

Dense southern yellow pine is used extensively in construction of
factories, warehouses, bridges, trestles, and docks in the form of
stringers, beams, posts, joists, and piling. Lumber of lower density
and strength finds many uses for building material, such as interior
finish, sheathing, subflooring, and joists, and for boxes and crates.
Southern yellow pine is used also for tight and slack cooperage.
When used for railroad ties, telephone and telegraph poles, and mine
timbers, it is treated with preservatives.

Pine, Sugar

Sugar pine (*Pinus lambertiana*) is sometimes called California
sugar pine. The range of sugar pine extends from the Coast and
Cascade Mountain Ranges of southern Oregon along the Coast Ranges
and the Sierra Nevada of California. Most of the sugar pine lumber
is produced in California and the remainder in southwestern Oregon.

The heartwood of sugar pine is buff or light brown, sometimes tinged with red. The sapwood is creamy white. The wood is straight grained, fairly uniform in texture, and easy to work with tools. It has very small shrinkage, is readily seasoned without warping or checking, and stays in place well. This species is light in weight, moderately low in strength, moderately soft, low in shock resistance, and low in stiffness.

Sugar pine is used almost entirely for lumber products. The largest amounts are used in boxes and crates, sash, doors, frames, blinds, general millwork, building construction, and foundry patterns. Like eastern white pine, sugar pine is suitable for use in nearly every part of a house because of the ease with which it can be cut, its ability to stay in place, and its good nailing properties. It is readily available in wide, thick pieces practically free from defects.

Pine, Virginia

Virginia pine (*Pinus virginiana*), known also as Jersey pine and scrub pine, grows from New Jersey and Virginia throughout the Appalachian region to Georgia and the Ohio Valley. The heartwood is orange and the sapwood nearly white and relatively thick. The wood is rated as moderately heavy, moderately strong, moderately hard, and moderately stiff and has moderately large shrinkage and high shock resistance. It is used for lumber, railroad ties, mine props, and fuel.

Pine, Western White

Western white pine (*Pinus monticola*) is also known as Idaho white pine or white pine. In the United States, it grows in western Montana, northern Idaho, and along the Cascade Mountains and Sierra Nevada through Washington and Oregon to central California. About four-fifths of the cut comes from Idaho with the remainder mostly from Washington; small amounts are cut in Montana and Oregon.

Heartwood of western white pine is cream colored to light reddish brown and darkens on exposure. The sapwood is yellowish white and generally from 1 to 3 inches wide. The wood is straight grained, easy to work, easily kiln-dried, and stays in place well after seasoning.

This species is moderately light in weight, moderately low in strength, moderately soft, moderately stiff, moderately low in shock resistance, and has moderately large shrinkage.

Practically all western white pine is sawed into lumber and used mainly for building construction, matches, boxes, patterns, and millwork products, such as sash, frames, doors, and blinds. In building construction, boards of the lower grades are used for sheathing, subflooring, and roof strips. High-grade material is made into siding of various kinds, exterior and interior trim, and knotty finish. It has practically the same uses as eastern white pine and sugar pine.

Port-Orford-Cedar

Port-Orford-cedar (*Chamaecyparis lawsoniana*) is sometimes known as Lawson cypress, Oregon cedar, and white cedar. It grows along the Pacific coast from Coos Bay, Oreg., southward to California. It does not extend more than 40 miles inland.

The heartwood of Port-Orford-cedar is light yellow to pale brown in color. Sapwood is thin and hard to distinguish. The wood has fine texture, generally straight grain, and a pleasant spicy odor. It is moderately light in weight, stiff, moderately strong and hard, and moderately resistant to shock. Port-Orford-cedar heartwood is highly resistant to decay. The wood shrinks moderately, has little tendency to warp, and stays in place well after seasoning.

A large proportion of the high-grade Port-Orford-cedar is used in the manufacture of battery separators and venetian-blind slats. Other uses are mothproof boxes, sash and door construction, flooring, interior finish, furniture, and boatbuilding.

Redcedar, Eastern

Eastern redcedar (*Juniperus virginiana*) grows throughout the eastern half of the United States, except in Maine, Florida, and a narrow strip along the gulf coast, and at the higher elevations in the Appalachian Mountain Range. Commercial production is principally in the southern Appalachian and Cumberland Mountain regions. Another species, southern redcedar (*J. silicicola*), grows over a limited area in the South Atlantic and Gulf Coastal Plains.

The heartwood of redcedar is bright red or dull red, and the thin sapwood is nearly white. The wood is moderately heavy, moderately low in strength, hard, and high in shock resistance, but low in stiffness. It has very small shrinkage and stays in place well after seasoning. The texture is fine and uniform. Grain is usually straight, except where deflected by knots, which are numerous. Eastern redcedar heartwood is very resistant to decay.

The greatest quantity of eastern redcedar is used for fence posts. Lumber is manufactured into chests, wardrobes, and closet lining. Other uses include flooring, pencils, scientific instruments, and small boats. Southern redcedar is used for the same purposes.

Redcedar, Western

Western redcedar (*Thuja plicata*) grows in northern California, western Oregon, western and northeastern Washington, northern Idaho, and northwestern Montana. It grows also along the Pacific coast northward to Alaska. Western redcedar is also called canoe cedar, giant arborvitae, shinglewood, and Pacific redcedar. Western redcedar lumber is produced principally in Washington, followed by Oregon, Idaho, and Montana.

The heartwood of western redcedar is reddish brown and the sapwood nearly white. The sapwood is narrow, often not over 1 inch in width. The wood is generally straight grained and has a uniform but rather coarse texture. It has very small shrinkage. This species is light in weight, moderately soft, low in strength when used as a beam or post, and low in shock resistance. Its heartwood is very resistant to decay.

Western redcedar is used principally for shingles, lumber, poles, posts, and piling. The lumber is used for exterior siding, interior finish, greenhouse construction, ship and boat building, boxes and crates, sash, doors, and millwork.

Redwood

Redwood (*Sequoia sempervirens*) is a very large tree growing on the coast of California. Another sequoia, giant sequoia (*Sequoia gigantea*), grows in a limited area in the Sierra Nevada of California. Other names for redwood are coast redwood, California redwood, and sequoia. Production of redwood lumber is limited to California.

The heartwood of redwood varies from a light cherry to a dark mahogany. The narrow sapwood is almost white. Typical old-growth redwood is moderately light in weight, moderately strong and stiff, and moderately hard. The wood is easy to work, generally straight grained, and shrinks and swells comparatively little. The heartwood has high decay resistance.

Most redwood lumber is used for building. It is remanufactured extensively into siding, sash, doors, blinds, finish, casket stock, and containers. Because of its durability, it is useful for cooling towers, tanks, silos, wood-stave pipe, and outdoor furniture. It is used in agriculture for buildings and equipment. Its use as timbers and large dimension in bridges and trestles is relatively minor. The wood splits readily and the manufacture of split products, such as posts and fence material, is an important business in the redwood area.

Spruce, Eastern

The term "eastern spruce" includes three species, red (*Picea rubens*), white (*P. glauca*), and black (*P. mariana*). White spruce and black spruce grow principally in the Lake States and New England, and red spruce in New England and the Appalachian Mountains. All three species have about the same properties, and in commerce no distinction is made between them. The wood dries easily, stays in place well, is moderately light in weight and easily worked, has moderate shrinkage, and is moderately strong, stiff, tough, and hard. The wood is light in color, and there is little difference between the heartwood and sapwood.

The largest use of eastern spruce is for pulpwood. It is used for framing material, general millwork, boxes and crates, ladder rails, and piano sounding boards.

Spruce, Engelmann

Engelmann spruce (*Picea engelmannii*) grows at high elevations in the Rocky Mountain region of the United States. This species is sometimes known by other names, such as white spruce, mountain spruce, Arizona spruce, silver spruce, and balsam. About two-thirds of the lumber is produced in the southern Rocky Mountain States. Most of the remainder comes from the northern Rocky Mountain States and Oregon.

The heartwood of Engelmann spruce is nearly white with a slight tinge of red. The sapwood varies from ¾ inch to 2 inches in width and is often difficult to distinguish from heartwood. The wood has medium to fine texture and is without characteristic taste or odor. It is generally straight grained. Engelmann spruce is rated as light in weight. It is low in strength as a beam or post. It is limber, soft, low in shock resistance, and has moderately small shrinkage. The lumber contains small knots.

Engelmann spruce is used principally for lumber and for mine timbers, railroad ties, and poles. It is used also in building construction in the form of dimension stock, flooring, sheathing, and studding. It has excellent pulp- and paper-making properties.

Spruce, Sitka

Sitka spruce (*Picea sitchensis*) is a tree of large size growing along the northwestern coast of North America from California to Alaska. It is generally known as Sitka spruce, although other names may be applied locally, such as yellow spruce, tideland spruce, western spruce, silver spruce, and west coast spruce. About two-thirds of the production of Sitka spruce lumber comes from Washington and one-third from Oregon. Small amounts are produced in California.

The heartwood of Sitka spruce is a light pinkish brown. The sapwood is creamy white and shades gradually into the heartwood; it may be 3 to 6 inches wide or even wider in young trees. The wood has a comparatively fine, uniform texture, generally straight grain, and no distinct taste or odor. It is moderately light in weight, moderately low in bending and compressive strength, moderately stiff, moderately soft, and moderately low in resistance to shock. It has moderately small shrinkage. On the basis of weight, it rates high in strength properties and can be obtained in clear, straight-grained pieces of large size.

Sitka spruce is used principally for lumber, pulpwood, and cooperage. Boxes and crates account for about one-half of the remanufactured lumber. Other important uses are furniture, planing-mill products, sash, doors, blinds, millwork, and boats. Sitka spruce has been by far the most important wood for aircraft construction. Other specialty uses are ladder rails and sounding boards for pianos.

Tamarack

Tamarack (*Larix laricina*) is a small- to medium-sized tree with a straight, round, slightly tapered trunk. In the United States it grows from Maine to Minnesota, with the bulk of the stand in the Lake States. It was formerly used in considerable quantity for lumber, but in recent years production for that purpose has been small.

The heartwood of tamarack is yellowish brown to russet brown. The sapwood is whitish, generally less than an inch wide. The wood is coarse in texture, without odor or taste, and the transition from springwood to summerwood is abrupt. The wood is intermediate in weight and in most mechanical properties.

Tamarack is used principally for pulpwood, lumber, railroad ties, mine timbers, fuel, fence posts, telegraph poles, and scaffolding poles. Lumber goes into framing material, tank construction, and boxes and crates.

Whitecedar, Northern and Atlantic

There are two species of whitecedar in the eastern part of the United States—northern whitecedar (*Thuja occidentalis*) and Atlantic whitecedar (*Chamaecyparis thyoides*). Northern whitecedar is also known as arborvitae, or simply cedar. Atlantic whitecedar is also known as juniper, southern whitecedar, swamp cedar, and boat cedar.

Northern whitecedar grows from Maine southward along the Appalachian Mountain Range and westward through the northern part of the Lake States. Atlantic whitecedar grows near the Atlantic coast from Maine to northern Florida and westward along the gulf coast to Louisiana. It is strictly a swamp tree.

Production of northern whitecedar lumber is probably greatest in Maine and the Lake States. Commercial production of Atlantic whitecedar centers in North Carolina and along the gulf coast.

The heartwood of whitecedar is light brown, and the sapwood is white or nearly so. The sapwood is usually thin. The wood is light in weight, rather soft and low in strength, and low in shock resistance. It shrinks little in drying. It is easily worked, holds paint well, and the heartwood is highly resistant to decay. The two species are used for similar purposes, mostly for poles, ties, lumber, and posts. Whitecedar lumber is used principally where high degree of durability is needed, as in tanks and boats, and for woodenware.

LITERATURE CITED

(1) BETTS, H. S.
   1945. [Individual Leaflets on Tree Species, With Production Figures.]
        U. S. Dept. Agr., Forest Serv. Amer. Woods Ser.
(2) BROWN, H. P., PANSHIN, A. J., AND FORSAITH, C. C.
   1949. TEXTBOOK OF WOOD TECHNOLOGY. VOL. I. STRUCTURE, IDENTIFI-
        CATION, DEFECTS, AND USES OF THE COMMERCIAL WOODS OF THE
        UNITED STATES. 652 pp., illus. New York.
(3) KOEHLER, A.
   1924. PROPERTIES AND USES OF WOOD. 354 pp., illus. New York.
(4) ———
   1943. GUIDE TO DETERMINING SLOPE OF GRAIN IN LUMBER AND VENEER.
        U. S. Forest Prod. Lab. Rpt. 1585, [56] pp., illus. [Processed.]
(5) MARKWARDT, L. J.
   1930. COMPARATIVE STRENGTH PROPERTIES OF WOODS GROWN IN THE UNITED
        STATES. U. S. Dept. Agr. Tech. Bul. 158, 39 pp., illus.
(6) ——— and HECK, G. E.
   1938. STANDARD TERMS FOR DESCRIBING WOOD. U. S. Forest Prod. Lab.
        Rept. R1169, [6] pp. [Processed.]
(7) WANGAARD, F. F.
   1950. MECHANICAL PROPERTIES OF WOOD. 377 pp., illus. New York.

PHYSICAL PROPERTIES OF WOOD

DECORATIVE FEATURES OF COMMON WOODS

The decorative value of wood depends upon its color, figure, luster, and the way in which it bleaches or takes fillers, stains, fumes, and transparent finishes (20).[2] Because of the combinations of color and the multiplicity of shades found in wood, it is impossible to give detailed descriptions of the colors of the various kinds. The sapwood of most species, however, is light in color, and in some species it is practically white. The white sapwood of certain species, such as maple, is preferable to the heartwood for specific uses. In some species, such as hemlock, the true firs, basswood, cottonwood, and beech, there is little or no difference in color between sapwood and heartwood, but in most species the heartwood is darker and fairly uniform in color. Table 2 describes in a general way the color of the heartwood of the more common kinds of woods.

Some types of figure are more pronounced in plainsawed lumber and others in quartersawed (table 2). Often lumber is neither strictly plainsawed nor strictly quartersawed but rather is intermediate, thereby losing some of its decorative effect.

In plainsawed boards and rotary-cut veneer, the annual growth rings frequently form ellipses and parabolas that make striking figures, especially when the rings are irregular in width and outline on the cut surface. On quartersawed surfaces, these rings form stripes, which are not especially ornamental unless they are irregular in width and direction. The relatively large rays, often referred to as flakes, form a conspicuous figure in quartersawed oak and sycamore. With interlocked grain, which slopes in alternate directions in successive layers from the center of the tree outward, quartersawed surfaces show a ribbon effect, either because of the difference in reflection of light from successive layers when the wood has a natural luster or because cross grain of varying degree absorbs stains unevenly. Much of this type of figure is lost in plainsawed lumber.

In open-grained hardwoods, the appearance of both plainsawed and quartersawed lumber can be varied greatly by the use of fillers of different colors. In softwoods, the annual growth layers can be made to stand out more by applying a stain.

Knots, pin wormholes, bird pecks, decay in isolated pockets, birdseye, mineral streaks, swirls in grain, and ingrown bark are decorative in some species when the wood is carefully selected for a particular architectural treatment.

IDENTIFICATION OF WOOD

Familiarity with most kinds of wood makes it possible to identify them by their general appearance. In the technical identification of wood and for instruction purposes, specific differences must be pointed out. Some woods, such as black walnut, can readily be identified by their color; others, such as Douglas-fir, cypress, and the cedars, can be distinguished by their odor. Many woods have a pronounced difference in color between sapwood and heartwood,

---

[2] Italic numbers in parentheses refer to Literature Cited, p. 63.

TABLE 2.—*Color and figure of common kinds of wood*

HARDWOODS

Species	Color of heartwood [1]	Type of figure in—	
		Plainsawed lumber or rotary-cut veneer	Quartersawed lumber or quarter-sliced veneer
Alder, red	Pale pinkish brown	Faint growth ring	Scattered large flakes, sometimes entirely absent.
Ash:			
Black	Moderately dark grayish brown.	Conspicuous growth ring; occasional burl.	Distinct, not conspicuous growth-ring stripe; occasional burl.
Oregon	Grayish brown, sometimes with reddish tinge.	____do____	Do.
White	____do____	____do____	Do.
Aspen	Light brown	Faint growth ring	None.
Basswood	Creamy white to creamy brown, sometimes reddish.	____do____	Do.
Beech, American	White with reddish tinge, to reddish brown.	____do____	Numerous small flakes up to ⅛ inch in height.
Birch:			
Paper	Light brown	____do____	None.
Sweet	Dark reddish brown	Distinct, not conspicuous growth ring; occasionally wavy.	Occasionally wavy.
Yellow	Reddish brown	____do____	Do.
Butternut	Light chestnut brown with occasional reddish tinge or streaks.	Faint growth ring	None.
Cherry, black	Light to dark reddish brown.	Faint growth ring; occasional burl.	Occasional burl.
Chestnut, American	Grayish brown	Conspicuous growth ring.	Distinct, not conspicuous growth-ring stripe.
Cottonwood	Grayish white to light grayish brown.	Faint growth ring	None.
Elm:			
American and rock	Light grayish brown, usually with reddish tinge.	Distinct, not conspicuous, with fine wavy pattern within each growth ring.	Faint growth-ring stripe.
Slippery	Dark brown with shades of red.	Conspicuous growth ring, with fine pattern within each growth ring.	Distinct, not conspicuous growth-ring stripe.
Hackberry	Light yellowish or greenish gray.	Conspicuous growth ring.	Distinct, not conspicuous growth-ring stripe.
Hickory	Reddish brown	Distinct, not conspicuous growth ring.	Faint growth-ring stripe.
Honeylocust	Cherry red	Conspicuous growth ring.	Distinct, no conspicuous growth-ring stripe.
Locust, black	Golden brown, sometimes with tinge of green.	Conspicuous growth rings.	Do.
Magnolia	Light to dark yellowish brown with greenish or purplish tinge.	Faint growth ring	None.

See footnotes at end of table.

TABLE 2.—*Color and figure of common kinds of wood*—Continued

HARDWOODS

| Species | Color of heartwood [1] | Type of figure in— | |
		Plainsawed lumber or rotary-cut veneer	Quartersawed lumber or quarter-sliced veneer
Maple: Black, bigleaf, red, silver, and sugar.	Light reddish brown____	Faint growth ring, occasionally birdseye, curly, and wavy.	Occasionally curly and wavy.
Oak:			
All species of red oak group.	Grayish brown, usually with fleshy tinge.	Conspicuous growth ring.	Pronounced flake; distinct, not conspicuous growth-ring stripe.
All species of white oak group.	Grayish brown, rarely with fleshy tinge.	____do _____	Do.
Sugarberry_____ _____	Light yellowish or greenish gray.	____do_____	Distinct, not conspicuous growth-ring stripe.
Sweetgum__	Reddish brown_____	Faint growth ring; occasionally irregular darker streaks in "figured" gum.	Distinct, not pronounced ribbon; occasionally irregular darker streaks in "figured" gum.
Sycamore_____	Flesh brown__ _____	Faint growth ring _____	Numerous pronounced flakes up to ¼ inch in height.
Tupelo: Black and water__	Pale to moderately dark brownish gray.	____do_____	Distinct, not pronounced ribbon.
Walnut, black_____	Chocolate brown occasionally with darker, sometimes purplish streaks.	Distinct, not conspicuous growth ring; occasionally wavy, curly, burl, and other types.	Distinct, not conspicuous growth-ring stripe; occasionally wavy, curly, burl, crotch, and other types.
Yellow-poplar_____	Light to dark yellowish brown with greenish or purplish tinge.	Faint growth ring_____	None.

SOFTWOODS

Baldcypress_____	Light yellowish brown to reddish brown.	Conspicuous irregular growth ring.	Distinct, not conspicuous growth-ring stripe.
Cedar:			
Alaska-_____	Yellow_____	Faint growth ring_____	None.
Atlantic white-_____	Light brown with reddish tinge.	Distinct, not conspicuous growth ring.	Do.
Eastern redcedar ____	Brick red to deep reddish brown.	Occasionally streaks of white sapwood alternating with heartwood.	Occasionally streaks of white sapwood alternating with heartwood
Incense-_____	Reddish brown_____	Faint growth ring_____	Faint growth-ring stripe.
Northern white-_____	Light to dark brown____	____do___ _____	Do.
Port-Orford-_____	Light yellow to pale brown.	____do_____	None.
Western redcedar_____	Reddish brown_____	Distinct, not conspicuous growth ring.	Faint growth-ring stripe.
Douglas-fir_____	Orange red to red; sometimes yellow.	Conspicuous growth ring.	Distinct, not conspicuous growth-ring stripe.
Fir:			
Balsam_____ _____	Nearly white _____	Distinct, not conspicuous growth ring.	Faint growth-ring stripe.

See footnotes at end of table.

TABLE 2.—*Color and figure of common kinds of wood*—Continued

SOFTWOODS

| Species | Color of heartwood [1] | Type of figure in— | |
		Plainsawed lumber or rotary-cut veneer	Quartersawed lumber or quarter-sliced veneer
Fir—Continued			
White	Nearly white to pale reddish brown.	Conspicuous growth ring.	Distinct, not conspicuous growth-ring stripe.
Hemlock:			
Eastern	Light reddish brown	Distinct, not conspicuous growth ring.	Faint growth-ring stripe.
Western	----do	----do	Do.
Larch, western	Russet to reddish brown	Conspicuous growth ring.	Distinct, not conspicuous growth-ring stripe.
Pine:			
Eastern white	Cream to light reddish brown.	Faint growth ring	None.
Lodgepole	Light reddish brown	Distinct, not conspicuous growth ring; faint "pocked" appearance.	Do.
Ponderosa	Orange to reddish brown	Distinct, not conspicuous growth ring.	Faint growth-ring stripe.
Red	----do	----do	Do.
Southern yellow [2]	----do	Conspicuous growth ring.	Distinct, not conspicuous growth-ring stripe.
Sugar	Light creamy brown	Faint growth ring	None.
Western white	Cream to light reddish brown.	----do	Do.
Redwood	Cherry to deep reddish brown.	Distinct, not conspicuous growth ring; occasionally wavy and burl.	Faint growth-ring stripe; occasionally wavy and burl.
Spruce:			
Black, Engelmann, red, white.	Nearly white	Faint growth ring	None.
Sitka	Light reddish brown	Distinct, not conspicuous growth ring.	Faint growth-ring stripe.
Tamarack	Russet brown	Conspicuous growth ring.	Distinct, not conspicuous growth-ring stripe.

[1] The sapwood of all species is light in color or virtually white unless discolored by fungus or chemical stains.

[2] Includes longleaf, loblolly, shortleaf, and slash pine.

whereas in others there is no difference in color. Pitch is found normally only in the pines, Douglas-fir, the spruces, larch, and tamarack. Frequently, it is necessary to refer to the finer details of the structure, such as the size of the rays, the size and arrangement of the pores, and the presence or absence of resin ducts, for accurate identification (*17, 18, 20, 21, 28, 29*).

GRAIN AND TEXTURE OF WOOD

The terms "grain" and "texture" are commonly used rather loosely in connection with wood. In fact, they do not have any definite meaning. Grain is often used in reference to the annual rings, as in fine grain and coarse grain, but it is also employed to indicate the direction of the fibers, as in straight grain, spiral grain, and curly grain. Painters refer to woods as open grained and close grained, meaning thereby the relative size of the pores, which determines whether the piece needs a filler. Texture is often used synonymously with grain, but usually it refers to the finer structure of the wood rather than to the annual rings. When the words "grain" or "texture" are used in connection with wood, the meaning intended should be made perfectly clear (*19*).

PLAINSAWED AND QUARTERSAWED LUMBER

Lumber can be cut from a log in two distinct ways: Tangent to the annual rings, producing "plainsawed" lumber in hardwoods and "flat-grained" or "slash-grained" lumber in softwoods; and radially to the rings or parallel to the rays, producing "quartersawed" lumber in hardwoods and "edge-grained" or "vertical-grained" lumber in softwoods (fig. 4). Usually so-called quartersawed or edge-grained lumber is not cut strictly parallel with the rays; and often in plainsawed boards the surfaces next to the edges are far from being tangent to the rings. In commercial practice, lumber with rings at angles of 45° to 90° with the surface is called quartersawed, and lumber with rings at angles of 0° to 45° with the surface is called plainsawed. Hardwood lumber in which the annual rings make angles of 30° to 60° with the faces is sometimes called "bastard sawn."

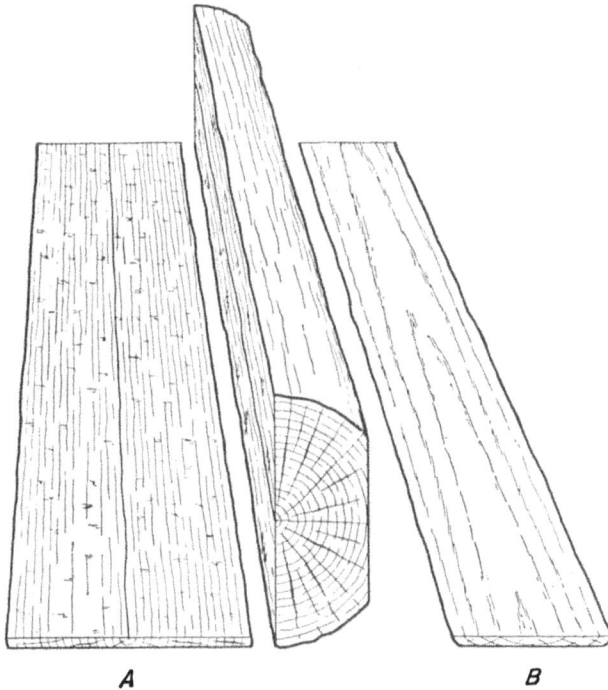

ZM 554 F

FIGURE 4.—Quartersawed (*A*) and plainsawed (*B*) boards cut from a log.

For many purposes either plainsawed or quartersawed lumber is satisfactory. Each type has certain advantages, however, that may be important in a particular use. Some of the advantages of plainsawed or flat-grained lumber are:

(1) It is cheaper, as a rule, because less time is required and less waste is involved in sawing it from the log.

(2) The figure patterns resulting from the annual rings and some other types of figure are brought out more conspicuously.

(3) Round or oval knots that may occur in plainsawed boards affect the surface appearance less than spike knots that may occur in quartersawed boards. Also, a board with a round or oval knot is weakened less by the knot than a board with a spike knot. A greater percentage of the boards from a log sawed to produce the maximum amount of plainsawed lumber will contain knots, however, than will boards from a log sawed to produce the maximum amount of quartersawed material.

(4) Shakes and pitch pockets when present extend through fewer boards.

(5) It does not collapse so easily in drying.

(6) It shrinks and swells less in thickness.

The principal advantages of quartersawed or edge-grained lumber are:

(1) It shrinks and swells less in width.

(2) It twists and cups less.

(3) It surface-checks and splits less in seasoning and in use.

(4) Raised grain caused by separation in the annual rings does not become so pronounced.

(5) It wears more evenly.

(6) Types of figure coming from pronounced rays, interlocked grain, and wavy grain are brought out more conspicuously.

(7) It does not allow liquids to pass into or through it so readily in some species.

(8) It holds paint better in some species.

(9) The sapwood appearing in boards is at the edges and its width is limited according to the width of the sapwood in the log.

WEATHERING OF WOOD

Boards exposed to the weather without a protective coating rapidly become weathered. Weathering may involve change in color, roughening and checking of the surface, and, if one side of the board only is fully exposed, cupping and tearing loose from fastenings, but it does not include decay. With all species, edge-grained boards check less conspicuously and cup less than flat-grained boards of the same species. Thin boards tend to cup more than thick boards do. To resist weathering well, boards should be no more than 8 times as wide as they are thick. Weathering may cause boards that are both too thin and too short, say less than 2 feet long, to split.

Actual erosion of wood from the surface due to weathering takes place very slowly; approximately a century is required for one-fourth inch of thickness to waste away. Twisting, another although less common effect of weathering, is caused primarily by uneven shrinkage resulting from spiral and interlocked grain; it is more pronounced in plainsawed than in quartersawed boards. Weathering, as a rule, changes all woods to a gray color, darker in some woods than in others, and attractive when accompanied by a silvery sheen, as it often is. Metal fastenings for wood that is exposed to the weather should be highly resistant to corrosion.

Woods that weather with—

- Light-gray color and silvery sheen.
  - Baldcypress
  - Cedar: Alaska-
  -      Port-Orford-

- Light-gray color and moderate sheen.
  - Aspen
  - Basswood
  - Birch
  - Cottonwood
  - Hemlock: Eastern
  -       Western
  - Hickory
  - Maple
  - Pine: Eastern white
  -    Ponderosa
  -    Sugar
  -    Western white
  - Spruce: Eastern
  -     Sitka
  - Sweetgum
  - Yellow-poplar

- Dark-gray color and little or no sheen.
  - Ash
  - Chestnut
  - Douglas-fir
  - Fir, commercial white
  - Larch, western
  - Oak: Red
  -    White
  - Pine, southern yellow
  - Redcedar, western
  - Redwood
  - Walnut, black

Woods on which weather checks are—

- Inconspicuous
  - Aspen
  - Baldcypress
  - Cedar: Alaska-
  -      Port-Orford-
  -      Western redcedar
  - Redwood
  - Yellow-poplar

- Conspicuous
  - Ash
  - Basswood
  - Birch
  - Chestnut
  - Cottonwood
  - Douglas-fir
  - Fir, commercial white
  - Hemlock: Eastern
  -       Western
  - Hickory
  - Larch, western
  - Maple
  - Oak: Red
  -    White
  - Pine: Eastern white
  -    Ponderosa
  -    Southern yellow
  -    Sugar
  -    Western white
  - Spruce: Eastern
  -     Sitka
  - Sweetgum
  - Walnut, black

	Slight_____	Baldcypress
		Cedar: Alaska-
		        Port-Orford-
		        Western redcedar
		Redwood

Woods that cup and tend to pull loose from fastenings when exposed to the weather	Distinct_____	Aspen
		Basswood
		Douglas-fir
		Fir, commercial white
		Hemlock: Eastern
		        Western
		Larch, western
		Pine: Eastern white
		        Ponderosa
		        Southern yellow
		        Sugar
		        Western white
		Spruce: Eastern
		        Sitka
		Yellow-popular

	Pronounced_____	Chestnut Walnut, black
	Very pronounced_____	Ash Birch Cottonwood Hickory Maple Oak: Red White Sweetgum
Woods likely to twist because of interlocked grain when exposed to the weather		Birch Cottonwood Elm Sweetgum Sycamore Tupelo: Black Water Many tropical hardwoods

Spiral grain, another cause of twisting, may occur sporadically in any species.

DECAY RESISTANCE OF HEARTWOOD OF NATIVE SPECIES

Wood kept either constantly dry or continuously submerged in water does not decay, regardless of species or of the presence of sapwood. A large proportion of the wood in use is kept so dry at all times that it lasts indefinitely. Moisture and temperature, which vary greatly with local conditions, are the principal factors affecting the rate of decay. When exposed to conditions that favor decay, wood in warm, humid areas of the United States deteriorates more rapidly than that in cool or dry areas. High altitudes, as a rule, are less favorable to decay than are low altitudes because the average temperatures are lower and the growing seasons for fungi, which cause decay, are shorter.

The natural decay resistance of all common native species of wood lies in the heartwood. When untreated, the sapwood of substantially all species has low resistance to decay and usually has a short life under decay-producing conditions. The decay resistance of heartwood in service is greatly affected by differences in the character of the wood, the attacking fungus, and the conditions of exposure. A widely different length of life may therefore be obtained from pieces of wood cut from the same species or even the same tree and used under apparently similar conditions. Further, in a few species, such as the spruces and the white firs (not Douglas-fir), the colors of the heartwood and of the sapwood are so similar that frequently the two cannot be easily distinguished.

Precise ratings of the decay resistance of the heartwood of different species are not possible because of differences within a species and the variety of service conditions to which wood is exposed. However, broad groupings of many of the native species, based on service records, laboratory tests, and general experience, are helpful in choosing woods for use under conditions favorable to decay. The following list shows (1) species with heartwood that is high in decay resistance, giving generally satisfactory service where decay hazards exist; and (2) species with heartwood that is moderate to low in decay resistanse, usually requiring some form of preservative treatment to give satisfactory service under conditions that favor decay.

GROUP 1.—*Species with heartwood of high resistance to decay*

Softwoods (conifers)	Hardwoods	
Baldcypress	Catalpa	Mulberry, red
Cedars	Chestnut	Osage-orange
Junipers	Locust, black	Walnut, black
Redwood	Mesquite	
Yew, Pacific		

GROUP 2.—*Species with heartwood of moderate to low resistance to decay*

Softwoods (conifers)	Hardwoods	
Douglas-fir	Alders	Hickory
Firs, true	Ashes	Honeylocust
Hemlocks	Aspens	Magnolia
Larch, western	Basswood	Maples
Pine, ponderosa	Beech	Oaks
Pine, white (several species)	Birches	Sassafras
	Butternut	Sweetgum
Pine, southern yellow (several species)	Cherry, black	Sycamore
	Cottonwoods	Tupelos
Spruces	Elms	Willows
Tamarack	Hackberry	Yellow-poplar

The heartwood of such species as Douglas-fir, eastern white pine, honeylocust, sassafras, southern yellow pine, sweetgum, western larch, and white oak is usually classified as moderate in decay resistance and generally gives good service under mild decay conditions. The heartwood of the other species of Group 2 is generally lower in decay resistance and frequently gives unsatisfactory service where decay hazards exist. The woods of lower decay resistance are improved by some form of preservative treatment for use under mild decay conditions; those with moderate decay resistance are usually treated for exposure to more severe decay hazards; and even the highly decay-resistant woods of Group 1 may require preservative treatment for important structural and other uses where high decay hazards exist and failure would require expensive repairs.

THERMAL CONDUCTIVITY OF WOOD

The thermal conductivity of wood is affected little by species, except as species differ with respect to factors that influence thermal conductivity. It does, however, vary with (1) direction of grain; (2) specific gravity; (3) moisture content and its distribution; (4) kind, quantity, and distribution of extractives; (5) proportion of springwood and summerwood; and (6) such features as checks, knots, and cross grain (23). In normal wood, the relative rate of heat flow is approxi-

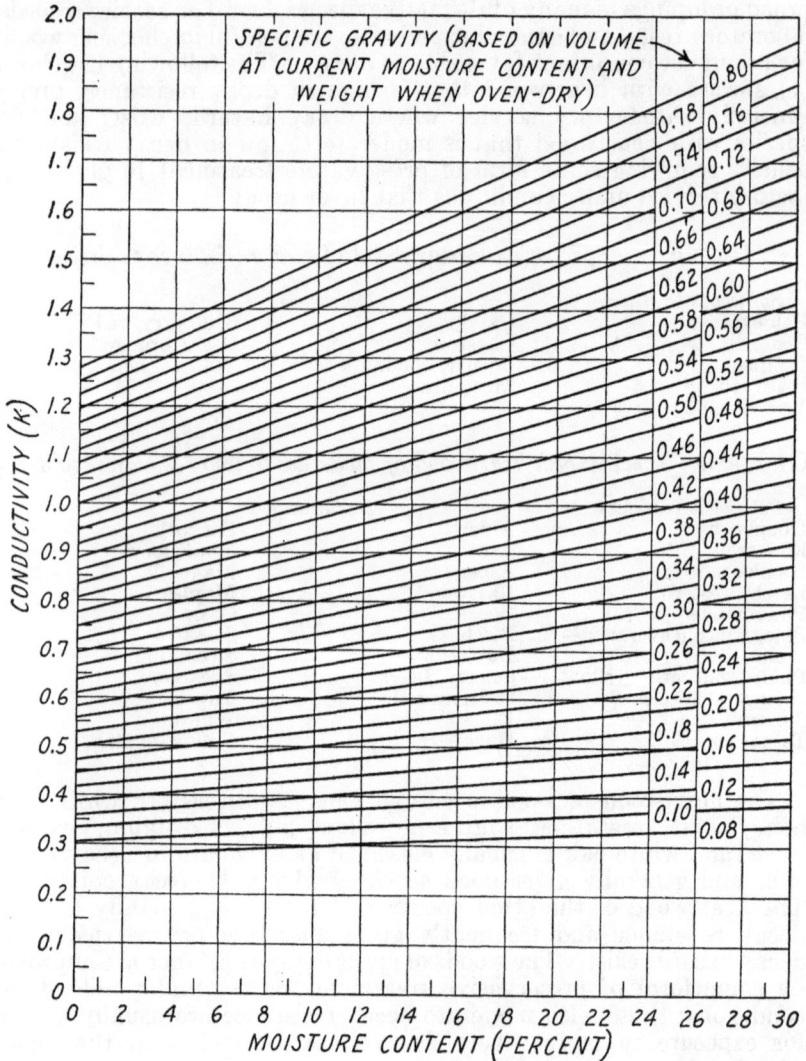

ZM 38147 F

FIGURE 5.—Relation between computed thermal conductivity and moisture content for wood having different specific gravity values.

mately the same in the radial and tangential direction, but thermal conductivity is generally 2¼ to 2¾ times faster along the grain than in the transverse directions. Since thermal conductivity increases with specific gravity, the lighter weight woods are the better insulators.

Thermal conductivity values for various species are given in tables 3 and 4. If the specific gravity (based on volume at current moisture content and weight when oven-dry) and the moisture content of the wood are known, the conductivity can also be determined by the chart shown in figure 5. Conductivity was computed from the formula

$$k=S[1.39+0.028M]+0.165$$

where k is conductivity, S is specific gravity based on volume at current moisture content and weight when oven-dry, and M is moisture content in percent.

TABLE 3.—*Thermal conductivity of oven-dry wood across the grain*

Species	Number of tests	Average specific gravity [1]	Average conductivity k [2]	Species	Number of tests	Average specific gravity [1]	Average conductivity k [2]
Hardwoods:				Softwoods—Continued			
Aspen, bigtooth	5	0.41	0.71	Douglas-fir	8	0.46	0.76
Balsa	4	.16	.41	Fir, white	2	.41	.71
Basswood, American	7	.38	.69	Hemlock, western	2	.46	.79
Elm, rock	1	.76	1.16	Larch, western	3	.57	.94
Maple, sugar	5	.68	1.13	Pine:			
Oak, red [3]	5	.67	1.19	Southern yellow [3]	7	.56	.94
Prima vera	1	.47	.82	White [3]	10	.40	.72
Tangile	4	.58	1.00	Redcedar, western	3	.34	.64
Softwoods:				Redwood	8	.40	.74
Baldcypress	5	.39	.75	Spruce, Engelmann	4	.34	.62

[1] Based on volume and weight when oven-dry.
[2] k is the quantity of heat expressed in British thermal units that flows in 1 hour through a 1-inch thickness of material, 1 square foot in area, when the temperature difference between the 2 surfaces is 1° F.
[3] Average of several species.

A more detailed discussion of the thermal conductivity and insulation values of wood and wood-base materials is given on pages 445–455.

THERMAL EXPANSION OF WOOD

Dry wood, like most other solids, expands on heating and contracts on cooling. The coefficient of linear thermal expansion (the increase in length per unit of length for a temperature rise of 1° F.) differs in the three different structural directions of wood. In the longitudinal direction (along the grain) the values are independent of the specific gravity of the wood and vary from about 1.7×10^{-6} to 2.5×10^{-6} per 1° F. for different species. Across the grain, or in the radial and tangential directions, the values, in general, vary directly with the specific gravity (based on weight and volume when oven-dry) (*35*). The data for several softwoods (Douglas-fir, Sitka spruce, redwood, and white fir) and two of the softer hardwoods (yellow-poplar and cottonwood) can be closely approximated by the simple relationships

$$a_t = 45 \times \text{specific gravity} \times 10^{-6} \text{ per } 1° \text{ F.}$$
$$a_r = 31 \times \text{specific gravity} \times 10^{-6} \text{ per } 1° \text{ F.}$$

where a_t is the coefficient of linear thermal expansion in the tangential direction and a_r is the coefficient in the radial direction. The values for two of the denser hardwoods (yellow birch and sugar maple) can be approximated from the relationships

$$a_t = 32 \times \text{specific gravity} \times 10^{-6} \text{ per } 1° \text{ F.}$$
$$a_r = 25 \times \text{specific gravity} \times 10^{-6} \text{ per } 1° \text{ F.}$$

The coefficient of linear thermal expansion of wood varies slightly with temperature but for all ordinary uses may be considered constant (*34*).

The coefficient of linear thermal expansion in the longitudinal direction of wood is from one-tenth to one-third of the values for the common metals, concrete, and glass. The values for wood in the transverse directions are generally larger than those in the longitudinal direction, but are usually less than those of other structural materials.

The coefficients of linear thermal expansion of resin-bonded birch laminates and plywood made from both resin-treated and untreated veneer (impreg, compreg, and normal laminates and plywood) were measured and compared with the values calculated by theoretical equations involving the values for wood, the values for the resin, and the modulus of elasticity values for wood in compression in the three different structural directions (*34*). The calculated values agreed closely with the measured values. The values for the coefficient of linear thermal expansion of resin-bonded, rotary-cut birch plywood in the sheet directions are about 3.0×10^{-6} per $1°$ F. parallel to the face ply and 4.2×10^{-6} per $1°$ F. across the face ply. These values are not much greater than the longitudinal value for the normal wood (2.0×10^{-6} per $1°$ F.).

The coefficient of thermal expansion of wood can be neglected in most structural designs, because the thermal expansion is much smaller than the swelling and shrinking of the wood that occurs under normal exposure conditions. Thermal expansion is of importance only in the case of special structures that are kept dry and are subjected to considerable temperature change.

Green wood behaves differently from dry wood when heated; it expands tangentially and shrinks radially. A discussion of this behavior and of the rate of temperature change in wood when heated would be too involved for this handbook, but information on these subjects may be found elsewhere (*24, 25, 26*).

ELECTRICAL PROPERTIES OF WOOD

The most important electrical properties of wood are its resistance to the passage of an electric current and its dielectric properties. The electrical resistance of wood, which is the reciprocal of the conductance, is utilized in electric meters for determining the moisture content of wood (*11*). It is of importance in connection with the use of wood poles and crossarms to carry high-voltage electrical powerlines and wood handles on linemen's tools. The dielectric properties—dielec-

TABLE 4.—*Thermal conductivity of wood across the grain at various moisture content values*

Species	Number of tests	Average specific gravity [1]	Average moisture content (percent)	Average conductivity k [2]
Hardwoods:				
Ash, white	7	0.56	15.6	1.19
Aspen, bigtooth	10	.41	12.1	.87
Balsa	6	.17	8.0	.44
Basswood, American	20	.37	9.3	.78
Beech, American	1	.59	11.1	1.17
Birch, yellow	11	.64	10.8	1.25
Cherry, black	1	.68	7.8	1.26
Chestnut, American	1	.42	11.0	.88
Elm:				
American	1	.54	9.4	1.06
Rock	9	.65	17.9	1.39
Greenheart	4	.85	18.4	1.78
Maple:				
Silver	1	.47	9.9	.95
Sugar	27	.66	11.7	1.30
Oak:				
Red [3]	12	.62	12.4	1.24
White [3]	18	.62	11.1	1.22
Pecan	1	.67	10.0	1.28
Primavera	9	.46	9.7	.93
Sweetgum	1	.55	11.0	1.10
Sycamore, American	1	.52	9.0	1.02
Tangile	6	.54	10.6	1.08
Tupelo, black	1	.46	10.0	.93
Softwoods:				
Baldcypress	30	.38	11.7	.82
Douglas-fir	52	.46	18.4	1.04
Fir, white [3]	2	.38	11.7	.82
Hemlock, western	5	.44	23.0	1.06
Juniper, bigberry	1	.44	16.0	.97
Larch, western	13	.46	12.6	.97
Pine:				
Southern yellow [3]	59	.53	13.8	1.11
White [3]	20	.36	9.8	.76
Redcedar, western	16	.32	13.3	.73
Redwood	15	.39	11.7	.83
Spruce	12	.35	13.0	.78

[1] Based on volume at current moisture content and weight when oven-dry.

[2] k is the quantity of heat expressed in British thermal units that flows in 1 hour through a 1-inch thickness of material, 1 square foot in area, when the temperature difference between the 2 surfaces is 1° F.

[3] Average of several species.

tric constant and power factor—assume importance when wood is heated in an oscillating high-frequency electric field, as in the curing of glues, and in connection with some moisture meters.

Electrical Resistance

The direct-current electrical resistance of wood (*14, 16, 31, 32*) varies greatly with moisture content, especially below the fiber saturation point of 30 percent, decreasing as the moisture content increases. It also varies with species, is greater across the grain than along it, and approximately doubles for each drop in temperature of 22.5° F. The electrical resistance varies inversely with the density of the wood. Fortunately, the electrical resistance is affected so much more by variations in moisture content that the effect of density of the wood on moisture content determinations by the resistivity method is not highly significant. The variation among species is probably caused by minerals or electrolytic materials in the wood or dissolved in water present in the wood.

Table 5 lists the electrical resistance along the grain of some common species, at different uniform values of moisture content, between needle electrodes driven each time to the same depth. The specific resistance, which is the resistance of a centimeter cube of wood, varies from 3×10^{17} to 3×10^{18} ohm-centimeters for oven-dry wood to 10^8 ohm-centimeters for wood at 16-percent moisture content (34).

Since the electrical resistance of wood may increase as much as 10^5 times with a moisture content change from 25 to 7 percent, the moisture content of wood used as an insulator, as in powerline poles or tool handles, is highly important. The leakage of current from powerlines increases with increase in the moisture content of the crossarms and poles. Salts should not be used as a preservative treatment for powerline poles, since the salt dissolved in the water of the wood causes an additional lowering of the electrical resistance.

Dielectric Properties

At a low moisture content, wood is normally classified as an electrical insulator, or dielectric, rather than as a conductor (30). A dielectric can be heated by using it as the medium between electrodes carrying charges of oscillating high-frequency electricity or placing it in an electrical field of like nature. This property is utilized in the curing of glue lines on wood by high-frequency electric current. The rate at which the wood is heated and electrical power is absorbed depends on the dielectric constant and the power factor. The resistivity of a dielectric can be determined by formula (30) from the relationship between the dielectric constant and the power factor, and when wood and glue lines are heated, the relative power absorbed in the wood and in the glue lines can be calculated from the resistivity values and dimensions of the wood and glue lines.

The dielectric constant of wood is the ratio of the capacitance of a condenser in which the wood is used as the dielectric to the capacitance of a similar condenser in which a vacuum dielectric is employed. The capacitance of a condenser is measured by the quantity of electricity put into it by a given voltage. The dielectric constant for a vacuum or dry air is unity, and the values for wood are all greater than unity. Since the structure and chemical composition of the cell walls of all species are essentially similar, it is not surprising that the dielectric constants are proportional to density at a given moisture content. The dielectric constant of wood increases with increase in density, and since the dielectric constant of water is approximately 81 as compared to 4.2 for oven-dry wood, it is evident that the dielectric constant of wood also increases with increase in moisture content. Variations in moisture content have a greater effect on the dielectric constant than

TABLE 5.—*The average electrical resistance along the grain in megohms, measured at 80° F. between 2 pairs of needle electrodes 1¼ inches apart and driven to a depth of 5/16 inch, of several species of wood at different values of moisture content*

Species	Electrical resistance in megohms when the moisture content in percent is—																		
	7	8	9	10	11	12	13	14	15	16	17	18	19	20	21	22	23	24	25
Hardwoods:																			
Ash, commercial white	12,000	2,190	690	250	105	55	28	14	8.3	5.0	3.2	2.0	1.32	0.89	0.63	0.50	0.44	0.40	0.40
Basswood	36,300	1,740	470	180	85	45	27	16	9.6	6.2	4.1	2.8	1.86	1.32	.93	.69	.51	.39	.31
Birch	87,000	19,950	4,470	1,290	470	200	96	53	30.2	18.2	11.5	7.6	5.13	3.55	2.51	1.78	1.32	.95	.70
Elm, American	18,200	2,000	350	110	45	20	12	7	3.9	2.3	1.5	1.0	.66	.48	.42	.40	.40	.40	.40
Gum, red	38,000	6,460	2,090	815	345	160	81	45	25.7	15.1	9.3	6.0	3.98	2.63	1.78	1.26	.87	.63	.46
Hickory, true		31,600	2,190	340	115	50	21	11	6.3	3.7	2.3	1.5	1.00	.71	.52	.44	.40	.40	.40
Khaya [1]	44,600	16,200	6,310	2,750	1,260	630	340	180	105.0	60.2	35.5	21.9	14.10	9.33	6.16	4.17	2.82	1.99	1.44
Magnolia	43,700	12,600	5,010	2,040	910	435	205	105	56.2	29.5	16.2	9.1	5.25	3.09	1.86	1.17	.74	.50	.37
Mahogany, American	20,900	6,760	2,290	870	380	180	85	43	22.4	12.3	7.2	4.4	2.69	1.66	1.07	.72	.49	.35	.26
Maple, sugar	77,400	13,800	3,160	690	250	105	53	29	16.6	10.2	6.8	4.5	3.16	2.24	1.62	1.23	.98	.75	.60
Oak: Commercial red [2]	14,400	4,790	1,590	630	265	125	63	32	18.2	11.3	7.3	4.6	3.02	2.09	1.45	.95	.80	.63	.50
Commercial white	17,400	3,550	1,100	415	170	80	42	22	12.6	7.2	4.3	2.7	1.70	1.15	.79	.60	.49	.44	.41
Shorea [3]	2,890	690	220	80	35	15	9	5	2.8	1.7	1.1	.7	.45	.30	.21	.16	.12	.09	.07
Tupelo, black [2]	31,700	12,600	5,020	1,820	725	275	120	58	27.6	13.0	6.9	3.7	2.19	1.38	.95	.63	.46	.33	.25
Walnut, black [2]	51,300	9,770	2,630	890	355	155	78	41	22.4	12.9	7.3	4.9	3.16	2.14	1.48	1.02	.72	.51	.38
Yellow-poplar [2]	24,000	8,320	3,170	1,260	525	250	140	76	43.7	25.2	14.5	8.7	5.76	3.81	2.64	1.91	1.39	1.10	.85
Softwoods:																			
Baldcypress	12,600	3,980	1,410	630	265	120	60	33	18.6	11.2	7.1	4.6	3.09	1.78	1.26	.91	.66	.51	.42
Douglas-fir (coast type)	22,400	4,780	1,660	630	265	120	60	33	18.6	11.2	7.1	4.6	3.09	2.14	1.51	1.10	.79	.60	.46
Fir: California red	31,600	6,760	2,000	725	315	150	83	48	28.8	18.2	11.8	7.6	5.01	3.31	2.29	1.58	1.15	.83	.63
White	57,600	15,850	3,980	1,120	415	180	83	46	26.9	16.6	11.0	6.6	4.47	3.02	2.14	1.55	1.12	.86	.62
Hemlock, western	22,900	5,620	2,040	850	400	185	98	51	28.2	16.2	10.0	6.0	3.89	2.52	1.58	1.05	.72	.51	.37
Larch, western	39,800	11,200	3,980	1,445	560	250	120	63	33.9	19.9	12.3	7.6	5.02	3.39	2.29	1.62	1.20	.87	.66
Pine: Eastern white	20,900	5,620	2,090	850	405	200	102	58	33.1	19.9	12.3	7.9	5.01	3.31	2.19	1.51	1.05	.74	.52
Longleaf	25,000	8,700	3,160	1,320	575	270	135	74	41.7	24.0	14.4	8.9	5.76	3.72	2.46	1.66	1.15	.79	.60
Ponderosa	39,800	8,910	3,310	1,410	645	300	150	81	44.7	25.1	14.8	9.1	5.62	3.55	2.34	1.62	1.15	.87	.69
Shortleaf	43,600	11,250	3,720	1,350	560	255	130	69	38.9	22.4	13.8	8.7	5.76	3.80	2.63	1.82	1.29	.93	.66
Sugar	22,900	5,250	1,660	645	280	140	76	44	25.9	15.9	10.0	6.6	4.36	3.02	2.09	1.48	1.05	.75	.56
Redwood	22,400	4,680	1,550	615	250	100	45	22	12.6	7.2	4.7	3.2	2.29	1.74	1.32	1.05	.85	.71	.60
Spruce, Sitka	22,400	5,890	2,140	830	365	165	83	44	25.1	15.5	9.8	6.3	4.27	3.02	2.14	1.58	1.17	.91	.71

[1] Known in the trade as "African mahogany."

[2] The values for this species were calculated from measurements on veneer.

[3] A Philippine hardwood, identified as tangile or some similar species.

do variations in density. The parallel-to-grain dielectric constant of wood is significantly greater than the corresponding perpendicular-to-grain constant. The dielectric constant of wood decreases with an increase in frequency of the oscillating current.

The power factor of wood is the ratio of the power absorbed in the wood per cycle of oscillation of an electric current to the total apparent power stored in the wood during that cycle. The power factor of wood is less dependent on density than is the dielectric constant but generally increases with increase in moisture content. At moisture content values of 0 to 7 percent, the parallel-to-grain power factors are greater than the corresponding perpendicular-to-grain power factors. The power factor varies with the frequency of the oscillating current and is greater at high frequencies when the wood is at a moisture content of 15 percent or lower.

EFFECT OF CHEMICALS ON WOOD

Wood is highly resistant to a number of chemicals. For this reason, it is used for various types of tanks, containers, and equipment in which chemicals are used and for structures near such equipment that may contact chemicals through spillage, leakage, or condensation.

Wood owes its extensive use in chemical equipment largely to its superiority over cast iron and ordinary steel in resistance to mild acids and solutions of acidic salts. Iron is far superior to wood in resistance to alkalies, however, and wood is therefore seldom used in contact with solutions that are more than weakly alkaline.

Experience has shown that the heartwood of cypress, southern yellow pine, Douglas-fir, and redwood is the most suitable for water tanks (13), and heartwood of the first three species for tanks for use where resistance to chemicals in appreciable concentrations is an important factor. The four species combine moderate to high resistance to water penetration with moderate to high natural resistance to decay and hydrolysis.

Chemicals may affect the strength of wood by three general types of action. The first involves swelling and the resultant weakening of the wood. This action is almost completely reversible, so that, if the swelling liquid or solution is removed by evaporation or by extraction followed by evaporation of the solvent, the original dimensions and strength are practically restored. The second type of action involves permanent changes in the wood, such as hydrolysis of the cellulose by acids or acid salts. The third type of action, which is also permanent, involves delignification of the wood and dissolving of hemicelluloses by alkalies. Occasionally dilute sodium carbonate solutions show some action of this type on the wood of cooling towers. There is also some evidence that iron salts catalyze oxidation of wood and precipitate toxic extractives, thus lowering its natural decay resistance.

The strength properties of water-swollen wood are in general considerably lower, and in some cases more than 50 percent lower (27, 36), than those for wood at 12 percent moisture content, which is about the average moisture content in normal use. This must be taken into account in designing tanks for water and dilute aqueous solutions or other structures in which the wood members are wet for an appreciable time.

Alcohols and other wood-swelling organic liquids that do not react chemically with wood reduce its compressive strength proportionally to the extent of swelling (12). A liquid such as acetone that swells wood 60 percent as much as water will cause 60 percent as much loss in strength as is caused by swelling in water. Nonswelling liquids, such as petroleum oils and creosote, have a negligible effect upon wood strength (see p. 420). Castor-oil hydraulic fluids swell wood to some extent, and reduce bending and compressive strength to an extent that depends upon the depth of penetration of the fluid and presumably the degree of swelling of the wood (10).

Acids have a hydrolytic effect upon the cellulose of wood and thus cause some permanent loss in strength (table 6). The weakening effect of alkalies (table 6) appears to be caused by a combination of excessive swelling and softening and dissolving of lignin and hemicelluloses. It appears however that acids with pH values above 2 or bases with pH values below 10 have little weakening effect upon wood at room temperature if length of exposure is moderate (6, 22).

TABLE 6.—*Ratios of loss in strength of wood specimens soaked for 4 weeks in acids and bases at room temperature, to loss of strength of water-soaked specimens* [1]

Species	Ratio [2] of modulus of rupture of specimens soaked in the following acids or bases of the indicated concentration [3] to that of specimens soaked in water													
	Hydrochloric acid		Sulfuric acid		Nitric acid		Lactic acid		Acetic acid		Sodium hydroxide		Ammonium hydroxide	
	2 percent	10 percent	2 percent	10 percent	2 percent	10 percent	2 percent	10 percent	2 percent	10 percent	2 percent	10 percent	2 percent	10 percent
Larch	1.09	0.81	1.11	1.00	1.07	0.95	1.21	1.10	1.07	1.04	0.96	0.52	1.06	0.92
Pine	1.08	.84	1.13	.84	1.07	.93	1.25	1.11	1.11	.88	.96	.51	1.00	.79
Spruce	1.00	.82	1.02	.94	.98	.96	1.00	1.04	1.09	.99	.93	.44	.98	.75
Beech	.96	.55	.97	.87	.83	.71	1.13	1.02	.97	.93	.59	.31	.67	.43
Oak	.93	.52	.97	.82	.82	.65	.99	1.05	.96	.87	.58	.29	.65	.36
Basswood	.88	.48	.87	.81	.76	.57	1.08	.96	1.09	1.00	.55	.21	.61	.33

[1] Values greater than unity indicate greater penetration and swelling by water than by the chemical solutions, thus causing greater loss of strength of water-soaked than of chemical-soaked specimens.
[2] Percent by weight.
[3] Data from "Technologie des Holzes" by F. Kollman (Berlin, 1936).

Highly acidic acid salts, such as zinc chloride, tend to hydrolyze wood if they are present in appreciable concentrations. Fortunately, the concentrations used in wood-preservative and fire-retardant treatments are sufficiently small, so that the strength properties other than impact resistance are not greatly affected under normal use conditions (8, 15). A significant loss in impact strength may occur, however, if higher concentrations are used. When used in hot arid regions, railroad ties containing relatively low concentrations of zinc chloride show signs of migration of the salt to the surface caused by occasional wetting and drying. This migration, combined with the high concentrations of salt relative to the small amount of water present, causes an acidic condition sufficient to make the wood brittle. None of the other common salt preservatives when used in high concentrations is as acidic as zinc chloride, so that their effect upon the strength of the wood can be disregarded in most cases (see p. 420).

Iron salts are frequently very acidic and show hydrolytic action on wood in the presence of free water. The softening and discoloration of the wood around corroded nails results from a partial hydrolysis of the wood. This action is especially pronounced in acidic woods, such as oak, and in woods containing considerable tannin and related compounds, such as redwood. It can be eliminated by using zinc-coated aluminum or copper nails.

A process commonly known as niggerizing has been used for some time to provide wood with increased acid resistance (3, 4, 5). In this process wood is treated with a coke-oven coal tar having a viscosity between 1,250 and 1,450 Saybolt seconds at 200° F., a flashpoint of 270° to 310°, and a fire point of 335° to 375°. Coal tar is the preferred treating material and heartwood of baldcypress or southern yellow pine the preferred wood for niggerizing (2). Acid-resistant paints have not proved suitable for protecting wood in continuous exposure to acids (1, 2). Tests on small laboratory specimens show that phenol resin treatment greatly increases the acid resistance of wood (33), but the difficulty and cost of treating by this method limit its use.

A new proprietary acid-resistant treating agent, somewhat similar to niggerizing agents, is claimed to increase materially the acid resistance of wood. Favorable service tests have been made on wood treated with this material (3, 4, 5, 7) for use in filter plates and frames; acid-storage tanks; "duck boards," walks, and platforms subject to acid spillage or drip; ducts for acid vapors; drainage boxes; and lumber, in general, used near operations involving the use of acid.

WEIGHT OF VARIOUS WOODS

Calculated weights of sawed or round timbers are necessarily approximate values owing to variations in moisture content, density, sapwood thickness, and the like that occur in different parts of the same timber. The calculated average weights obtained by the methods given in this section are not 100-percent accurate, but are more accurate than the weights commonly given in grading rules as a basis for estimating timber transportation costs or other exacting transactions. The methods described here are also useful for roughly determining truck capacity needed to haul a given lot of timbers or the possibility of driving or towing logs. There is enough difference between the weights of sawed and round timbers to require separate methods for estimating their average weights.

Sawed Timbers

Table 7 gives the average weight per cubic foot of sawed timbers of various species at moisture content values of 8 and 15 percent, and the average weight of 1,000 board-feet when air-dry (15-percent moisture content). Factors for adjusting values for each 1 percent change in moisture content are given.

TABLE 7.—*Average weights [1] of sawed hardwood and softwood timbers at moisture contents of 8 and 15 percent, with adjustment factors for changes in moisture content values*

Species	Weight per cubic foot			Weight per 1,000 board-feet air-dry (15-percent moisture content)	
	Based on weight and volume at a moisture content of 15 percent	Based on weight and volume at a moisture content of 8 percent	Factor [2] for adjusting value for each 1-percent change in moisture content	Actual board-feet	Dressed (1 by 8 dressed to $^{25}\!/_{32}$ by 7½)
Hardwoods:	*Pounds*	*Pounds*		*Pounds*	*Pounds*
Alder, red	28.8	28.0	0.112	2,400	1,760
Apple	48.5	47.6	.133	4,040	2,960
Ash:					
Black	35.3	34.3	.142	2,940	2,150
Blue	40.7	39.2	.208	3,390	2,480
Green	40.7	39.4	.179	3,390	2,480
White	42.7	41.5	.175	3,560	2,610
Aspen:					
Bigtooth	27.3	26.6	.104	2,270	1,660
Quaking	27.0	26.1	.129	2,250	1,650
Basswood, American	26.0	25.5	.075	2,170	1,590
Beech, American	44.3	43.2	.162	3,690	2,700
Birch:					
Alaska paper	38.8	38.0	.117	3,230	2,370
Paper	38.9	38.2	.095	3,240	2,370
Sweet	47.2	46.0	.175	3,930	2,880
Yellow	43.4	42.4	.142	3,620	2,650
Buckeye, yellow	25.5	24.8	.104	2,120	1,550
Butternut	27.4	26.4	.145	2,280	1,670
California-laurel	39.5	38.2	.183	3,290	2,410
Cherry, black	36.1	34.8	.183	3,010	2,200
Chestnut, American	30.5	29.5	.145	2,540	1,860
Chinkapin, golden	32.3	31.3	.145	2,690	1,970
Cottonwood:					
Black	24.5	23.8	.104	2,040	1,490
Eastern	28.9	28.0	.125	2,410	1,770
Cucumbertree	34.3	33.3	.142	2,860	2,090
Dogwood:					
Flowering	51.5	50.7	.120	4,290	3,140
Pacific	45.9	44.9	.142	3,820	2,800
Elm:					
American	36.3	35.5	.117	3,020	2,210
Cedar	45.9	44.6	.187	3,820	2,800
Rock	44.2	42.7	.208	3,680	2,700
Slippery	37.8	36.7	.154	3,150	2,310
Hackberry	37.4	36.2	.175	3,120	2,290
Hickory:					
Mockernut	51.4	50.4	.145	4,280	3,130
Pecan	46.5	45.0	.212	3,870	2,830
Pignut	53.4	52.6	.120	4,450	3,260
Shagbark	51.2	50.3	.129	4,270	3,130
Shellbark	49.0	47.8	.170	4,080	2,990
Water	44.6	42.3	.329	3,720	2,720
Holly, American	39.8	38.9	.133	3,320	2,430
Honeylocust	45.3	43.6	.250	3,770	2,760
Hophornbeam, eastern	50.0	48.8	.167	4,170	3,050

See footnotes at end of table.

TABLE 7.—*Average weights* [1] *of sawed hardwood and softwood timbers at moisture contents of 8 and 15 percent, with adjustment factors for changes in moisture content values*—Continued

Species	Weight per cubic foot			Weight per 1,000 board-feet air-dry (15-percent moisture content)	
	Based on weight and volume at a moisture content of 15 percent	Based on weight and volume at a moisture content of 8 percent	Factor [2] for adjusting value for each 1-percent change in moisture content	Actual board-feet	Dressed (1 by 8 dressed to $^{25}\!/_{32}$ by 7½)
Hardwoods—Continued	*Pounds*	*Pounds*		*Pounds*	*Pounds*
Locust, black	49.0	46.7	0.324	4,080	2,990
Madrone, Pacific	45.6	44.6	.150	3,800	2,780
Magnolia, southern	35.5	34.4	.162	2,960	2,170
Mangrove	70.0	68.3	.241	5,830	4,270
Maple:					
Bigleaf	34.2	33.2	.145	2,850	2,090
Black	40.9	39.8	.150	3,410	2,500
Red	37.0	35.6	.195	3,080	2,260
Silver	33.9	32.8	.154	2,820	2,070
Sugar	44.5	43.4	.154	3,710	2,720
Mountain-laurel	48.8	47.7	.158	4,070	2,980
Oak, red:					
Black	44.0	43.0	.150	3,670	2,690
Cherrybark	47.2	45.8	.200	3,930	2,880
Laurel	45.5	44.7	.108	3,790	2,780
Northern red	43.8	42.5	.187	3,650	2,670
Southern red	41.1	40.1	.142	3,420	2,500
Water	44.6	43.8	.120	3,720	2,720
Willow	44.6	43.8	.112	3,720	2,720
Oak, white:					
Bur	45.0	43.5	.208	3,750	2,750
Post	47.9	47.1	.120	3,990	2,920
Swamp chestnut	48.5	47.9	.087	4,040	2,960
White	46.8	45.6	.167	3,900	2,860
Osage-orange	57.3	54.9	.337	4,770	3,490
Palmetto, cabbage	29.6	29.0	.087	2,470	1,810
Persimmon, common	50.8	49.7	.158	4,230	3,100
Poplar, balsam	23.2	22.5	.100	1,930	1,410
Sugarberry	36.5	35.3	.167	3,040	2,230
Sweetgum	36.4	35.5	.133	3,030	2,220
Sycamore, American	35.7	34.7	.137	2,970	2,180
Tupelo:					
Black	36.4	35.5	.129	3,030	2,220
Water	35.1	34.0	.162	2,920	2,140
Walnut, black	38.6	37.0	.233	3,220	2,360
Willow, black	27.6	26.9	.104	2,300	1,680
Yellow-poplar	30.3	29.2	.150	2,520	1,850
Softwoods:					
Baldcypress	32.6	31.4	.167	2,720	1,990
Cedar:					
Alaska-	31.6	30.4	.170	2,630	1,930
Atlantic white-	23.8	23.0	.120	1,980	1,450
Eastern redcedar	33.5	32.2	.187	2,790	2,040
Incense-	25.5	24.2	.183	2,120	1,550
Northern white-	21.8	20.8	.145	1,820	1,330
Port-Orford-	30.1	28.9	.175	2,510	1,840
Western redcedar	23.4	22.4	.137	1,950	1,430

See footnotes at end of table.

TABLE 7.—*Average weights* [1] *of sawed hardwood and softwood timbers at moisture contents of 8 and 15 percent, with adjustment factors for changes in moisture content values*—Continued

Species	Weight per cubic foot			Weight per 1,000 board-feet air-dry (15-per-cent moisture content)	
	Based on weight and volume at a moisture content of 15 percent	Based on weight and volume at a moisture content of 8 percent	Factor [2] for adjusting value for each 1-percent change in moisture content	Actual board-feet	Dressed (1 by 8 dressed to $2\frac{5}{32}$ by $7\frac{1}{2}$)
Softwoods—Continued					
Douglas-fir:	*Pounds*	*Pounds*		*Pounds*	*Pounds*
Coast type	34.3	33.1	0.170	2,860	2,090
Intermediate type	31.8	30.8	.137	2,650	1,940
Rocky Mountain type	30.5	29.2	.179	2,540	1,860
Fir:					
Alpine	22.5	21.3	.167	1,870	1,370
Balsam	26.9	26.4	.071	2,240	1,640
California red	28.3	27.2	.158	2,360	1,730
Grand	28.3	27.3	.145	2,360	1,730
Noble	27.1	26.2	.129	2,260	1,660
Pacific silver	28.1	27.3	.117	2,340	1,710
White	26.7	25.8	.129	2,220	1,630
Hemlock:					
Eastern	29.0	28.0	.150	2,420	1,770
Western	29.6	28.7	.129	2,470	1,810
Juniper, alligator	36.7	35.4	.179	3,060	2,240
Larch, western	39.4	38.2	.170	3,280	2,400
Pine:					
Eastern white	25.4	24.2	.167	2,120	1,550
Jack	30.3	29.2	.158	2,520	1,850
Lodgepole	29.2	28.2	.142	2,430	1,780
Pitch	34.9	33.8	.150	2,910	2,130
Ponderosa	28.6	27.5	.162	2,380	1,740
Red	31.4	30.4	.142	2,620	1,920
Southern yellow:					
Loblolly	36.3	35.2	.154	3,020	2,210
Longleaf	41.6	40.3	.179	3,470	2,540
Shortleaf	35.7	34.6	.154	2,970	2,180
Slash	43.9	42.6	.179	3,660	2,680
Sugar	26.0	24.9	.162	2,170	1,590
Western white	28.0	27.1	.129	2,330	1,710
Redwood (old-growth)	28.6	27.4	.175	2,380	1,740
Spruce:					
Black	28.8	27.8	.142	2,400	1,760
Engelmann	24.1	23.2	.129	2,010	1,470
Red	28.4	27.2	.175	2,370	1,740
Sitka	28.1	27.1	.145	2,340	1,710
White	29.4	28.7	.104	2,450	1,790
Tamarack	37.6	36.3	.187	3,130	2,290
Yew, Pacific	45.7	44.0	.250	3,810	2,790

[1] Based on the weights and volumes of 2- by 2-inch, clear specimens from the top 4 feet of 16-foot butt logs of typical trees.

[2] To adjust value to any desired moisture content up to the fiber saturation point, add factor to value to be adjusted for each 1 percent increase in moisture content; subtract factor from value to be adjusted for each 1 percent decrease in moisture content. These factors take shrinkage or swelling with moisture changes into consideration.

In any lot of lumber of a given species in an air-dry condition at 15-percent moisture content, the weight per cubic foot will rarely vary more than 10 percent from the figure given in table 7. The greatest changes in weight are those which occur in the early stages of drying of green wood. Changes in the moisture content of air-dry wood are attended by only relatively small changes in weight per cubic foot, owing to the countereffect of change in volume as a result of accompanying shrinkage and swelling.

The values given in table 7 for weight per 1,000 board-feet at 15-percent moisture content were determined by multiplying the values per cubic foot at 15 percent by 83.3. Weights per 1,000 board-feet in column 5 apply to actual board-feet and not to 1,000 board-feet lumber scale. Rough lumber is generally oversized and dressed lumber undersized with respect to thickness. Therefore, the values in column 5 usually need to be adjusted for actual shipments of lumber. The adjustment for 1- by 8-inch boards dressed to $2\frac{5}{32}$ inch in thickness and $7\frac{1}{2}$ inches in width is as follows:

$$\frac{2\frac{5}{32} \times 7\frac{1}{2}}{1 \times 8} = 0.7324.$$

Values in column 5 (actual board-feet at 15-percent moisture content) multiplied by the constant 0.7324 give the weights of the dressed lumber shown in column 6 for various species. Similarly, constants for any dressed size may be worked out and the weight per 1,000 board-feet computed. An adjustment for rough oversized lumber is made in a like manner, that is, actual size divided by nominal size.

Round Timbers

The weight per unit volume of green round timbers, such as logs, pulpwood, posts, poles, and piling, may be estimated by means of tables 8 and 9 when the average specific gravity of the wood and the moisture content of its sapwood and heartwood are known or are obtainable from standard reference tables. Table 8 gives the percentage of sapwood in round timbers for various thicknesses and diameters. Table 9 gives the weight per cubic foot of green wood at various specific gravity and moisture content values.

All four tables are necessary for estimating the weight per cubic foot of round timbers, because in round timbers the proportions of sapwood and heartwood in the total volume often differ widely. Furthermore, the sapwood generally contains more water than the heartwood, and both the sapwood and heartwood contain more moisture in the butt logs than in the top logs.

The following example illustrates how to determine the approximate weight per cubic foot of green round timber by using tables 8, 9, 12, and 40.

For black tupelo, the average specific gravity is found from table 12 to be 0.46. The moisture content of the sapwood can be determined by actual measurement or estimated from table 40 as 115 percent.

The moisture content of the heartwood can be determined by actual measurement or estimated from table 40 as 87 percent.

Measure the average diameter of the timber and average width of sapwood. If, for example, the average diameter is 10 inches and the average sapwood thickness is 1¾ inches, then from table 8 the percentage of the volume of the round timber occupied by the sapwood is found to be 58 percent. The percentage of the volume occupied by the heartwood will therefore be 42 percent (100 percent minus 58 percent).

TABLE 8.—*Sapwood, in percent of volume, of round timbers*

Sapwood thickness (inches)	Percent of sapwood in timber in which the average diameter in inches is—												
	4	5	6	7	8	9	10	11	12	13	14	15	16
¼	23	19	16	14	12	11	10	9	8	8	7	7	6
½	44	36	31	27	23	21	19	17	16	15	14	13	12
¾	61	51	44	38	34	31	28	25	23	22	20	19	18
1	75	64	56	49	44	40	36	33	31	28	27	25	23
1¼	86	75	66	59	53	48	44	40	37	35	33	31	29
1½	94	84	75	67	61	56	51	47	44	41	38	36	34
1¾	98	91	83	75	68	63	58	54	50	47	44	41	39
2	100	96	89	82	75	69	64	60	56	52	49	46	44
2¼	------	99	94	87	81	75	70	65	61	57	54	51	48
2½	------	100	97	92	86	80	75	70	66	62	59	56	53
2¾	------	------	99	95	90	85	80	75	71	67	63	60	57
3	------	------	100	98	94	89	84	79	75	71	67	64	61
3¼	------	------	------	99	96	92	88	83	79	75	71	68	65
3½	------	------	------	100	98	95	91	87	83	79	75	72	68
3¾	------	------	------	------	100	97	94	90	86	82	78	75	72
4	------	------	------	------	100	99	96	93	89	85	82	78	75
4¼	------	------	------	------	------	100	98	95	91	88	85	81	78
4½	------	------	------	------	------	100	99	97	94	91	87	84	81
4¾	------	------	------	------	------	------	100	98	96	93	90	87	83
5	------	------	------	------	------	------	100	99	97	95	92	89	86

In table 9 under a specific gravity of 0.46 for a sapwood moisture content of 115 percent, the weight per cubic foot is found to be 61.7 pounds. Under the same specific gravity value and a moisture content of 87 percent, the weight of the heartwood is estimated to be halfway between that given for moisture content values of 86 percent and 88 percent, or 53.7 pounds per cubic foot. (Moisture content values in the left column may be applied to either sapwood or heartwood.)

To find the weight in pounds per cubic foot of the round timber, multiply the weight of sapwood by the percentage of sapwood divided by 100, and similarly for heartwood.

Thus: 61.7 × 58/100=35.8 pounds
53.7 × 42/100=22.6 pounds

The total weight of the round timber per cubic foot is 35.8 plus 22.6, or 58.4 pounds.

TABLE 9.—*Weight in pounds per cubic foot of green wood at various values of specific gravity and moisture content*

Moisture content of wood (percent)	Weight in pounds per cubic foot when the specific gravity [1] is—																				
	0.30	0.32	0.34	0.36	0.38	0.40	0.42	0.44	0.46	0.48	0.50	0.52	0.54	0.56	0.58	0.60	0.62	0.64	0.66	0.68	0.70
30	24.3	26.0	27.6	29.2	30.8	32.4	34.1	35.7	37.3	38.9	40.6	42.2	43.8	45.4	47.0	48.7	50.3	51.9	53.5	55.2	56.8
32	24.7	26.4	28.0	29.7	31.3	32.9	34.6	36.2	37.9	39.5	41.2	42.8	44.5	46.1	47.8	49.4	51.1	52.7	54.4	56.0	57.7
34	25.1	26.8	28.4	30.1	31.8	33.4	35.1	36.8	38.5	40.1	41.8	43.5	45.2	46.8	48.5	50.2	51.8	53.5	55.2	56.9	58.5
36	25.5	27.2	28.9	30.6	32.2	33.9	35.6	37.3	39.0	40.7	42.4	44.1	45.8	47.5	49.2	50.9	52.6	54.3	56.0	57.7	59.4
38	25.8	27.6	29.3	31.0	32.7	34.4	36.2	37.9	39.6	41.3	43.1	44.8	46.5	48.2	49.9	51.7	53.4	55.1	56.8	58.6	60.3
40	26.2	28.0	29.7	31.4	33.2	34.9	36.7	38.4	40.2	41.9	43.7	45.4	47.2	48.9	50.7	52.4	54.2	55.9	57.7	59.4	61.2
42	26.6	28.4	30.1	31.9	33.7	35.4	37.2	39.0	40.8	42.5	44.3	46.1	47.8	49.6	51.4	53.2	54.9	56.7	58.5	60.3	62.0
44	27.0	28.8	30.6	32.3	34.1	35.9	37.7	39.5	41.3	43.1	44.9	46.7	48.5	50.3	52.1	53.9	55.7	57.5	59.3	61.1	62.9
46	27.3	29.2	31.0	32.8	34.6	36.4	38.3	40.1	41.9	43.7	45.6	47.4	49.2	51.0	52.8	54.7	56.5	58.3	60.1	62.0	63.8
48	27.7	29.6	31.4	33.2	35.1	36.9	38.8	40.6	42.5	44.3	46.2	48.0	49.9	51.7	53.6	55.4	57.3	59.1	61.0	62.8	64.6
50	28.1	30.0	31.8	33.7	35.6	37.4	39.3	41.2	43.1	44.9	46.8	48.7	50.5	52.4	54.3	56.2	58.0	59.9	61.8	63.6	65.5
52	28.5	30.4	32.2	34.1	36.0	37.9	39.8	41.7	43.6	45.5	47.4	49.3	51.2	53.1	55.0	56.9	58.8	60.7	62.6	64.5	66.4
54	28.8	30.8	32.7	34.6	36.5	38.4	40.4	42.3	44.2	46.1	48.0	50.0	51.9	53.8	55.7	57.7	59.6	61.5	63.4	65.3	67.3
56	29.2	31.2	33.1	35.0	37.0	38.9	40.9	42.8	44.8	46.7	48.7	50.6	52.6	54.5	56.5	58.4	60.4	62.3	64.2	66.2	68.1
58	29.6	31.5	33.5	35.5	37.5	39.4	41.4	43.4	45.4	47.3	49.3	51.3	53.2	55.2	57.2	59.2	61.1	63.1	65.1	67.0	69.0
60	30.0	31.9	33.9	35.9	37.9	39.9	41.9	43.9	45.9	47.9	49.9	51.9	53.9	55.9	57.9	59.9	61.9	63.9	65.9	67.9	69.9
62	30.3	32.3	34.4	36.4	38.4	40.4	42.5	44.5	46.5	48.5	50.5	52.6	54.6	56.6	58.6	60.7	62.7	64.7	66.7	68.7	70.8
64	30.7	32.7	34.8	36.8	38.9	40.9	43.0	45.0	47.1	49.1	51.2	53.2	55.3	57.3	59.4	61.4	63.4	65.5	67.5	69.6	71.6
66	31.1	33.1	35.2	37.3	39.4	41.4	43.5	45.6	47.6	49.7	51.8	53.9	55.9	58.0	60.1	62.2	64.2	66.3	68.4	70.4	72.5
68	31.4	33.5	35.6	37.7	39.8	41.9	44.0	46.1	48.2	50.3	52.4	54.5	56.6	58.7	60.8	62.9	65.0	67.1	69.2	71.3	73.4
70	31.8	33.9	36.1	38.2	40.3	42.4	44.6	46.7	48.8	50.9	53.0	55.2	57.3	59.4	61.5	63.6	65.8	67.9	70.0	72.1	74.3
72	32.2	34.3	36.5	38.6	40.8	42.9	45.1	47.2	49.4	51.5	53.7	55.8	58.0	60.1	62.3	64.4	66.5	68.7	70.8	73.0	75.1
74	32.6	34.7	36.9	39.1	41.3	43.4	45.6	47.8	49.9	52.1	54.3	56.5	58.6	60.8	63.0	65.1	67.3	69.5	71.7	73.8	76.0
76	32.9	35.1	37.3	39.5	41.7	43.9	46.1	48.3	50.5	52.7	54.9	57.1	59.3	61.5	63.7	65.9	68.1	70.3	72.5	74.7	76.9
78	33.3	35.5	37.8	40.0	42.2	44.4	46.7	48.9	51.1	53.3	55.5	57.8	60.0	62.2	64.4	66.6	68.8	71.1	73.3	75.5	77.8
80	33.7	35.9	38.2	40.4	42.7	44.9	47.2	49.4	51.7	53.9	56.2	58.4	60.7	62.9	65.1	67.4	69.6	71.9	74.1	76.4	78.6
82	34.1	36.3	38.6	40.9	43.2	45.4	47.7	50.0	52.2	54.5	56.8	59.1	61.3	63.6	65.9	68.1	70.4	72.7	75.0	77.2	79.5
84	34.4	36.7	39.0	41.3	43.6	45.9	48.2	50.5	52.8	55.1	57.4	59.7	62.0	64.3	66.6	68.9	71.2	73.5	75.8	78.1	80.4
86	34.8	37.1	39.5	41.8	44.1	46.4	48.7	51.1	53.4	55.7	58.0	60.4	62.7	65.0	67.3	69.6	72.0	74.3	76.6	78.9	81.2
88	35.2	37.5	39.9	42.2	44.6	46.9	49.3	51.6	54.0	56.3	58.7	61.0	63.3	65.7	68.0	70.4	72.7	75.1	77.4	79.8	82.1
90	35.6	37.9	40.3	42.7	45.1	47.4	49.8	52.2	54.5	56.9	59.3	61.7	64.0	66.4	68.8	71.1	73.5	75.9	78.2	80.6	83.0
92	35.9	38.3	40.7	43.1	45.5	47.9	50.3	52.7	55.1	57.5	59.9	62.3	64.7	67.1	69.5	71.9	74.3	76.7	79.1	81.5	83.9

94	36.3	38.7	41.2	43.6	46.0	48.4	50.8	53.3	55.7	58.1	60.5	62.9	65.4	67.8	70.2	72.6	75.1	77.5	79.9	82.3	84.7
96	36.7	39.1	41.6	44.0	46.5	48.9	51.4	53.8	56.3	58.7	61.2	63.6	66.0	68.5	70.9	73.4	75.8	78.3	80.7	83.2	85.6
98	37.1	39.5	42.0	44.5	46.9	49.4	51.9	54.4	56.8	59.3	61.8	64.2	66.7	69.2	71.7	74.1	76.6	79.1	81.5	84.0	86.5
100	37.4	39.9	42.4	44.9	47.4	49.9	52.4	54.9	57.4	59.9	62.4	64.9	67.4	69.9	72.4	74.9	77.4	79.9	82.4	84.9	87.4
105	38.4	40.9	43.5	46.1	48.6	51.2	53.7	56.3	58.8	61.4	64.0	66.5	69.1	71.6	74.2	76.8	79.3	81.9	84.4	87.0	89.5
110	39.3	41.9	44.6	47.2	49.8	52.4	55.0	57.7	60.3	62.9	65.5	68.1	70.8	73.4	76.0	78.6	81.2	83.9	86.5	89.1	91.7
115	40.2	42.9	45.6	48.3	51.0	53.7	56.3	59.0	61.7	64.4	67.1	69.8	72.4	75.1	77.8	80.5	83.2	85.9	88.5	91.2	93.9
120	41.2	43.9	46.7	49.4	52.2	54.9	57.7	60.4	63.1	65.9	68.6	71.4	74.1	76.9	79.6	82.4	85.1	87.9	90.6	93.4	96.1
125	42.1	44.9	47.7	50.5	53.4	56.2	59.0	61.8	64.6	67.4	70.2	73.0	75.8	78.6	81.4	84.2	87.0	89.9	92.7	95.5	98.3
130	43.1	45.9	48.8	51.7	54.5	57.4	60.3	63.1	66.0	68.9	71.8	74.6	77.5	80.4	83.2	86.1	89.0	91.9	94.7	97.6	100.5
135	44.0	46.9	49.9	52.8	55.7	58.7	61.6	64.5	67.5	70.4	73.3	76.3	79.2	82.1	85.1	88.0	90.9	93.8	96.8	99.7	102.6
140	44.9	47.9	50.9	53.9	56.9	59.9	62.9	65.9	68.9	71.9	74.9	77.9	80.9	83.9	86.9	89.9	92.9	95.8	98.8	101.8	104.8
145	45.9	48.9	52.0	55.0	58.1	61.2	64.2	67.3	70.3	73.4	76.4	79.5	82.6	85.6	88.7	91.7	94.8	97.8	100.9	104.0	107.0
150	46.8	49.9	53.0	56.2	59.3	62.4	65.5	68.6	71.8	74.9	78.0	81.1	84.2	87.4	90.5	93.6	96.7	99.8	103.0	106.1	109.2

¹ Based on weight when oven-dry and volume when green.

Working Qualities of Wood

In selecting wood for a given purpose, the ease with which it may be worked is sometimes a factor, especially when handtools are to be used. No test has been devised for definitely classifying woods as to workability with handtools, but a classification based on the experience of the Forest Products Laboratory together with the general reputation of the wood is given in table 10.

Woodworking nowadays is largely done with machines rather than with handtools. Different woods vary in their machining properties just as they do in other properties. In addition, several machining operations are involved, all of which must be considered in appraising the machinability of any given wood. Table 11 summarizes the results of a long series of machining tests with hardwoods (9). The comparisons are based on quality of surface produced, which is the most important consideration when woodworking is done by machine. Comparable data on softwoods are not available.

TABLE 10.—*Classification of certain hardwood and softwood species according to ease of working with handtools* [1]

HARDWOODS

Group 1—Easy to work	Group 2—Relatively easy to work	Group 3—Least easy to work
Alder, red	Birch, paper	Ash, commercial white
Basswood	Cottonwood	Beech
Butternut	Magnolia	Birch
Chestnut	Sweetgum	Cherry
Yellow-poplar	Sycamore	Elm
	Tupelo:	Hackberry
	Black	Hickory, true and pecan
	Water	Honeylocust
	Walnut, black	Locust, black
		Maple
		Oak:
		Commercial red
		Commercial white

SOFTWOODS

Cedar:	Baldcypress	Douglas-fir
Atlantic white-	Fir:	Larch, western
Incense-	Balsam	Pine, southern yellow
Northern white-	White	
Port-Orford-	Hemlock:	
Western redcedar	Eastern	
Pine:	Western	
Eastern white	Pine, lodgepole	
Ponderosa	Redcedar, eastern	
Sugar	Redwood	
Western white	Spruce:	
	Eastern	
	Sitka	

[1] The groupings in the table indicate the approximate order of ease of working based on the experience of the Forest Products Laboratory and the general reputation of the wood. Direct comparison of species within a group and comparison of hardwoods and softwoods is not intended.

TABLE 11.—*Some machining and related properties of hardwoods*

Kind of wood	Planing— perfect pieces	Shaping— good to excellent pieces	Turning— good to excellent pieces	Boring— good to excellent pieces	Mortising— fair to excellent pieces	Sanding— good to excellent pieces	Steam bending— unbroken pieces	Nail splitting— pieces free from complete splits	Screw splitting— pieces free from complete splits
	Percent	*Percent*	*Percent*	*Percent*	*Percent*	*Percent*	*Percent*	*Percent*	*Percent*
Ash	75	51	79	94	62	75	67	65	71
Basswood	64	9	68	75	51	17	2	79	68
Beech		21	90	99	93	49	75	42	58
Birch	63	53	80	98	97	34	72	32	48
Buckeye		6	58	75	18		9		
Chestnut	74	24	87	91	72	64	56	66	60
Cottonwood	21	3	70	70	52	19	44	82	78
Elm	33	11	65	94	75	66	74	80	74
Hackberry	74	10	77	99	70		94	63	63
Hickory		19	84	100	98	80	76	35	63
Magnolia	65	25	79	69	32	37	85	73	76
Mahogany	80	68	89	100	100		41	68	78
Maple:									
Hard	54	62	82	99	95	38	57	27	52
Soft	41	22	76	80	36	37	59	58	61
Oak:									
Chestnut		23	90	100	100	75	85	49	70
Red	91	21	84	99	100	81	86	66	78
White	87	28	85	95	100	83	91	69	74
Pecan	88	31	89	100	100		78	47	69
Sweetgum	51	21	86	92	58	23	67	69	69
Sycamore	22	8	85	98	96	21	29	79	74
Tupelo:									
Black	48	23	75	82	24	21	42	65	63
Water		43	79	62	35	34	46	64	63
Walnut, black	62	34	91	100	98		78	50	59
Willow	52	5	58	71	24	24	73	89	62
Yellow-poplar	70	12	81	87	63	19	58	77	67
Average	61	25	79	89	70	45	62	62	67

LITERATURE CITED

(1) ANONYMOUS.
1929. RESISTANT WOODS SATISFY MOST CORROSION REQUIREMENTS. Chem. & Metall. Engin. 36 (9): 567–568, illus.
(2) ———
1942. WOOD FOR CHEMICAL EQUIPMENT. Chem. Engin. 49 (9): 104.
(3) ———
1946. REPORT OF COMMITTEE 7–10. DIVERSIFIED USES OF TREATED WOOD. Amer. Wood-Preservers' Assoc. Proc. 42: 255–265, illus.
(4) ———
1947. REPORT OF COMMITTEE 7–10. DIVERSIFIED USES OF TREATED WOOD. Amer. Wood-Preservers' Assoc. Proc. 43: 317–330, illus.
(5) ———
1948. REPORT OF COMMITTEE 7–10. DIVERSIFIED USES OF TREATED WOOD. Amer. Wood-Preservers' Assoc. Proc. 44: 343–352, illus.
(6) ALLIOTT, E. A.
1926. EFFECTS OF ACIDS ON THE MECHANICAL STRENGTH OF TIMBER: A PRELIMINARY STUDY. Soc. Chem. Indus. Jour. 45 (53): 463T–466T, illus.

(7) BESCHER, R. H.
    1947. ACID-PROOFING OF WOOD. Forest Prod. Res. Soc. Proc. 1: 120–123, illus.
(8) BETTS, H. S., AND NEWLIN, J. A.
    1915. STRENGTH TESTS OF STRUCTURAL TIMBERS TREATED BY COMMERCIAL WOOD-PRESERVING PROCESSES. U. S. Dept. Agr. Bul. 286, 15 pp., illus.
(9) DAVIS, E. M.
    1942. MACHINING AND RELATED PROPERTIES OF SOUTHERN HARDWOODS. U. S. Dept. Agr. Tech. Bul. 824, 42 pp., illus.
(10) DROW, J. T.
    1945. EFFECT OF HYDRAULIC-EQUIPMENT OILS ON THE BENDING AND COMPRESSIVE STRENGTH OF SITKA SPRUCE. U. S. Forest Prod. Lab. Rpt. 1520, 8 pp., illus. [Processed.]
(11) DUNLAP, M. E., AND BELL, E. R.
    1951. ELECTRICAL MOISTURE METERS FOR WOOD. U. S. Forest Prod. Lab. Rpt. 1660, 10 pp., illus.
(12) ERICKSON, H. D., AND REES, L. W.
    1940. EFFECT OF SEVERAL CHEMICALS ON THE SWELLING AND THE CRUSHING STRENGTH OF WOOD. Jour. Agr. Res. 60 (9): 593–603, illus.
(13) HARTE, C. R.
    1936. WOOD TANKS. EQUIPMENT FOR THE CHEMICAL PROCESS INDUSTRIES. Indus. and Engin. Chem. 28 (2): 176–179, illus.
(14) HASSELBLATT, M.
    1926. VAPOR PRESSURE AND ELECTRIC CONDUCTIVITY OF WOOD AT DIFFERENT MOISTURE CONTENTS. Ztschr. f. Anorgan. u. Allg. Chem. 154: 375, illus.
(15) HATT, W. K.
    1906. EXPERIMENTS ON THE STRENGTH OF TREATED TIMBER. U. S. Dept. Agr. Forest Serv. Cir. 39, 31 pp.
(16) HIRUMA, J.
    1913. EXPERIMENTS ON THE ELECTRICAL RESISTANCE OF WOOD. Bul. Forest Expt. Sta., Meguro, Tokyo, 10: 59–65.
(17) KOEHLER, A.
    1917. GUIDEBOOK FOR THE IDENTIFICATION OF WOODS USED FOR TIES AND TIMBERS. U. S. Dept. Agr. Misc. RL–1, 79 pp., illus.
(18) ——
    1922. IDENTIFICATION OF TRUE MAHOGANY AND CERTAIN SO-CALLED MAHOGANIES. U. S. Dept. Agr. Bul. 1050, 18 pp., illus.
(19) ——
    1924. PROPERTIES AND USE OF WOOD. 354 pp., illus. New York.
(20) ——
    1926. IDENTIFICATION OF FURNITURE WOODS. U. S. Dept. Agr. Misc. Cir. 66, 78 pp., illus.
(21) ——
    1932. IDENTIFICATION OF LONGLEAF PINE TIMBERS. South. Lumberman 145 (1841): 36–37, illus.
(22) KOLLMANN, F.
    1936. TECHNOLOGIE DES HOLZES. 764 pp., illus. Berlin.
(23) MACLEAN, J. D.
    1941. THERMAL CONDUCTIVITY OF WOOD. Heating, Piping and Air Conditioning 13 (6): 380–391, illus.
(24) ——
    1946. TEMPERATURES OBTAINED IN TIMBERS WHEN THE SURFACE TEMPERATURE IS CHANGED AFTER VARIOUS PERIODS OF HEATING. U. S. Forest Prod. Lab. Rpt. R1609, 30 pp., illus. [Processed.]
(25) ——
    1951. RATE OF DISINTEGRATION OF WOOD UNDER DIFFERENT HEATING CONDITIONS. Amer. Wood-Preservers' Assoc. Proc. 47: 155–169, illus.
(26) ——
    1952. EFFECT OF TEMPERATURE ON THE DIMENSIONS OF GREEN WOOD. Amer. Wood-Preservers' Assoc. Proc. 48: 136–157, illus.

(27) MARKWARDT, L. J., AND WILSON, T. R. C.
    1935. STRENGTH AND RELATED PROPERTIES OF WOODS GROWN IN THE
        UNITED STATES. U. S. Dept. Agr. Bul. 479, 99 pp., illus.
(28) RECORD, S. J.
    1934. IDENTIFICATION OF THE TIMBERS OF TEMPERATE NORTH AMERICA,
        INCLUDING ANATOMY AND CERTAIN PHYSICAL PROPERTIES OF
        WOOD. 196 pp., illus. New York.
(29) RENO, J., AND KUKACHKA, B. F.
    1950. WOOD IDENTIFICATION CHART. Wood 5 (1): 25-27, illus.; (2):
        28–29; (3): 22–23, illus.; (4): 24–25; (5): 26–27; (6): 26–27;
        (7): 22–23; (8): 20–21; (9): 24–25. Chicago.
(30) SKAAR, C.
    1948. DIELECTRIC PROPERTIES OF WOOD AT SEVERAL RADIO FREQUENCIES.
        N. Y. State Col. Forestry, 36 pp., illus.
(31) STAMM, A. J.
    1927. ELECTRICAL RESISTANCE OF WOOD AS A MEASURE OF THE MOISTURE
        CONTENT. Indus. and Engin. Chem. 19 (9): 1021–1025, illus.
(32) ———
    1930. ELECTRICAL CONDUCTIVITY METHOD FOR DETERMINING THE MOIS-
        TURE CONTENT OF WOOD. Indus. and Engin. Chem., Analyt.
        Ed. 2 (3): 240–244, illus.
(33) ——— AND SEBORG, R. M.
    1939. RESIN-TREATED PLYWOOD. Indus. and Engin. Chem. 31 (7):
        897–902, illus.
(34) WEATHERWAX, R. C., AND STAMM, A. J.
    1945. ELECTRICAL RESISTIVITY OF RESIN-TREATED WOOD, LAMINATED
        HYDROLIZED WOOD, AND PAPER-BASE PLASTICS. Elect. Engin.
        64: 833, illus.
(35) ——— AND STAMM, A. J.
    1946. COEFFICIENTS OF THERMAL EXPANSION OF WOOD AND WOOD PROD-
        UCTS. U. S. Forest Prod. Lab. Rpt. 1487, 24 pp., illus. [Proc-
        essed.]
(36) WILSON, T. R. C.
    1932. STRENGTH-MOISTURE RELATIONS FOR WOOD. U. S. Dept. Agr.
        Tech. Bul. 282, 88 pp., illus.

STRENGTH VALUES OF CLEAR WOOD AND RE-LATED FACTORS

STRENGTH VALUES

The strength properties for any species of wood, or the properties that enable wood to resist applied forces, are truly representative only when obtained from tests on small, clear pieces of wood, because the effect of such things as knots, cross grain, checks and splits, and compression wood is then eliminated. Table 12 gives these strength properties for some of the commercially important species; additional information is presented in U. S. Department of Agriculture Technical Bulletins 158 and 479 (15, 16).[3] The data are based on tests of specimens 2 by 2 inches or smaller in cross section and of different lengths, depending on the test. Standard testing procedures of the American Society for Testing Materials were followed (1).

Since there is a considerable difference in the strength of small, clear pieces of wood when green and when air-dry, strength values are given for both conditions. The normal increase in strength with loss of moisture shown in table 12, however, does not hold for large pieces, because the development of checks in and around knots and of shakes and checks along the neutral axis during seasoning usually offsets the increase in strength caused by drying. For allowable working stresses for various species, see page 137.

Common and Botanical Names of Species, Column 1

The species listed in table 12 are grouped for convenience into hardwoods and softwoods. (The terms are not correlated with the hardness or softness of the wood.) Many of the species have numerous common names, and frequently one common name is applied to several species. In order to eliminate confusion, the common and botanical names in the tables of this report conform to the standard nomenclature of the U. S. Forest Service.

Moisture Content, Column 2

Moisture content is the weight of the water contained in the wood expressed as a percentage of the weight of the oven-dry wood.

The moisture content given for green wood is the average for specimens taken from the pith to the circumference of the log. It includes the moisture found in both the heartwood and the sapwood and is the approximate moisture content of the living tree. In many species, there is much more moisture in the sapwood than in the heartwood.

The moisture content of seasoned wood when tested varied somewhat among the different species. To facilitate comparison of the strength properties, the test values were adjusted to conform to the uniform moisture content of 12 percent, in accordance with standard procedures outlined in the section on moisture content.

---

[3] Italic numbers in parentheses refer to Literature Cited, p. 103.

Specific Gravity, Column 3

Specific gravity is the ratio of the weight of a given volume of wood to that of an equal volume of water at a standard temperature. Since the weight of wood in a given volume changes with the shrinkage and swelling caused by changes in moisture content, specific gravity is an indefinite quantity unless the conditions under which it is obtained are specified. The 2 specific gravity values presented for each species are based on the weight of the wood when oven-dry and its volume when green and at 12-percent moisture content.

Static Bending, Columns 4, 5, 6, 7, and 8

Fiber Stress at Proportional Limit

The fiber stress at proportional limit in static bending or flexure is the computed stress in the wood specimen at which the strain (or deflection) becomes no longer proportional to the stress (or load). It is therefore the stress in the specimen at which the load-deflection curve departs from a straight line.

Modulus of Rupture

The modulus of rupture is a measure of the ability of a beam to support a slowly applied load for a short time. It is an accepted criterion of strength, although it is not a true stress, since the formula by which it is computed is only valid to the proportional limit.

Modulus of Elasticity

The modulus of elasticity of wood is a measure of its stiffness or rigidity. For a beam, the modulus of elasticity is a measure of its resistance to deflection. The modulus of elasticity as determined from bending tests includes deflection due to shear distortion. The moduli of elasticity in compression parallel to grain may be taken as 10 percent higher than the figures in column 6.

Work in Bending to Proportional Limit

Work to proportional limit in static bending is a measure of the energy absorbed by a beam when it is stressed to the proportional limit. It is a comparative property that indicates the ability of the wood to absorb shock without permanent damage.

Work in Bending to Maximum Load

Work to maximum load in static bending represents the ability of the timber to absorb shock with some permanent deformation and more or less injury to the timber. It is a measure of the combined strength and toughness of wood under bending stresses.

Impact Bending, Column 9

Height of Drop Causing Complete Failure

In the impact bending test, a hammer of given weight is dropped upon a beam from successively increased heights until complete rupture occurs. The height of the maximum drop, or the drop that causes failure, is a comparative figure representing the ability of wood to absorb shocks that cause stresses beyond the proportional limit. This ability of wood is important in tool handles, baseball bats, and other articles subjected to frequent shocks.

Compression Parallel to Grain, Columns 10 and 11

Fiber Stress at Proportional Limit

The fiber stress at proportional limit in compression parallel to the grain is the stress in the specimen at the limit of proportionality between stress and strain. It is the stress at which the stress-strain curve for compression specimens having a ratio of length to least dimension of less than 11 departs from a straight line.

Maximum Crushing Strength

The maximum crushing strength is the maximum stress sustained by a compression specimen having a ratio of length to least dimension of less than 11 under a load slowly applied parallel to the grain. This property permits evaluation of the strength of posts or short blocks.

Compression Perpendicular to Grain, Column 12

Fiber Stress at Proportional Limit

The fiber stress at proportional limit in compression perpendicular to the grain is the maximum across-the-grain stress of a few minutes' duration that can be applied through a plate covering only a portion of a timber surface without causing injury to the timber.

Shearing Strength Parallel to Grain, Column 13

Shearing strength is a measure of the ability of timber to resist slipping of one part upon another along the grain.

Tensile Strength Perpendicular to Grain, Column 14

Tensile strength perpendicular to the grain is a measure of the resistance of wood to forces acting across the grain that tend to split a member.

Hardness, Columns 15 and 16

Hardness represents the resistance of wood to wear and marring. Values are presented for end-grain surfaces and side-grain surfaces (average of radial and tangential values). It is measured by the load required to imbed a 0.444-inch ball to one-half its diameter in the wood.

TABLE 12.—*Strength properties of some commercially important woods grown in the United States*

[Results of tests on small, clear specimens in the green and air-dry condition [1]]

Common and botanical names of species (1)	Moisture content (2)	Specific gravity [2] (3)	Static bending — Fiber stress at proportional limit (4)	Static bending — Modulus of Rupture (5)	Static bending — Modulus of Elasticity (6)	Static bending — Work to Proportional limit (7)	Static bending — Work to Maximum load (8)	Impact bending—height of drop causing complete failure (50-pound hammer) (9)	Compression parallel to grain — Fiber stress at proportional limit (10)	Compression parallel to grain — Maximum crushing strength (11)	Compression perpendicular to grain—fiber stress at proportional limit (12)	Shear parallel to grain—maximum shearing strength (13)	Tension perpendicular to grain—maximum tensile strength (14)	Hardness—load required to embed a 0.444-inch ball to ½ its diameter — End (15)	Hardness — Side (16)
	Pct.		*P.s.i.*	*P.s.i.*	*1,000 p.s.i.*	*In.-Lb. per cu. in.*	*In.-Lb. per cu. In.*	*In.*	*P.s.i.*	*P.s.i.*	*P.s.i.*	*P.s.i.*	*P.s.i.*	*Lb.*	*Lb.*
HARDWOODS															
Alder, red (*Alnus rubra*)	98	0.37	3,800	6,500	1,170	0.70	8.0	22	2,620	2,960	310	770	390	550	440
	12	.41	6,900	9,800	1,380	1.85	8.4	20	4,530	5,820	540	1,080	420	980	590
Ash:															
Black (*Fraxinus nigra*)	85	.45	2,600	6,000	1,040	.41	12.1	33	1,690	2,300	430	860	490	590	520
	12	.49	7,200	12,600	1,600	1.57	14.9	35	4,520	5,970	940	1,570	700	1,150	850
Green (*F. pennsylvanica*)	48	.53	5,300	9,500	1,400	1.14	11.8	35	3,560	4,200	910	1,260	590	960	870
	12	.56	8,900	14,100	1,660	2.72	13.4	32	5,120	7,080	1,620	1,910	700	1,630	1,200
Oregon (*F. latifolia*)	48	.50	4,200	7,600	1,130	.92	12.2	39	2,760	3,510	650	1,190	590	850	790
	12	.55	7,000	12,700	1,360	2.08	14.4	33	4,100	6,040	1,540	1,790	720	1,430	1,160
White (*F. americana*)	42	.55	5,100	9,600	1,460	1.04	16.6	38	3,190	3,990	810	1,380	590	1,010	960
	12	.60	8,900	15,400	1,770	2.60	17.6	43	5,790	7,410	1,410	1,950	940	1,720	1,320
Aspen, quaking (*Populus tremuloides*)	94	.35	3,200	5,100	860	.69	6.4	22	1,670	2,140	220	660	230	280	300
	12	.38	5,600	8,400	1,180	1.53	7.6	21	3,040	4,250	460	850	260	510	350
Basswood, American (*Tilia americana*)	105	.32	2,700	5,000	1,040	.40	5.3	16	1,690	2,220	210	600	280	290	250
	12	.37	5,900	8,700	1,460	1.37	7.2	16	3,800	4,730	450	990	350	520	410
Beech, American (*Fagus grandifolia*)	54	.56	4,300	8,600	1,380	.85	11.9	43	2,550	3,550	670	1,290	720	970	850
	12	.64	8,700	14,900	1,720	2.63	15.1	41	4,880	7,300	1,250	2,010	1,010	1,590	1,300
Birch:															
Paper (*Betula papyrifera*)	65	.48	3,000	6,400	1,170	.45	16.2	49	1,640	2,360	340	840	380	470	560
	12	.55	6,900	12,300	1,590	1.80	16.0	34	3,610	5,690	740	1,210		890	910

Species	(1)	(2)	(3)	(4)	(5)	(6)	(7)	(8)	(9)	(10)	(11)	(12)	(13)	(14)	(15)
Sweet (*B. lenta*)	53	.60	4,800	9,400	1,650	.94	15.7	48	2,680	3,740	580	1,240	430	1,070	970
	12	.65	10,100	16,900	2,170	2.72	18.0	47	6,330	8,540	1,340	2,240	950	1,960	1,470
Yellow (*B. alleghaniensis*)	67	.55	4,200	8,300	1,500	.70	16.1	48	2,620	3,380	530	1,110	430	810	780
	12	.62	10,100	16,600	2,010	2.89	20.8	55	6,130	8,170	1,190	1,880	920	1,480	1,260
Butternut (*Juglans cinerea*)	104	.36	2,900	5,400	970	.52	8.2	24	2,020	2,420	270	760	430	410	390
	12	.38	5,700	8,100	1,180	1.59	8.2	24	4,200	5,110	570	1,170	440	570	490
Cherry, black (*Prunus serotina*)	55	.47	4,200	8,000	1,310	.80	12.8	33	2,940	3,540	440	1,130	570	750	660
	12	.50	9,000	12,300	1,490	3.11	11.4	29	5,960	7,110	850	1,700	560	1,470	950
Chestnut, American (*Castanea dentata*)	122	.40	3,100	5,600	930	.59	7.0	24	2,080	2,470	380	800	440	530	420
	12	.43	6,100	8,600	1,230	1.78	6.5	19	3,780	5,320	760	1,080	460	720	540
Cottonwood:															
Black (*Populus trichocarpa*)	132	.32	2,900	4,800	1,070	.44	5.0	20	1,760	2,160	200	600	270	280	250
	12	.35	5,300	8,300	1,260	1.25	6.7	22	3,270	4,420	370	1,020	330	540	350
Eastern (*P. deltoides*)	111	.37	2,900	5,300	1,010	.49	7.3	21	1,740	2,280	240	680	410	380	340
	12	.40	5,700	8,500	1,370	1.39	7.4	20	3,490	4,910	470	930	580	580	430
Cucumbertree (*Magnolia acuminata*)	80	.44	4,200	7,400	1,560	.66	10.0	30	2,810	3,140	410	990	440	600	520
	12	.48	8,000	12,300	1,820	1.98	12.2	35	4,840	6,310	710	1,340	660	950	700
Elm:															
American (*Ulmus americana*)	89	.46	3,900	7,200	1,110	.81	11.8	38	1,920	2,910	440	1,000	590	680	620
	12	.50	7,600	11,800	1,340	2.53	13.0	39	4,030	5,520	850	1,510	660	1,110	830
Cedar (*U. crassifolia*)	60	.59	3,900	9,200	1,170	.75	20.0	60	2,350	3,730	600	1,320	690	1,160	1,100
	12	.64	5,700	13,500	1,480	1.24	18.6	59	2,600	6,020	950	2,240	1,220	1,640	1,320
Rock (*U. thomasii*)	48	.57	4,600	9,500	1,190	1.05	19.8	54	2,970	3,780	750	1,270	—	980	940
	12	.63	8,000	14,800	1,540	2.45	19.2	56	4,700	7,050	1,520	1,920	640	1,510	1,320
Slippery (*U. rubra*)	85	.48	4,000	8,000	1,230	.82	15.4	47	2,790	3,320	510	1,110	530	750	660
	12	.53	7,700	13,000	1,490	2.35	16.9	45	4,760	6,360	1,010	1,630	850	1,120	860
Winged (*U. alata*)	59	.60	4,000	9,200	1,210	.81	21.7	73	1,980	3,700	630	1,300	1,210	1,100	1,140
	12	.66	6,300	14,800	1,650	1.36	23.1	69	2,920	6,780	1,020	2,370	630	1,850	1,540
Hackberry (*Celtis occidentalis*)	65	.49	2,900	6,500	950	.58	14.5	48	2,070	2,650	490	1,070	580	760	700
	12	.53	5,900	11,000	1,190	1.72	12.8	43	3,710	5,440	1,100	1,590	—	1,110	880
Hickory, pecan:															
Bitternut (*Carya cordiformis*)	66	.60	5,500	10,300	1,400	1.22	20.0	66	4,330	4,570	990	1,240	—	—	—
	12	.66	9,300	17,100	1,790	2.73	18.2	66	—	9,040	2,070	—	—	—	—
Nutmeg (*C. myristicaeformis*)	74	.56	4,900	9,100	1,290	1.06	22.8	54	3,620	3,980	940	1,030	—	—	—
	12	.60	8,100	16,600	1,700	2.04	25.1	—	—	6,910	1,930	—	—	—	—
Pecan (*C. illinoensis*)	63	.60	5,200	9,800	1,370	1.18	14.6	53	3,100	3,990	960	1,480	680	1,270	1,310
	12	.66	9,100	13,700	1,730	2.81	13.8	44	5,180	7,850	2,130	2,080	—	1,930	1,820
Water (*C. aquatica*)	80	.61	6,000	10,700	1,560	1.29	18.8	56	3,240	4,660	1,090	1,440	—	—	—
	12	.62	10,200	17,800	2,020	2.88	19.3	53	5,400	8,600	1,910	—	—	—	—

See footnotes at end of table.

Table 12.—*Strength properties of some commercially important woods grown in the United States*—Continued

[Results of tests on small, clear specimens in the green and air-dry condition [1]]

Common and botanical names of species	Moisture content	Specific gravity[2]	Static bending — Fiber stress at proportional limit	Static bending — Modulus of Rupture	Static bending — Modulus of Elasticity	Work to — Proportional limit	Work to — Maximum load	Impact bending — height of drop causing complete failure (50-pound hammer)	Compression parallel to grain — Fiber stress at proportional limit	Compression parallel to grain — Maximum crushing strength	Compression perpendicular to grain — fiber stress at proportional limit	Shear parallel to grain — maximum shearing strength	Tension perpendicular to grain — maximum tensile strength	Hardness — End	Hardness — Side
(1)	(2)	(3)	(4)	(5)	(6)	(7)	(8)	(9)	(10)	(11)	(12)	(13)	(14)	(15)	(16)
	Pct		*P.s.i.*	*P.s.i.*	*1,000 p.s.i.*	*In.-Lb. per cu. in.*	*In.-Lb. per cu. in.*	*In.*	*P.s.i.*	*P.s.i.*	*P.s.i.*	*P.s.i.*	*P.s.i.*	*Lb.*	*Lb.*
HARDWOODS—continued															
Hickory, true:															
Mockernut (*Carya tomentosa*)	59	.64	6,300	11,100	1,570	1.38	26.1	88	3,900	4,480	1,000	1,280	-----	-----	-----
	12	.72	11,900	19,200	2,220	3.41	22.6	77		8,940	2,140	1,740	-----	-----	-----
Pignut (*C. glabra*)	54	.66	6,200	11,700	1,650	1.34	31.7	89	3,950	4,810	1,140	1,370	-----	-----	-----
	12	.75	11,300	20,100	2,260	3.23	30.4	74		9,190	2,450	2,150	-----	-----	-----
Shagbark (*C. ovata*)	60	.64	5,900	11,000	1,570	1.28	23.7	74	3,430	4,580	1,040	1,520	-----	-----	-----
	12	.72	10,700	20,200	2,160	3.01	25.8	67		9,210	2,170	2,430	-----	-----	-----
Shellbark (*C. laciniosa*)	61	.62	5,600	10,500	1,340	1.36	29.9	104	2,740	3,920	1,000	1,190	-----	-----	-----
	12	.69	8,900	18,100	1,890	2.29	23.6	88		8,000	2,220	2,110	-----	-----	-----
Honeylocust (*Gleditsia triacanthos*)	63	.60	5,600	10,200	1,290	1.40	12.6	88	3,320	4,420	1,420	1,660	930	1,440	1,390
	12	--------	8,800	14,700	1,630	2.74	13.3	47	5,250	7,500	2,280	2,250	900	1,860	1,580
Locust, black (*Robinia pseudoacacia*)	40	.66	8,800	13,800	1,850	2.36	15.4	47	6,120	6,800	1,430	1,760	770	1,640	1,570
	12	.69	12,800	19,400	2,050	4.62	18.4	44	6,800	10,180	2,260	2,480	640	1,580	1,700
Magnolia, southern (*Magnolia grandiflora*)	105	.46	3,600	6,800	1,110	.67	15.4	57	2,160	2,700	570	1,040	610	780	740
	12	.50	6,800	11,200	1,400	1.90	12.8	54	3,420	5,460	1,060	1,530	740	1,280	1,020
Maple:															
Bigleaf (*Acer macrophyllum*)	72	.44	4,400	7,400	1,100	1.02	8.7	29	2,510	3,240	550	1,110	600	760	620
	12	.48	6,600	10,700	1,450	1.66	7.8	23	4,790	5,950	930	1,730	540	1,330	850
Black (*A. nigrum*)	65	.52	4,100	7,900	1,330	.70	12.8	28	2,800	3,270	740	1,130	720	940	840
	12	.57	8,300	13,300	1,620	2.39	12.5	48	4,600	6,680	1,250	1,820	670	1,700	1,180
Red (*A. rubrum*)	63	.49	3,800	7,700	1,390	.71	11.4	32	2,360	3,280	500	1,150		780	700
	12	.54	8,700	13,400	1,640	2.84	12.5	32	4,650	6,540	1,240	1,850		1,430	950

Silver (*A. saccharinum*)	66	.44	3,100	5,800	940	.61	11.0	29	1,930	2,490	460	1,050	560	670	590
	12	.47	6,200	8,900	1,140	1.90	8.3	25	4,360	5,220	910	1,480	500	1,140	700
Sugar (*A. saccharum*)	58	.56	5,100	9,400	1,550	1.03	13.3	40	2,850	4,020	800	1,460	--------	1,070	970
	12	.63	9,500	15,800	1,830	2.76	16.5	39	5,390	7,830	1,810	2,330	--------	1,840	1,450
Oak, red:															
Black (*Quercus velutina*)	80	.56	4,600	8,200	1,180	1.02	12.2	40	2,720	3,470	870	1,220	--------	1,000	1,060
	12	.61	7,900	13,900	1,640	2.15	13.7	41	4,750	6,520	1,150	1,910	--------	1,380	1,210
Cherrybark (*Q. falcata* var. *pagodaefolia*)	78	.61	6,500	10,800	1,790	1.32	14.7	54	3,820	4,620	940	1,320	800	1,270	1,240
	12	.68	11,200	18,100	2,280	3.09	18.3	49	6,350	8,740	1,540	2,000	840	1,570	1,480
Laurel (*Q. laurifolia*)	84	.56	4,500	7,900	1,390	.86	11.2	39	2,650	3,170	710	1,180	770	1,020	1,000
	12	.63	7,700	12,600	1,690	2.02	11.8	39	4,640	6,980	1,310	1,830	790	1,230	1,210
Northern red (*Q. rubra*)	80	.56	4,100	8,300	1,350	.73	13.2	44	2,360	3,440	760	1,210	750	1,060	1,000
	12	.63	8,500	14,300	1,820	2.33	14.5	43	4,580	6,760	1,250	1,780	800	1,580	1,290
Pin (*Q. palustris*)	75	.58	4,000	8,300	1,320	.71	14.0	48	--------	3,680	880	1,290	800	1,000	1,070
	12	.63	8,000	14,000	1,730	2.22	14.8	45	4,620	6,820	1,260	2,080	1,050	1,600	1,510
Scarlet (*Q. coccinea*)	65	.60	4,500	10,400	1,480	.81	15.0	54	2,840	4,090	1,030	1,410	700	1,170	1,200
	12	.67	9,700	17,400	1,910	2.92	20.5	53	5,550	8,330	1,380	1,890	870	1,690	1,400
Southern red (*Q. falcata*)	90	.52	4,200	6,900	1,140	.93	8.0	29	2,220	3,030	680	930	480	910	860
	12	.59	6,000	10,900	1,490	1.44	9.4	26	2,910	6,090	1,080	1,390	510	1,020	1,060
Water (*Q. nigra*)	81	.56	5,600	8,900	1,550	1.14	11.1	39	3,260	3,740	770	1,240	820	1,050	1,010
	12	.63	8,900	15,400	2,020	2.24	21.5	44	3,960	6,770	1,260	2,020	920	1,400	1,190
Willow (*Q. phellos*)	94	.56	4,400	7,400	1,290	.88	8.8	35	2,340	3,000	750	1,180	760	1,020	980
	12	.69	9,300	14,500	1,900	2.61	14.6	42	4,380	7,040	1,400	1,650	--------	1,420	1,460
Oak, white:															
Bur (*Quercus macrocarpa*)	70	.58	3,600	7,200	880	.89	10.7	44	2,380	3,290	840	1,350	800	1,160	1,110
	12	.64	6,400	10,300	1,030	2.37	9.8	29	3,580	6,060	1,480	1,820	680	1,410	1,370
Chestnut (*Q. prinus*)	72	.57	4,600	8,000	1,370	.90	9.4	35	2,890	3,520	660	1,210	690	970	890
	12	.66	9,000	13,300	1,590	2.88	11.0	40	4,420	6,830	1,040	1,490	--------	1,250	1,130
Overcup (*Q. lyrata*)	83	.57	4,000	8,000	1,150	.79	12.6	44	2,270	3,370	540	1,320	730	1,010	960
	12	.63	5,500	12,600	1,420	1.20	15.7	38	2,610	6,200	810	2,000	940	1,410	1,190
Post (*Q. stellata*)	69	.60	5,000	8,100	1,090	1.31	11.0	44	2,840	3,480	1,060	1,280	790	1,160	1,130
	12	.67	7.600	13,200	1,510	2.25	13.2	46	3,700	6,600	1,760	1,840	780	1,350	1,360
Swamp chestnut (*Q. michauxii*)	76	.60	4,800	8,500	1,350	1.00	12.8	45	3,000	3,540	710	1,260	670	1,100	1,110
	12	.67	7,300	13,900	1,770	1.68	12.0	41	4,400	7,270	1,370	1,990	690	1,290	1,240
Swamp white (*Q. bicolor*)	74	.64	5,400	9,900	1,590	1.05	14.5	50	3,580	4,360	940	1,300	860	1,200	1,160
	12	.72	10,200	17,700	2,050	2.88	19.2	49	5,830	8,600	1,470	2,000	830	1,680	1,620
White (*Q. alba*)	68	.60	4,700	8,300	1,250	1.08	11.6	42	3,090	3,560	830	1,250	770	1,120	1,060
	12	.68	8,200	15,200	1,780	2.27	14.8	37	4,760	7,440	1,320	2,000	800	1,520	1,360

See footnotes at end of table.

TABLE 12.—*Strength properties of some commercially important woods grown in the United States*—Continued

[Results of tests on small, clear specimens in the green and air-dry condition [1]]

Common and botanical names of species	Moisture content	Specific gravity [2]	Static bending					Impact bending—height of drop causing complete failure (50-pound hammer)	Compression parallel to grain		Compression perpendicular to grain—fiber stress at proportional limit	Shear parallel to grain—maximum shearing strength	Tension perpendicular to grain—maximum tensile strength	Hardness—load required to embed a 0.444-inch ball to ½ its diameter	
			Fiber stress at proportional limit	Modulus of—		Work to—			Fiber stress at proportional limit	Maximum crushing strength				End	Side
				Rupture	Elasticity	Proportional limit	Maximum load								
	(2)	(3)	(4)	(5)	(6)	(7)	(8)	(9)	(10)	(11)	(12)	(13)	(14)	(15)	(16)
	Pct.		*P.s.i.*	*P.s.i.*	*1,000 p.s.i.*	*In.-Lb. per cu. in.*	*In.-Lb. per cu. in.*	*In.*	*P.s.i.*	*P.s.i.*	*P.s.i.*	*P.s.i.*	*P.s.i.*	*Lb.*	*Lb.*
HARDWOODS—continued															
Sugarberry (*Celtis laevigata*)	62	0.47	3,200	6,600	810	0.78	12.0	33	1,990	2,800	580	1,050	660	840	740
	12	.51	6,200	9,900	1,140	2.18	11.2	36	3,970	5,620	1,240	1,280	----	1,280	960
Sweetgum (*Liquidambar styraciflua*)	115	.46	3,500	7,100	1,200	.60	10.1	36	2,040	3,040	380	996	540	670	600
	12	.52	6,600	12,500	1,640	1.59	11.9	32	3,670	6,320	660	1,600	760	1,080	850
Sycamore, American (*Platanus occidentalis*)	83	.46	3,300	6,500	1,060	.60	7.5	26	2,400	2,920	450	1,000	630	700	610
	12	.49	6,400	10,000	1,420	1.66	8.5	26	3,710	5,380	860	1,470	720	920	770
Tupelo:															
Black (*Nyssa sylvatica*)	55	.46	4,000	7,000	1,030	.91	8.0	30	2,490	3,040	600	1,100	570	790	640
	12	.50	7,300	9,600	1,200	2.54	6.2	22	3,470	5,520	1,150	1,340	500	1,240	810
Water (*N. aquatica*)	97	.46	4,200	7,300	1,050	.98	8.3	30	2,690	3,370	590	1,190	600	800	710
	12	.50	7,200	9,600	1,260	2.41	6.9	23	4,280	5,920	1,070	1,590	700	1,200	880
Walnut, black (*Juglans nigra*)	81	.51	5,400	9,500	1,420	1.16	14.6	37	3,520	4,300	600	1,220	570	960	900
	12	.55	10,500	14,600	1,680	3.70	10.7	34	5,780	7,580	1,250	1,370	690	1,050	1,010
Yellow-poplar (*Liriodendron tulipifera*)	83	.40	3,400	6,000	1,220	.55	7.5	26	2,070	2,660	300	790	510	480	440
	12	.42	6,200	10,100	1,580	1.39	8.8	24	3,730	5,540	560	1,190	540	670	540
SOFTWOODS															
Baldcypress (*Taxodium distichum*)	91	.42	4,200	6,600	1,180	.91	6.6	25	3,100	3,580	500	810	300	440	390
	12	.46	7,200	10,600	1,440	2.15	8.2	24	4,740	6,360	900	1,000	270	660	510

Species	Trees / M.C.	Sp. gr.	(col L)	(col K)	(col J)	(col I)	(col H)	(col G)	(col F)	(col E)	(col D)	(col C)	(col B)	(col A)
Cedar:														
Alaska- (Chamaecyparis nootkatensis)	38	.42	6,400	1,140	.77	9.2	27	2,500	3,050	430	840	330	540	440
	12	.44	11,100	1,420	2.06	10.4	29	5,210	6,310	770	1,130	360	790	580
Atlantic white- (C. thyoides)	55	.31	4,700	750	.51	5.9	18	1,660	2,390	300	690	180	400	290
	12	.32	6,800	930	1.46	4.1	13	2,740	4,700	500	800	220	520	350
Eastern redcedar (Juniperus virginiana)	35	.44	7,000	650	1.08	15.0	35	2,540	3,570	860	1,010	330	760	650
	22	.47	8,800	880	1.01	8.3	22	----	6,020	1,140	----	----	900	900
Incense- (Libocedrus decurrens)	108	.35	6,200	840	.94	6.4	17	2,940	3,150	460	830	280	570	390
	12	.37	8,000	1,040	1.67	5.4	17	4,760	5,200	730	880	270	830	470
Northern white- (Thuja occidentalis)	55	.29	4,200	640	.60	5.7	15	1,490	1,990	290	620	240	320	280
	12	.31	6,500	800	1.72	4.8	12	2,630	3,960	380	850	240	450	320
Port-Orford- (Chamaecyparis lawsoniana)	43	.40	6,200	1,420	.65	7.4	21	2,770	3,130	350	830	180	460	400
	12	.42	11,300	1,730	1.97	9.1	28	5,890	6,470	760	1,080	400	730	560
Western redcedar (Thuja plicata)	37	.31	5,100	920	.63	5.0	17	2,470	2,750	340	710	230	430	270
	12	.33	7,700	1,120	1.44	5.8	17	4,360	5,020	610	860	220	660	350
Douglas-fir:														
Coast type (Pseudotsuga menziesii)	38	.45	7,600	1,570	.75	7.6	26	3,130	3,860	440	930	300	570	500
	12	.48	12,200	1,950	1.77	9.8	31	5,850	7,430	870	1,160	340	900	710
Intermediate type (P. menziesii)	48	.41	6,800	1,350	.63	6.6	22	2,570	3,300	480	840	300	510	450
	12	.44	11,200	1,640	1.87	8.8	27	5,540	6,720	920	1,130	340	710	600
Rocky Mountain type (P. menziesii var. glauca)	38	.40	----	1,180	.65	6.8	20	2,540	3,000	450	880	350	450	400
	12	.43	9,600	1,400	1.60	6.4	26	4,660	6,060	820	1,070	330	740	630
Fir:														
Balsam (A. balsamea)	117	.34	4,900	960	.52	4.7	16	2,080	2,400	210	610	180	290	290
	12	.36	7,600	1,230	1.23	5.1	20	3,970	4,530	380	710	180	510	400
California red (A. magnifica)	116	.37	6,000	1,210	.59	6.4	21	2,350	2,850	360	800	380	430	360
	12	.39	10,800	1,540	1.48	8.9	24	4,220	5,650	650	1,090	390	880	500
Grand (A. grandis)	94	.37	6,100	1,300	.58	5.6	22	2,640	3,020	340	760	240	420	360
	12	.40	9,300	1,630	1.22	7.5	28	4,420	5,430	620	930	240	660	490
Noble (A. procera)	36	.35	5,800	1,270	.61	6.0	19	2,420	2,740	340	750	230	330	290
	12	.38	10,000	1,580	1.59	8.8	23	4,960	5,550	640	980	220	690	410
Pacific silver (A. amabilis)	66	.35	----	1,260	.60	8.0	21	2,380	2,670	290	670	240	360	310
	12	.38	9,400	1,530	1.40	9.3	24	4,660	5,550	490	1,050	----	620	430
White (A. concolor)	115	.35	5,700	1,030	.84	5.1	22	2,390	2,710	370	750	290	380	330
	12	.37	9,300	1,380	1.72	6.7	17	3,590	5,350	600	930	260	730	440
Hemlock:														
Eastern (Tsuga canadensis)	111	.38	6,400	1,070	.76	6.7	21	2,600	3,080	440	850	230	500	400
	12	.40	8,900	1,200	1.79	6.8	21	4,020	5,410	800	1,060	----	810	500
Western (T. heterophylla)	74	.38	6,100	1,220	.57	6.8	22	2,480	2,990	390	810	310	520	430
	12	.42	10,100	1,490	1.82	7.5	26	5,340	6,210	680	1,170	310	940	580

See footnotes at end of table.

TABLE 12.—*Strength properties of some commercially important woods grown in the United States*—Continued

[Results of tests on small, clear specimens in the green and air-dry condition [1]]

Common and botanical names of species (1)	Moisture content (2)	Specific gravity [2] (3)	Static bending					Impact bending-height of drop causing complete failure (50-pound hammer) (9)	Compression parallel to grain		Compression perpendicular to grain-fiber stress at proportional limit (12)	Shear parallel to grain-maximum shearing strength (13)	Tension perpendicular to grain-maximum tensile strength (14)	Hardness—load required to embed a 0.444-inch ball to ½ its diameter	
			Fiber stress at proportional limit (4)	Modulus of—		Work to—			Fiber stress at proportional limit (10)	Maximum crushing strength (11)				End (15)	Side (16)
				Rupture (5)	Elasticity (6)	Proportional limit (7)	Maximum load (8)								
	Pct.		*P.s.i.*	*P.s.i.*	*1,000 p.s.i.*	*In.-Lb. per cu. in.*	*In.-Lb. per cu. in.*	*In.*	*P.s.i.*	*P.s.i.*	*P.s.i.*	*P.s.i.*	*P.s.i.*	*Lb.*	*Lb.*
SOFTWOODS—continued															
Larch, western (*Larix occidentalis*)	58	0.51	4,600	8,200	1,530	0.81	10.3	29	3,010	3,990	420	900	330	580	510
	12	.55	8,300	13,900	1,960	1.99	12.6	35	5,620	8,110	980	1,410	430	1,120	830
Pine.															
Eastern white (*Pinus strobus*)	73	.34	3,000	4,900	990	.53	5.2	17	2,040	2,440	280	680	250	300	290
	12	.35	5,700	8,600	1,240	1.51	6.8	18	3,670	4,800	510	900	310	480	380
Jack (*P. banksiana*)	60	.40	3,400	6,000	1,070	.63	7.2	26	2,360	2,950	310	750	360	440	400
	12	.43	5,800	9,900	1,350	1.43	8.3	27	3,550	5,660	600	1,170	420	800	570
Lodgepole (*P. contorta*)	65	.38	3,000	5,500	1,080	.49	5.6	20	2,110	2,610	310	680	220	320	330
	12	.41	6,700	9,400	1,340	1.97	6.8	20	4,310	5,370	750	880	290	530	480
Ponderosa (*P. ponderosa*)	91	.38	3,100	5,000	970	.59	5.1	20	2,070	2,400	360	680	290	310	310
	12	.40	6,300	9,200	1,266	1.85	6.6	17	4,060	5,270	740	1,160	400	550	450
Red (*P. resinosa*)	92	.41	3,200	5,800	1,280	.46	6.1	26	2,160	2,730	280	690	300	350	340
	12	.44	7,000	11,000	1,630	1.77	9.9	26	4,160	6,070	650	1,210	460	740	560
Pine, southern yellow:															
Loblolly (*P. taeda*)	81	.47	4,100	7,300	1,410	.68	8.2	30	2,550	3,490	480	850	260	420	450
	12	.51	7,800	12,800	1,800	1.92	10.4	30	4,820	7,080	980	1,370	470	750	690
Longleaf (*P. palustris*)	63	.54	5,200	8,700	1,600	.95	8.9	35	3,430	4,300	590	1,040	330	550	590
	12	.58	9,300	14,700	1,990	2.44	11.8	34	6,150	8,440	1,190	1,500	470	920	870
Shortleaf (*P. echinata*)	81	.46	3,900	7,300	1,390	.63	8.2	30	2,500	3,430	440	850	320	410	440
	12	.51	7,700	12,800	1,760	1.93	11.0	33	5,090	7,070	1,000	1,310	470	750	690

Species															
Slash (*P. elliottii*)	66	.56	5,100	8,900	1,580	1.02	9.5	36	3,040	4,340	680	1,000	400	600	630
	12	.61	9,800	15,900	2,060	2.76	12.6	36	6,280	9,100	1,390	1,730	570	1,080	1,010
Pine:															
Sugar (*P. lambertiana*)	137	.35	3,400	5,100	940	.70	5.4	17	2,330	2,530	350	680	270	320	310
	12	.36	5,700	8,000	1,200	1.53	5.5	18	4,140	4,770	590	1,050	350	530	380
Virginia (*P. virginiana*)	88	.45	4,000	7,300	1,220	.75	10.9	34	2,500	3,420	390	890	400	590	540
	12	.48	7,100	13,000	1,520	1.86	13.7	32	3,820	6,710	910	1,350	380	980	740
Western white (*P. monticola*)	54	.36	3,400	5,200	1,170	.56	5.0	19	2,430	2,650	290	640	260	310	310
	12	.38	6,200	9,500	1,510	1.47	8.8	23	4,480	5,620	540	850		440	370
Redwood, old-growth (*Sequoia sempervirens*)	112	.38	4,800	7,500	1,180	1.18	7.4	21	3,700	4,200	520	800	260	570	410
	12	.40	6,900	10,000	1,340	2.04	6.9	19	4,560	6,150	860	940	240	790	480
Spruce:															
Black (*Picea mariana*)	38	.38	2,900	5,400	1,060	.45	7.4	24	1,540	2,570	180	660	100	430	370
	12	.40	5,800	10,300	1,530	1.34	10.5	23	4,520	5,320	650	1,030		700	520
Engelmann (*P. engelmannii*)	80	.32	2,600	4,500	960	.42	5.1	16	1,850	2,190	250	590	240	310	260
	12	.34	5,500	8,700	1,280	1.34	6.4	18	3,580	4,770	540	1,030	350	560	350
Red (*P. rubens*)	43	.38	3,400	5,800	1,190	.58	6.9	18	2,380	2,650	340	760	220	410	350
	12	.41	6,800	10,200	1,520	1.73	8.4	25	4,610	5,890	580	1,080	350	640	490
Sitka (*P. sitchensis*)	42	.37	3,300	5,700	1,230	.53	6.3	24	2,240	2,670	340	760	250	430	350
	12	.40	6,700	10,200	1,570	1.62	9.4	25	4,780	5,610	710	1,150	370	760	510
White (*P. glauca*)	50	.37	3,300	5,600	1,070	.60	6.0	22	2,130	2,570	290	690	220	350	320
	12	.40	6,500	9,800	1,340	1.76	7.7	20	3,700	5,470	570	1,080	360	610	480
Tamarack (*Larix laricina*)	52	.49	4,200	7,200	1,240	.84	7.2	28	2,930	3,480	480	860	260	400	380
	12	.53	8,000	11,600	1,640	2.19	7.1	23	4,780	7,160	990	1,280	400	670	590

[1] The values in the first line for each species are from tests of green material; those in the second line are from tests of seasoned material adjusted to an average air-dry condition of 12-percent moisture content.

[2] Based on weight when oven-dry and volume when green or at 12-percent moisture content.

ELASTIC PROPERTIES OF WOOD

Certain elastic properties of wood that are necessary in design, particularly when plywood is used, are presented in table 13 (2).

The modulus of elasticity of wood perpendicular to the grain is designated as E_T when the direction in which the deformation takes place is tangential to the annual growth rings and as E_R when the direction is radial to the annual growth rings. These properties were evaluated for a few species, and the results are presented as ratios E_T/E_L and E_R/E_L in table 13, where E_L is the modulus of elasticity parallel to the grain. These ratios vary among species and are considerably affected by differences in specific gravity and moisture content. For species not listed in the table, approximate values of $E_T/E_L = 0.05$ and $E_R/E_L = 0.10$ may be used. Values of E_L may be obtained by adding 10 percent to the modulus of elasticity value in column 6 of table 12.

The modulus of elasticity in shear, called the modulus of rigidity, must be associated with shear deformation in 1 of the 3 mutually perpendicular planes defined by the L (longitudinal), T (tangential), and R (radial) directions and with shear stresses in the other 2 planes. The symbol for modulus of rigidity has subscripts denoting the plane of deformation. Thus, the modulus of rigidity G_{LT} refers to shear deformations in the LT plane resulting from shear stresses in the LR and RT planes. Values of these moduli for a few species are given in table 13, and their variations among species and with changes in specific gravity and moisture content are shown. For species not listed in the table, approximate values of $G_{LT}/E_L = 0.06$, $G_{LR}/E_L = 0.075$, and $G_{RT}/E_L = 0.018$ may be used.

The Poisson's ratio relating to the contraction in the T (tangential) direction under a tensile stress acting in the L (longitudinal) direction and thus normal to the RT plane is designated as μ_{LT}. The symbols μ_{LR}, μ_{RT}, μ_{RL}, μ_{TR}, and μ_{TL} have similar significance, the first letter of the subscript in each relating to the direction of stress and the second to the direction of lateral deformation under that stress. The Poisson's ratios appear to be independent of specific gravity but are variously affected by differences in moisture content. Information on Poisson's ratios for wood is meager, and values for only a few species are given in table 13.

COMPRESSION PERPENDICULAR TO GRAIN

Effect of Bearing Area on Stress

Proportional limit stresses in compression perpendicular to the grain given in table 12 are from tests in which the load was applied through a plate 2 inches wide placed on the central 2 inches of a specimen 6 inches long. The supporting action of the wood fibers adjacent to the plate in the unloaded area caused an increase in stress over that which would have resulted if the entire area of the specimen had been covered. The test results indicate that the stress obtained in the standard test procedure is about 50 percent greater than that obtained in loading the entire area of a 2-inch cube. As the ratio of loaded to unloaded

TABLE 13.—*Elastic constants of various species*

Species	Specific gravity [1]	Moisture content [1] (percent)	Young's modulus ratios		Modulus of rigidity ratios			Poisson's ratios					
			E_T/E_L	E_R/E_L	G_{LR}/E_L	G_{LT}/E_L	G_{RT}/E_L	μ_{LR}	μ_{LT}	μ_{RT}	μ_{TR}	μ_{RL}	μ_{TL}
Ash	0.801	13.6	0.064	0.109	0.057	0.041	0.0165	0.533	0.653	0.656	0.386	0.0582	0.0421
Balsa [2]	.131	9.4	.015	.046	.054	.037	.005	.229	.488	.665	.231	.018	.009
Birch, yellow	.64	13.3	.050	.078	.074	.068	.017	.426	.451	.697	.426	.043	.024
Do	.65	12.6	.061	.090				.295	.441	.447	.368	.030	.023
Douglas-fir	.506	7.5	.050	.068	.064	.078	.007	.292	.449	.390	.374	.036	.029
Do	.506	12.9	.032	.056				.274	.496	.560	.396	.018	.019
Do	.506	19.9											
Do	.45	11.2											
Khaya	.45	11.3	.050	.111	.086	.066	.021	.297	.641	.604	.264	.033	.032
Mahogany	.50	11.7	.064	.107	.082	.055	.028	.314	.533	.600	.326	.033	.034
Quipo [3]	.137	11.2	.055	.182	.070	.068	.032	.216	.666	.455	.128	.047	.032
Sitka spruce	.378	7.1	.050	.089	.064	.061	.044	.375	.436	.468	.248	.034	.022
Do	.378	12.8	.043	.078			.003	.372	.467	.435	.245	.040	.025
Do	.378	16.3	.036	.064				.374	.504	.527	.278	.030	.020
Do	.378	21.6	.029	.053				.371	.539	.512	.278	.022	.017
Do	.39	6.8											
Do	.39	11.1											
Sweetgum	.536	11.2	.050	.115	.089	.061	.021	.325	.403	.682	.309	.044	.023
Do	.530	10.2											
Walnut	.593	11.0	.056	.106	.085	.062	.0209	.495	.632	.718	.367	.0520	.0360
Yellow-poplar	.376	10.7	.043	.092	.075	.069	.011	.318	.392	.703	.329	.030	.019

[1] Specific gravity based on volume at test and weight when oven-dry. The values shown to 2 decimal places and the corresponding moisture content values are tentative and subject to revision.

[2] The balsa in these tests varied in specific gravity from 0.06 to 0.22, in which range E_L is given approximately by the equation $E_L = 5,500,000 \times \text{specific gravity} - 200,000$.

[3] The quipo in these tests varied in specific gravity from 0.08 to 0.20, in which range E_L is given approximately by the equation $E_L = 3,260,000 \times \text{specific gravity} - 170,000$.

area is decreased, further increases in proportional limit stresses are obtained. This factor is further discussed in arriving at safe working stresses for bearing perpendicular to the grain, wherein higher working stresses are permitted for bearing areas less than 6 inches in length (p. 163).

Effect of Ring Placement on Stress

Investigations at the Forest Products Laboratory and elsewhere have shown that properties in compression perpendicular to the grain are affected by the angle of growth rings with respect to the direction of load (fig. 6) and that properties of hardwoods and softwoods are affected differently by the orientation of growth rings. In softwoods, the lowest stress at proportional limit is obtained when the growth rings are at an angle of 45° to the applied load, and it is approximately one-half to two-thirds of the value obtained with the rings at 0°, or parallel, to the direction of load. Stress values at the 90° ring position are approximately the same as those at the 0° position. At a 45° ring orientation, modulus of elasticity values are decreased somewhat more than are proportional limit stresses but are about one-fourth greater for the 90° than for the 0° orientation. For hardwoods, the 0° and 45° positions give about the same values for proportional limit stress and modulus of elasticity, but values at 90° are higher by about one-third.

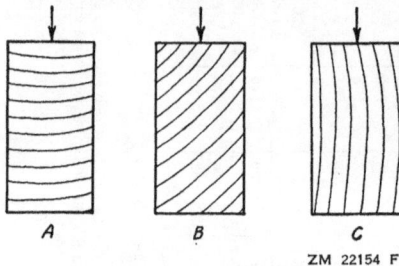

FIGURE 6.—The direction of the load in relation to the direction of the annual growth rings: A, 90°, or perpendicular; B, 45°; C, 0°, or parallel.

ZM 22154 F

COMPRESSIVE STRENGTH ON SURFACES INCLINED TO THE GRAIN

A formula for determining the compressive strength of wood on surfaces at an angle to the grain, known as the Hankinson formula, is recommended for general use in timber framing. The formula is:

$$N=\frac{PQ}{P\sin^2\theta+Q\cos^2\theta}$$

in which N represents the allowable unit stress on the inclined surface; Q, the unit stress in compression perpendicular to the grain; P, the unit stress in compression parallel to grain; and θ, the angle between the direction of the load and the direction of the grain (fig. 7).

FIGURE 7.—The angle θ between the direction of the load and the direction of the grain.

ZM 22153 F

For safe working stresses, the unit stress constants in the formula would be the working stresses in compression parallel and perpendicular to the grain for the species.

SHEAR PARALLEL TO GRAIN

In the block shear tests (1) upon which the values in table 12 are based, a 2- by 2-inch portion that projects ¾ inch from a block 2 inches wide and 2½ inches long is sheared from the block. The specimen is supported at the base, so that a ⅛-inch offset exists between the adjacent edges of the support and the loading plate. The results obtained are influenced to some extent by a nonuniform stress distribution throughout the depth of the shear area and by components of stress perpendicular to the grain. The values presented, however, indicate the relative shearing strength of the various species and may be considered as reliable estimates of true shearing strength.

TENSION PARALLEL TO GRAIN

Relatively few data are available on the tensile strength of various species parallel to the grain. In the absence of sufficient tensile test data on which to base values for use in design, the modulus of rupture values are also used for tension. While it is recognized that those values are conservative, the pronounced effect of stress concentration, slope of grain, and other factors upon tensile strength make the use of conservative values desirable. Furthermore, it is practically impossible to devise attachments that permit the tensile strength of the full cross section of a wooden member to be developed.

TORSIONAL PROPERTIES

The torsional strength of wood is seldom needed in design, but data are available that permit calculation of torsional properties if they are required.

The torsional deformation of wood is related to the moduli of rigidity in the longitudinal-radial (LR), longitudinal-tangential (LT), and radial-tangential (RT) planes. These moduli were discussed under elastic properties of wood. When a wood member is twisted about an axis parallel to the grain, the modulus in the RT plane is not involved. Since the moduli in the LR and LT planes do not differ greatly, the "mean modulus of rigidity," taken as one-sixteenth of the modulus of elasticity parallel to the grain, may be used safely in standard formulas for computing the strength and rigidity of wood members twisted in this manner.

For solid wood members, the allowable ultimate torsional shear stress may be taken as the shear stress parallel to the grain in table 12, and two-thirds of this value may be used as the allowable torsional shear stress at the proportional limit. Information on critical buckling and twisting due to torsional instability is presented in a bulletin on Design of Wood Aircraft Structures (23).

TOUGHNESS

Although a number of the properties given in table 12 indicate the ability of wood to absorb shock or impact loads, a simple test has been developed with a Forest Products Laboratory machine by which the toughness of relatively small samples can be determined (21). This test affords a means of comparing species and a basis for selecting stock of known properties, particularly when used in conjunction with specific gravity. Toughness values for some species of wood are listed in table 14. These data are based on specimens ⅝ inch square and 10 inches long, supported over a span of 8 inches. That size was used as a standard by the Forest Products Laboratory until 1949 when, by international agreement, a specimen 2 centimeters (0.79 inch) square by 28 centimeters (11 inches) long was adopted (1).

FATIGUE RESISTANCE

The fatigue resistance of wood is a measure of the ability of wood to sustain repeated, reversed, or vibrational loads without failure. Its resistance to fatigue, like that of metals, is an important consideration in design and may be adversely affected by abrupt changes in cross section, bolt holes, slope of grain, knots, and other strength-reducing factors.

Tests to date indicate that wood, a fibrous material, is less sensitive to repeated loads than are more crystalline structural materials, such as metals. Endurance limits are likely to be higher in proportion to the ultimate strength values for wood than for some of the metals. For example, tests in fatigue in tension parallel to the grain indicate an endurance load for 30 million cycles of stress of about 40 percent of the strength of air-dry wood as determined by a standard static test. Similar results were obtained on tests of air-dry wood with glued scarf joints. The minimum repeated load for each cycle was one-tenth of the maximum for each cycle.

Tests of small cantilever bending specimens of air-dry solid wood and plywood (8), subjected to fully reversed bending stresses, indicated an endurance limit for 30 million cycles of stress of about 30 percent of the modulus of rupture of the specimens.

Fatigue tests in repeated bending of green, standard-size wood static bending specimens indicate that, if maximum repeated stresses are below the proportional limit (60 percent of the modulus of rupture or less), the specimens will withstand 30 million repetitions of stress without failure and with only superficial damage at points of loading.

Tests of small rotating beam specimens (3) indicate an endurance load for wood at 30 million cycles of stress of about 28 percent of the modulus of rupture for specimens of this type.

TABLE 14.—*Average toughness values for some species of wood*

Species	Moisture content	Specific gravity [1]	Toughness		Species	Moisture content	Specific gravity [1]	Toughness	
			Radial [2]	Tangential [3]				Radial [2]	Tangential [3]
	Percent		*In.-lb.*	*In.-lb.*		*Percent*		*In.-lb.*	*In.-lb.*
Birch:					Juniper— Continued				
Paper	10	0.56	180	180	Utah	51	0.48	130	120
Yellow	12	.65	260	330		12	.48	100	100
Catalpa, northern	66	.40	180	180	Larch, western	61	.51	140	210
	12	.41	100	120		12	.55	110	180
Cedar:					Maple, sugar	14	.64	190	190
Alaska-	10	.48	110	120	Oak, pin	12	.64	230	220
Western red-cedar	9	.33	40	70	Pine:				
Chestnut, Japanese	90	.53	190	220	Eastern white	99	.33	60	80
	14	.54	140	200		12	.34	60	60
Douglas-fir:					Jack	49	.41	100	200
Coast type	43	.46	120	200		12	.42	80	130
	12	.48	110	200	Red	111	.40	110	190
Intermediate type	36	.43	80	110		12	.43	80	150
	10	.46	90	150	Southern yellow:				
Elm, winged	56	.60	350	360	Loblolly	86	.47	140	180
	12	.66	250	290		12	.51	90	150
Fir:					Longleaf	90	.54	180	230
California red	132	.36	70	90		13	.57	90	140
	12	.39	60	90	Shortleaf	88	.48	140	190
Corkbark	55	.31	40	50		13	.50	80	120
	10	.31	40	50	Slash	78	.55	180	240
Noble	62	.36	------	130		12	.59	110	170
	11	.40	100	------	Virginia	124	.45	180	250
	12	.39	------	120	Redwood:				
Hemlock:					Old-growth	103	.39	60	110
Eastern	12	.41	60	90		11	.39	50	80
Western	11	.38	60	90	Second-growth	138	.32	60	80
	12	.43	------	130		12	.33	40	50
Juniper:					Spruce:				
Rocky Mountain	52	.39	100	90	Engelmann	60	.34	60	80
	12	.41	70	70	Sitka	11	.39	------	140
					Willow, black	153	.38	160	190
						11	.40	110	120
					Yellow-poplar	110	.43	170	160

[1] Based on oven-dry weight and volume at test.
[2] Load applied to radial face of specimen.
[3] Load applied to tangential face of specimen.

Tests on a glued, double-shear type of specimen (9) indicate an endurance load for 30 million cycles of stress of about 40 percent of the strength of the joint. The specimens were loaded repeatedly so that the minimum repeated stress for each cycle was one-tenth of the maximum for each cycle.

VARIABILITY IN STRENGTH

Variability, or variation in properties, is common to all materials, including rope, steel, concrete, and wood, but the degree of variability, or spread of property values, differs markedly with the type of material. Since wood is a natural material and subject to numerous constantly changing influences, such as moisture, soil conditions, and growing space, it may show a decided variation in properties even in clear material.

The property values tabulated in table 12 were obtained by averaging individual test values for each tree and then averaging the individual tree averages to arrive at the value for the species. Obviously, some values obtained in individual tests were higher than the average, and others were lower. Normally, the fact that the piece selected may be stronger, harder, or stiffer than the average is of less concern to the user than if it is weaker, although this may not be true if light-weight material is selected for a specific purpose. It is desirable, however, that some indication of the spread of property values be given. If a frequency distribution curve were presented for each property and species, it would show a large number of property values from the individual tests grouped around the average, and a smaller number varying widely from higher to lower levels.

Magnitude of variation can also be indicated by standard deviation. Standard deviation is the square root of the mean of the squares of individual deviations from the average value. Since it is affected by every value, it is a better measure of variability than the range of all or any part of the values. The result is in units for each property, and to facilitate comparisons, it is divided by the average value to get a percentage figure that is termed the "coefficient of variation." Coefficient of variation values for strength properties of green wood are presented in the following tabulation:

	Coefficient of variation [1] (percent)
Specific gravity	10
Shrinkage:	
Radial	15
Tangential	14
Volumetric	16
Static bending:	
Fiber stress at proportional limit	22
Modulus of rupture	16
Modulus of elasticity	22
Work to proportional limit	38
Work to maximum load	34
Impact bending, height of drop causing complete failure	25
Compression parallel to grain:	
Fiber stress at proportional limit	24
Maximum crushing strength	18
Modulus of elasticity	29
Compression perpendicular to grain, fiber stress at proportional limit	28
Shear parallel to grain, maximum shearing strength	14
Tension perpendicular to grain, maximum tensile strength	25
Hardness:	
End	17
Side	20
Toughness	34

[1] Values given are based on results of tests of approximately 50 species of green wood. Values for wood in the air-dry condition (12-percent moisture content) may be assumed to be approximately of the same magnitude.

In a normal frequency distribution, about two-thirds of the individual values lie within the range encompassed by the sum and the difference of the average value and the product of the coefficient of variation times the average value. Also, 95 percent of the individual results will lie in the range of the sum and the difference of the average value and twice the same product. Thus, if the average value is 50 and the coefficient of variation is 6 percent, then two-thirds of the pieces tested could be expected to have values between 47 and 53, and 95 percent would have values between 44 and 56. Variability from the average must be considered in establishing working stresses for a species; this is discussed further on page 137.

MOISTURE CONTENT AND STRENGTH

Wood increases in strength as it dries (26). With small, clear pieces, the strength in endwise compression, for example, is about twice as great for a moisture content of 12 percent as for green wood, and drying to about 5-percent moisture content will sometimes triple this property. Increase in strength does not begin, however, until the fiber saturation point is reached. The fiber saturation point, which is at approximately 30-percent moisture content, is reached when the free water in the cell cavities has been evaporated and the cell walls are still saturated.

Not all the strength properties of wood increase with a decrease in moisture content; in fact, properties that represent toughness, or shock resistance, sometimes actually decrease as the wood dries. This is because dried wood will not bend so far as green wood before failure, although it will sustain a greater load, and because toughness is dependent upon both strength and pliability.

Frequently, strength values are obtained for pieces at different moisture-content values, and the results are recorded as obtained. It is then necessary to adjust some values before a true comparison of results can be made. An approximate adjustment for the property in question can be made by using the values given in the following tabulation (22). These values represent the average increase (or decrease) in wood strength properties for a 1-percent decrease (or increase) in moisture content.

	Change per 1-percent change in moisture content (percent)
Static bending:	
Fiber stress at proportional limit	5
Modulus of rupture	4
Modulus of elasticity	2
Work to proportional limit	8
Work to maximum load	.5
Impact bending, height of drop causing complete failure	.5
Compression parallel to grain:	
Fiber stress at proportional limit	5
Maximum crushing strength	6
Compression perpendicular to grain, fiber stress at proportional limit	5.5
Shear parallel to grain, maximum shearing strength	3
Tension perpendicular to grain, maximum tensile strength	1.5
Hardness:	
End	4
Side	2.5

A more accurate adjustment can be made by using the following "exponential formula," devised by the Forest Products Laboratory:

$$\text{Log } S_3 = \log\ S_1 + \left(\frac{M_1 - M_3}{M_1 - M_2}\right) \log \frac{S_2}{S_1}$$

where S_1 and M_1 are one pair of corresponding strength and moisture content values as found from test, S_2 and M_2 are another pair, and S_3 is the strength value adjusted to the moisture content M_3.

If one strength value is for green wood, the moisture content that must be used is that corresponding to the intersection of straight lines giving strength-moisture content relations when strength values are plotted as ordinates and moisture content values as abscissas on semilogarithmic paper (26). This value, which is somewhat lower than the fiber saturation point, is designated as M_p. As previously stated, the strength undoubtedly increases when the moisture is lowered below the fiber saturation point, but, because of limitations in test procedure, it is not practical to obtain a strength-moisture content relation starting at this point. Experimentally determined values of M_p for certain species follow:

	M_p (percent)		M_p (percent)
Ash, white	24	Pine, longleaf	21
Birch, yellow	27	Pine, red	24
Chestnut, American	24	Redwood	21
Douglas-fir	24	Spruce, red	27
Hemlock, western	28	Spruce, Sitka	27
Larch, western	28	Tamarack	24
Pine, loblolly	21		

For other species, a value of 25 percent is assumed for M_p.

The exponential formula is not applicable to results of tests on wood in which there is a large variation in moisture content from one part of the cross section to another.

Examples of Application of Formula

Case 1.—Strength values S_1 and S_2 from matched specimens tested green, M_1, and air-dry, M_2 are known, and the strength, S_3, at another air-dry moisture content, M_3, is desired. Then

$$\text{Log } S_3 = \log S_1 + \left(\frac{M_1 - M_3}{M_1 - M_2}\right) \log \frac{S_2}{S_1}$$

Example: Tests of matched specimens of Sitka spruce gave values of maximum crushing strength of 2,600 and 5,770 pounds per square inch, respectively, for green wood and wood at 8.9-percent moisture content. What is the strength value for a moisture content of 12 percent?

$$\text{Log } S_3 = \log 2{,}600 + \frac{27 - 12}{27 - 8.9} \log \frac{5{,}770}{2{,}600}$$

$$S_3 = 5{,}030 \text{ pounds per square inch.}$$

Case 2.—The strength value, S_2, at some air-dry moisture content, M_2, only is known, and the strength value, S_3, at another air-dry moisture content, M_3, is desired. A good estimation of the true strength value S_3 can be obtained from the formula

$$\text{Log } S_3 = \log S_2 + (M_2 - M_3) \frac{\log \dfrac{S_{12}}{S_g}}{(M_p - 12)}$$

where S_g and S_{12} are values pertaining to green wood and wood at 12-percent moisture content, respectively, as found in table 12 for the species in question.

Example: Some tests of eastern hemlock at 8-percent moisture content gave a value of modulus of rupture of 10,000 pounds per square inch. Estimate the value that would have been obtained had the tests been made at 13-percent moisture content.

$$\text{Log } S_3 = \log 10,000 + (8 - 13) \frac{\log \dfrac{8,900}{6,400}}{25 - 12}$$

$$S_3 = 8,810 \text{ pounds per square inch.}$$

(Since M_p has not been determined experimentally for eastern hemlock, the value of 25 percent is taken.)

Case 3.—The strength value at some air-dry moisture content, M_2, only is known, and the strength value at another air-dry moisture content, M_3, is desired. The species is not included in table 12 as in case 2, and hence, only an approximation of the true strength value can be obtained. The formula is

$$\text{Log } S_3 = \log S_2 + (M_2 - M_3) \frac{\log \dfrac{S_{12}}{S_g}}{25 - 12}$$

Average values of S_{12} over S_g, based on many different species and for use in the preceding formula, are given in the following tabulation:

	S_{12}/S_g	
	Hardwoods (average of 113 species)	Softwoods (average of 54 species)
Static bending:		
Fiber stress at proportional limit	1. 80	1. 81
Modulus of rupture	1. 59	1. 61
Modulus of elasticity	1. 31	1. 28
Work to proportional limit	2. 49	2. 56
Work to maximum load	1. 05	1. 13
Impact bending, height of drop causing complete failure	. 89	1. 03
Compression parallel to grain:		
Fiber stress at proportional limit	1. 74	1. 86
Maximum crushing strength	1. 95	1. 97
Compression perpendicular to the grain, fiber stress at proportional limit	1. 84	1. 96
Hardness:		
End	1. 55	1. 67
Side	1. 33	1. 40
Shear parallel to the grain, maximum shearing strength	1. 43	1. 37
Tension perpendicular to the grain, maximum tensile strength	1. 20	1. 23

Example: A piece of Central American mahogany at 9-percent moisture content tested 11,700 pounds per square inch in modulus of rupture. Estimate the value for the same piece at 12-percent moisture content.

$$\text{Log } S_3 = \log 11,700 + (9-12)\frac{\log 1.59}{25-12}$$
$$S_3 = 10,510 \text{ pounds per square inch.}$$

SPECIFIC GRAVITY AND STRENGTH

The substance of which wood is composed is actually heavier than water, its specific gravity being about 1.5 regardless of the species of wood. In spite of this fact, the dry wood of most species floats in water, and it is thus evident that a large part of the volume of a piece of wood is occupied by cell cavities and pores. Variations in the size of these openings and in the thickness of the cell walls cause some species to have more wood substance than others and therefore to have higher specific gravity values. Specific gravity thus is an excellent index of the amount of wood substance a piece of dry wood contains and hence is an index of its strength properties. It should be noted, however, that specific gravity values are also affected by gums, resins, and extractives, which contribute little to strength.

The relationships of specific gravity and various strength properties of wood are expressed as equations in the following tabulation, which is based on average results of strength tests of more than 160 species:

	Specific gravity-strength relation [1]	
	Green wood	*Air-dry wood (12-percent moisture content)*
Static bending:		
Fiber stress at proportional limit___p. s. i__	$10,200G^{1.25}$	$16,700G^{1.25}$
Modulus of rupture_____p. s. i__	$17,600G^{1.25}$	$25,700G^{1.25}$
Work to maximum load__in.-lb. per cu. in__	$35.6G^{1.75}$	$32.4G^{1.75}$
Total work_____in.-lb. per cu. in__	$103G^2$	$72.7G^2$
Modulus of elasticity_____1,000 p. s. i__	$2,360G$	$2,800G$
Impact bending, height of drop causing complete failure_____in__	$114G^{1.75}$	$94.6G^{1.75}$
Compression parallel to grain:		
Fiber stress at proportional limit__p. s. i__	$5,250G$	$8,750G$
Maximum crushing strength_____p. s. i__	$6,730G$	$12,200G$
Modulus of elasticity_____1,000 p. s. i__	$2,910G$	$3,380G$
Compression perpendicular to grain, fiber stress at proportional limit_____p. s. i__	$3,000G^{2.25}$	$4,630G^{2.25}$
Hardness:		
End_____lb__	$3,740G^{2.25}$	$4,800G^{2.25}$
Side_____lb__	$3,420G^{2.25}$	$3,770G^{2.25}$

[1] The properties and values should be read as equations; for example, modulus of rupture for green wood = $17,600G^{1.25}$, where G represents the specific gravity of oven-dry wood, based on the volume at the moisture condition indicated.

Some properties, such as maximum crushing strength in compression parallel to the grain, increase approximately in proportion to the increase in specific gravity, whereas other properties increase at a more rapid rate. Because these general relations vary considerably, they cannot be expected to give exact strength values, although they do give reliable estimates.

Strength values for pieces of wood within a species vary by a higher power of the specific gravity values than do average strength values of different species. If the exponents to specific gravity in the preceding tabulation are increased by 0.25, they will indicate the variation within a species.

TEMPERATURE AND STRENGTH

Immediate Effect of Temperature

In general, when the temperature of wood is raised above normal, it tends to become weaker in most strength properties, and when the temperature is lowered, it becomes stronger. This effect on strength is immediate, and its magnitude depends upon the moisture content of the wood and, when the temperature is elevated, upon the time of exposure. For most structural uses under ordinary atmospheric conditions, wood exposed to temperatures above normal, if the exposure is for a limited period and the temperature is not excessive, can be expected to recover essentially all of its original strength when the temperature is reduced to normal. Experiments indicate that air-dry wood can probably be exposed to temperatures up to nearly 150° F. for a year or more without an important permanent loss in most strength properties, but its strength while heated will be temporarily reduced as compared to the strength at normal temperature.

The approximate immediate effect of temperature on most static strength properties of dry wood (12-percent moisture content) within the range of 0° to 150° F. can be estimated as an increase or a decrease in the strength at 70° of about $\frac{1}{8}$ to $\frac{1}{2}$ percent for each 1° decrease or increase in temperature $(4, 5, 7, 19, 20)$. The change in properties will be greater if the moisture content is high and less if the moisture content is low. In some geographical locations fairly high temperatures are commonly experienced, but the accompanying relative humidity is ordinarily quite low. Wood exposed to such conditions will generally have a low moisture content, and the immediate effect of the high temperature is not so pronounced as in locations where wood has a higher moisture content.

Figure 8, *A* shows the approximate temperature and moisture content relations that apply for modulus of rupture in bending.

Tests of wood conducted at about −300° F. show that the important strength properties of dry wood in bending and in compression, including stiffness and shock resistance, are much higher at the extremely low temperature than at normal temperature.

Permanent Effect of Exposure to High Temperatures

When wood is exposed to temperatures of 150° F. or more for extended periods of time, it will be permanently weakened $(12, 13, 14)$, even though the temperature is subsequently reduced and the wood is used at normal temperatures The permanent or nonrecoverable strength loss will depend upon a number of factors, including the moisture content and temperature of the wood, the heating medium and time of exposure, and to some extent upon the species and the size of the piece. In the following discussion it should be understood

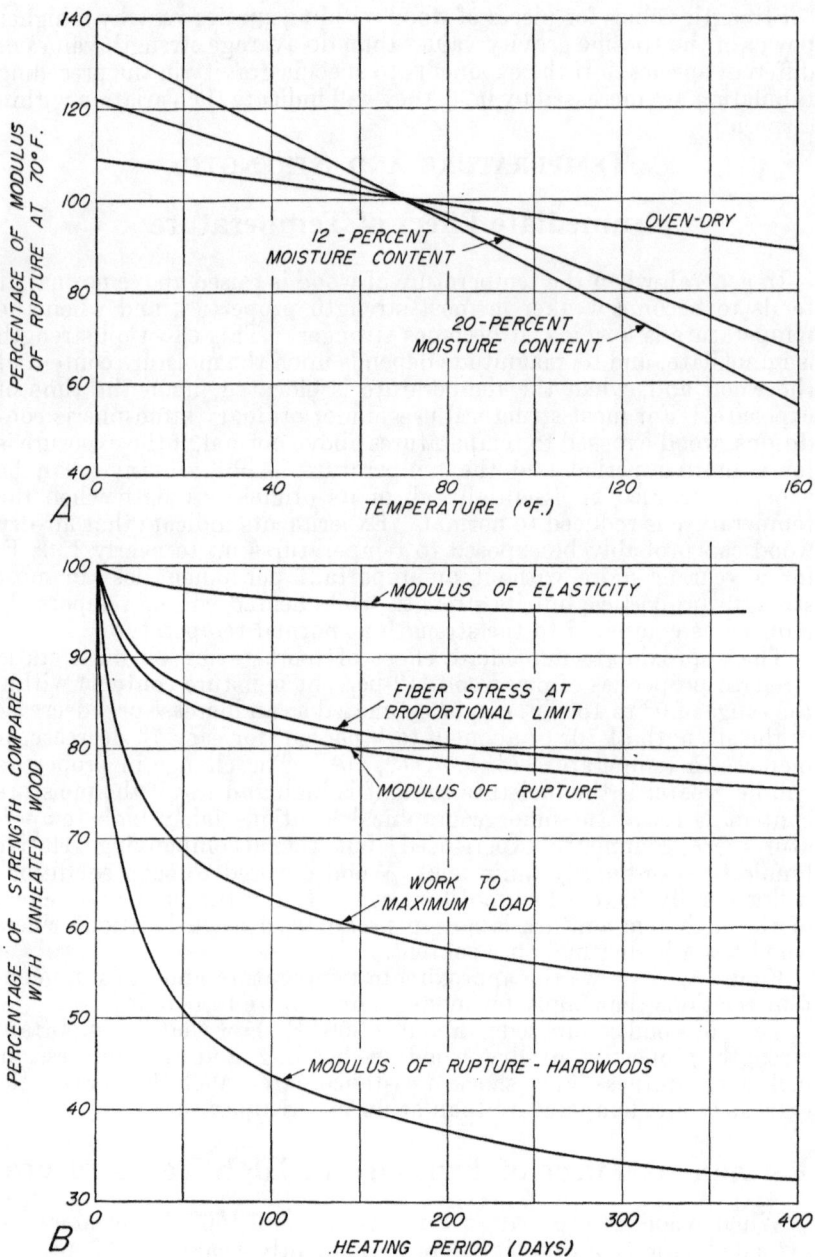

ZM 92547

FIGURE 8.—*A*, Approximate immediate effect of temperature on modulus of rupture at several moisture content conditions.   *B*, Effect of heating on bending properties of wood exposed to water at 200° F. for various periods and then tested at normal temperature and 12-percent moisture content.   All curves are for softwoods unless otherwise indicated.

that losses in strength are permanent losses, measured at normal temperature, after exposure to high-temperature conditions for various periods. Reductions in strength would be substantially higher if measured at the elevated temperature.

Broadly, the available data indicate that exposure of wet wood for a period of 1 year to a temperature of 180° F. will result in a substantial permanent loss in strength, and in wet wood an important strength loss will occur even at 150°. Exposure to higher temperatures will result in increasingly greater strength losses in shorter periods of time as the temperature is increased. Strength properties are affected differently by exposure to high temperatures (fig. 8, B). The shock resistance of wood (work to maximum load) is the first bending strength property affected to a measurable degree regardless of the species, temperature, or heating medium. Modulus of elasticity, a measure of stiffness, is least affected.

The effect of high temperatures on various species is different, but, in general, hardwoods are affected to a considerably greater extent than softwoods (fig. 8, B).

The heating medium used has a considerable bearing on the amount of reduction of strength that results from exposure of wood at a particular temperature for various periods of time. At all temperatures, exposure to hot water causes somewhat less strength loss in a given period of time than steam; the effect of exposure to a hot press is considerably less than it is for hot water, and the least effect results from exposure to hot, dry air.

The important interrelation of temperature and time of exposure is illustrated in figure 9, which shows the great reduction in modulus of rupture of wood soaked in hot water as the temperature and time of exposure are increased. Similar curves showing different rates of deterioration would apply for other properties and heating media. The relatively greater effect on hardwoods is clearly evident. For instance, the modulus of rupture of a hardwood shows an insignificant reduction at 260° F. if exposed for 2 hours, but a reduction of about 13 percent if exposed for 4 hours and 22 percent if exposed for 8 hours.

Lowering the temperature to 250° F. results in no measurable reduction in modulus of rupture after 2 hours of exposure, only about 4 percent reduction after 4 hours, and about 14 percent after 8 hours. Thus, 4 hours of exposure to hot water at 260° may be considered essentially equivalent to 8 hours of exposure at 250° so far as the effect on modulus of rupture is concerned. The effect of time of exposure becomes very large with a substantial difference in temperature. For example, 32 hours of exposure to hot water at 250° is no more harmful than only 2 hours of exposure at 300°.

The effect of time and temperature on shock resistance is greater than on modulus of rupture. When a softwood is exposed to hot water for 10 hours at a temperature of 300° F., the reduction in modulus of rupture would approximate 30 percent, whereas in work to maximum load, a measure of shock resistance, the corresponding reduction would be more than 40 percent.

The effect of exposure to high temperatures on the strength of wood is cumulative. For example, if wood at a particular moisture content is exposed 6 different times to a temperature of 180° F. for 1 month

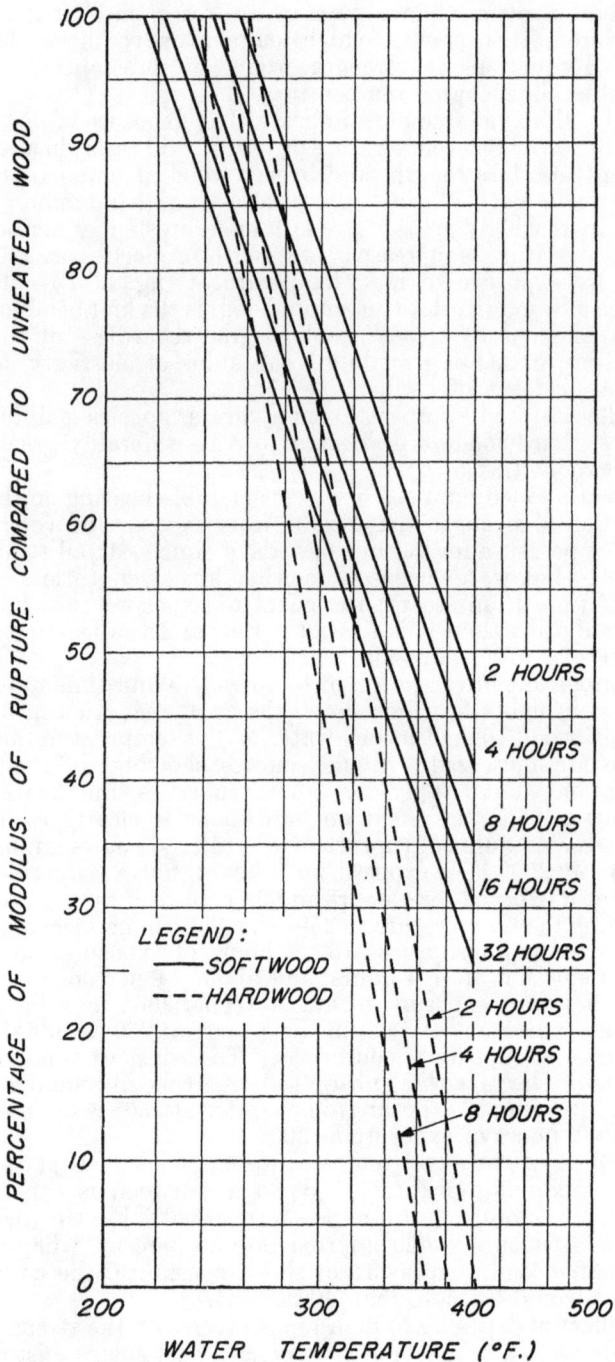

FIGURE 9.—Variation in modulus of rupture with temperature for softwoods and hardwoods soaked in hot water for various periods. Strength tests were made at normal temperature and 12-percent moisture content.

each, the overall effect would be approximately the same as for a single exposure of 6 months.

The shape and size of wood pieces can be expected to influence the overall temperature effect in relation to the heating medium, exposure time, moisture content, and the strength properties considered. If the exposure is for only a short time, so that the inner parts of a large piece do not reach the temperature of the surrounding medium, the immediate effect on the strength properties of the inner parts will be less than for the outer parts. On the other hand, the outer fibers of a piece stressed in bending are subjected to the greatest load and will ordinarily govern the ultimate strength of the piece; hence, the fact that the inner part is at a lower temperature may be of little advantage.

For long-extended exposures, it can be assumed that the entire piece reaches the temperature of the heating medium and will, therefore, be subject to permanent, nonrecoverable strength losses throughout the volume of the piece, regardless of size. It should be recognized, however, that in ordinary construction service, such as in buildings, the temperature of the air surrounding the wood varies substantially during the day and the seasons of the year and that the wood itself, or at least any substantial part of a member, is not likely to reach the maximum temperature of the surrounding air. This is true particularly of the larger or structural members.

Duration of Stress

The duration of stress, or the time during which a load acts on a wood member, is an important factor in determining the load that a member can safely carry. For members that carry a load continuously for a long period of time, such as posts or beams in a warehouse, the load-carrying capacity is much less than that determined from the strength properties of table 12. A wood member continuously loaded for 10 years can carry only about 60 percent of the load required to produce failure in a standard strength test of only a few minutes' duration (fig. 26) (27).

Conversely, if the loading rate is very rapid, as it might be in certain aircraft members during flight or maneuvers, the load-carrying capacity is increased. This increase is approximately 25 percent over that indicated by standard strength tests when the load is applied in 1 second. As an approximate indication of variation in strength of wood members with duration of stress, the strength may be said to increase or decrease 7 to 8 percent as the duration of loading time is decreased or increased by a factor of 10 (10). The duration of stress must be considered in establishing safe working stresses for timber and is further discussed on page 160.

Strength-Reducing Factors

Knots

As a knot appears on a sawed surface, it is merely a section of the entire knot, and its shape depends upon the direction of the cut. When a knot is sawed through at right angles to its length, a round knot results; when cut diagonally, an oval knot; and when sawed lengthwise, a spike knot.

Knots are further classified as intergrown or encased (fig. 10). As long as a limb remains alive, there is continuous growth at the junction of the limb and the trunk of the tree, and the resulting knot is called intergrown; after the limb has died, additional growth on the trunk encloses the dead limb, and an encased knot results. The encased knot and the fibers of the trunk are not continuous, and consequently, the distortion of the grain around the knot is less than for intergrown knots. Encased knots and knotholes are accompanied by less cross grain than are intergrown knots and hence have no more effect on the bending strength of lumber than do intergrown knots.

Knots decrease strength because their grain is at a large angle to the grain of the piece, and because the grain around them is distorted. Also, checking may occur in and around them in drying. The weakening effect of knots is greater on the tensile than on the compressive strength.

In structural beams, the effect of knots on the bending strength is largely dependent upon the location of the knot (17, 24). For example, in a beam simply supported, knots on the lower side are placed in tension, those on the upper side in compression, and those at or near the neutral axis in horizontal shear. On the lower side at the point of maximum stress, a knot has a marked effect on the maximum load a beam will sustain; knots on the compression side are somewhat less serious. In any location knots have little effect on shear. The stiffness of beams is also affected but little by knots.

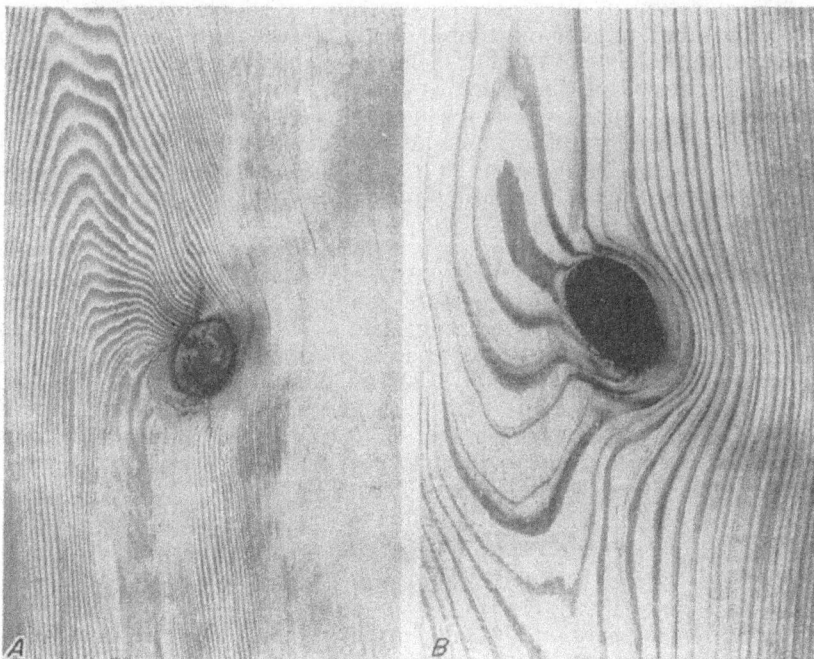

M 22604 F

FIGURE 10.—A, Encased knot; B, intergrown knot.

In long columns, in which stiffness is the controlling factor, knots are not of importance. In short or intermediate columns, the reduction in strength caused by knots is approximately proportional to the size of the knot. However, large knots have a somewhat greater relative effect than small knots.

Knots in round timbers, such as poles and piling, have less effect on strength than knots in sawed timbers. Although the grain is irregular around knots in both forms of timber, its angle with the surface is less in naturally round than in sawed timber. Furthermore, the round piece is often sufficiently enlarged at a knot to compensate, at least partially, for the effect of the irregular grain.

Observation of trees and poles bent or broken by windstorms or in felling or handling, as well as observations in tests of round timber, shows that first failure often occurs between, rather than at knots.

Cross Grain

The fibers in a cross-grained piece of wood lie at an angle to the axis of the piece (25). The principal types of cross grain are spiral grain and diagonal grain (fig. 11). Other less important types are wavy, dipped, interlocked, and curly grain.

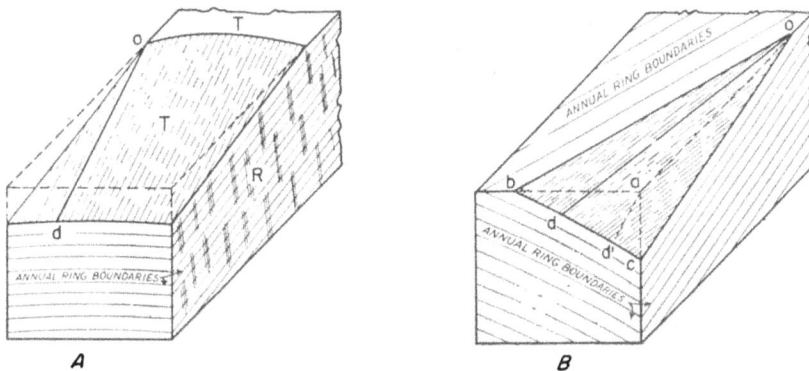

ZM 48832 F

FIGURE 11.—Interior views of two pieces of wood showing inclination of the annual rings (diagonal grain) and inclination of fibers in the plane of the annual rings (spiral grain). The line do indicates the direction of the fibers in each piece with respect to a longitudinal corner. A, A piece of wood with radial (R) and tangential (T) surfaces and with left-handed spiral grain. B, A piece of wood without radial and tangential surfaces and with right-handed spiral grain. If spiral grain were not present, the direction of the fibers would be parallel to a line from d' to o, d' being at the base of a perpendicular line from the corner a to the annual ring bc.

Spiral grain (fig. 12) is caused by the fibers growing in a winding or spiral course about the bole of a tree instead of in a vertical course.

Spiral grain is not always readily detected by ordinary visual inspection. The best test for spiral grain is to split a sample section from the piece at right angles to the annual rings. Another method of determining the presence of spiral grain is to note the alinement of

FIGURE 12.—Spiral-grained and straight-grained forest trees.

pores, rays, and resin ducts on flat-grained faces. Checks on a flat-sawed surface also follow the fibers and when present indicate the direction of grain.

Diagonal grain results from sawing a piece of lumber at an angle other than parallel with the bark. One of the frequent causes is sawing a log having a pronounced taper parallel with the axis of the tree rather than parallel with the bark. It also occurs in sawing crooked or swelled-butt logs and from careless sawing. Diagonal grain is easily detected by noting the slope of the annual rings on an edge-grain surface. Its presence but not the slope is evidenced on a tangential surface by the numerous hyperbolas produced by the intersection of growth rings with the surface.

The slope of grain is usually expressed by the ratio between a 1-inch deviation of the grain from the edge or axis of a piece and the distance within which this deviation occurs (6). When a piece contains both spiral and diagonal grain, the combined slope must be considered rather than the greater of the two slopes. The combined slope is determined by taking the square root of the sum of the squares of the slopes of the two types of cross grain. For example, if these slopes are 1 in 18 and 1 in 12, the combined slope is

$$\sqrt{\left(\frac{1}{18}\right)^2+\left(\frac{1}{12}\right)^2}=\frac{1}{10} \text{ or a slope of 1 in 10}$$

Because of practical considerations in commercial stress grading, slope of grain limitations in structural timbers are measured in the face having the greatest slope of grain. This fact is taken into account in the reduction factors applied for slope of grain in establishing working stresses for various grades (p. 149).

Relationships between strength properties and the actual combined slope of grain, determined for several species in tests by the Forest Products Laboratory, are given in table 15.

TABLE 15.—*Strength of wood members with various grain slopes compared to strength of a straight-grained member, expressed as percentages*

Maximum slope of grain in member	Static bending		Impact bending —height of drop causing complete failure (50-pound hammer)	Compression parallel to grain—maximum crushing strength
	Modulus of rupture	Modulus of elasticity		
	Percent	*Percent*	*Percent*	*Percent*
Straight-grained	100	100	100	100
1 in 25	96	97	95	100
1 in 20	93	96	90	100
1 in 15	89	94	81	100
1 in 10	81	89	62	99
1 in 5	55	67	36	93

Compression Wood

Compression wood is an abnormal type of wood that is characteristic of softwood trees or logs in which the pith is off center, as might occur in leaning trees. Compression wood may occur in all softwood (coniferous) species. It is denser and harder than the normal wood, is characterized by wide annual growth rings that are usually eccentric, and includes what appears to be exceptionally large proportions of summerwood growth (fig. 13, *A*). The contrast in color between springwood and summerwood, however, is usually less in compression wood than in normal wood.

Compression wood has larger shrinkage along the grain than normal wood, and low strength for its weight. When compression wood is present in the same board with normal wood, the unequal longitudinal shrinkage causes internal stresses that result in warping, sometimes in rough lumber, but more often in planed, ripped, or resawed lumber. The seriousness of the effects of compression wood is generally related to its widely varying forms. The most serious effects occur in pronounced compression wood that can be readily detected by ordinary visual examination, as compared to borderline forms that merge with normal wood and frequently are only detected by microscopical examination. Warping is serious in millwork and cabinets in which compression wood is limited to small tolerances. Moldings that are securely fastened in place usually are not seriously affected by warping caused by compression wood. Its reduced strength as compared with normal wood, however, requires that compression wood be restricted to only small amounts for uses such as aircraft in which safety of life and property are involved.

M 90614 F

FIGURE 13.—*A*, The darker areas shown are compression wood; *B*, compression failure is shown by the irregular lines across the grain.

Tension Wood

Tension wood is an abnormal type of wood that occurs in many hardwood species, mainly on the upper side of leaning trees but sometimes distributed entirely around some trees. Its presence in lumber causes excessive longitudinal shrinkage and warping when moderate to large numbers of its abnormal fibers are present. In addition, these abnormal fibers hold together tenaciously, so that sawed surfaces usually have projecting fibers and planed surfaces often are torn or have raised grain. The undesirable characteristics vary in seriousness in relation to the presence of small to large numbers of tension wood fibers.

Compression Failures

Excessive bending of standing trees from wind or snow, felling trees across boulders, logs, or irregularities in the ground, or the rough handling of logs or lumber may produce excessive compression stresses along the grain that cause minute compression failures. In some instances, such failures are visible on the surface of a board as minute lines or zones formed by the crumpling or buckling of the cells (fig.

13, *B*), although usually they appear only as white lines or may even be invisible to the naked eye. Their presence frequently is indicated by fiber breakage on end grain (fig. 14).

The strength of a piece containing visible compression failures is seriously reduced, especially in tensile strength and shock resistance; such a piece may break suddenly along the compression failure when subjected to shock or tension, in which event the failure will have a brittle appearance. Even when compression failures are so small that they are not visible with favorable lighting, they are known to reduce shock resistance seriously and cause brittle fractures (*18*).

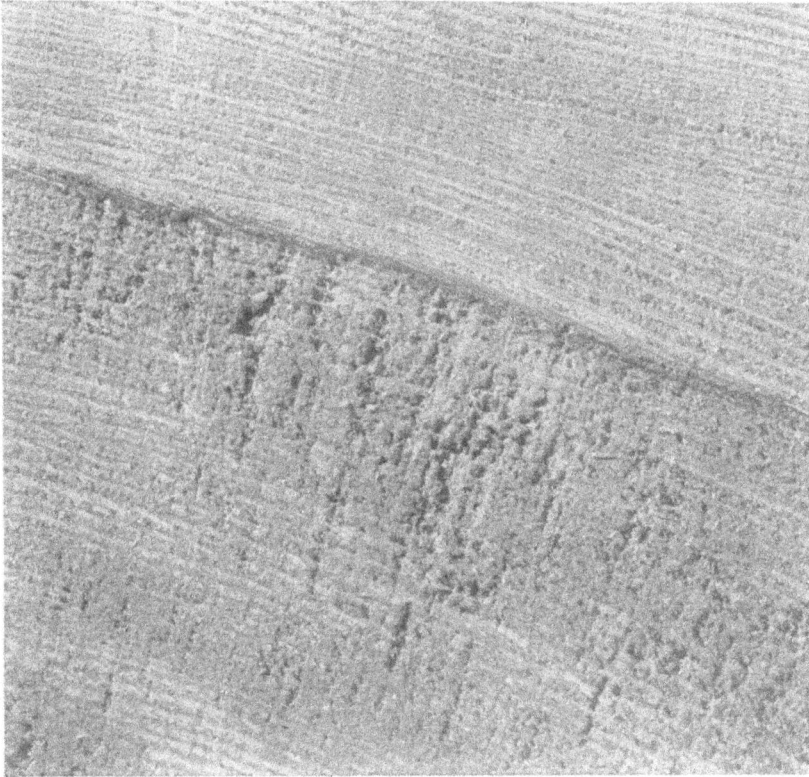

M 81195 F

FIGURE 14.—End-grain surfaces of spruce lumber showing fiber breakage caused by compression failures confined to the part below the dark line. Slightly magnified.

Molds and Staining Fungi

Molds and staining fungi (p. 381) do not ordinarily appreciably affect the strength of wood. Where molding or staining of wood is intensive, the shock resistance may be greatly reduced, but even then such properties as strength in compression parallel to the grain and bending strength are usually but little affected.

Although molds and stains themselves are usually not important as regards strength, conditions that favor the development of these fungi are likewise ideal for the growth of wood-destroying fungi (p. 382). Since decay even in its incipient stage may greatly reduce some strength properties, stained lumber should be regarded with suspicion when strength is an important factor.

Decay

Unlike the sap-staining fungi, the decay-producing fungi cause a serious reduction in strength. Even apparently sound wood adjacent to obviously decayed parts may contain incipient decay that is decidedly weakening, especially in shock resistance.

Different wood-destroying fungi affect wood differently. The fungi that cause an easily recognized pitting of the wood, for example, may be less injurious to the strength than those that, in the early stages, give a slight discoloration of the wood as the only visible effect.

No method is known for estimating from the appearance of decayed wood the amount of reduction in its strength. Therefore, when strength is an important consideration, the safe procedure is to discard every piece that contains even a small amount of decay (p. 383), with the exception of pieces in which decay occurs in a knot but does not extend into the surrounding wood.

Insect Damage

Insect damage may occur in standing trees, logs, and unseasoned or seasoned lumber. Damage in the standing tree is difficult to control, but otherwise insect damage can be largely eliminated by proper control methods.

Insect holes are generally classified as pinholes, grub holes, and powderpost (p. 386). The powderpost larvae by their irregular burrows may destroy most of the interior of a piece, while the surface shows only small holes, and the strength of the piece may be reduced virtually to zero.

No method is known for estimating from the appearance of insect-damaged wood the amount of reduction in strength, and, when strength is an important consideration, the safe procedure is to discard every piece that contains insect holes.

Pitch Pockets

A pitch pocket is a well-defined opening extending parallel to the annual rings, substantially flat toward the pith and curved on the bark side, which contains more or less free resin. Pith pockets are confined to the pines, spruces, Douglas-fir, tamarack, and western larch.

The effect of pitch pockets on strength depends upon their number, size, and location in the piece; their effect on strength has often been overestimated. A large number of pitch pockets indicates a lack of bond between annual growth layers, and a piece containing them should be inspected for shakes, or separations along the grain.

Mineral Streaks

Maple, hickory, white ash, and a number of other species are often damaged by small holes made by sapsuckers. These holes or bird pecks are often placed in horizontal rows, sometimes encircling the tree, and a brown or a black discoloration known as a "mineral streak" has its origin at such holes. Holes for tapping maple trees are also a source of mineral streaks. The streaks are caused by oxidation and other chemical changes in the wood.

Bird pecks and mineral streaks are not important defects as regards strength, although they do impair the appearance of the wood. However, if several bird pecks occur in a row across the outer surface of a piece of wood that is to be used in a bent product, such as a handle, the holes would appreciably weaken the piece.

Seasoning Degrade

Seasoning degrade (p. 323) that influences strength is confined mostly to surface and end checking. Some loss in strength, especially in shock resistance, may occur when wood is dried at excessively high temperatures, although such wood may show no visible signs of degrade. Even improperly seasoned wood, however, may be considerably higher in strength than green wood, since properly seasoned wood in the form of small, clear specimens may more than double in strength in seasoning from a green to a dry condition; the increase depends upon the strength property and the final moisture content.

Different drying conditions are required for different species of wood, because the strength of wood is affected by the temperature and the time of exposure in relation to its moisture content (p. 89). Therefore, suitable kiln-drying schedules have been worked out for the various species. These kiln-drying schedules result in no significant difference in the strength properties of the wood as compared to those obtained when careful air-drying practices, where temperature is not a factor, are followed. Proper kiln-drying offers better control of drying conditions than is usually the case with air-drying, and less degrade due to drying can be expected.

VALUE OF TIMBER FROM LIVE AND FROM DEAD TREES

Sound timber from trees killed by insects, blight, wind, or fire is undoubtedly as good for any structural purpose as that from live trees, provided additional injury from further insect attack, staining, decay, or seasoning degrade has not occurred.

If a tree stands on the stump too long after its death, the sapwood is likely to decay or to be attacked severely by wood-boring insects, and in time the heartwood will be similarly affected. Such deterioration occurs also in logs that have been cut from live trees and improperly cared for afterwards. Because of variations in climatic and local weather conditions and in other factors that affect deterioration, the length of the period during which dead timber may stand or lie in the forest without serious deterioration varies.

Tests on wood from trees that had stood as long as 15 years after being killed by fire demonstrated that this wood was as sound and as strong as wood from live trees. Also, logs of some of the more durable species have had thoroughly sound heartwood after lying on the ground in the forest for many years.

On the other hand, decay may cause great loss of strength within a very brief time, both in trees standing dead on the stump and in logs that have been cut from live trees and allowed to lie on the ground. Consequently, the important consideration is not whether the trees from which timber products are cut are alive or dead, but whether the products themselves are free from decay or other defects that would render them unsuitable for use.

Specifications for some timber products sometimes require that only live trees be used; this requirement, however, is difficult to enforce unless inspection is made in the forest, because wood cut from fire-killed, wind-fallen, or otherwise naturally killed trees before weathering, seasoning, discoloration, decay, insect attack, or similar change has occurred cannot be distinguished from wood taken from live trees. All specifications should omit the live-tree requirement and depend entirely upon inspection to determine the suitability of the timber for use.

RESIN

Resin is formed in some of the conifers, especially the southern yellow pines. It is common in amounts up to 6 percent of the dry weight of the wood, and pieces with a resin content up to 50 percent are sometimes found.

Tests at the Forest Products Laboratory on southern yellow pine indicate that resin will slightly increase some strength properties, but the effect is too small to be of any particular significance; that is, any additional strength resulting from the presence of resin is very small as compared to that which would normally result from an equal additional weight of wood substance. An excessive amount of resin is sometimes associated with an injury that may have greatly reduced the strength.

TURPENTINING

Some species of southern yellow pine trees are frequently tapped for turpentine, and if the turpentining is properly handled, the trees are not unduly damaged for lumber purposes. Comparative tests on mature turpentined and unturpentined trees have shown that (1) turpentined timber is as strong as unturpentined timber of the same weight, (2) the weight and shrinkage of wood are not affected by turpentining, and (3) practically speaking, properly turpentined trees contain neither more nor less resin than unturpentined trees.

EXTRACTIVES

Extractives, the part of a piece of wood that dissolves when the piece is placed in an inert solvent, occur in many species and are especially abundant in redwood, western redcedar, and black locust (11). The three species mentioned are also relatively high in certain strength properties for the amount of wood substance they contain, and tests at the Forest Products Laboratory have shown that the presence of extractives is probably accountable. The extent to which extractives affect the strength is apparently dependent upon the amount and nature of extractives, the moisture condition of the piece, and the mechanical property under consideration.

Of the strength properties examined, maximum crushing strength in compression parallel with the grain showed the greatest increase from the infiltration of extractives in the change of sapwood into heartwood, the modulus of rupture came next, and shock resistance showed the least. In fact, under some conditions, shock resistance appears to be actually lowered by extractives. That the extractives found in different species may affect the strength differently is indicated by the fact that they change the strength of western redcedar less than the strength of black locust, although black locust has a smaller percentage of extractives (11).

LITERATURE CITED

(1) AMERICAN SOCIETY FOR TESTING MATERIALS.
  1952. STANDARD METHODS FOR TESTING SMALL CLEAR SPECIMENS OF TIMBER. Amer. Soc. Testing Mater. Standard D. 143–52. Pt. 4: 720–757, illus.

(2) DOYLE, D. V., DROW, J. T., AND MCBURNEY, R. S.
  1945. ELASTIC PROPERTIES OF WOOD: THE YOUNG'S MODULI, MODULI OF RIGIDITY, AND POISSON'S RATIOS OF BALSA AND QUIPO. U. S. Forest Prod. Lab. Rpt. 1528, 16 pp., illus. [Processed.] (See supplementary Rpts. 1528A through 1528H for elastic properties of other species.)

(3) FULLER, F. B., AND OBORG, T. T.
  1943. FATIGUE CHARACTERISTICS OF NATURAL AND RESIN-IMPREGNATED COMPRESSED LAMINATED WOODS. Jour. Aeronaut. Sci. 10 (3): 81–85, illus.

(4) GEORGE, H. O.
  1933. EFFECT OF LOW TEMPERATURES ON THE STRENGTH OF WOOD. N. Y. State Col. Forestry, Syracuse Univ., Tech. Pub. 43, 18 pp., illus.

(5) GREENHILL, W. L.
  1936. STRENGTH TESTS PERPENDICULAR TO THE GRAIN OF TIMBER AT VARIOUS TEMPERATURES AND MOISTURE CONTENTS. Austral. Council Sci. & Indus. Res., Jour. 9 (4): 265–276, illus.

(6) KOEHLER, A.
  1943. GUIDE TO DETERMINING SLOPE OF GRAIN IN LUMBER AND VENEER. U. S. Forest Prod. Lab. Rpt. 1585, 25 pp., illus. [Processed.]

(7) KOLLMAN, F.
  1941. MECHANICAL CHARACTERISTICS OF VARIOUS MOIST WOODS WITHIN TEMPERATURE GRADIENTS FROM −200° TO +200° C. Translation, U. S. Natl. Advisory Com. Aeronaut. Tech. Memo 984, 37 pp., illus.

(8) KOMMERS, W. J.
  1943. FATIGUE BEHAVIOR OF WOOD AND PLYWOOD SUBJECTED TO REPEATED AND REVERSED BENDING STRESSES. U. S. Forest Prod. Lab. Rpt. 1327, 11 pp., illus. [Processed.]

(9) LEWIS, W. C.
  1951. FATIGUE OF WOOD AND GLUED JOINTS USED IN LAMINATED CONSTRUCTION. Forest Prod. Res. Soc. Proc. 5: 221–229, illus.

(10) LISKA, J. A.
  1950. EFFECT OF RAPID LOADING ON THE COMPRESSIVE AND FLEXURAL STRENGTH OF WOOD. U. S. Forest Prod. Lab. Rpt. 1767, 10 pp., illus. [Processed.]

(11) LUXFORD, R. F.
  1931. EFFECT OF EXTRACTIVES ON THE STRENGTH OF WOOD. Jour. Agr. Res. 42: 801–826, illus.

(12) MacLean, J. D.
    1953. EFFECT OF STEAMING ON THE STRENGTH OF WOOD.   Amer. Wood-
        Preservers' Assoc. Proc. 49: 88–112, illus.
(13) ———
    1951. RATE OF DISINTEGRATION OF WOOD UNDER DIFFERENT HEATING
        CONDITIONS.   Amer. Wood-Preservers' Assoc. Proc. 47: 155–168,
        illus.
(14) ———
    1945. EFFECT OF HEAT ON THE PROPERTIES AND SERVICEABILITY OF WOOD.
        U. S. Forest Prod. Lab. Rpt. R1471, 12 pp., illus. [Processed.]
(15) Markwardt, L. J.
    1930. COMPARATIVE STRENGTH PROPERTIES OF WOODS GROWN IN THE
        UNITED STATES.   U. S. Dept. Agr. Tech. Bul. 158, 39 pp., illus.
(16) ——— and Wilson, T. R. C.
    1935. STRENGTH AND RELATED PROPERTIES OF WOODS GROWN IN THE
        UNITED STATES.   U. S. Dept. Agr. Tech. Bul. 479, 99 pp., illus.
(17) Newlin, J. A., and Johnson, R. P. A.
    1924. STRUCTURAL TIMBERS: DEFECTS AND THEIR INFLUENCES ON
        STRENGTH.   Amer. Soc. Testing Mater. Proc. 24, Pt. II; 975–989,
        illus.
(18) Pillow, M. Y.
    1949. STUDIES OF COMPRESSION FAILURES AND THEIR DETECTION IN
        LADDER RAILS.   U. S. Forest Prod. Lab. Rpt. D1733, 10 pp.,
        illus. [Processed.]
(19) Sulzberger, P. H.
    1943. EFFECT OF TEMPERATURES ON STRENGTH PROPERTIES OF WOOD,
        PLYWOOD, AND GLUED JOINTS.   Austral. Council Sci. Indus. &
        Res. Jour. 16 (4): 263–265.
(20) Thunell, B.
    1940. EFFECT OF TEMPERATURES ON THE BENDING STRENGTH OF SWEDISH
        PINE WOOD.   (Sweden) Govt. Testing Inst. Med. 80: 38 pp.,
        illus. (English summary).
(21) U. S. Forest Products Laboratory.
    1941. FOREST PRODUCTS LABORATORY'S TOUGHNESS TESTING MACHINE.
        U. S. Forest Prod. Lab. Rpt. 1308, 13 pp., illus. [Processed.]
(22) ———
    1941. MOISTURE CONTENT-STRENGTH ADJUSTMENTS FOR WOOD.   U. S.
        Forest Prod. Lab. Rpt. 1313, 9 pp., illus. [Processed.]
(23) ———
    1951. DESIGN OF WOOD AIRCRAFT STRUCTURES.   ANC Bul. 18, 234 pp.,
        illus. (Issued by Subcommittee on Air Force-Navy-Civil Air-
        craft Design Criteria, Aircraft Committee, Munitions Board.)
(24) Wangaard, F. F.
    1950. MECHANICAL PROPERTIES OF WOOD.   377 pp., illus. New York.
(25) Wilson, T. R. C.
    1921. EFFECT OF SPIRAL GRAIN ON THE STRENGTH OF WOOD.   Jour.
        Forestry 19: 740–747, illus.
(26) ———
    1932. STRENGTH-MOISTURE RELATIONS FOR WOOD.   U. S. Dept. Agr.
        Tech. Bul. 282, 88 pp., illus.
(27) Wood, L. W.
    1947. BEHAVIOR OF WOOD UNDER CONTINUED LOADING.   Engin. News-
        Rec. 139 (24): 108–111, illus.

GRADES AND SIZES OF LUMBER

A log when sawed into lumber often yields boards of widely varying quality. To enable a user to buy the quality that best suits his purpose, the product of a log is divided into several lumber grades, each having a relatively narrow range in quality.

Except as noted later, the grade of a piece of lumber is based on the number, character, and location of features that may lower the strength, durability, or utility value of the lumber. Among the more common of such features are knots, checks, pitch pockets, shake, and stain, some of which are a natural part of the tree. The best grades are free or practically free from them, but the other grades comprising the great bulk of lumber contain fairly numerous knots and other features that affect quality. These features, however, do not prevent such lumber from giving satisfaction for many widespread uses.

AGENCIES PROMULGATING GRADING RULES

Name and address	*Woods covered by grading rules* [1]
Hardwood Dimension Manufacturers' Association, 230 Heyburn Bldg., Louisville, Ky.	Hardwoods (hardwood dimension lumber, squares and laminated stock) (*2*).[2]
Maple Flooring Manufacturers' Association, 35 E. Wacker Dr., Chicago, Ill.	Maple (northern hard), beech, birch (flooring) (*3*).
National Hardwood Lumber Association, 59 E. Van Buren St., Chicago, Ill.	Hardwoods, cypress, aromatic redcedar, hardwood veneer (*4*).
National Oak Flooring Manufacturers' Association, 814 Sterick Bldg., Memphis, Tenn.	Oak and pecan (flooring) (*5*).
California Redwood Association, 576 Sacramento St., San Francisco, Calif.	Redwood (*1*).
Northeastern Lumber Manufacturers' Association, 271 Madison Ave., New York, N. Y.	Balsam fir, northern white pine, eastern hemlock, and eastern spruce (*6*, *7*).
Northern Hemlock & Hardwood Manufacturers' Association, Oshkosh, Wis.	Northern white pine, Norway pine, jack pine, eastern spruce, balsam, eastern hemlock, tamarack, northern whitecedar (*8*, *9*).
Northern Pine Manufacturers' Association, 4329 Oakland Ave., Minneapolis, Minn.	Northern white pine, Norway pine, eastern spruce, and jack pine.
Southern Cypress Manufacturers' Association, Barnett National Bank Bldg., Jacksonville, Fla.	Southern cypress (upland and tidewater types) (*10*).
Southern Pine Inspection Bureau,[3] National Bank of Commerce Bldg., New Orleans, La.	Southern yellow pine and longleaf yellow pine (*11*).

[1] Species names in this list are those used in the grading rules of the different agencies.
[2] Italic numbers in parentheses refer to Literature Cited, p. 135.
[3] Affiliated with the Southern Pine Association.

Name and address	*Woods covered by grading rules* [1]
West Coast Bureau of Lumber Grades & Inspection,[4] 1410 S. W. Morrison St., Portland, Oreg.	Douglas-fir (coast region), west coast hemlock, Sitka spruce, western redcedar, white fir, noble fir (*13*).
Western Pine Association, Yeon Bldg., Portland, Oreg.	Ponderosa pine, Idaho white pine, sugar pine. lodgepole pine, western larch, Douglas-fir, white fir, Engelmann spruce, ·incense cedar, and western redcedar (*14*).
Red Cedar Shingle Bureau, White Bldg., Seattle, Wash.	Western redcedar (shingles).

[4] Affiliated with the West Coast Lumbermen's Association.

HARDWOOD LUMBER GRADING

Hardwood Lumber (All Species)

Most hardwood boards are not used in their entirety but are cut into smaller pieces to make furniture or some other fabricated product.

The rules that are considered standard in grading hardwood lumber in the United States are those adopted by the National Hardwood Lumber Association (*4*).[4] In these rules the grade of a piece of hardwood lumber is determined by the proportion of the piece that can be cut into a certain number of smaller pieces of material clear on one side and not less than a certain size. In other words, the grade classification is based upon the amount of usable lumber in the piece rather than upon the number or size of growth features that characterize softwood yard grades. This usable material, commonly termed "cuttings," must have one face clear and the reverse face sound, which means free from such things as rot, heart center, and shake that materially impair the strength of the cutting. The lowest grade requires only that cuttings be sound.

The highest grade of hardwood lumber is termed "Firsts" and the next grade "Seconds." Firsts and Seconds, or, as they are generally written, "FAS," nearly always are combined in one grade. The third grade is termed "Selects," followed by No. 1 Common, No. 2 Common, Sound Wormy, No. 3A Common, and No. 3B Common.

A brief summary of the standard hardwood lumber grades follows. This summary should not be regarded as a complete set of grading rules, as there are numerous details, exceptions, and special rules for certain species that are not included. The complete official rules of the National Hardwood Lumber Association should be followed as the only full description of existing grades. The summary is intended only as a preliminary guide in distinguishing between the general qualities to be expected under the various grades.

[4] This association publishes a booklet that contains detailed grading rules for various hardwood products, such as lumber, car stock, and vehicle stock. The association maintains bonded lumber inspectors in various hardwood producing and consuming centers who issue inspection certificates on shipments. The correctness of the grades as shown on these certificates is guaranteed by the association.

Summary of Standard Grades of the National Hardwood Lumber Association

Woods Included [1]

Alder, red
Ash
Aspen
Basswood
Beech
Birch
Boxelder
Buckeye
Butternut
Cedar, red
Cedar, Spanish
Cherry
Chestnut
Cottonwood
Cypress
Elm:
- Rock (or cork)
- Soft

Gum:
- Black
- Red and sap
- Tupelo

Hackberry
Hardwoods (Philippine)
Hardwoods (Tropical American other than mahogany and Spanish cedar)
Hickory
Locust
Magnolia
Mahogany:
- African
- Cuban and San Dominican
- Philippine

Maple:
- Hard (or sugar)
- Soft

Oak:
- Red
- White

Pecan
Poplar
Sycamore
Walnut
Willow

[1] Species names are those used in grading rules of the association. Two of the woods included, namely, cedar (eastern redcedar) and cypress (baldcypress), are not hardwoods. Cypress lumber has a different set of grading rules from those used for the hardwoods.

Standard Lengths

Standard lengths are 4, 5, 6, 7, 8, 9, 10, 11, 12, 13, 14, 15, and 16 feet, but not more than 50 percent of odd lengths will be admitted.

Standard Thicknesses

Standard thicknesses for hardwood lumber are given in table 16.

TABLE 16.—*Standard thicknesses for hardwood lumber*

Rough (inches)	Surfaced 1 side (S1S)	Surfaced 2 sides (S2S)	Rough (inches)	Surfaced 1 side (S1S)	Surfaced 2 sides (S2S)
	Inches	*Inches*		*Inches*	*Inches*
3/8	1/4	3/16	2 1/2	2 5/16	2 1/4
1/2	3/8	5/16	3	2 13/16	2 3/4
5/8	1/2	7/16	3 1/2	3 5/16	3 1/4
3/4	5/8	9/16	4	3 13/16	3 3/4
1	7/8	13/16	4 1/2	(1)	(1)
1 1/4	1 1/8	1 1/16	5	(1)	(1)
1 1/2	1 3/8	1 5/16	5 1/2	(1)	(1)
2	1 13/16	1 3/4	6	(1)	(1)

[1] Finished size not specified in rules. Thickness subject to special contract.

Description of Standard Hardwood Grades

A description of the standard hardwood grades is given in table 17. The highest grade, Firsts, calls for pieces that will allow 91 2/3 percent of their surface measure to be cut into clear-face material; that is,

457249°—58——8

TABLE 17.—*Standard hardwood grades* [1]

Grade and lengths allowed (feet)	Widths allowed	Surface measure of pieces	Amount of each piece that must work into clear-face cuttings	Maximum cuttings allowed	Minimum size of cuttings required
	Inches	*Square feet*	*Percent*	*Number*	
Firsts: [2] 8 to 16 (will admit 30 percent of 8- to 11-foot, ½ of which may be 8- and 9-foot).	6+	4 to 9	91⅔	1	4 inches by 5 feet, or 3 inches by 7 feet.
		10 to 14	91⅔	2	
		15+	91⅔	3	
Seconds: [2] 8 to 16 (will admit 30 percent of 8- to 11-foot, ½ of which may be 8- and 9-foot).	6+	4 and 5	83⅓	1	Do.
		6 and 7	83⅓	1	
		6 and 7	91⅔	2	
		8 to 11	83⅓	2	
		8 to 11	91⅔	3	
		12 to 15	83⅓	3	
		12 to 15	91⅔	4	
		16+	83⅓	4	
Selects: 6 to 16 (will admit 30 percent of 6- to 11-foot, ⅙ of which may be 6- and 7-foot).	4+	2 and 3	91⅔	1	Do.
		4+	(3)		
No. 1 Common: 4 to 16 (will admit 10 percent of 4- to 7-foot, ½ of which may be 4- and 5-foot).	3+	1	100	0	4 inches by 2 feet, or 3 inches by 3 feet.
		2	75	1	
		3 and 4	66⅔	1	
		3 and 4	75	2	
		5 to 7	66⅔	2	
		5 to 7	75	3	
		8 to 10	66⅔	3	
		11 to 13	66⅔	4	
		14+	66⅔	5	
No. 2 Common: 4 to 16 (will admit 30 percent of 4- to 7-foot, ⅓ of which may be 4- and 5-foot).	3+	1	66⅔	1	3 inches by 2 feet.
		2 and 3	50	1	
		2 and 3	66⅔	2	
		4 and 5	50	2	
		4 and 5	66⅔	3	
		6 and 7	50	3	
		6 and 7	66⅔	4	
		8 and 9	50	4	
		10 and 11	50	5	
		12 and 13	50	6	
		14+	50	7	
No. 3A Common: 4 to 16 (will admit 50 percent of 4- to 7-foot, ½ of which may be 4- and 5-foot).	3+	1+	[4] 33⅓	(5)	Do.
No. 3B Common: 4 to 16 (will admit 50 percent of 4- to 7-foot, ½ of which may be 4- and 5-foot).	3+	1+	[6] 25	(5)	1½ inches by 2 feet.

[1] Inspection to be made on the poorer side of the piece, except in Selects.

[2] Firsts and Seconds are combined as 1 grade (FAS). The percentage of Firsts required in the combined grade varies from 20 to 40 percent, depending on the species.

[3] Same as Seconds.

[4] This grade also admits pieces that grade not below No. 2 Common on the good face and have the reverse face sound.

[5] Not specified.

[6] The cuttings must be sound; clear face not required.

not more than 8⅓ percent of each piece can be wasted in making the required cuttings. In the grade of Seconds 83⅓ percent of the surface measure of the pieces must yield clear-face cuttings.[5] Both Firsts and Seconds require pieces not less than 6 inches wide and 8 feet long. In the grade Selects the minimum width is 4 inches and the minimum length 6 feet. Both Firsts and Seconds and the face side of Selects must, in addition to cutting requirements, meet specified requirements as to knots, holes, and other imperfections. The cutting requirements of Selects are 91⅔ percent of clear face in pieces with 2 and 3 surface feet. In larger pieces the cutting requirements are the same as for Seconds on the face side. The reverse side of the cuttings in Selects must be sound or the reverse side of the piece not below No. 1 Common.

The next 2 grades, No. 1 Common and No. 2 Common, call for lumber not less than 3 inches wide and 4 feet long and require 66⅔ percent and 50 percent of clear-face cuttings, respectively.[6] The minimum size of cuttings in these 2 grades is reduced from 4 inches by 5 feet or 3 inches by 7 feet in Firsts, Seconds, and Selects to 3 inches by 2 feet. The number of allowable cuttings in grades No. 2 Common and better are limited by the hardwood grading rules (4) according to the surface measure of each piece.

A grade of hardwood lumber called Sound Wormy has the same requirements as No. 1 Common except that wormholes and limited sound knots and other imperfections are allowed in the cuttings.

The grade of 3A Common admits pieces that will furnish 33⅓ percent of clear face in cuttings not less than 3 inches wide and 2 feet long. This grade will also admit pieces that grade not below No. 2 Common on the good face and have the reverse face of the cutting sound. The lowest grade, No. 3B Common, allows pieces that will cut 25 percent in sound material not less than 1½ inches wide by 2 feet long.

Hardwood Flooring

Hardwood flooring is generally graded under the rules of the Maple Flooring Manufacturers' Association (3) and the rules of the National Oak Flooring Manufacturers' Association (5). Tongued-and-grooved and end-matched hardwood flooring is commonly furnished. Square-edge and square-end strip flooring is also available, as well as parquetry flooring suitable for laying on a mastic base or on an ordinary subfloor.

The Maple Flooring Manufacturers' Association grading rules cover flooring manufactured from sugar maple (hard maple), beech, and birch.

Each species has the three grades designated as First grade, Second grade, and Third grade. Combination grades of Second and Better and Third and Better are sometimes made. There are also four

---

[5] Boards 6 to 15 feet surface measure will admit of 1 additional cutting to yield 91⅔ percent of clear face.

[6] Exceptions in No. 1 Common are pieces with 1 foot surface measure and 2 feet surface measure, which require 100 percent of clear face and 75 percent of clear face, respectively, and in No. 2 Common pieces with 1 foot surface measure, which require 66 percent of clear face.

special grades—White Clear Northern Hard Maple, Brown Clear Northern Hard Maple, Red Clear Northern Beech, and Red Clear Northern Birch, which are made up of special stock selected for uniformity of color. First grade flooring must have one face practically free from all imperfections. Variations in the natural color of the wood are allowed. Second grade flooring admits tight, sound knots and slight imperfections but must lay without waste. Third grade flooring has no restrictions as to imperfections in the grain but must be of such a character that it can be properly laid and will give a good serviceable floor. The standard thickness of maple, beech, and birch flooring is $^{25}\!/_{32}$ inch. Face widths are 1½, 2, 2¼, and 3¼ inches. Standard lengths are 2 feet and longer in First and Second grade flooring and 1¼ feet and longer in Third grade flooring.

The grading rules of the National Oak Flooring Manufacturers' Association mainly cover quartersawed and plainsawed oak flooring. Quartersawed flooring has three grades—Clear, Sap Clear, and Select. Plainsawed flooring has four grades—Clear, Select, No. 1 Common, and No. 2 Common. The Clear grade in both plainsawed and quartersawed flooring must have one face practically free from surface imperfections except three-eighths inch of bright sap. Color is not considered in any grade. Sap Clear quartersawed flooring must have one face practically clear but will admit unlimited bright sap. Select flooring (plainsawed or quartersawed) may contain sap and will admit a few features such as pin wormholes and small tight knots. No. 1 Common plainsawed flooring must contain material that will make a sound floor without cutting. No. 2 Common may contain grain and surface imperfections of all kinds but must lay a serviceable floor. Standard thicknesses of oak flooring are $^{25}\!/_{32}$, ½, and ⅜ inch. Standard face widths are 1½, 2, 2¼, and 3¼ inches. Lengths in upper grades are 2 feet and up with a required average of 4½ feet in a shipment. In the lower grades lengths are 1¼ feet and up with a required average of 2½ or 3 feet.

The most recent rules of the National Oak Flooring Manufacturers' Association also include specifications for flooring of pecan, hard maple, beech, and birch. The grades of pecan flooring are: First grade, practically clear but unselected for color; First grade red, practically clear but made from all heartwood; First grade white, practically clear but made from all bright sapwood; Second grade, admits sound tight knots, pin wormholes, streak, and slight machining imperfections; Second grade red, similar to Second grade but must be made from all heartwood; Third grade, must make a sound floor without cutting; and Fourth grade, must lay a serviceable floor. The standard sizes for pecan flooring are the same as those for oak flooring.

The rules of the National Oak Flooring Manufacturers' Association for hard maple, beech, and birch flooring are the same as those of the Maple Flooring Manufacturers' Association.

Hardwood Dimension Stock

Hardwood dimension stock, as used by furniture and cabinet manufacturers, is generally graded under the rules of the Hardwood Dimension Manufacturers' Association (2). These rules apply primarily to dimension stock cut from kiln-dried rough lumber and cover three classes of material—kiln dried glued dimension, solid dimension flat stock, and solid dimension squares. Each class may be rough, semifinished, or finished. Solid dimension flat stock has five grades— Clear Two Faces, Clear One Face, Paint, Core, and Sound. Solid dimension squares have three grades—Clear, Select, and Sound.

SOFTWOOD LUMBER GRADING

Softwood lumber, unlike hardwood lumber, is graded under a number of different association rules. Not only are the different kinds of softwoods graded under different rules, but the same softwoods in a number of cases are graded under different association rules.

The first industry-sponsored grading rules for softwoods were established before 1900 and were comparatively simple because the sawmills marketed their lumber locally and grades had only local significance. As new sources were developed and lumber was transported to distant points, each producing region continued to establish its own grading rules with the result that lumber from the various regions differed in size, grade name, and, most important of all, in grade quality. When several different species graded under different rules came into keen competition in the chief consuming areas, confusion and dissatisfaction were inevitable.

In order to eliminate unnecessary differences in the grading rules of the various softwood lumber manufacturers' associations and to improve and simplify these rules, American lumber standards for softwood lumber were formulated in the period from 1919 to 1925. These standards were the result of a number of conferences organized by the U. S. Department of Commerce and attended by representatives of lumber manufacturers, distributors, wholesalers, retailers, engineers, architects, and contractors. The standards themselves are issued in pamphlet form as simplified practice recommendations of the U. S. Department of Commerce (12).

The agencies promulgating grading rules for commercial use have based their rules on the American lumber standards for softwood lumber. These include California Redwood Association, Southern Cypress Manufacturers' Association, Southern Pine Association, West Coast Lumbermen's Association, Western Pine Association, Northern Hemlock & Hardwood Manufacturers' Association, and Northeastern Lumber Manufacturers' Association.

Nomenclature of Commercial Softwoods

The names of lumber adopted by the trade as American lumber standards are not always identical with the names of trees adopted as official by the Forest Service. Table 18 has therefore been prepared to show the American lumber standards commercial name for lumber corresponding to the Forest Service tree name used in this handbook. Other names sometimes used locally but not in contracts are also shown, as are accepted botanical names.

TABLE 18.—*Nomenclature of commercial softwoods*

CEDARS AND JUNIPERS

Official Forest Service tree name used in this handbook	Name adopted as standard for lumber under American lumber standards	Other names sometimes used	Botanical name
Alaska-cedar	Alaska cedar	Yellow cedar, Sitka cypress, yellow cypress.	*Chamaecyparis nootkatensis.*
Alligator juniper	Western juniper	Juniper	*Juniperus deppeana.*
Atlantic whitecedar	Southern white cedar	White cedar, swamp cedar.	*Chamaecyparis thyoides.*
Eastern redcedar	Eastern red cedar	Red cedar, cedar, juniper.	*Juniperus virginiana.*
Incense-cedar	Incense cedar	Cedar, white cedar.	*Libocedrus decurrens.*
Northern whitecedar	Northern white cedar	Arborvitae, cedar, swamp cedar, white cedar.	*Thuja occidentalis.*
Port-Orford-cedar	Port Orford cedar	Lawson's cypress, Oregon cedar, white cedar.	*Chamaecyparis lawsoniana.*
Rocky Mountain juniper.	Western juniper	Juniper	*Juniperus scopulorum.*
Utah juniper	----do	Juniper, white cedar.	*J. osteosperma.*
Western juniper	----do	Juniper, cedar.	*J. occidentalis.*
Western redcedar	Western red cedar	Red cedar, cedar, western cedar.	*Thuja plicata.*

CYPRESS

Baldcypress	Red cypress (coast type), yellow cypress (inland type), white cypress (inland type).	Cypress, tidewater red cypress, Gulf coast red cypress, Louisiana red cypress, bald cypress, black cypress.	*Taxodium distichum.*

DOUGLAS-FIR

Douglas-fir	Douglas fir	Red fir, Oregon fir, Douglas spruce, yellow fir, Puget Sound pine.	*Pseudotsuga menziesii.*

TRUE FIRS

Balsam fir	Balsam fir	Balsam, eastern fir	*Abies balsamea.*
California red fir	White fir	Red fir	*A. magnifica.*
Fraser fir	Balsam fir	Balsam, eastern fir	*A. fraseri.*
Grand fir	White fir	Yellow fir	*A. grandis.*
Noble fir	Noble fir	Red fir	*A. procera.*
Pacific silver fir	White fir	Red fir, fir	*A. amabilis.*
Subalpine fir	----do	Balsam	*A. lasiocarpa.*
White fir	----do	Colorado white fir	*A. concolor.*

HEMLOCKS

Carolina hemlock	Eastern hemlock	Hemlock, hemlock spruce, spruce pine.	*Tsuga caroliniana.*
Eastern hemlock	----do	----do	*T. canadensis.*
Mountain hemlock	Mountain hemlock	Weeping spruce, Alpine spruce, hemlock spruce.	*T. mertensiana.*
Western hemlock	West coast hemlock	Hemlock, hemlock spruce, Pacific hemlock, Alaska pine.	*T. heterophylla.*

Table 18.—*Nomenclature of commercial softwoods*—Continued

LARCH

Official Forest Service tree name used in this handbook	Name adopted as standard for lumber under American lumber standards	Other names sometimes used	Botanical name
Western larch..........	Western larch............	Tamarack, larch..........	*Larix occidentalis.*

PINES

Official Forest Service tree name used in this handbook	Name adopted as standard for lumber under American lumber standards	Other names sometimes used	Botanical name
Eastern white pine......	Northern white pine....	White pine, cork pine, soft pine, northern pine, pumpkin pine.	*Pinus strobus.*
Jack pine...............	Jack pine...............	Scrub pine...............	*P. banksiana.*
Loblolly pine...........	Southern yellow pine...	Old-field pine, slash pine, shortleaf pine, Virginia pine, sap pine, yellow pine, North Carolina pine.	*P. taeda.*
Lodgepole pine.........	Lodgepole pine..........	Scrub pine, spruce pine...	*P. contorta.*
Longleaf pine...........	Longleaf yellow pine,[1] southern yellow pine.	Southern pine, yellow pine, hard pine, Georgia pine, pitch pine, heart pine, fat pine.	*P. palustris.*
Pitch pine.............	Southern yellow pine...	-------------------------	*P. rigida.*
Ponderosa pine.........	Ponderosa pine.........	Bull pine, Arizona white pine, western soft pine, western pine.	*P. ponderosa.*
Red pine..............	Norway pine...........	Hard pine, northern pine.	*P. resinosa.*
Shortleaf pine..........	Southern yellow pine...	Yellow pine, spruce pine, oldfield pine, Arkansas soft pine, North Carolina pine.	*P. echinata.*
Slash pine.............	Longleaf yellow pine,[1] southern yellow pine.	Swamp pine, pitch pine..	*P. elliottii.*
Sugar pine.............	Sugar pine.............	Big pine..................	*P. lambertiana.*
Virginia pine..........	Southern yellow pine....	-------------------------	*P. virginiana.*
Western white pine.....	Idaho white pine	White pine, soft pine......	*P. monticola.*

REDWOOD

Official Forest Service tree name used in this handbook	Name adopted as standard for lumber under American lumber standards	Other names sometimes used	Botanical name
Redwood...............	Redwood...............	Sequoia, coast redwood...	*Sequoia sempervirens.*

SPRUCES

Official Forest Service tree name used in this handbook	Name adopted as standard for lumber under American lumber standards	Other names sometimes used	Botanical name
Black spruce...........	Eastern spruce..........	-------------------------	*Picea mariana.*
Blue spruce............	Engelmann spruce......	-------------------------	*P. pungens.*
Engelmann spruce......	....do................	White spruce, silver spruce, balsam, mountain spruce.	*P. engelmannii.*
Red spruce............	Eastern spruce..........	-------------------------	*P. rubens.*
Sitka spruce...........	Sitka spruce	Spruce, tideland spruce, western spruce, yellow spruce, silver spruce.	*P. sitchensis.*
White spruce..........	Eastern spruce..........	-------------------------	*P. glauca.*

[1] The commercial requirements for longleaf yellow pine lumber are that not only must it be produced from the species *Pinus elliottii* and *P. palustris*, but each piece must average either on one end or the other not less than 6 annual rings per inch and not less than ⅓ summerwood. Longleaf yellow pine lumber is sometimes designated as pitch pine in the export trade.

TABLE 18.—*Nomenclature of commercial softwoods*—Continued

Official Forest Service tree name used in this handbook	Name adopted as standard for lumber under American lumber standards	Other names sometimes used	Botanical name
		TAMARACK	
Tamarack_____	Tamarack_____	Larch, hackmatack, red larch, black larch.	*Larix laricina.*
		YEW	
Pacific yew_____	Pacific yew_____	Yew, western yew, mountain mahogany.	*Taxus brevifolia.*

General Classification of Softwood Lumber

Softwood lumber is divided into three main classes—yard lumber, structural lumber (often referred to under the general term "timber"), and factory and shop lumber. The following classification of softwood lumber gives the grade names in general use by lumber manufacturers' associations for the various classes of lumber.

Grades

SOFTWOOD LUMBER (This classification applies to rough or dressed lumber; sizes given are nominal.)

YARD LUMBER (lumber less than 5 inches thick, intended for general building purposes; grading based on use of the entire piece).

- Finish (4 inches and under thick and 16 inches and under wide). — A. B. C. D.
- Common boards (less than 2 inches thick and 1 or more inches wide). — No. 1. No. 2. No. 3. No. 4. No. 5.
- Common dimension (2 inches and under 5 inches thick and 2 or more inches wide).
  - Planks (2 inches and under 4 inches thick and 8 or more inches wide). — No. 1. No. 2. No. 3.
  - Scantling (2 inches and under 5 inches thick and less than 8 inches wide). — No. 1. No. 2. No. 3.
  - Heavy joists (4 inches thick and 8 or more inches wide). — No. 1. No. 2. No. 3.

STRUCTURAL LUMBER (lumber 5 or more inches thick and wide, except joists and planks; grading based on strength and on use of entire piece).

- Joists and planks (2 to 4 inches thick and 4 or more inches wide).
- Beams and stringers (5 or more inches thick and 8 or more inches wide).
- Posts and timbers (5 by 5 inches and larger).

FACTORY AND SHOP LUMBER (grading based on area of piece suitable for cuttings of certain size and quality).

- Factory plank graded for door, sash, and other cuttings 1¼ or more inches thick and 5 or more inches wide.
- Shop lumber graded for general cut-up purposes.

Association grading rules should be referred to for standard grades and sizes.

Yard Lumber—Size Standards

Standard lengths of yard lumber are multiples of 1 foot as specified in the manufacturers' association grading rules. In practice, however, even foot lengths are the rule.

Standard thicknesses and widths of various yard lumber products are given in table 19. In commercial practice the dressed dimensions are considered minimum dimensions, and some association rules provide for thicker and wider sizes than American lumber standards.

TABLE 19.—*Summary of American standard thicknesses and widths* [1] *for softwood yard lumber, including finish, boards, dimension, siding, flooring, ceiling, partition, shiplap, and dressed and matched lumber*

Item	Thickness		Widths	
	Nominal [2]	Minimum dressed	Nominal	Minimum dressed
	Inches	*Inches*	*Inches*	*Inches*
Finish: Select or Common	3/8	5/16	2	1 5/8
	1/2	7/16	3	2 5/8
	5/8	9/16	4	3 1/2
	3/4	11/16	5	4 1/2
	1	25/32	6	5 1/2
	1 1/4	1 1/16	7	6 1/2
	1 1/2	1 5/16	8	7 1/4
	1 3/4	1 7/16	9	8 1/4
	2	1 5/8	10	9 1/4
	2 1/2	2 1/8	11	10 1/4
	3	2 5/8	12	11 1/4
	3 1/2	3 1/8	14	13
	4	3 1/2	16	15
Bevel siding	1/2	7/16 butt, 3/16 tip	4	3 1/2
	9/16	15/32 butt, 3/16 tip	5	4 1/2
	5/8	9/16 butt, 3/16 tip	6	5 1/2
	3/4	11/16 butt, 3/16 tip	8	7 1/4
	1	3/4 butt, 3/16 tip	10	9 1/4
			12	11 1/4
Bungalow siding	3/4	1 1/16 butt, 3/16 tip	8	7 1/4
			10	9 1/4
			12	11 1/4
Rustic and drop siding (shiplapped, 3/8-inch lap)	5/8	9/16	4	3 1/8
	1	3/4	5	4 1/8
			6	5 1/16
Rustic and drop siding (shiplapped, 1/2-inch lap)	5/8	9/16	4	3
	1	3/4	6	5
			8	6 11/16
			10	8 11/16
			12	10 11/16
Rustic and drop siding (dressed and matched)	5/8	9/16	4	3 1/4
	1	3/4	5	4 1/4
			6	5 3/16
			8	6 11/16
			10	8 11/16
Flooring	3/8	5/16	2	1 1/2
	1/2	7/16	3	2 3/8
	5/8	9/16	4	3 1/4
	1	25/32	5	4 1/4
	1 1/4	1 1/16	6	5 3/16
	1 1/2	1 5/16		

See footnotes at end of table.

TABLE 19.—*Summary of American standard thicknesses and widths* [1] *for softwood yard lumber, including finish, boards, dimension, siding, flooring, ceiling, partition, shiplap, and dressed and matched lumber*—Continued

Item	Thickness		Widths	
	Nominal [2]	Minimum dressed	Nominal	Minimum dressed
	Inches	*Inches*	*Inches*	*Inches*
Ceiling	3/8	5/16	3	2 3/8
	1/2	7/16	4	3 1/4
	5/8	9/16	5	4 1/4
	3/4	11/16	6	5 3/16
Partition	1	23/32	3	2 3/8
			4	3 1/4
			5	4 1/4
			6	5 3/16
Stepping	1	25/32		
	1 1/4	1 1/16	8	7 1/4
	1 1/2	1 5/16	10	9 1/4
	2	1 5/8	12	11 1/4
Boards [3]	1	25/32	2	1 5/8
	1 1/4	1 1/16	3	2 5/8
	1 1/2	1 5/16	4	3 5/8
			5	4 5/8
			6	5 1/2
			7	6 1/2
			8	7 1/2
			9	8 1/2
			10	9 1/2
			11	10 1/2
			12	11 1/2
			14	13 1/2
			16	15 1/2
Shiplap, 3/8-inch lap	1	25/32	4	3 1/8
			6	5 1/8
			8	7 1/8
			10	9 1/8
			12	11 1/8
			14	13 1/8
			16	15 1/8
Shiplap, 1/2-inch lap	1	25/32	4	3
			6	5
			8	7
			10	9
			12	11
			14	13
			16	15
Center matched, 1/4-inch tongue	1	25/32	4	3 1/4
	1 1/4	1 1/16	6	5 3/16
	1 1/2	1 5/16	8	7
			10	9
			12	11
Dimension, plank, and joists	2	1 5/8	2	1 5/8
	2 1/2	2 1/8	3	2 5/8
	3	2 5/8	4	3 5/8
	3 1/2	3 1/8	6	5 1/2
	4	3 5/8	8	7 1/2
			10	9 1/2
			12	11 1/2
			14	13 1/2
			16	15 1/2
			18	17 1/2

See footnotes at end of table.

TABLE 19.—*Summary of American standard thicknesses and widths* [1] *for softwood yard lumber, including finish, boards, dimension, siding, flooring, ceiling, partition, shiplap, and dressed and matched lumber*—Continued

Item	Thickness		Widths	
	Nominal	Minimum dressed	Nominal	Minimum dressed
	Inches	*Inches*	*Inches*	*Inches*
Timbers_____	[5] 5	½ off _____	[6] 5	½ off
	2	1⅝_____	4	[7] 3
Factory flooring, heavy roofing, decking, and sheet piling [4]_____	2½	2⅛_____	6	[7] 5
	3	2⅝_____	8	[8] 7
	4	3⅝_____	10	[8] 9
	5	4⅝_____	12	[8] 11

[1] The thicknesses apply to all widths and the widths to all thicknesses except as modified. In tongued-and-grooved flooring and in tongued-and-grooved shiplapped ceiling of ⁵⁄₁₆-, ⁷⁄₁₆-, and ⁹⁄₁₆-inch dressed thicknesses, the tongue or lap shall be ³⁄₁₆ inch wide, with the overall widths ³⁄₁₆ inch wider than the face widths shown in this table. In all other worked lumber of dressed thickness 1¹⁄₁₆ inch to 1⁵⁄₁₆ inches the tongue shall be ¼ inch wide or wider in tongued-and-grooved lumber, and the lap ⅜ inch wide or wider in shiplapped lumber, and the overall widths shall be not less than the dressed face widths shown in this table, plus the widths of the tongue lap.

[2] For nominal thicknesses under 1 inch, the board measure count is based on the nominal surface dimensions (width by length). With these exceptions, the nominal thicknesses and widths in this table are the same as the board measure or count sizes.

[3] In some regions lumber thicker than 1½ inch is graded according to board rules. When proper provision is made for this in the applicable grading rules, such lumber may be regarded as American standard lumber.

[4] In worked lumber of nominal thickness of 2 inches and over, the tongue shall be ⅜ inch wide in tongued-and-grooved lumber and the lap ½ inch wide in shiplapped lumber, with the overall widths ⅜ inch and ½ inch wider, respectively, than the face widths shown in this table.

[5] 5 and thicker.

[6] 5 and wider.

[7] Face width when shiplapped; when tongued and grooved the face width is ⅛ inch greater; when grooved for splines the face width is ½ inch greater.

[8] Face width when shiplapped or tongued and grooved; when grooved for splines the face width is ½ inch greater.

Yard Lumber—Grade Standards

Ordinary building lumber is graded by lumber manufacturers' associations as finish in grades A, B, C, and D; as common boards in grades No. 1, No. 2, No. 3, No. 4, and No. 5; and as common dimension in grades No. 1, No. 2, and No. 3. The general requirements of these grades as used by lumber manufacturers' associations are summarized in the following paragraphs.

Finish Lumber—Grade Qualities

Finish grades [7] of lumber provide for good appearance and finishing qualities; grades A and B are suitable for natural finishes and grades C and D for paint finishes. In a few species where there is a pronounced difference in color between heartwood and sapwood and where high natural resistance to decay is required, a grade of clear heartwood is available.

[7] Detailed descriptions and use recommendations for the various grades are given in "Lumber Grade-use Guide," issued by the National Lumber Manufacturers' Association, Washington, D. C.

Grade A is practically clear wood. It is manufactured for such items as finish, flooring, ceiling, partition, and siding. A large number of manufacturers do not segregate the grade even in these items, and some of the lumber associations do not recognize the quality as a separate grade. When the grade is not segregated, it is combined with B grade and sold as B and Better. Grade A lumber is used almost entirely for interior and exterior trim and for flooring. The demand is small and confined largely to high-class construction, such as office buildings and the higher cost residences.

Grade B allows a few small imperfections. In practice these small imperfections mainly take the form of minor skips in manufacture and small checks or stain due to seasoning, and, depending on the species, small pitch areas, pin knots, or the like. Grade A pieces in the mixed grade are practically clear, but the average board contains 1 to 2 small imperfections. Grade B and Better is the highest quality segregated in a number of woods. In construction it is the grade most commonly used for high-class interior and exterior trim, paneling, and cabinet-work, especially where these are to receive a natural finish. It is the principal grade used for flooring in homes, offices, and public buildings. In industrial uses it meets the special requirements for large-sized, practically clear stock.

Grade C allows a limited number of small imperfections that can be easily covered with paint. Specifically, the number of these per board averages about twice that of B and Better, and the proportion of small knots is greater than in B and Better. Grade C lumber is especially adapted to use where the highest class paint finish is desired. It is therefore popular for cornice and other exterior parts of dwellings, porch flooring, porch columns, trim for bedrooms and kitchens, built-in kitchen fixtures, and siding for the better class of structures. It is used to some extent for natural finishes in medium-priced dwellings and offices.

Grade D allows any number of surface imperfections that do not detract from the appearance of the finish when painted. In practice the number of such surface features per board averages 3 to 5 times as many as in B and Better. Certain natural and manufacturing imperfections are not much more numerous in grade D than in grade C, but the number and size of the knots in grade D are considerably greater than in grade C, and usually the back is of somewhat lower quality. Commercial grading permits an occasional coarse knot or hole in grade D that may be cut out with restrictions as to waste. Grade D is used in construction for the same uses as grade C. It goes into moderate- or low-priced houses, furnishing a medium-priced lumber for casing, cornice work, shelving, and built-in fixtures that are to be painted. It is also used extensively for millwork and molding and is adaptable to industrial uses requiring short-length clear lumber.

The knots occurring in grade B and Better are predominantly under one-half inch in diameter and have smooth, hard surfaces. A small proportion of the knots in grade C are as large as 1 inch in diameter, and a few are not of the best quality. A few knots in grade D lumber are more than 1 inch in diameter and in quality are slightly soft, rough, or loose.

Depending on the species, the highest commercially recognized grade may be C and Better or D and Better, but no such combination of grades, except B and Better, is recommended in American lumber standards.

Seasoning faults, such as checks, either in flat surfaces or at the ends of boards, are among the more frequent imperfections in the finish grades. Imperfect seasoning often causes a lowering of grade, but the number of such occurrences is considerably reduced at plants of careful manufacturers.

Pitch pocket is relatively common in the finish grades of several species but occurs less frequently than knots in all the important species except one. The variation among grades in the number and size of pitch pockets is not so marked as in the case of knots. The frequency of pitch pockets as compared with other forms of pitch varies considerably among the species.

Among the other factors in the finish grades are stain and chipped and torn grain.

Common Boards and Dimension—Grade Qualities

Common grades of boards contain features that detract from appearance, but the boards are suitable for general-utility and construction purposes. The differences between the various board grades are due to the character more than to the number of such features as knots and pitch. The number of knots and like features in a board averages in different species about 5 to 20 per 8 board-feet regardless of grade. No. 1 and No. 2 boards are for use without waste. No. 3, No. 4, and No. 5 boards permit a limited amount of waste.

No. 1 boards are usually required to be sound and tight-knotted stock, and the size of knots is limited. Such lumber is usually considered watertight. In most species practically all boards in the grade contain knots, although in some species pitch is the predominant feature of the grade. The size of the knots varies with the species. From one-half to three-fourths of the knots are usually intergrown; the remaining knots are encased, a small proportion of which are unsound, broken, or checked.

No. 1 boards are used for siding, cornices, shelving, and paneling in medium- and low-priced homes, and for sheathing and roof boards in the more expensive type of buildings. They are well adapted to coverage for farm buildings, especially where a weathertight structure is required. In this grade it is difficult to entirely conceal the knots with paint. The No. 1 board is a general-utility item in industrial use. It is used extensively for door and window frames and for backing and concealed parts of furniture and fixtures. Figure 15 illustrates the appearance of No. 1 boards of four species.

No. 2 boards allow large and coarse features, such as knots, that may be considered graintight. In practice, a small amount of through-shake, through-pitch pocket, and decay is permitted in the grade. The proportion of large knots is greater than in No. 1, and whereas 33 to 75 percent of the knots are intergrown, 10 percent or more are usually unsound, loose, or otherwise partially open. Some commercial grading rules allow knotholes in the grade, provided they are

M 90499 F

FIGURE 15.—Typical No. 1 boards of four species.

strictly limited as to size and number. No. 2 boards are used primarily as coverage where the wood is not painted or otherwise finished. Subfloors, sheathing, and concrete forms are typical uses for the grade. Dressed and matched, it is used for rough flooring in inexpensive farm and factory buildings, garages, and warehouses. The popular type of knotty finish is largely selected from this grade. Typical No. 2 boards of four species are shown in figure 16.

No. 3 boards (fig. 17) allow larger and coarser knots than No. 2 boards and also occasional knotholes. A larger part of large knots and increased amounts of shake, decay, and holes distinguish No. 3 boards from No. 2 boards. No. 3 boards are used in construction for concrete forms, sheathing, subfloors, roof boards, barn boards, and temporary construction. They fill a demand from builders of less exacting type of buildings for a cheaper material than No. 2 grade. Industrially they are extensively used for boxes and crates, and in some species for foundry flasks.

Grades No. 4 and 5 boards are usually provided for in grading rules, but grade No. 5 is not produced in some species. No. 4 is low-quality lumber admitting the coarsest features, such as decay and holes. The only requirement as to the quality of No. 5 lumber is that it must be usable.

No. 4 boards (fig. 18) are not graded for use as a whole. They are used for sheathing, subfloors, and roof boards in the cheaper types of construction, but their most important industrial outlet is for boxes and crates.

No. 5 boards are seldom shipped far from the mill and are therefore not commonly available at retail yards in nonforested regions. They are used for rough and temporary coverage of buildings, and for boxes, crates, and dunnage.

The fact that all lumber manufacturers' associations do not divide their lumber into five grades and that there are differences in inherent properties of the various species makes it impossible to consider the common grades of corresponding name for the different woods as interchangeable in use.

In most species one-third to two-thirds of the knots in the common grades of boards are intergrown, whereas in the finish grades intergrown knots comprise only a small proportion of the total. Intergrown knots may check if large, but they are integral parts of the wood and will not loosen or drop out.

In some species a large number of the encased knots remain tight, but in other species many of them loosen. The loosening becomes more pronounced in large knots and as the lumber becomes drier. As a rule encased knots comprise a larger percentage of the total number of knots in the small-knotted select grades than they do in the larger knotted grades of boards.

A spike knot is formed by cutting a branch lengthwise rather than crosswise as is the case with a round knot. Spike knots occur infrequently in grades better than No. 2.

Wood that contains small areas of clearly evident decay is not permitted in grades better than No. 2.

Common dimension lumber is manufactured in Nos. 1, 2, and 3 dimension grades mainly, although stress grades are available in several species.

M 90433 F

FIGURE 16.—Typical No. 2 boards of four species.

M 90434 F

FIGURE 17.—Typical No. 3 boards of four species.

457249°—58——9

M 90435 F

FIGURE 18.—Typical No. 4 boards of four species.

Grading is based principally on the requirements of framing for buildings. The dimension grades are best adapted to use where stiffness is the controlling factor, as in joists and studs or where the size of the member is determined by common building practice rather than by design to carry definite live and dead loads.[8] Rafters for houses are good examples of members whose size is generally determined by common practice rather than by special design.

No. 1 dimension is a sound grade allowing knots limited in size depending on the size of the piece. The location of the knots in the piece is important in grading dimension. Features, such as pitch, torn grain, checks, and stain, that do not materially affect the strength of the piece are not limited. Wane is limited to provide good nailing on 1 side and 2 edges. The grade is for joists, rafters, and scaffolding in light framing and for less exacting items in heavier framing. Likewise, it is used for planking for warehouses, platforms, and other heavy-duty flooring, where the wearing surface rather than load-carrying capacity is the important factor. The structural grades of joists and plank or stress grades of dimension should be used for flooring designed to carry heavy loads.

No. 2 dimension admits larger and coarser features than No. 1 but each piece must be usable in its full dimensions for ordinary framing purposes. It is used in construction for joists and rafters in medium-priced light-frame construction and for plates, sills, studding, and other vertical load-bearing members in high-priced light construction. It is used as flooring to provide wearing surface where the load is not carried by the wood floor. Industrially it is used for crating and other uses where short lengths are required.

No. 3 dimension admits all features of No. 2 but in a more serious degree. In general it is suitable for general-utility purposes in low-priced and temporary light-frame construction. In small buildings, where the members are short, No. 3 dimension may be cut and used with considerable economy.

A grade of No. 4 dimension is made in some woods, but is low-quality lumber of limited utility.

Structural Lumber

Structural lumber is graded on the basis of strength and is intended for use where working stresses are required. A full discussion of structural grades, sizes, and appropriate working stresses appears on pages 137–164.

Factory and Shop Lumber

Factory and shop lumber is divided into two classes from the standpoint of use—factory plank and shop lumber—each of which has a different set of grades. These grades are based on the percentage of the area of each board or plank that will furnish cuttings of specified sizes and qualities. Association grading rules give standard grades

---

[8] Where strength rather than stiffness is the controlling factor, as in heavy construction, the structural grades of joists and planks or stress grades of dimension should be used.

for the various species, which differ somewhat in cutting requirements, and standard sizes.

Factory plank is 1¼ inches or more in thickness and is used largely for door and sash cuttings. The No. 1 cuttings for factory plank as a rule must be free of knots and other features that affect quality on both sides. The No. 2 cuttings may contain any one of the following seven features: a limited amount of blue or brown stain, a small tight knot, a small pitch pocket or streak, small seasoning checks, and slightly torn grain. The cuttings are of various lengths and widths, depending on the door or sash parts for which they are used.

Shop lumber is used for general cut-up purposes. Two types of cuttings, (a) and (b), are recognized; (a) cuttings must be clear on both faces, (b) cuttings equal in grade on one face to B and Better finish lumber. The (a) cuttings for shop lumber as a rule must be at least 9½ inches wide and 18 inches long, and the (b) cuttings at least 5 inches wide and 3 feet long. Shop lumber grades in some woods are available in only 1-inch stock. Thick stock in these woods for cut-up purposes is bought under factory plank rules.

STANDARD LUMBER PATTERNS

Figure 19 shows six typical patterns of lumber: Flooring, standard match; ceiling, edge beading; drop siding, shiplapped; bevel siding; dressed and matched, center matched; and shiplap.

With softwood flooring, "standard match" means that the upper lip of the groove is thicker than the lower. The thickness of the lower lip is the same for all standard thicknesses of flooring, and hence the difference between upper and lower lips becomes more pronounced in the heavier thicknesses. Ceiling, which is thinner than 1 inch, is usually machined with a V. Partition usually has the bead and V, also, but on both sides, and it is thicker than ceiling. Drop siding is usually made from 1-inch lumber and probably is made in more patterns than any other product except molding. Some of these patterns are shiplapped, and others are tongued and grooved. Bevel siding is made by resawing 4/4- or 5/4-inch lumber on an angle. Square-edged lumber in either boards, timbers, or dimension, of course, forms only rectangles of different dimensions. Boards are frequently dressed and matched (D&M), in which event the tongue and groove are in the center, making the pieces center matched. For some uses it is considered preferable to shiplap boards.

LUMBER ITEMS USUALLY CARRIED IN RETAIL YARDS

The small retail yards throughout the United States carry softwoods required for ordinary construction purposes and sometimes small stocks of 1 or 2 hardwoods in the grades suitable for finishing or cabinetwork. Any particular hardwood desired, however, may be obtained by special order through the retail lumberyard. Hardwoods are used in building chiefly for interior trim, cabinets, and flooring. In modern practice, trim items are cut to size in standard pattern at millwork plants and are sold in such form by the retail yards. Cabinets are usually made by the millwork plant ready for installation on

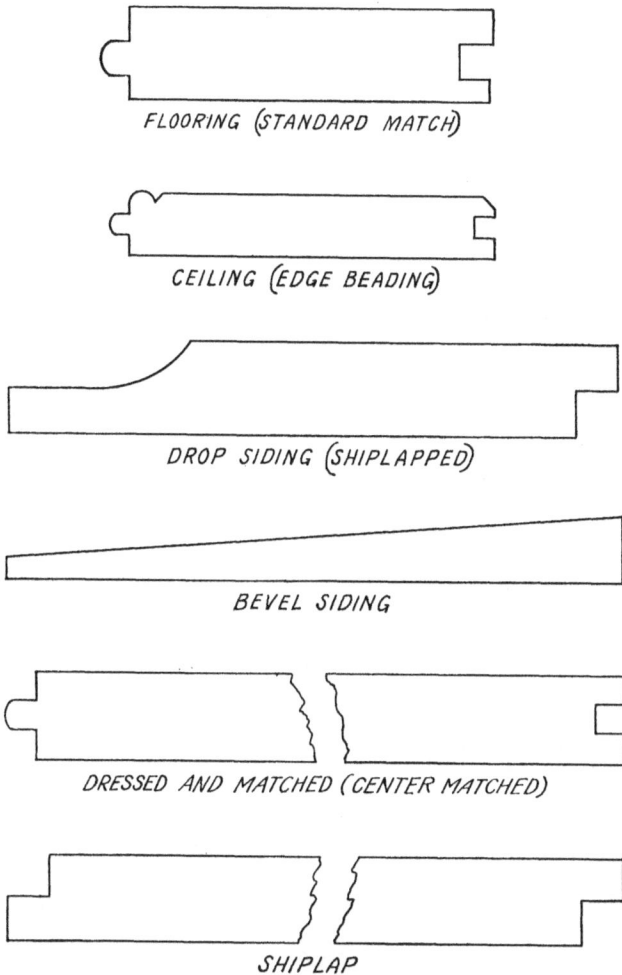

FLOORING (STANDARD MATCH)

CEILING (EDGE BEADING)

DROP SIDING (SHIPLAPPED)

BEVEL SIDING

DRESSED AND MATCHED (CENTER MATCHED)

SHIPLAP

ZM 22047 F

FIGURE 19.—Six typical patterns of lumber.

the job by the carpenter. Hardwood flooring invariably is a planing-mill product and is available to the buyer only in standard patterns.

The assortment of species in general construction items carried by retail yards depends largely upon geographical location. For instance, in the Pacific Northwest local yards will, as a rule, stock Douglas-fir, spruce, ponderosa pine, western hemlock, and western redcedar. An Iowa yard may stock eastern hemlock, ponderosa pine, southern yellow pine, and Douglas-fir. For some species only certain items of various species may be available; for example, southern yellow pine or Douglas-fir may be stocked only in the form of dimension, eastern hemlock only in sheathing and 1-inch boards, cypress and redwood in finish boards or in siding. A New York market might stock eastern spruce and hemlock, eastern white pine, cypress, ponderosa pine, southern yellow pine, and Douglas-fir. In the eastern part of the United States lumber from the Pacific coast is readily available because of low-cost water transportation via the Panama Canal.

Wholesalers do not ordinarily stock lumber. However, some large wholesalers located in extensive lumber-consuming districts have yards stocked with many kinds of lumber, both of hardwood and softwood. These wholesalers supply the varied needs of retail yards, wood-using factories, and larger contractors, although purchases are sometimes made directly from the lumber mills.

Lumber is sold in a number of standard general-purpose items and also in certain special-purpose items. Retail yards carry all of the general-purpose items, but as a rule only the more important of the special-purpose items. Some of these lumber items come in one group of grades and some in another. There are not many items made in the complete range of grades. Among the typical special-purpose items are stepping, casing and base, silo staves, molding, battens, window and door jambs, and porch columns. Descriptions of lumber items commonly carried by all retail yards follow.

Common Boards

Common boards, so-called, or boards are a general-purpose item available at all yards in one or more of the kinds of wood most frequently used in building. Boards are usually of nominal 1-inch thickness dressed 2 sides to $2\frac{5}{32}$ inch thick. Sometimes thinner stock is offered, but it should not be classed as standard lumber. The standard nominal widths are 4, 6, 8, 10, and 12 inches. Although boards are manufactured in all grades from No. 1 to No. 5, No. 1, No. 2, and No. 3 are the grades most generally available in retail yards.

Boards are sold square edge, dressed and matched (tongued and grooved) or with a shiplap joint. The largest uses for common boards are subfloors, sheathing, barn boards, roofing boards, rough siding, and concrete forms, and, in some species and grades, knotty finish.

Dimension

Dimension is primarily framing lumber, such as joists, rafters, and studding. It also comprises the planking used for heavy floors. Strength, stiffness, and uniformity of size are essential requirements. It is stocked in all yards, frequently in only one of the general-purpose construction woods, such as pine, fir, hemlock, or spruce. It is usually nominal 2 inches thick dressed 1 or 2 sides to $1\frac{5}{8}$ inches, nominal 4, 6, 8, 10, or 12 inches wide, and 4 to 18 feet long in multiples of 2 feet. Dimension thicker than 2 inches and longer than 18 feet is manufactured but only in comparatively small quantities. The grades in most common use are No. 1 and No. 2.

Finish

Finish, except for such special types as knotty pine and pecky cypress, is of select quality and is made to some extent in practically all softwoods, although a local lumberyard usually stocks only 1 or

2 kinds. It is manufactured in all finish grades but principally B and Better and C. It is usually nominal 1 inch, dressed 2 sides to $^{25}\!/_{32}$ inch. The widths most usually stocked are 4 to 12 inches in even inches. Finish goes principally into interior and exterior trim, although it has numerous other uses.

Siding

Siding, as the name implies, is made specifically for purposes of exterior coverage. It is of two principal types, bevel siding and drop siding; the latter also known as rustic siding or barn siding. Bevel siding is ordinarily stocked only in the finish grades of white pine, ponderosa pine, western redcedar, cypress, or redwood. Drop siding is sometimes stocked both in finish and in sound, tight-knotted utility grades of southern yellow pine, Douglas-fir, and hemlock, in addition to the woods mentioned under bevel siding. Bevel siding may be of the narrow type, $3\frac{1}{2}$-, $4\frac{1}{2}$-, or $5\frac{1}{2}$-inch face width, or of the wide type, $7\frac{1}{4}$-, $9\frac{1}{4}$-, or $11\frac{1}{4}$-inch face width. The narrow type is $^{7}\!/_{16}$, $^{15}\!/_{32}$, or $^{9}\!/_{16}$ inch thick, the wide type $^{9}\!/_{16}$ or $\frac{3}{4}$ inch thick. Drop siding is ordinarily 5-, 6-, or 8-inch nominal width and dressed to a thickness of $\frac{3}{4}$ inch. It may be either dressed and matched or shiplapped.

Ceiling and Partition

Ceiling and partition, which are usually in the finish grades, are not made or stocked as commonly as in former years. Although, as their name indicates, they are manufactured for a specific use, they are often used for a variety of purposes where a finished appearance is required and where a simple pattern is preferred to plain unbroken surfaces. Ceiling and partition may be $2\frac{3}{8}$-, $3\frac{1}{4}$-, $4\frac{1}{4}$-, and $5\frac{5}{16}$-inches face width. Ceiling is $^{5}\!/_{16}$, $^{7}\!/_{16}$, $^{9}\!/_{16}$, and $^{11}\!/_{16}$ inch thick, and partition is $^{23}\!/_{32}$ inch thick.

Flooring

Flooring is made chiefly in the harder softwood species, such as Douglas-fir, western larch, and southern yellow pine, and in hardwoods, such as maple and oak. At least one of the softwoods and one of the hardwoods are usually stocked in most yards. Flooring is usually nominal 1-inch dressed to $^{25}\!/_{32}$ inch and 3- and 4-inch nominal width. Thicker flooring is available for heavy-duty floors both in hardwoods and softwoods. Thinner flooring is available in hardwoods, especially for re-covering old floors. Vertical and flat grain (also called quartersawed and plainsawed, respectively) are manufactured in both softwoods and hardwoods, and many dealers carry both types in stock.

Vertical-grained flooring shrinks and swells less than flat-grained flooring, is more uniform in texture, wears more uniformly, and the joints do not open so much.

Softwood flooring is usually available in B and Better grade, C Select, or D Select. The chief grades in maple are Clear No. 1 and No. 2. The grades in quartersawed oak are Clear, Sap Clear, and Select, and in plainsawed Clear, Select, and No. 1 Common. The quartersawed has the same advantages that vertical-grained softwood flooring has. In addition the silver or flaked grain of quartersawed flooring is frequently preferred to the figure of plainsawed flooring. Beech, birch, and, for fancy parquetry flooring, walnut and mahogany are also used.

Casing and Base

Casing and base is a standard item in the more important softwoods and is stocked in most yards in at least one species. The chief grade is B and Better. It is made and graded to meet the requirements of interior trim for dwellings. It is usually nominal 1-inch stock, dressed to ¾ inch, of 5-, 6-, and 7-inch nominal widths. Hardwoods for the same purposes, such as oak and birch, may be carried in stock in the retail yard or may be obtained on special order at the local planing mill.

Shingles

Shingles are made from western redcedar, cypress (baldcypress), and redwood. Western redcedar furnishes the greater proportion. The shingle grades of western redcedar are No. 1, No. 2, and No. 3; cypress, No. 1—bests, primes, economy, and clippers; and redwood, No. 1 and No. 2. The western redcedar No. 1 grade, tidewater red cypress No. 1, and redwood No. 1 meet the requirements of No. 1 grade of commercial standards as promulgated by the Department of Commerce. These highest grades, which are usually the most economical grades for permanent construction, are all clear, all heartwood, and all edge grained.

Shingles that are all heartwood give greater resistance to decay than do shingles that contain sapwood. Edge-grained shingles are less likely to warp than flat-grained shingles; thick-butted shingles less than thin shingles; and narrow shingles less than wide shingles. The standard thicknesses of shingles are described as 4/2, 5/2¼, and 5/2 meaning, respectively, 4 shingles to 2 inches of butt thickness, 5 shingles to 2¼ inches of butt thickness, and 5 shingles to 2 inches of butt thickness. Lengths may be 16, 18, or 24 inches. Random widths and specified widths ("dimension" shingles) are available in western redcedar, redwood, and cypress (baldcypress). Random-width shingles are usually packed by the square, dimension by the thousand shingles.

Lath

Lath are made in nearly all the important softwoods. The chief grades are No. 1 and No. 2.

GRADE-MARKED LUMBER AND MOISTURE CONTENT PROVISIONS

Grade-marked and trade-marked lumber is now available in most softwood species and by lumber items. Each piece of such lumber typically is stamped with its proper grade, a number identifying the mill where it was made, and the mark of the inspection agency promulgating the grading rules, or with a certificate of inspection. The grade designation stamped on a board indicates the quality at the time the piece was graded. Subsequent exposure to unfavorable storage conditions, improper drying, or careless handling may cause the material to fall below its original grade.

Lumber may be purchased under moisture content provisions. The allowable moisture content is lower in the thinner lumber and in the finish grades. In one specification of kiln-dried 4/4- and 5/4-inch lumber of C and Better quality, for instance, the moisture content must not exceed 12 percent in 90 percent of the pieces, and the remainder must not exceed 15-percent moisture content. For thicker finish lumber and for kiln-dried boards and dimension the allowable moisture content values are from 15 to 20 percent. Specifications for air-dried lumber are expressed similarly with the allowable moisture content values ranging from 16 percent for 4/4-inch C and Better up to 19 percent for 8/4-inch dimension.

POINTS TO CONSIDER WHEN ORDERING SOFTWOOD LUMBER OR TIMBER

The following, based on a Federal specification, lists some of the points to consider when ordering softwood lumber or timber.

1. *Quantity.*—Feet, board measure, number of pieces if of definite size and length.

2. *Size.*—Thickness in inches—nominal and actual if surfaced on faces. Width in inches—nominal and also actual if surfaced on edges. Length in feet—may be nominal average length, limiting lengths, or a single uniform length.

3. *Grade.*—As indicated in grading rules of lumber manufacturers' associations.

4. *Species of wood.*—Douglas-fir, cypress, etc.

5. *Product*—Flooring, siding, timbers, boards, etc.

6. *Condition of seasoning.*—Air-dry, kiln-dry, commercially shipping dry, etc. (Definite interpretation of requirements in this respect necessitates specifying in terms of moisture content and how it is to be determined.)

7. *Surfacing.*—Indicate whether rough (unplaned) or dressed (surfaced) stock is desired. S1S means surfaced on 1 side; S2S, surfaced on 2 sides; S1S1E, surfaced on 1 side and 1 edge; S4S, surfaced on 4 sides.

8. *Association rules.*—Southern Pine Association, Western Pine Association, etc.

STANDARD LUMBER ABBREVIATIONS

The following standard lumber abbreviations are in common use in contracts and other documents arising in the transactions of purchase and sale of lumber.

AD—air-dried.
A. d. f.—after deducting freight.
A. l.—all lengths.

ALS—American lumber standards.
Av. or avg.—average.
Av. w.—average width.
Av. l.—average length.
A. w.—all widths.
B1S—beaded one side.
B2S—beaded two sides.
BBS—box bark strips.
B&B or B & Btr.—B and better.
B&S—beams and stringers.
Bd.—board.
Bd. ft.—board-foot (or board-feet); that is, an area of 1 square foot by 1
    inch thick.
Bdl.—bundle.
Bdl. bk. s.—bundle bark strips.
Bev.—bevel.
B/L—bill of lading.
Bm.—board measure.
Btr.—better.
CB1S—center bead one side.
CB2S—center bead two sides.
CF—cost and freight.
CG2E—center groove two edges.
CIF—cost, insurance, and freight.
CIFE—cost, insurance, freight, and exchange.
Clg.—ceiling.
Clr.—clear.
CM—center matched; that is, the tongued-and-grooved joints are worked
    along the center of the edges of the piece.
Com.—common.
CS—calking seam.
Csg.—casing.
Ctg.—crating.
Cu. ft.—cubic foot or feet.
CV1S—center V one side.
CV2S—center V two sides.
DB. Clg.—double-beaded ceiling (E&CB1S).
DB. Part.—double-beaded partition (E&CB2S).
DET—Double end trimmed.
D&CM—dressed (1 or 2 sides) and center matched.
D&H—dressed and headed; that is, dressed 1 or 2 sides and worked to
    tongued-and-grooved joints on both the edge and the ends.
D&M—dressed and matched; that is, dressed 1 or 2 sides and tongued and
    grooved on the edges. The match may be center or standard.
D&SM—dressed (1 or 2 sides) and standard matched.
D2S&CM—dressed two sides and center matched.
D2S&M—dressed two sides and (center or standard) matched.
D2S&SM—dressed two sides and standard matched.
Dim.—dimension.
Dkg.—decking.
D/S or D/Sdg.—drop siding.
E—edge.
EB1S—edge bead one side.
EB2S—edge bead two sides.
E&CB1S—edge and center bead 1 side; surfaced 1 or 2 sides and with a
    longitudinal edge and center bead on a surfaced face.
E&CB2S—edge and center bead 2 sides; all 4 sides surfaced and with a
    longitudinal edge and center bead on the 2 faces.
ECM—ends center matched.
E&CV1S—edge and center V 1 side; surface 1 or 2 sides and with a longi-
    tudinal edge and center V-shaped groove on a surfaced face.
E&CV2S—edge and center V two sides.
EG—edge (vertical) grain.
EE—eased edges.
EM—end matched—either center or standard.
ESM—ends standard matched.

EV1S—edge V one side.
EV2S—edge V two sides.
Fac.—factory.
FAS—First and Seconds—a combined grade of the two upper grades of hardwoods.
FAS—free alongside (named vessel).
F. bk.—flat back.
FBM—foot or feet board measure.
Fcty.—factory (lumber).
FG—flat (slash) grain.
Flg.—flooring.
FOB—free on board (named point).
FOHC—free of heart center or centers.
F. o. k.—free of knots.
Frm.—framing.
Frt.—freight.
Ft.—foot or feet. Also one accent (′).
Feet b. m.—feet board measure.
Feet s. m.—feet surface measure.
GM—grade marked.
G/R or G/Rfg.—grooved roofing.
HB—hollow back.
Hdl.—handle (stock).
Hdwd.—hardwood.
H&M—hit and miss.
H or M—hit or miss.
Hrt.—heart.
Hrt. CC—heart cubical content.
Hrt. FA—heart facial area.
Hrt. G—heart girth.
Hrtwd.—heartwood.
1s&2s.—Ones and Twos—a combined grade of the hardwood grades of Firsts and Seconds.
In.—inch or inches. Also two accent marks (″).
J&P—joists and planks.
KD—kiln-dried.
K. d.—knocked down.
Lbr.—lumber.
LCL—less than carload.
LFT or LIN. ft.—linear foot (or feet); that is 12 inches.
Lgr.—longer.
Lgth.—length.
Lin.—Linear.
Lng.—lining.
LR.—log run.
Lr. MCO—log run, mill culls out.
Lth.—lath.
M—thousand.
MBM—thousand (feet) board measure.
MC—moisture content.
MCO—mill culls out.
Merch.—merchantable.
M. l.—mixed lengths.
Mldg.—molding.
MR—mill run.
M. s. m.—thousand (feet) surface measure.
M. w.—mixed widths.
No.—number.
N1E—nosed one edge.
N2E—nosed two edges.
Og.—Ogee.
Ord.—order.
P.—planed.
Par.—paragraph.
Part.—partition.
Pat.—pattern.

Pc.—piece.
Pcs.—Pieces.
PE—plain end.
Pky.—pecky.
Pln.—plain, as plainsawed.
PO—purchase order.
P&T—post and timbers.
Qtd.—quartered—when referring to hardwoods.
Rdm.—random.
Reg.—regular.
Res.—resawed.
Rfg.—roofing.
Rfrs.—roofers.
Rgh.—rough.
Rip.—ripped.
R/L—random lengths.
Rnd.—round.
R. Sdg.—rustic siding.
R/W—random widths.
R/W&L—random widths and lengths.
S&E—surfaced 1 side and 1 edge.
S1E—surfaced one edge.
S2E—surfaced two edges.
S1S—surfaced one side.
S2S—surfaced two sides.
S1S1E—surfaced 1 side and 1 edge.
S2S1E—surfaced 2 sides and 1 edge.
S1S2E—surfaced 1 side and 2 edges.
S4S—surfaced four sides.
S4S&CS—surfaced four sides with a calking seam on each edge.
S&M—surfaced and matched; that is, surfaced 1 or 2 sides and tongued and
    grooved on the edges.   The match may be center or standard.
S2S&SM—surfaced two sides and standard matched.
S2S&CM—surfaced two sides and center matched.
S2S&M—surfaced two sides and center or standard matched.
S2S&S/L—surfaced two sides and shiplapped.
Sap.—sapwood.
SB—standard bead.
Sd.—seasoned.
Sdg.—siding.
Sel.—select.
SE Sdg.—square-edge siding.
SE&S—square edge and sound.
S. f.—surface foot; that is, an area of 1 square foot.
Sftwd.—softwood.
Sh. D.—shipping dry.
Ship.—shiplap.
S. m.—surface measure.
SM—standard matched.
Smkd.—smoked (dried).
Smk. stnd.—smoke stained.
S. n. d.—sap no defect.
Snd.—sound.
Sq.—square.
Sqrs.—squares.
Std.—standard.
Stnd.—stained.
Stk.—stock.
SW—sound wormy.
T&G—tongued and grooved.
TB&S—top, bottom, and sides.
Tbrs.—timbers.
Thickness—4/4, 5/4, 6/4, 8/4, etc.=1 inch, 1¼ inches, 1½ inches, 2 inches, etc.
V1S—V 1 side; that is, a longitudinal V-shaped groove on 1 face of a piece
    of lumber.

V2S—V on 2 sides; that is, a longitudinal V-shaped groove on 2 faces of a piece of lumber.
VG—vertical grain.
W. a. l.—wider, all lengths.
Wth.—width.
Wdr.—wider.
Wt.—weight.

LITERATURE CITED

(1) CALIFORNIA REDWOOD ASSOCIATION.
 1951. STANDARD SPECIFICATIONS FOR GRADES OF CALIFORNIA REDWOOD LUMBER. 61 pp., illus. San Francisco.
(2) HARDWOOD DIMENSION MANUFACTURERS' ASSOCIATION.
 1953. STANDARDS AND GRADING RULES FOR THE MEASUREMENT AND INSPECTION OF GLUED AND SOLID HARDWOOD DIMENSION LUMBER FOR FURNITURE AND INDUSTRIAL WOOD PARTS. 15 pp. Louisville, Ky.
(3) MAPLE FLOORING MANUFACTURERS' ASSOCIATION.
 1953. GRADING RULES FOR NORTHERN HARD MAPLE, BEECH, AND BIRCH FLOORING. 11 pp. Chicago.
(4) NATIONAL HARDWOOD LUMBER ASSOCIATION.
 1953. RULES FOR THE MEASUREMENT AND INSPECTION OF HARDWOOD LUMBER, CYPRESS, VENEERS, AND THIN LUMBER. 104 pp. Chicago.
(5) NATIONAL OAK FLOORING MANUFACTURERS' ASSOCIATION.
 1949. OFFICIAL FLOORING GRADING RULES, OAK, PECAN, BEECH, BIRCH, HARD MAPLE. 10 pp. Memphis, Tenn.
(6) NORTHEASTERN LUMBER MANUFACTURERS' ASSOCIATION.
 1950. STANDARD GRADING RULES FOR EASTERN SPRUCE AND BALSAM FIR. 18 pp. New York.
(7) ———
 1948. STANDARD GRADING RULES FOR NORTHERN WHITE PINE AND NORWAY PINE. 20 pp. New York.
(8) NORTHERN HEMLOCK AND HARDWOOD MANUFACTURERS' ASSOCIATION.
 1947. OFFICIAL GRADING RULES FOR NORTHERN WHITE PINE, NORWAY PINE, JACK PINE, EASTERN SPRUCE, BALSAM, ASPEN. 55 pp. Oshkosh, Wis.
(9) NORTHERN HEMLOCK AND HARDWOOD MANUFACTURERS' ASSOCIATION; APPLACHIAN HARDWOOD MANUFACTURERS INC.
 1950. OFFICIAL GRADING RULES FOR EASTERN HEMLOCK LUMBER. 76 pp. Oshkosh, Wis.
(10) SOUTHERN CYPRESS MANUFACTURERS' ASSOCIATION.
 1950. STANDARD SPECIFICATIONS FOR GRADES OF TIDEWATER RED CYPRESS. 64 pp., illus. Jacksonville, Fla.
(11) SOUTHERN PINE INSPECTION BUREAU.
 1948. STANDARD GRADING RULES FOR SOUTHERN PINE LUMBER. 124 pp., illus. New Orleans, La.
(12) U. S. DEPARTMENT OF COMMERCE.
 1953. AMERICAN LUMBER STANDARDS FOR SOFTWOOD LUMBER. Simplified Pract. Recom. R16–53. 26 pp. [Processed.]
(13) WEST COAST BUREAU OF LUMBER GRADES AND INSPECTION.
 1947. STANDARD GRADING AND DRESSING RULES FOR DOUGLAS FIR, SITKA SPRUCE, WEST COAST HEMLOCK, WESTERN RED CEDAR LUMBER. 19 pp., illus. Portland, Oreg.
(14) WESTERN PINE ASSOCIATION.
 1953. STANDARD GRADING RULES FOR PONDEROSA PINE, SUGAR PINE, IDAHO WHITE PINE, LODGEPOLE PINE, LARCH, DOUGLAS FIR, WHITE FIR, ENGELMANN SPRUCE, INCENSE CEDAR, WESTERN RED CEDAR LUMBER. 147 pp. Portland, Oreg.

STRESS GRADES AND WORKING STRESSES

Individual pieces of lumber, as they come from the saw, represent a wide range in strength as well as other properties. The orderly marketing of lumber, therefore, requires the establishment of grades that permit the procurement of any required quality of lumber in any desired quantity. Structural or stress grades of lumber are established on the basis of features that relate to strength and strength uses. They afford lumber of assured minimum strength to which working stress values for structural design can be assigned. Factors that affect stress grades and working stresses and their application in the grading and use of structural lumber are discussed in this section.

FACTORS INFLUENCING STRENGTH

The most important factors that influence the strength of structural lumber are the strength and variability of the clear wood, the moisture content, the duration of load, and the size, number, and location of strength-reducing features, such as knots, cross grain, and checks and splits.

Strength of Clear Wood

Quality of Wood

The strength of the clear wood of any species varies over a considerable range (7).[9] Pieces in the lower part of the range of strength for a species are undesirable for strength uses; on the other hand, recognition of the higher strength of the better wood is desirable.

Strength is closely related to the weight or the density of the wood. Higher working stresses can be assigned to lumber if pieces of low strength are eliminated by excluding those that are of exceptionally light weight and by using the rate of growth and the percentage of summerwood to select pieces of superior strength from certain species. Selection for rate of growth requires that the number of annual rings per inch be within a specified range. Selection for density requires, in addition to a specified rate of growth, a minimum percentage of summerwood.

Decay

Decay in any form is severely restricted or prohibited in strength grades because its extent is difficult to determine and its effect on strength is often greater than visual observation would indicate.

Heartwood and Sapwood

Heartwood and sapwood of the same species are of equal strength, and no heartwood requirement need be made when strength alone is the criterion. Since heartwood of some species is more resistant to

---

[9] Italic numbers in parentheses refer to Literature Cited, p. 164.

decay than their sapwood, heartwood may be required if the untreated wood is to be exposed to a decay hazard. On the other hand, sapwood takes preservative treatment more readily and should not be limited in lumber that is to be treated.

Moisture Content of Wood

When wood is seasoned, the direct effect of the loss of moisture is the stiffening and strengthening of the wood fibers. In large timbers, however, this increase in strength may be largely offset by checking or splitting that may occur in seasoning. Checks lessen resistance to shear or to tension across the grain. Some net increase of strength with drying is recognized in structural lumber not more than 4 inches in thickness that is subjected to bending stresses (joists or beams) and in posts or columns of any thickness. Glued laminated structural members generally are not subject to such checking, and larger increases of strength are recognized (p. 250).

Duration of Load

Both the elastic limit and the ultimate strength of wood are higher under short-time loading than under long-time loading (8). Wood is thus able to absorb overloads of considerable magnitude for short periods or smaller overloads for longer periods. This property is important in strength uses because it directly affects working stresses. A substantial reduction factor is applied to stress values obtained from laboratory tests in which the load duration is a few minutes, to convert them to working stresses suitable for long-time loading. Conversely, stresses for shorter periods of loading in service are increased above those for long-time loading.

Strength-Reducing Features

Structural lumber typically contains knots, checks, and other features that reduce its strength. The size and extent of these features are taken into account in all structural grades.

Slope of Grain

In zones of cross grain, where the direction of the wood fibers is not parallel to the edges of the piece, longitudinal tensile and compressive stresses have components acting across the grain, the direction in which wood is least strong. Cross-grained pieces are undesirable also because they tend to twist with changes in their moisture content. Stresses caused by shrinkage in seasoning are greater in structural lumber than in small, clear specimens and are increased in zones of sloping or distorted grain. To provide a factor of safety, the reduction of strength due to cross grain in structural lumber is taken as about twice the reduction observed in tests of small, clear specimens.

Knots

Knots interrupt the direction of grain and cause localized cross grain with steep slopes. Intergrown or live knots resist some kinds of stress, but encased knots or knotholes are obviously of little or no value. On the other hand, distortion of grain is greater around an intergrown knot than around an encased or dead knot. As a result, overall strength effects are roughly equalized, and no distinction is made in stress grading between live knots, dead knots, and knotholes.

Knots reduce tensile strength more than compressive or shear strength.

The effect of knots on stiffness is small. Although the zone of distorted grain around a knot has less stiffness than straight-grained wood, such zones generally comprise only a minor part of the volume of a piece of lumber, and stiffness reflects the character of all parts.

The strength effect of a knot depends on the proportion of the cross section of the piece of lumber occupied by the knot. Limits on knot sizes are therefore made in accordance with the width of the face in which the knot appears. The effect of knots upon strength also depends upon their location and upon the distribution of stress in the piece. Compression members are stressed about equally throughout, and no limitation related to location of knots is imposed. In structural members subjected to bending, stresses are greatest in the middle part of the length and are greater at the top and bottom edges than at midheight. These facts are recognized in differing limitations on the sizes of knots in different locations. Since bending members may be used with either edge uppermost, and since the effect of knots on strength is greater in tension than in compression, knot requirements suitable for tensile stress are applied to both edges.

As knot size increases, the distortion of grain around the knot is more than proportionally increased. For that reason, sizes of knots in wide faces wider than 12 inches and narrow faces wider than 6 inches are permitted to increase only in proportion to the square root of the increase of the width of the face, whereas, below the 12- and 6-inch limits, the sizes permitted are generally proportioned to the width of the face.

Cluster knots are prohibited because neither the sizes of the individual knots nor those of the cluster are good measures of the extent of distortion of grain and the resultant effect on strength.

If two or more knots occur in the same cross section of a piece of structural lumber there is a cumulative effect on most of the strength properties. Two or more knots of maximum size are not permitted in the same cross section.

Knots in glued laminated structural members are not continuous as in sawed structural lumber, and different methods are used for evaluating their effect on strength (p. 251).

Shakes

Shakes in members subjected to bending reduce the resistance to shear and are therefore limited most closely in those parts of a bending member where shearing stresses are highest. In members subjected only to longitudinal compression, shakes do not greatly affect strength; they are limited because of appearance and because they permit entrance of moisture that may make conditions favorable to decay. Smaller shakes are permitted in green than in seasoned lumber, because they may be extended as the lumber dries.

Checks and Splits

Whereas shakes indicate a weakness of bond between annual rings that is presumed to extend lengthwise without limit, checks and splits are rated on the more restricted basis of the area of actual opening. An end split is considered the same as an end check that extends through the full thickness of the piece. The effects of checks and splits upon strength and the principles of their limitation are the same as for shakes.

Wane

Requirements of appearance, fabrication, or bearing or nailing surface generally impose stricter limitations on wane than does strength. Wane is therefore limited in structural lumber on those bases.

Pitch Pockets

Pitch pockets ordinarily have so little effect on structural lumber that they can be disregarded in grading for strength. The presence of a large number of pitch pockets, however, may indicate a weakness of bond between annual rings, and the piece should be carefully inspected for shakes.

Holes

Since the strength effect of a knot depends more upon the distortions of the surrounding grain than upon the knot itself, knotholes are considered to have the same effect as knots, Holes due to other causes are not accompanied by distortion of grain, and the same limitations that apply to knotholes are sufficient.

Combination of Strength-Reducing Features

Where combinations of strength-reducing features are present, their combined effects are taken into account in structural grading. Lumber containing such combinations may be accepted if above average in density and if the combinations are not too serious. Grading rules commonly prohibit any serious combination.

PRINCIPLES OF STRESS GRADING

Basic Stress

Basic stress is a generalized working strength value for the clear wood of a species. It is reduced from the average laboratory value obtained in strength tests in order to conform more closely to the conditions of use of structural lumber. This reduction provides a factor of safety, but it does not allow for the effect of any knots or other strength-reducing features. Basic stress is thus independent of grade and provides a measure of the inherent strength of the clear wood of a species, from which any grade of that species can be evaluated.

Strength Ratio

The strength ratio of a piece of structural lumber or of a number of similar pieces grouped in a structural grade represents the strength remaining after allowance is made for the knots, cross grain, or other strength-reducing features present. Thus, a strength ratio of 75 percent applies to a piece or a grade in which the maximum reduction in strength from that of clear wood is 25 percent. If a grade permits a range in sizes of strength-reducing features, the strength ratio is based on the largest size permitted. Strength reductions from knots and cross grain generally are not cumulative; thus, if the ratio from knots is 75 percent and that from cross grain is 69 percent, the combined strength ratio is 69 percent.

Strength ratios are applied to stresses in transverse bending, tension parallel to grain, compression parallel to grain, and horizontal shear in beams. Modulus of elasticity and compression perpendicular to grain are little affected by such features as knots, and strength ratios of 100 percent are assumed for all grades. A piece or a grade may have the same or differing strength ratios in bending, tension, compression, or shear. Commonly, strength ratios in the various properties for a single grade are adjusted to be approximately in balance under the loading for which that grade is designed.

Working Stress

The basic stress for a species, when multiplied by the strength ratio for a piece or a grade, becomes the working stress for structural design with that piece or grade of the species.

USE CLASSES OF STRUCTURAL LUMBER

The effects of knots, cross grain, and other strength-reducing features on the strength of a timber vary with the loads placed on the timber in use. Also, the effect of seasoning varies with the size of the timber. Consequently, efficiency in grading necessitates classifying structural lumber according to its size and use as follows:

Beams and stringers.—Pieces of rectangular cross section, 5 by 8 inches (nominal dimensions) and up, graded with respect to their strength in bending when loaded on the narrow face.

Joists and planks.—Pieces of rectangular cross section, 2 to 4 inches thick and 4 or more inches wide (nominal dimensions), graded primarily with respect to their strength in bending edgewise or flatwise but also used where tensile or compressive strength is important. Dimension lumber (2 inches in nominal thickness) is sometimes placed in stress grades separate from the grades of the thicker joists and planks.

Posts and timbers.—Pieces of square or approximately square cross section, 5 by 5 inches (nominal dimensions) and larger, graded primarily for use as posts or columns but adapted to miscellaneous uses in which strength in bending is not especially important.

Combined uses.—The principles of stress grading make possible the assignment of strength ratios in compression or tension to the joist and plank grades, in bending to the post and timber grades, or in bending flatwise to the beam and stringer grades. The most common of such combinations is in the assignment of compression or tension values to joist and plank grades used in the fabrication of chord or web members in wood trusses.

Round timbers.—Round timbers, such as poles or piles (p. 429), are not generally graded to a specific strength rating, although their strength is affected by such features as spiral grain and knots in much the same way as is the strength of sawed lumber. Logs or poles are sometimes used in the framing of structures where strength considerations govern their design.

Boards.—The increasing use of 1-inch boards in lightweight trusses or other structural elements, or in glued laminated structural members, frequently requires that they be stress graded. Test data show that a knot reduces the strength of a board in proportion to the part of the width occupied by the knot. The sum of sizes of knots is limited in the same way as in joists and planks (p. 144). Effects upon strength from cross grain are about the same as in thicker lumber. Since any strength-reducing features present appear in one or both wide faces, the narrow faces, or edges, need not be considered in grading.

GENERAL REQUIREMENTS FOR STRESS GRADES

Size Standards

The minimum dressed [10] sizes (green) of structural lumber are given in the grading rule books of the various lumber manufacturers' associations for industrial species. In general, these sizes are as follows, although the grading rule books should be consulted for a given species:

Beams and stringers.—Nominal thicknesses, 5 or more inches; nominal widths, 8 or more inches; standard sizes, S1S, S1E, S2S, or S4S, ½ inch off each way.

Joists and planks.—Nominal thicknesses, 2 inches to but not including 5 inches; nominal widths, 4 or more inches; standard thicknesses, S1S or S2S, ⅜ inch off; standard widths, S1E or S2E, 4 inches, ⅜ inch off, and 6 inches or more wide, ½ inch off.

Posts and timbers.—Nominal sizes, 5 by 5 inches and larger; standard sizes, S1S, S1E, S2S, or S4S, ½ inch off each way.

Rough lumber should be sawed to full nominal dimensions, except that occasional slight variation in sawing is permissible (*6*). No shipment should contain more than 20 percent of pieces of minimum dimension due to such variation in sawing. In any one shipment, at least 80 percent of the pieces should be of full nominal dimensions, and the remainder should be not more than one-sixteenth inch scant.

---

[10] Dressing is specified as surfacing on 1 side (S1S), 2 sides (S2S), 1 edge (S1E), 2 edges (S2E), or both sides and both edges (S4S); ½ or ⅜ inch off means ½ or ⅜ inch less than nominal.

General Quality of Lumber

No piece of exceptionally light weight should be permitted. All lumber should be well manufactured.

Structural grades prohibit or severely restrict decay in any form. Unsound knots and limited amounts of decay in its early stages are permitted in some of the lower grades, which generally are not used as primary structural elements. Sapwood decay, resulting from improper seasoning practices, is not admitted under this provision.

Slope of Grain

Slope of grain is measured and limited at whatever point in the length of a structural timber that shows the greatest slope. It is measured over a distance sufficiently great to define the general slope, disregarding such short local deviations as those around knots.

Knots and Holes

Knotholes and holes from causes other than knots are measured and limited as provided for knots. Cluster knots and knots in groups should not be permitted.

Beams and Stringers

The size of a knot on a narrow face of a beam or stringer is taken as its width between lines enclosing the knot and parallel to the edges of the piece (fig. 20), except that, when a knot on a narrow face of a side-cut piece extends into the adjacent one-fourth of the width of a wide face, its least dimension is taken as its size. A corner knot in a piece containing the pith should be measured either by its width on the narrow face between lines parallel to the edges or by its smallest diameter on the wide face, whichever is greater.

The size of a knot on a wide face is measured by its smallest diameter. A knot at the edge of a wide face is limited to the same size as a knot on a narrow face but is measured by its smallest diameter.

ZM 22545 F

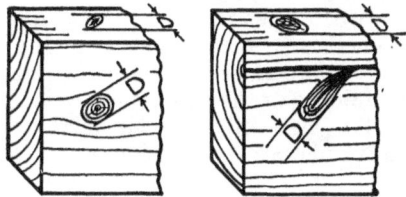

FIGURE 20.—Measurement of knots in beams and stringers.

The sizes of knots on narrow faces and at the edges of wide faces may increase proportionately from the size permitted in the middle one-third of the length to twice that size at the ends of the piece (fig. 21), except that the size of no knot should exceed the size permitted at the centerline of the wide face. The size of knots on wide faces may increase proportionately from the size permitted at the edge to the size permitted at the centerline.

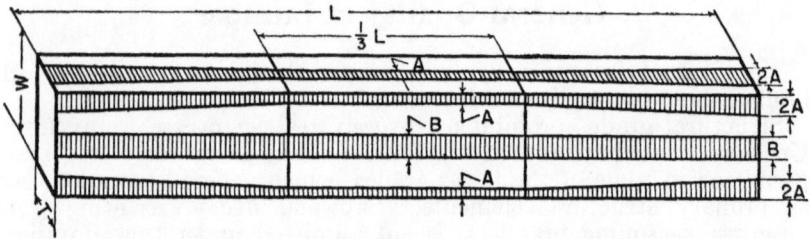

ZM 22546 F

FIGURE 21.—Maximum size of knots permitted in vaiious parts of beams and stringers: *A*, Maximum size on narrow face or edge of wide face, middle third of length with a gradual increase to 2*A* at ends of piece; *B*, maximum size at centerline of wide face; *L*, length; *W*, width; *T*, thickness.

The sum of the sizes of all knots within the middle one-half of the length of any face, when measured as specified for the face under consideration, should not exceed four times the size of the largest knot allowed on that face. Two or more knots of maximum permissible size should not be allowed in the same 6 inches of length on any face of beams and stringers intended for stress in tension or compression.

Joists and Planks

The size of a knot on a narrow face of a joist or plank is taken as its width between lines enclosing the knot and parallel to the edges of the piece (fig. 22). A narrow-face knot that appears also on the wide face of a side-cut joist (but does not contain the intersection of those faces) is measured and graded only on the wide face.

The size of a knot on a wide face is the average of its largest and smallest diameters; the size of a spike knot is the average of its length and its greatest width (fig. 22).

A corner knot or a knot extending entirely across the width of a narrow face in a side-cut joist is measured on its end or ends between lines parallel to the edges of the piece and is graded with respect to the face on which it is measured (fig. 22). A corner knot in a joist containing the pith is measured either by its width on the narrow face between lines parallel to the edges or by its smallest diameter on the wide face, whichever is most restrictive.

If joists and planks are of square cross section, each of the four faces should be graded as a narrow face.

The sizes of knots on narrow faces and at the edges of wide faces may increase proportionately from the size permitted in the middle one-third of the length to twice that size at the ends of the piece, except that the size of no knot should exceed the size permitted at the centerline of the wide face (p. 152). The sizes of knots on wide faces may increase proportionately from the size permitted at the edge to the size permitted at the centerline.

The sum of the sizes of all knots in any 6 inches of length of the piece should not exceed twice the size of the largest permitted knot. Two or more knots of maximum permissible size should not be allowed in the same 6 inches of length on any face.

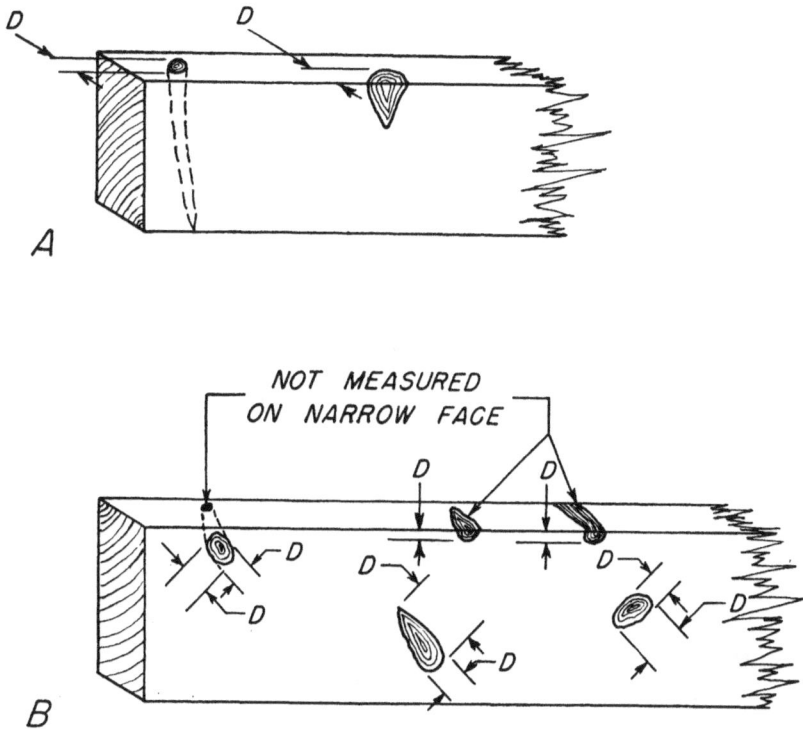

FIGURE 22.—Measurement of knots in side-cut joists and planks: *A*, Narrow face; *B*, wide face.

Posts and Timbers

The size of a knot on any face of a post or timber is taken as the diameter of a round knot, the lesser of the two diameters of an oval knot, and the greatest diameter perpendicular to the length of a spike knot (fig. 23). A corner knot is measured wherever the measurement will represent the true diameter of the branch causing the knot. In compression members with greater width than thickness, the sizes of knots in both narrow and wide faces are referred to the width of the wide face for determination of the strength ratio.

The sum of the sizes of all knots in any 6 inches of length of a post or timber should not exceed twice the size of the largest permitted knot. Two or more knots of maximum permissible size appearing on a face should not be allowed in the same 6 inches of length.

One-Inch Boards

Knots in 1-inch boards are measured by the average of the widths on the 2 opposite faces of the board, each width being taken perpendicular to the length of the board. Narrow-face knots are ignored, since they appear in one or both wide faces. A knot appearing in both wide faces is considered to be only one knot. Sizes of knots are

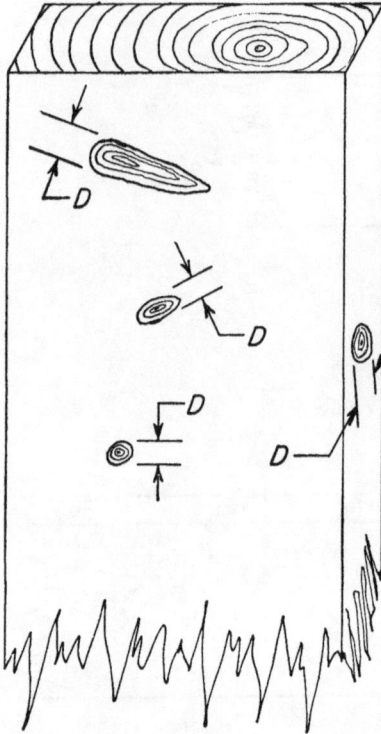

FIGURE 23.—Measurement of knots in posts and timbers or othe compression members.

ZM 91872 F

limited to the sizes permitted in the center of the wide faces of joists and planks (p. 152). The sum of sizes should be limited in the same way as in joists and planks.

Shakes, Checks, and Splits

Joists and Planks or Beams and Stringers

Limitations on shakes, checks, or splits in joists and planks or beams and stringers are based on the reduction of area effective in resisting horizontal shear. Since shear stress in most joists or beams is greatest near the ends, the restrictions are applied only for a distance from each end equal to three times the height of the piece. (Height equals width of wide face.) Since shear stress is greatest near the neutral axis, the restrictions also are applied only in the middle one-half of the height of the piece, and only the shakes, checks, and splits in this section are measured.

Shakes are measured at the ends of the piece, and the size permitted is determined by the width of the narrow face of the piece. The size of a shake is the distance between lines enclosing the shake and parallel to the wide faces of the piece (fig. 24).

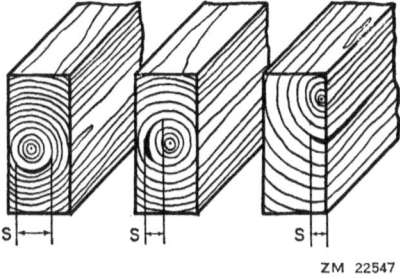

FIGURE 24.—Measurement of shakes in beams and stringers or joists and planks.

ZM 22547 F

Checks and splits are limited to the same nominal dimensions as shakes (table 24) but are measured differently.   The size of a check is its average depth of penetration into the piece measured from and perpendicular to the surface of the wide face.   The size of an end split is one-third of its average length.

The grading of any combination of shakes, checks, and splits in the middle one-half of the height is based on the grader's judgment of the probable effects of seasoning or loading in service on the combination.   Where a combination of two checks in opposite faces, a check and a split, a check and a shake, or a split and a shake may later become a single horizontal shear plane, the sum of the sizes in the combination should be restricted to the allowable size of shake. Where such a combination is not additive in this way, only the largest single characteristic is considered.

Where 2-inch dimension is to be used in light building construction, in which the shear stress is not critical, a more liberal provision on end splits can be made.   It is suggested that the length of an end split in such lumber be limited to 1½ times the width of the wide face in a grade with a shear strength ratio of 50 percent, or three-quarters of the width of the wide face in a grade with a shear strength ratio of 75 percent.

Posts and Timbers

Shakes, checks, and splits have little effect on the strength of a column.   They are limited in posts and timbers for the sake of appearance and, in exterior exposures, to reduce the opportunity for moisture to enter and decay to start.   It is suggested that shakes (and side checks) be limited to one-third the width of unseasoned lumber or one-half the width of dry pieces of all post and timber grades and be measured in a direction perpendicular to the plane of the face in which they appear (fig. 25).   The width may be considered to be either of the two dimensions of a rectangular cross section.   The length of an end split should not exceed 1½ times the allowable size of a shake or side check, but this length may be permitted in addition to the allowable sum of shakes and side checks.

Requirements for Continuous Spans

If beams and stringers or joists and planks are to be used as continuous beams, the restrictions for knots in the middle one-third of their length should be applied to the middle two-thirds of the length

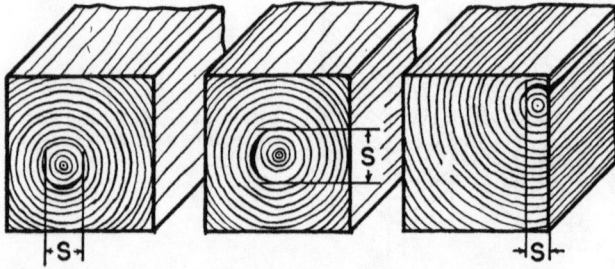

ZM 22548 F

FIGURE 25.—Measurement of shakes in posts and timbers.

of pieces continuous over 2 spans and the full length of pieces continuous over 3 or more spans. Restrictions for checks applied only in zones at the ends of the piece in simple spans should be applied to zones of the same length and height anywhere in the length of continuous spans.

OPTIONAL REQUIREMENTS FOR STRESS GRADES

Certain optional requirements may be applied to stress grades if the buyer and the seller agree. Some do not directly affect strength.

Heartwood

If heartwood is desired for decay resistance in untreated lumber, a requirement may be made in terms of a minimum amount or percentage of heartwood on the girth, or on each face, side, or edge, measured at the point where the greatest amount of sapwood occurs. Lumber intended for preservative treatment should permit unlimited amounts of sapwood.

Wane

The amounts of wane ordinarily permitted from considerations of appearance and bearing or nailing surface are not large enough to have a serious effect on the strength of structural lumber. It is suggested that wane in the higher structural grades be limited to one-fourth the nominal width of the face on which it appears, or that an optional requirement of "free from wane" be specified if especially high-grade timbers are desired for special purposes.

Density—Douglas-Fir and Southern Yellow Pine

Dense Douglas-fir or dense southern yellow pine averages on one end or the other of each piece not less than six annual rings per inch and one-third or more of summerwood (the darker, harder part of the annual ring) measured on a representative radial line (5, 6) but excluding the rapidly grown wood near the pith in boxed-heart pieces. The contrast in color between springwood and summerwood must be distinct. Douglas-fir pieces that average not less than 5 annual rings per inch and southern yellow pine pieces that average less than 6 annual rings per inch meet the requirements of dense Douglas-fir and southern yellow pine if they average one-half or more of summerwood. In case of disagreement, 2 radial lines are chosen, and the decision is based on the average of measurements made on these 2 lines.

Lumber qualifying as dense Douglas-fir or dense southern yellow pine is given basic stress values in bending, tension, compression parallel to grain, and compression perpendicular to grain that are about one-sixth higher than those for wood of these species without such qualifications.

Close Grain—Douglas-Fir, Southern Yellow Pine, and Redwood

Close-grained Douglas-fir or close-grained southern yellow pine averages on one end or the other of each piece not less than 6 nor more then 30 annual rings per inch, measured on a representative radial line (5, 6), but excluding the rapidly grown wood near the pith in boxed-heart pieces. Pieces averaging at least 5 or more than 30 rings per inch are accepted if they consist of one-third or more of summerwood. In case of disagreement, 2 radial lines are chosen, and the decision is based on the average of these 2 lines.

Close-grained redwood averages in each piece not less than 8 nor more than 40 annual rings per inch (2).

Lumber qualifying as close-grained Douglas-fir, southern yellow pine, and redwood is given basic stress values in bending, tension, compression parallel to grain, and compression perpendicular to grain that are about one-fifteenth higher than those for wood of these species without such qualifications.

Medium Grain

Medium-grained southern yellow pine or Douglas-fir has not less than four annual rings per inch, measured on a representative radial line as with close grain. Since wood of faster growth than four rings per inch in these species often has reduced strength as compared to slower grown wood, most stress grades, except those based on close grain or density, carry the requirement for medium grain.

VALUES OF STRENGTH RATIOS

Tables 20 through 24 can be used to estimate the strength ratio of a known piece or grade or to determine specification limits for a grade with a desired strength ratio. Limiting sizes of the various strength-reducing features shown in these tables are related to the methods of measurement described in the preceding sections.

Slope of Grain

Table 20 gives strength ratios for stress in extreme fiber in bending and compression parallel to grain for various slopes of grain

TABLE 20.—*Beams and stringers or joists and planks, and posts and timbers: strength ratios corresponding to various slopes of grain*

| Slope of grain | Maximum strength ratio | | Slope of grain | Maximum strength ratio | |
	Beams and stringers or joists and planks [1]	Posts and timbers [2]		Beams and stringers or joists and planks [1]	Posts and timbers [2]
	Percent	*Percent*		*Percent*	*Percent*
1 in 6		56	1 in 15	76	100
1 in 8	53	66	1 in 16	80	
1 in 10	61	74	1 in 18	85	
1 in 12	69	82	1 in 20	100	
1 in 14	74	87			

[1] Strength ratios for stress in extreme fiber in bending or in tension parallel to grain.
[2] Strength ratios for stress in compression parallel to grain.

Knots

Strength ratios for various combinations of size of knot and width of face are given in tables 21, 22, and 23.

The use of these tables is illustrated by the following example. The sizes of knots permitted in an 8- by 16-inch (nominal) beam in a grade having a strength ratio of 70 percent for stress in extreme fiber in bending are desired. The smallest ratio in the column for 8-inch face in table 21 that equals or exceeds 70 percent is opposite 2⅛ inches in the size-of-knot column. A similar ratio in the column for 16-inch face in table 22 is opposite 4¼ inches. Hence, the permissible sizes are 2⅛ inches on the 8-inch face and 4¼ inches on the 16-inch face.

Shakes, Checks, Splits

Strength ratios for various combinations of width of face with size of shakes, checks or splits are given in table 24.

RECOMMENDED BASIC STRESS VALUES [11]

Table 25 gives basic stress values for structural design with clear lumber under long-time service at full design load. They are based on the strength of green lumber and are applicable, with certain adjustments, to lumber of any degree of seasoning. Adjustments for seasoning, short-time loading, and other factors are discussed on pages 158–163.

[11] Acknowledgment is made to the Ottawa and Vancouver Laboratories, Forest Products Division, Forest Service, Canada, who have made available their data on strength values in species common to Canada and the United States and have materially contributed to the discussions leading to the present system of basic stresses.

TABLE 21.—*Beams and stringers or joists and planks—knots on narrow face within middle third of length of piece: strength ratios* [1] *corresponding to various combinations of size of knot and width of face*

Size of knot (inches)	Strength ratio when nominal face width is—									
	2 in.	3 in.	4 in.	5 in.	6 in.	8 in.	10 in.	12 in.	14 in.	16 in.
	Pct.	Pct.	Pct.	Pct.	Pct.	Pct.	Pct.	Pct.	Pct.	Pct.
¼	90	93	95	96	96	97	97	97	98	98
⅜	83	89	92	93	94	95	96	96	96	97
½	77	85	88	91	92	93	94	95	95	95
⅝	71	81	85	88	90	92	92	93	94	94
¾	65	76	82	86	88	90	91	92	92	93
⅞	58	72	79	83	86	88	89	90	91	91
1	52	68	76	81	84	86	88	89	89	90
1⅛	46	64	73	78	82	84	86	87	88	89
1¼	26	60	70	76	80	83	84	86	87	88
1⅜	--------	56	67	73	78	81	83	84	85	86
1½	--------	51	63	71	76	79	81	83	84	85
1⅝	--------	47	60	68	74	77	80	81	83	84
1¾	--------	35	57	66	71	75	78	80	81	83
1⅞	--------	30	54	63	69	73	76	78	80	81
2	--------	25	51	61	67	72	75	77	79	80
2⅛	--------	--------	48	58	65	70	73	75	77	79
2¼	--------	--------	45	56	63	68	71	74	76	77
2⅜	--------	--------	36	53	61	66	70	72	74	76
2½	--------	--------	32	51	59	64	68	71	73	75
1⅝	--------	--------	29	48	57	63	67	70	72	74
2¾	--------	--------	25	46	55	61	65	68	70	72
2⅞	--------	--------	--------	--------	53	59	63	67	69	71
3	--------	--------	--------	--------	51	57	62	65	68	70
3⅛	--------	--------	--------	--------	--------	55	60	64	66	68
3¼	--------	--------	--------	--------	--------	54	59	62	65	67
3⅜	--------	--------	--------	--------	--------	52	57	61	64	66
3½	--------	--------	--------	--------	--------	50	55	59	62	65
3⅝	--------	--------	--------	--------	--------	--------	54	58	61	63
3¾	--------	--------	--------	--------	--------	--------	52	56	59	62
3⅞	--------	--------	--------	--------	--------	--------	50	55	58	61
4	--------	--------	--------	--------	--------	--------	--------	53	57	60
4⅛	--------	--------	--------	--------	--------	--------	--------	52	55	58
4¼	--------	--------	--------	--------	--------	--------	--------	50	54	57
4⅜	--------	--------	--------	--------	--------	--------	--------	--------	53	56
4½	--------	--------	--------	--------	--------	--------	--------	--------	51	54
4⅝	--------	--------	--------	--------	--------	--------	--------	--------	50	53
4¾	--------	--------	--------	--------	--------	--------	--------	--------	--------	52
4⅞	--------	--------	--------	--------	--------	--------	--------	--------	--------	51

[1] Strength ratios for stress in extreme fiber in bending.

Strength ratios of 45 percent or higher are based on the relation of the size of knot to the nominal width of face; lower ratios are based on the relation of the size of knot to the actual width of face. In tabulating the sizes of knots, the rule used in approximating decimals by common fractions with ⅛-inch intervals has the effect of subtracting ½₄ inch from the size of each knot. For example, the strength ratios listed for a 3½-inch knot are those which actually would obtain for a size of 3¹¹⁄₂₄ inches. In view of the allowance thus introduced and the fact that nominal instead of actual widths of face are used, the strength ratios given should be taken as maximum values.

Sizes of knots permitted for a given strength ratio in narrow faces wider than 6 inches in beams and stringers are increased only in proportion to the square root of the increase of width of the face. Below these limits, the sizes permitted are proportioned to the width of the face.

TABLE 22.—*Beams and stringers or joists and planks—knots along centerline of wide face at any point in length of piece; posts and timbers—knots at any point on any face; and 1-inch boards—knots anywhere in wide face: strength ratios [1] corresponding to various combinations of size of knot and width of face*

Size of knot [2] (inches)	Strength ratio when nominal face width is—												
	3 in.	4 in.	5 in.	6 in.	8 in.	10 in.	12 in.	14 in.	16 in.	18 in.	20 in.	22 in.	24 in.
	Pct.	Pct.	Pct.	Pct.	Pct.	Pct.	Pct.	Pct.	Pct.	Pct.	Pct.	Pct.	Pct.
¼	93	95	96	96	97	98	98	98	98	99	99	99	99
½	85	88	91	92	94	95	96	96	97	97	97	97	97
¾	76	82	86	88	91	93	94	94	95	95	95	96	96
1	68	76	81	84	88	90	92	93	93	93	94	94	94
1¼	60	70	76	80	85	88	90	91	91	92	92	93	93
1½	51	63	71	76	82	85	88	89	89	90	91	91	91
1¾	35	57	66	71	79	83	86	87	88	88	89	89	90
2	25	51	61	67	75	80	84	85	86	87	87	88	88
2¼		45	56	63	72	78	82	83	84	85	86	86	87
2½		32	51	59	69	75	79	81	82	83	84	85	85
2¾		25	46	55	66	73	77	79	80	82	82	83	84
3			36	51	63	70	75	77	79	80	81	82	83
3¼			31	47	60	68	73	75	77	78	79	80	81
3½			25	37	57	65	71	73	75	76	78	79	80
3¾				33	54	63	69	71	73	75	76	77	78
4				28	50	60	67	69	71	73	74	76	77
4¼					47	58	65	67	70	71	73	74	75
4½					41	55	63	66	68	70	71	73	74
4¾					37	53	61	64	66	68	70	71	72
5					34	50	59	62	64	66	68	70	71
5¼					31	48	57	60	62	65	66	68	69
5½					27	45	54	58	61	63	65	66	68
5¾						40	52	56	59	61	63	65	66
6						37	50	54	57	59	61	63	65
6¼						35	48	52	55	58	60	62	63
6½						32	46	50	53	56	58	60	62
6¾						29	42	48	52	54	57	59	60
7						27	39	46	50	53	55	57	59
7¼							37	42	48	51	53	56	57
7½							35	40	46	49	52	54	56
7¾							33	38	42	48	50	53	55
8							31	36	40	46	49	51	53
8¼							29	34	38	42	47	49	52
8½							26	32	37	40	45	48	50

[1] Beams and stringers or joists and planks—strength ratio for stress in extreme fiber in bending; posts and timbers—strength ratio for stress in compression parallel to grain; one-inch boards—strength ratio for all stresses.

Strength ratios of 45 percent or higher are based on the relation of the size of knot to the nominal width of face; lower ratios are based on the relation of the size of knot to the actual width of face. In tabulating the sizes of knots, the rule used in approximating decimals by common fractions with ⅛-inch intervals has the effect of subtracting ¹⁄₂₄ inch from the size of each knot. For example, the strength ratios listed for a 3½-inch knot are those which actually would obtain for a size of 3¹¹⁄₂₄ inches. In view of the allowance thus introduced and the fact that nominal instead of actual widths of face are used, the strength ratios given should be taken as maximum values.

Sizes of knots permitted for a given strength ratio in wide faces wider than 12 inches in beams and stringers and joists and planks, or faces wider than 12 inches in posts and timbers, are increased only in proportion to the square root of the increase of width of the face. Below these limits, the sizes permitted are proportioned to the width of the face.

[2] Ratios corresponding to other sizes of knot can be found by interpolation.

Table 23.—*Joists and planks—knots at edge of wide face within middle ⅓ of length of piece: strength ratios [1] corresponding to various combinations of size of knot and width of face*

Size of knot (inches)	Strength ratio when nominal face width is—													
	2 in.	3 in.	4 in.	5 in.	6 in.	8 in.	10 in.	12 in.	14 in.	16 in.	18 in.	20 in.	22 in.	24 in.
	Pct.	Pct.	Pct.	Pct.	Pct.	Pct.	Pct.	Pct.	Pct.	Pct.	Pct.	Pct.	Pct.	Pct.
¼	80	86	90	92	93	95	96	97	97	97	97	97	97	98
⅜	70	79	84	87	89	92	93	94	95	95	95	96	96	96
½	60	72	78	82	85	89	91	92	93	93	94	94	94	95
⅝	50	65	73	78	81	86	89	90	91	92	92	93	93	93
¾	32	58	68	74	78	83	86	88	89	90	91	91	91	92
⅞		52	63	69	74	80	84	87	88	89	89	89	90	90
1		46	58	65	71	77	82	85	86	87	87	88	89	89
1⅛		34	53	61	67	75	79	83	84	85	86	86	87	88
1¼		29	49	57	64	72	77	81	82	83	84	85	86	86
1⅜			40	54	60	69	75	79	80	82	83	83	84	85
1½			36	50	57	67	73	77	79	80	81	82	83	84
1⅝			32	47	54	64	71	75	77	78	80	81	81	82
1¾			28	40	51	62	69	74	75	77	78	79	80	81
1⅞				36	.48	59	67	72	74	75	77	78	79	80
2				33	45	57	65	70	72	74	75	76	77	78
2⅛				30	40	55	63	68	70	72	74	75	76	77
2¼				27	37	52	61	67	69	71	72	73	75	76
2⅜					34	50	59	65	67	69	71	72	73	74
2½					32	48	57	63	66	68	69	71	72	73
2⅝					29	46	55	62	64	66	68	69	71	72
2¾					27	41	53	60	63	65	66	68	69	71
2⅞					25	39	51	58	61	63	65	67	68	69
3						37	50	57	60	62	64	65	67	68
3⅛						35	48	55	58	60	62	64	66	67
3¼						33	46	54	57	59	61	63	64	66
3⅜						31	42	52	55	58	60	62	63	65
3½						29	40	51	54	56	58	60	62	63
3⅝						27	39	49	52	55	57	59	61	62
3¾						26	37	48	51	54	56	58	60	61
3⅞							36	46	50	52	55	57	58	60
4							34	45	48	51	53	55	57	59
4⅛							33	42	47	50	52	54	56	58
4¼							31	40	46	48	51	53	55	57
4⅜							29	39	43	47	50	52	54	55
4½							28	37	41	46	48	51	53	54
4⅝							27	36	40	45	47	50	51	53
4¾							25	35	39	42	46	48	50	52
4⅞								34	37	41	45	47	49	51
5								32	36	40	42	46	48	50

[1] Strength ratios for stress in extreme fiber in bending.

Strength ratios of 45 percent or higher are based on the relation of the size of knot to the nominal width of face; lower ratios are based on the relation of the size of knot to the actual width of face. In tabulating the sizes of knots, the rule used in approximating decimals by common fractions with ⅛-inch intervals has the effect of subtracting 1/24 inch from the size of each knot. For example, the strength ratios listed for a 3½-inch knot are those which actually would obtain for a size of 3¹/24 inches. In view of the allowance thus introduced and the fact that nominal instead of actual widths of face are used, the strength ratios given should be taken as maximum values.

Sizes of knots permitted for a given strength ratio at the edges of wide faces wider than 12 inches in joists and planks are increased only in proportion to the square root of the increase of width of the face. Below these limits sizes are proportioned to the square of the remaining width of the face after the width of the knot is subtracted.

TABLE 24.—*Beams and stringers or joists and planks—shakes or checks in middle ½ of height: strength ratios[1] corresponding to various combinations of size of shake or check and width of end of piece*

Size of shake or check[2] (inches)	Green lumber										Seasoned lumber									
	Strength ratio when nominal end width of piece is—										Strength ratio when nominal end width of piece is—									
	2 in.	3 in.	4 in.	5 in.	6 in.	8 in.	10 in.	12 in.	14 in.	16 in.	2 in.	3 in.	4 in.	5 in.	6 in.	8 in.	10 in.	12 in.	14 in.	16 in.
	Pct.	Pct.	Pct.	Pct.	Pct.	Pct.	Pct.	Pct.	Pct.	Pct.	Pct.	Pct.	Pct.	Pct.	Pct.	Pct.	Pct.	Pct.	Pct.	Pct.
¼	90	93	95	96	96	97	98	98	98	99	100	100	100	100	100	100	100	100	100	100
⅜	83	89	92	93	94	96	97	97	98	98	94	100	100	100	100	100	100	100	100	100
½	77	85	88	91	92	94	95	96	97	97	87	95	100	100	100	100	100	100	100	100
⅝	71	81	85	88	90	93	94	95	96	96	80	91	96	99	99	100	100	100	100	100
¾	65	76	82	86	88	91	93	94	95	96	73	86	93	97	97	99	100	100	100	100
⅞	58	72	79	83	86	90	92	93	94	95	66	81	89	94	94	97	99	100	100	100
1	52	68	76	81	84	88	90	92	93	94	59	77	85	91	92	95	98	100	100	100
1⅛	46	64	73	78	82	86	89	91	92	93	52	72	82	88	90	94	97	100	100	100
1¼	26	60	70	76	80	85	88	90	91	92	45	67	78	85	87	92	96	99	100	100
1⅜		56	67	73	78	83	87	89	90	92	20	62	75	82	85	90	95	98	100	100
1½		51	63	71	76	82	85	88	90	91		58	71	80	83	87	93	98	100	100
1⅝		47	60	68	74	80	84	87	89	90		53	68	77	80	85	92	96	99	100
1¾		35	57	66	71	79	83	86	88	89		48	64	74	78	83	91	95	98	99
1⅞		30	54	63	69	77	82	85	87	88		34	61	71	76	81	89	93	97	98
2		25	51	61	67	75	80	84	86	88		29	57	68	73	80	88	92	96	97
2⅛			48	58	65	74	79	83	85	87			54	66	71	78	86	91	95	96
2¼			45	56	63	72	78	82	84	86			50	63	69	76	83	89	94	95
2⅜			36	53	61	71	77	81	83	85			47	60	66	74	82	88	93	94
2½			32	51	59	69	75	79	82	84			36	57	64	73	81	87	92	93
2⅝			29	48	57	68	74	78	81	83			32	54	62	71	79	86	91	93
2¾			25	46	55	66	73	77	80	82			28	52	59	67	76	85	90	92
2⅞					53	65	72	76	79	81			25	49	57	64	74	82	89	90
3					51	63	70	75	77	80					52	60	71	80	87	88
3¼						60	68	73	75	78					48	57	68	78	85	86
3½						57	65	71	73	77						53	65	75	83	85
3¾						54	63	69	72	75						50	62	73	81	83
4						50	60	67	70	74							59	71	79	81
4¼							58	65	68	72								68	77	79
4½							55	63	66	71									75	
4¾							53	61												

Width											
5	50		59	65	69				57	66	73	78
5¼			57	63	67				54	64	71	76
5½			54	61	66				51	61	69	74
5¾			52	59	64					59	67	72
6			50	57	63					57	65	71
6¼				56	61					54	63	69
6½				54	60					52	61	67
6¾				52	58					50	59	65
7				50	56						57	64
7¼					55						55	62
7½					53						53	60
7¾					52						51	58
8					50							56
8¼												55
8½												53
8¾												51
9												49

¹ Strength ratio for stress in horizontal shear.

Strength ratios of 45 percent or higher are based on the relation of the size of shake or check to the nominal width of face, lower ratios on the relation of the size of shake or check to the actual width of face.   In tabulating the sizes of shakes or checks, the rule used in approximating decimals by common fractions with ⅛-inch intervals has the effect of subtracting ¹⁄₆₄ inch from the size of each shake or check.   For example, the strength ratios listed for a 3½-inch shake or check are those that actually would obtain for a size of 3³¹⁄₆₄ inches.   In view of the allowance thus introduced and the fact that nominal instead of actual widths of face are used, the strength ratios in this table should be taken as maximum values.

² Ratios for sizes of shake or check other than those listed between 3 and 9 inches can be found by interpolation.

Derivation of basic stresses involves consideration of a number of factors representing the various circumstances that affect the adequacy of structural lumber. Some of these factors are well understood or can be accurately evaluated from available data; others are definable only in the light of engineering judgment and experience. It is therefore not possible to arrive at basic stresses by mathematical calculation alone. Experience and judgment, backed by pertinent test data, must be employed to formulate basic stresses that are applicable to certain specific conditions. With these specific conditions defined, means may be suggested by which the stress values can be adjusted by the engineer to meet other circumstances of design that may be encountered.

TABLE 25.—*Basic stresses for clear lumber under long-time service at full design load,[1] for use in determining working stresses according to grade of lumber and other applicable factors*

SOFTWOOD LUMBER

Species [2]	Extreme fiber in bending or tension parallel to grain	Maximum horizontal shear	Compression perpendicular to grain [3]	Compression parallel to grain L/d = 11 or less	Modulus of elasticity in bending
(1)	(2)	(3)	(4)	(5)	(6)
	P. s. i.	*P. s. i.*	*P. s. i.*	*P. s. i.*	*1,000 p. s. i.*
Baldcypress (cypress)	1,900	150	220	1,450	1,200
Cedar:					
Alaska-	1,600	130	185	1,050	1,200
Atlantic white- (southern whitecedar) and northern white-	1,100	100	130	750	800
Port-Orford	1,600	130	185	1,200	1,500
Western redcedar	1,300	120	145	950	1,000
Douglas-fir:					
Coast type	2,200	130	235	1,450	1,600
Coast type, close-grained	2,350	130	250	1,550	1,600
Rocky Mountain type	1,600	120	205	1,050	1,200
All types, dense	2,550	130	275	1,700	1,600
Fir:					
Balsam	1,300	100	110	950	1,000
California red, grand, noble, and white	1,600	100	220	950	1,100
Hemlock:					
Eastern	1,600	100	220	950	1,100
Western (west coast hemlock)	1,900	110	220	1,200	1,400
Larch, western	2,200	130	235	1,450	1,500
Pine:					
Eastern white (northern white), ponderosa, sugar, and western white (Idaho white)	1,300	120	185	1,000	1,000
Jack	1,600	120	160	1,050	1,100
Lodgepole	1,300	90	160	950	1,000
Red (Norway pine)	1,600	120	160	1,050	1,200
Southern yellow	2,200	160	235	1,450	1,600
Dense	2,550	160	275	1,700	1,630
Redwood	1,750	100	185	1,350	1,200
Close-grained	1,900	100	195	1,450	1,200
Spruce:					
Englemann	1,100	100	130	800	800
Red, white, and Sitka	1,600	120	185	1,050	1,200
Tamarack	1,750	140	220	1,350	1,300

TABLE 25.—*Basic stresses for clear lumber under long-time service at full design load,[1] for use in determining working stresses according to grade of lumber and other applicable factors*—Continued

HARDWOOD LUMBER

Species [2]	Extreme fiber in bending or tension parallel to grain	Maximum horizontal shear	Compression perpendicular to grain [3]	Compression parallel to grain L/d=11 or less	Modulus of elasticity in bending
(1)	(2)	(3)	(4)	(5)	(6)
	P. s. i.	*P. s. i.*	*P. s. i.*	*P. s. i.*	*1,000 p. s. i.*
Ash:					
Black	1,450	130	220	850	1,100
Commercial white	2,050	185	365	1,450	1,500
Aspen, bigtooth and quaking	1,300	100	110	800	800
Beech, American	2,200	185	365	1,600	1,600
Birch, sweet and yellow	2,200	185	365	1,600	1,600
Cottonwood, eastern	1,100	90	110	800	1,000
Elm:					
American and slippery (soft elm)	1,600	150	185	1,050	1,200
Rock	2,200	185	365	1,600	1,300
Hickory, true and pecan	2,800	205	440	2,000	1,800
Maple, black and sugar (hard maple)	2,200	185	365	1,600	1,600
Oak, commercial red and white	2,050	185	365	1,350	1,500
Sweetgum (gum, red gum, sap gum)	1,600	150	220	1,050	1,200
Tupelo, black (black gum) and water	1,600	150	220	1,050	1,200
Yellow-poplar (poplar)	1,450	130	160	1,050	1,200

[1] These stresses are based on the strength of green lumber and are applicable, with certain adjustments, to lumber of any degree of seasoning or lumber used under any conditions of duration of load.

[2] Species names approved by U. S. Forest Service. Commercial designations are shown in parentheses.

[3] Values given in previous editions of this handbook presumed some drying and were therefore at a higher level than these for green lumber.

The following basic principles were assumed in arriving at the values in table 25: (1) Each structural member to carry its own load; (2) consideration of lower range of strength values (7); (3) competent design and fabrication (3); (4) reliable stress grading; (5) supervision to prevent gross overloading; (6) adequate maintenance; and (7) a reasonable factor of safety.

The basic stress for a species is multiplied by the strength ratio appropriate for a stress grade to give the working stress for structural design with that grade of the species. Grade strength ratios are determined from the size of knots or other strength-reducing features permitted by the rules of the grade, evaluated in accordance with tables 20 to 24. Strength ratios for use in evaluating working stresses from the basic stresses in bending, tension parallel to grain, and compression parallel to grain (columns 2 and 5, table 25) are obtained from tables 20, 21, 22, and 23. Strength ratios for use in evaluating working stresses from the basic stresses in shear (column 3, table 25) are obtained from table 24. Strength in compression perpendicular to grain and modulus of elasticity (columns 4 and 6, table 25) are but little affected by grade, and a strength ratio of 100 percent is assumed for all grades; in other words, the basic stress is also the working stress for use in design.

Increases for density and close grain, as outlined on page 148, are incorporated in the basic stresses of table 25 for dense and for close-grained material.

The values for compression parallel to grain in table 25 apply to posts, columns, or struts whose unsupported length does not exceed 11 times the least dimension of the cross section. For more slender members, suitable intermediate- or long-column formulas (p. 216) are to be used.

ADJUSTMENT OF WORKING STRESSES

The principal modifications made in working stresses for various causes are summarized in table 26.

TABLE 26.—*Modification of working stresses by grade and use factors* [1]

Working stress	Size classification	Working stress modified by—				
		Grade	Rate of growth	Density	Seasoning	Duration of load
(1)	(2)	(3)	(4)	(5)	(6)	(7)
Extreme fiber in bending and tension parallel to grain.	Joists and planks	Yes	Yes	Yes	Yes [2]	Yes.
	Beams and stringers	Yes	Yes	Yes	No	Yes.
Horizontal shear	Joists and planks	Yes	No	No	Yes [2]	Yes.
	Beams and stringers	Yes	No	No	Yes	Yes.
Compression perpendicular to grain.	Joists and planks	No	Yes	Yes	Yes	Yes.
	Beams and stringers					
	Posts and timbers					
Compression parallel to grain	Joists and planks	Yes	Yes	Yes	Yes [2]	Yes.
	Posts and timbers	Yes	Yes	Yes	Yes	Yes.
Modulus of elasticity	Joists and planks	No	No	No	Yes	No.[3]
	Beams and stringers					
	Posts and timbers					

[1] Modification for grade (column 3) is accomplished by application of the grade strength ratio. Modifications for rate of growth and density (columns 4 and 5) are shown under the appropriate species in the basic stresses of table 25. Modifications for seasoning and duration of load (columns 6 and 7) are to be made by the designer to fit the particular conditions for which the design is made.

See p. 161 for a discussion of possible adjustments of working stress for decay hazard.

[2] Modification for seasoning in current commercial practice is accomplished by liberalizing grade limitation rather than increasing working stress.

[3] If deflection under long-time load must be limited, it is common practice to provide for the long-time increase of deformation by using one-half of the modulus of elasticity value in table 25 in calculating the stiffness of the member. Modulus of elasticity, when considered as a measure of the basic stiffness of the wood, is not affected by duration of load.

Seasoning Effects

The strength of clear wood increases as its moisture content decreases below the fiber saturation point, but in sawed structural lumber this increase of strength is largely offset by seasoning degrade, such as checks and honeycombing. Basic stress values in table 25 are suitable for green lumber or lumber that is completely submerged in use, provided no decay hazard exists (p. 161). Certain increases above these values can be made for lumber that will be continuously dry in use, as in most covered structures. Larger increases for drying are recognized in glued laminated lumber.

No increase for drying is made in working stresses for beams and stringers.

The effect of seasoning on strength in bending or in tension parallel to grain of joists and planks depends upon grade, the improvement being greater for the higher than for the lower grades. For lumber 2 to 4 inches in nominal thickness (joists and planks) that will be continuously dry in use, the strength ratio for bending stress in the extreme fiber or for stress in tension or in compression parallel to grain is increased by one-half of its excess over 50 percent. If, for example, the strength ratio for green lumber from tables 20, 21, 22, and 23 in a grade of joists and planks is 68 percent, the strength ratio for dry wood is 77 percent $\left(68+\dfrac{68-50}{2}=77\right)$. Since many joists remain continuously dry in use, it is common practice to take advantage of the increase in strength from drying by increasing permissible sizes of knots or other features rather than by increasing the working stress; under this circumstance, the working stress may require reduction if lumber with knots of these sizes is to be used under wet conditions.

Basic stresses for use in obtaining the working stresses for all grades of 1-inch boards used under dry conditions may be increased from the values of table 25 by one-quarter in bending or tension parallel to grain, one-eighth in horizontal shear, one-half in compression perpendicular to grain, three-eighths in compression parallel to grain, or one-tenth in modulus of elasticity. Similar increases are included in the values given in table 36 for basic stresses of glued laminated wood under dry conditions, and the values in that table may also be used for 1-inch boards.

Strength in compression parallel to grain is less affected by drying degrade than is bending strength, and the increase in this strength property on drying is significant in all grades of compression members. A strength increase of 10 percent above the values of table 25 for columns of all lengths may be taken for drying regardless of size or grade.[12] If existing grades of joists and planks have a bonus for drying in the permitted knot sizes, a 10-percent increase in the basic stress values for compression parallel to grain (table 25) may also be made.

Table 24 shows the increased sizes of shakes or checks permitted in seasoned lumber as compared to green lumber.

Working stresses in compression perpendicular to grain may be increased by 50 percent above the values in table 25 for lumber that will be continuously dry in use.

An increase of 10 percent in modulus of elasticity from the values of table 25 may be assumed for seasoned lumber.

---

[12] Since a column of thick cross section seasons very slowly in use, care must be taken in applying this increase that the column is not overloaded before an appreciable amount of seasoning can take place in the outer fibers. Seasoning progresses inward from all exposed surfaces, and a moderate depth of seasoning exerts a considerable influence on the strength of the column.

Although drying results in an increase of strength in many structural timbers, designers of timber structures should consider that the sizes assumed in design may be substantially reduced by the shrinkage that usually accompanies drying. Increases of working stresses for drying are based on strength properties related to the actual dimensions of the dry timbers.

Duration of Load

Design stresses derived from the basic stress values of table 25 are suitable for long-time loading at full load. If the conditions of use permit shorter periods of maximum load to be assumed, the working stresses (except modulus of elasticity) can be increased in accordance with the curve shown in figure 26. These increases apply to either continuous or intermittent loading with a cumulative duration as indicated.

In applying a stress increase for conditions of less than full duration of load, the safe stress for the permanent part of a combined loading must not be exceeded. If the assumed loading conditions involve an infrequent large load and a more frequent smaller load, the working stresses and sizes of structural members should be safe for each of the assumed loads. It is possible for sizes of members to be governed by the permanent load or by some semipermanent load below the maximum load.

Working stress values, except modulus of elasticity, may be used without allowance for occasional impacts up to 100 percent of the

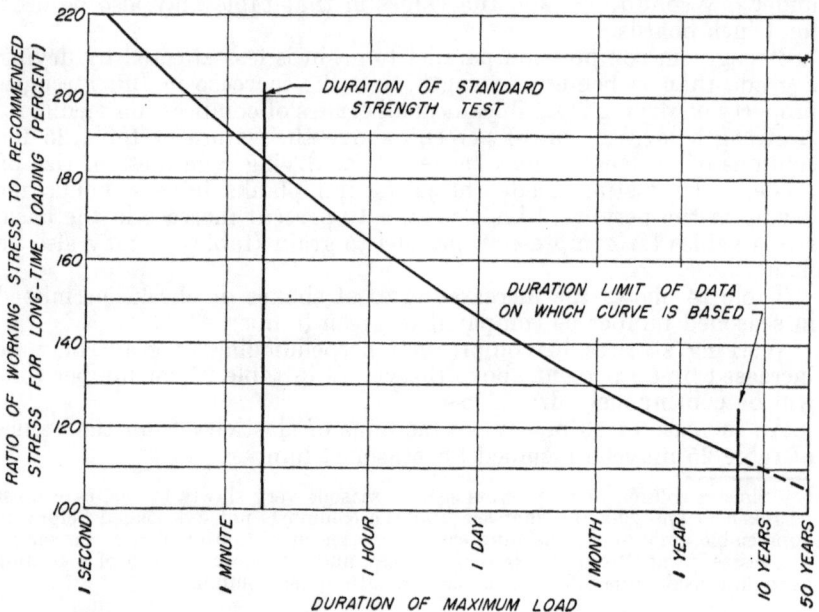

ZM 89380 F

FIGURE 26.—Relation of working stress to duration of load.

static load, provided that the sizes of structural members are safe for the static loadings. Frequently repeated impacts, however, have fatigue effects (p. 82).

In designing for wind loads in addition to dead and live loads, the working stresses may be increased 50 percent above the level of the basic stresses in table 25, provided the resulting structural members are not smaller than those designed for dead and live load alone.

Wood under continuing load takes on a continuing increment of deformation, usually very slow, but persistent over long periods of time. The "set" or "sag" of beams is an illustration. The deformation is greatest when green timbers season under load. It may increase with time to as much as twice the initial deformation without endangering the safety of the timber. This effect is most important if the long-continued load or part of the load is at or near the maximum design level. If deformation or deflection under long periods of loading must be limited, the necessary extra stiffness can be provided by setting the initial deformation limit at one-half the long-time deformation limit. Values for modulus of elasticity in table 25 will give the initial deflection of a beam, which will increase under long-continued load. The increase may be somewhat greater under varying temperature and moisture conditions than under uniform conditions.

Modulus of elasticity, when used to calculate deflections or deformations, does not increase with time. When used, however, in calculating the strength of a long column, a reduction factor of 3 is applied to the modulus of elasticity (p. 217) to give a working value that is safe for long-time loading. Though the design of long columns is predicted on an initial deflection, most long columns are in a state of unstable equilibrium without deflection. A member in unstable equilibrium is sensitive to long-time loading; that is, it may remain in the unstable equilibrium for a period of time and then suddenly deflect.

Since the reduction factor of 3 on modulus of elasticity has been found satisfactory for long-time-loading conditions, the working stress for a long column based on that factor may receive the same increases for shorter durations of load that are given to other working stresses. In the case of an intermediate-length column (p. 216), the increase for short-time loading is most conveniently made after the long-time-loading stress for the column is calculated from the appropriate long-time working stresses for compression parallel to grain and modulus of elasticity.

Decay

Since there is no satisfactory way of numerically appraising the effect of decay on strength of wood (4), decay is excluded from most structural grades. No working stress can be assigned with assurance to timber containing decay. Decay confined to knots and not present in wood surrounding them is permitted in some structural grades. Limited decay of pocket type is also permitted in the lowest dimension grades of some species. Structural lumber exposed to the hazard of decay should be inspected at frequent and regular intervals. If decay is detected in or near highly stressed areas, the member should be replaced. Special attention given to such features as drainage and ventilation will help reduce or eliminate the necessity of removing lumber because of decay. Treated wood or the heartwood of species of high natural decay resistance should be used wherever conditions are favorable to decay in order to prolong service life.

When untreated wood is used under conditions favorable to decay, some loss of strength may occur before the decay is detected. A reduction of design stress may be in order so that the structure may safely carry its design loads for the time until decay is detected and the affected material can be removed. The amount of reduction is a matter of choice for the structural designer. A general range from a few percent up to 25 percent is adequate. Stiffness is affected less than strength, and shock resistance is affected most.

Reductions of working stress for decay hazard are intended to provide only temporary protection until repairs or replacements can be made. It should be emphasized again that there is no safe working stress for decayed wood.

Preservative-Treated Timber

It may be necessary, when establishing allowable working stresses for preservative-treated timber, to take into account possible reductions in strength that may result from the treating process (see p. 420). Results of tests of treated timber show reductions in stress in extreme fiber in bending and in compression perpendicular to grain ranging from a few percent up to 25 percent, depending on the treating conditions. Compression parallel to grain is affected less and modulus of elasticity very little. The effect on resistance to horizontal shear can be estimated by inspection for shakes and checks after treatment.

These reductions in strength can be minimized by restricting temperatures, heating periods, and pressures as much as is consistent in obtaining the absorption and penetration required for proper treatment.

Treating conditions specified by the American Wood-Preservers' Association (1) should never be exceeded. The maximum temperatures and periods permitted by those specifications are seldom necessary in the treatment of partially seasoned structural lumber.

Temperature

Stress values in table 25 are applicable to lumber used under ordinary ranges of temperature. Special allowance should be made for lumber subjected to abnormally high temperatures, particularly for long periods of time (p. 89).

Other Conditions

Recommended stress values contain factors of safety that are applicable to the usual conditions of design. The overall factor of safety can be increased by reducing the design stress, or it can be reduced by increasing the design stress. In structural design, the engineer may weigh the factor of safety against the economics of the specific project that is to be built. If an individual timber is in a critical location where its failure might threaten the integrity of a major structure or endanger human life, he may reduce its working stress and thereby reduce the hazard of failure. Conversely, in situations where failure of a member can result only in property damage of limited extent and repair can easily be made, he may be justified in economizing through the use of stresses higher than those of table 25. The engineer's freedom to increase working stresses is subject, of course, to restrictions in any applicable codes or other building regulations.

The values in table 25 for compression perpendicular to grain apply to bearings 6 inches or more in length located anywhere in the length of a structural member and to bearings of any length located at the ends of beams or joists. For bearings shorter than 6 inches located 3 inches or more from the end of a member, the stresses may be increased in accordance with the following factors:

Length of bearing (inches):	Factor of increase	Length of bearing (inches) Con.	Factor of increase
½	1. 75	3	1. 13
1	1. 38	4	1. 10
1½	1. 25	6 or more	1. 00
2	1. 19		

For stress under a washer or other round bearing area, the same factor may be taken as for a bearing whose length equals the diameter of the washer.

MANUFACTURERS' STANDARD GRADES

Standard stress grades are sponsored by various agencies of lumber manufacturers. This sponsorship consists of (1) publication of detailed descriptions and grading rules for the various grades, (2) responsibility for inspection service either at the mill or at destination, and issue of a certificate of inspection, and (3) application of a copyrighted grade mark to each graded piece. The following associations and agencies offer some or all of these services:

California Redwood Association.
National Hardwood Lumber Association.
Northeastern Lumber Manufacturers Association.
Northern Hemlock and Hardwood Manufacturers Association.
Southern Cypress Manufacturers Association.
Southern Pine Inspection Bureau.
West Coast Bureau of Lumber Grades and Inspection.
Western Pine Association.

All published stress grades are listed with their recommended working stresses in the current issue of "National Design Specification for Stress-Grade Lumber and Its Fastenings," published by the National Lumber Manufacturers Association (*3*).

Recommendations of working stresses made by lumber manufacturers' associations or agencies are worked out in accordance with the principles of stress grading set forth in this section. Manufacturers' recommendations are generally for the condition defined as "continuously dry, such as in most covered structures" (*3*), and thus include increases for drying (p. 158). Since many buildings may carry full maximum design load for only short periods throughout a service life of many years, an increase of 10 percent for less-than-long-time loading is also included. This loading condition is designated as normal loading (*3*) and assumes that the continuous or cumulative duration of full maximum load will not exceed about 10 years during the service life of a permanent structure (fig. 26). The manufacturers' recommendations provide for removal of the 10-percent increase for normal loading in the case of structures or structural elements on which maximum load is applied for long periods. Specific increases for snow, wind, earthquake, or other short-time loads, in accordance with figure 26, are also recommended by the manufacturers (*3*).

LITERATURE CITED

(1) AMERICAN WOOD-PRESERVERS' ASSOCIATION.
    1952. MANUAL OF RECOMMENDED PRACTICE. Standard Specif. (A loose-leaf handbook, revised annually.)
(2) CALIFORNIA REDWOOD ASSOCIATION.
    1952. STANDARD SPECIFICATIONS FOR GRADES OF CALIFORNIA REDWOOD LUMBER. 61 pp., illus. San Francisco.
(3) NATIONAL LUMBER MANUFACTURERS ASSOCIATION.
    1953. NATIONAL DESIGN SPECIFICATION FOR STRESS-GRADE LUMBER AND ITS FASTENINGS. 67 pp., illus.
(4) SCHEFFEF, T. C., WILSON, T. R. C., LUXFORD, R. F., AND HARTLEY, C.
    1941. EFFECT OF CERTAIN HEART ROT FUNGI ON THE SPECIFIC GRAVITY AND STRENGTH OF SITKA SPRUCE AND DOUGLAS-FIR. U. S. Dept. Agr. Tech. Bul. 779, 24 pp., illus.
(5) SOUTHERN PINE INSPECTION BUREAU.
    1948. STANDARD GRADING RULES FOR SOUTHERN PINE LUMBER. 124 pp., illus. New Orleans, La.
(6) WEST COAST BUREAU OF LUMBER GRADES AND INSPECTION.
    1947. STANDARD GRADING AND DRESSING RULES FOR DOUGLAS-FIR, SITKA SPRUCE, WEST COAST HEMLOCK, WESTERN RED CEDAR LUMBER. 199 pp., illus. Portland, Oreg.
(7) WOOD, L. W.
    1951. VARIATION OF STRENGTH PROPERTIES IN WOODS USED FOR STRUCTURAL PURPOSES. U. S. Forest Prod. Lab. Rpt. R1780, 11 pp., illus. [Processed.]
(8) ————
    1951. RELATION OF STRENGTH OF WOOD TO DURATION OF LOAD. U. S. Forest Prod. Lab. Rpt. R1916, 5 pp., illus. [Processed.]

TIMBER FASTENINGS

The strength and stability of any structure are in great measure dependent upon the fastenings that hold its parts together. Among structural materials, wood is exceptional because of the facility with which structural parts made of it can be joined together with a wide variety of fastenings, including nails, spikes, screws, bolts, lag screws, drift pins, and metal connectors of various shapes. For utmost strength and service, of course, each type of fastening requires joint designs adapted to the varying strength properties of wood along and across the grain and to the dimensional changes that occur with changes in moisture content.

The allowable loads given in this section for various fastenings are applicable under conditions of long-continued or permanent loading. For loads of short duration, when the wood determines the load capacity, the tabulated values and the equations for computing allowable loads can be modified in accordance with the provisions set forth in the section on working stresses. The allowable loads given for a particular loading condition for all fastenings are for one unit. Loads for more than one unit, of either the same size or miscellaneous sizes, are the sum of the loads permitted for each unit, provided that the spacings, end distances, and edge margins are sufficient to develop the full strength of each unit.

NAILS

Nails are the most common mechanical fastenings used in temporary and permanent construction (*32, 34*).[13] There are many types, sizes, and forms of standard nails (fig. 27) and, in addition, many special-purpose nails. The formulas and tables presented here for allowable loads apply for bright, smooth, common steel wire nails driven into wood with no visual evidence of splitting. For nails other than common wire nails, the allowable loads can be adjusted in accordance with the provisions presented in subsequent paragraphs or by a factor that the designer considers to conform with the character of the construction.

In general, nails give stronger joints when driven into the side grain (perpendicular to the wood fibers) than into the end grain of wood (*13, 29*). Also, nails should preferably be used so that their lateral resistance, rather than direct withdrawal resistance, is utilized. The withdrawal resistance of nails is greatly affected by such factors as type of nail point, type of shank, time the nail remains in the wood, surface coatings, and moisture content changes in the wood.

Withdrawal Resistance

The resistance of a nail to direct withdrawal from a piece of wood is intimately related to the density or specific gravity of the wood (*31*), the diameter of the nail, and the depth of penetration. The surface condition of the nail at the time of driving also influences the immediate withdrawal resistance.

---

[13] Italic numbers in parentheses refer to Literature Cited, p. 201.

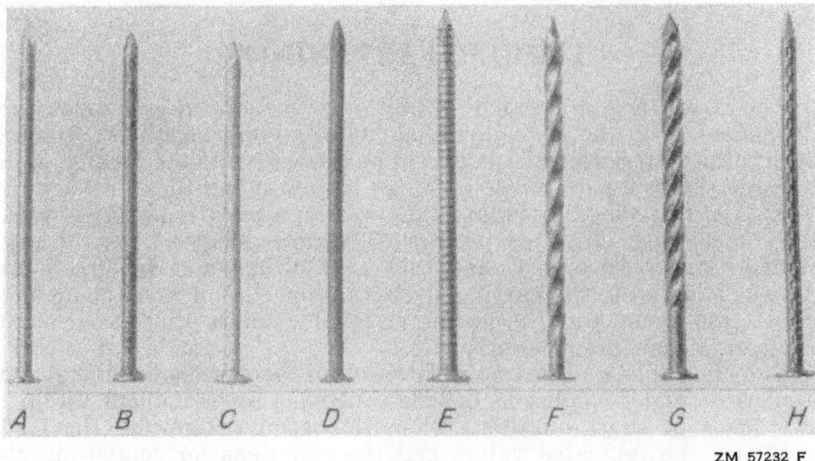

ZM 57232 F

FIGURE 27.—Nails with various surface conditions: A, Bright, smooth wire
nail; B, cement-coated; C, zinc-coated; D, chemically etched; E, annular
grooved; F, spirally grooved; G, spirally grooved and barbed; and H, barbed
nail.

For bright, common wire nails driven into the side grain of seasoned
wood that will remain dry or unseasoned wood that will remain wet,
the allowable withdrawal load is given by the formula:

$$p = 1{,}150 \ G^{5/2} D$$

in which p represents the allowable load (one-sixth of the ultimate
load) per lineal inch of penetration in the member holding the nail
point; G, the specific gravity of the wood based on oven-dry weight
and volume; and D, the diameter of the nail in inches. The average
specific gravity raised to the five-halves power is listed in table 27
for a number of the common hardwood and softwood species. The
diameters of various penny or gage sizes of nails are given in table
28.

The relationship expressed by this equation is general, but certain
species give test values that are somewhat higher or lower than the
equation values. Usually common knowledge of the characteristics
of the nail, and of the species with particular reference to its splitting
tendencies, will aid in deciding whether the withdrawal resistance
will fall above or below the general equation values.

Although the general equation for nail withdrawal resistance indicates that the dense, heavy woods offer greater nail-holding power
than the lighter weight ones, lighter species are not disqualified for
uses requiring high holding power (25). As a rule, the less dense
species do not split so readily as the dense ones (33) and thus offer
an opportunity for increasing the diameter, length, and number of
the nails to compensate for the wood's lower holding power.

In practically all species, nails driven into green wood and pulled
before any seasoning takes place offer about the same withdrawal
resistance as nails driven into seasoned wood and pulled soon after
driving. If, however, common smooth-shank nails are driven into
green wood that is allowed to season or into seasoned wood that is

subjected to cycles of wetting and drying before the nails are pulled, they lose a major part of their withdrawal resistance. In seasoned wood that is not subjected to appreciable moisture content changes, the withdrawal resistance of nails may also diminish with lapse of time. On the other hand, tests indicate that if, under some conditions of moisture variation and time, the wood fibers are affected or the nail rusts, withdrawal resistance is very erratic and may be regained or even increased over the immediate withdrawal resistance. Under all these conditions of use, the withdrawal resistance of nails differs

TABLE 27.—*Factors for calculating the allowable strength of mechanical fastenings in seasoned wood*

HARDWOODS

Species	Specific gravity G and powers [1]				Allowable lateral load constant (K)			Basic stresses for bolts [2]		Species group for connector loads [5]
	G	$G^{3/2}$	G^2	$G^{5/2}$	Nails [3]	Screws [3]	Lag screws [4]	Parallel to grain	Perpendicular to grain	
Ash:										
Black	0.53	0.39	0.28	0.20	1,250	2,900	1,900	850	300	
Commercial white	.61	.48	.37	.29	1,700	4,000	2,200	1,450	500	4
Aspen:										
Bigtooth	.41	.26	.17	.11	900	2,100	1,700	800	150	1
Quaking	.40	.25	.16	.10	900	2,100	1,700	800	150	1
Basswood, American	.40	.25	.16	.10	900	2,100	1,700			1
Beech, American	.67	.55	.45	.37	1,700	4,000	2,200	1,600	500	4
Birch:										
Sweet	.71	.60	.50	.42	1,700	4,000	2,200	1,600	500	4
Yellow	.66	.54	.44	.35	1,700	4,000	2,200	1,600	500	4
Chestnut, American	.45	.30	.20	.14	900	2,100	1,700			2
Cottonwood:										
Black	.37	.22	.14	.08	900	2,100	1,700			1
Eastern	.43	.28	.18	.12	900	2,100	1,700	800	150	1
Elm:										
American	.55	.41	.30	.22	1,250	2,900	1,900	1,050	250	3
Rock	.66	.54	.44	.35	1,700	4,000	2,200	1,600	500	4
Slippery	.57	.43	.32	.25	1,250	2,900	1,900	1,050	250	3
Hackberry	.56	.42	.31	.23	1,250	2,900				
Hickory:										
Pecan	.65	.52	.42	.34	1,700	4,000	2,200	2,000	600	4
True	.74	.64	.55	.47	1,700	4,000	2,200	2,000	600	4
Magnolia, southern	.53	.39	.28	.20	1,250	2,900				
Maple:										
Black	.62	.49	.38	.31	1,700	4,000	2,200	1,600	500	4
Red	.55	.41	.30	.22	1,250	2,900	1,900			3
Silver	.51	.36	.26	.19	1,250	2,900	1,900			3
Sugar	.68	.56	.46	.38	1,700	4,000	2,200	1,600	500	4
Oak:										
Commercial red	.66	.54	.44	.35	1,700	4,000	2,200	1,350	500	4
Commercial white	.71	.60	.50	.42	1,700	4,000	2,200	1,350	500	4
Sweetgum	.53	.39	.28	.20	1,250	2,900	1,900	1,050	300	3
Sycamore, American	.54	.40	.29	.21	1,250	2,900	1,900			3
Tupelo:										
Black	.55	.41	.30	.22	1,250	2,900	1,900	1,050	300	3
Water	.52	.37	.27	.19	1,250	2,900	1,900	1,050	300	3
Yellow-poplar	.43	.28	.18	.12	900	2,100	1,700	950	220	2

See footnotes at end of table.

TABLE 27.—*Factors for calculating the allowable strength of mechanical fastenings in seasoned wood*—Continued

SOFTWOODS

Species	Specific gravity G and powers [1]				Allowable lateral load constant (K)			Basic stresses for bolts [2]		Species group for connector loads [5]
	G	$G^{3/2}$	G^2	$G^{5/2}$	Nails [3]	Screws [3]	Lag screws [4]	Parallel to grain	Perpendicular to grain	
Baldcypress	0.48	0.33	0.23	0.16	1,125	2,700	1,700	1,450	300	2
Cedar:										
Alaska	.46	.31	.21	.14	1,125	2,700	1,700	1,050	2r0	2
Atlantic white	.35	.21	.12	.07	900	2,100	1,500	750	1r0	
Northern white	.32	.18	.10	.06	900	2,100	1,500	750	1 0	
Port-Orford	.44	.29	.19	.13	1,125	2,700	1,700	1,200	2r0	2
Western redcedar	.34	.20	.12	.07	900	2,100	1,500	950	200	1
Douglas-fir:										
Coast type [6]	.51	.36	.26	.19	1,375	3,300	1,900	1,450	320	3
Rocky Mountain type	.45	.30	.20	.14	1,125	2,700	1,700	1,050	2r0	2
Fir:										
Balsam	.41	.26	.17	.11	900	2,100	1,500	950	1r0	1
Commercial white	.41	.26	.17	.11	900	2,100	1,500	950	300	1
Hemlock:										
Eastern	.43	.28	.18	.12	900	2,100	1,500	950	300	1
Western	.44	.29	.19	.13	1,125	2,700	1,700	1,200	300	2
Larch, Western	.59	.45	.35	.27	1,375	3,300	1,900	1,450	320	3
Pine:										
Eastern white	.37	.22	.14	.08	900	2,100	1,500	1,000	2r0	1
Lodgepole	.43	.28	.18	.12	900	2,100	1,500	9r0	220	
Ponderosa	.42	.27	.18	.11	900	2,100	1,r00	1,000	2r0	1
Red	.51	.36	.26	.19	1,12r	2,r00	1,r00	1,0r0	220	2
Southern yellow [6]	.59	.45	.35	.27	1,375	3,r00	1,r00	1,4r0	320	3
Sugar	.38	.23	.14	.09	900	2,100	1,r00	1,000	2r0	1
Western white	.42	.27	.18	.11	900	2,100	1,500	1,000	2r0	1
Redwood (old-growth)	.42	.27	.18	.11	1,125	2,700	1,700	1,3r0	2r0	2
Spruce:										
Englemann	.35	.21	.12	.07	900	2,100	1,500	8r0	1r0	1
Red	.41	.26	.17	.11	900	2,100	1,500	1,0r0	2r0	2
Sitka	.42	.27	.18	.11	900	2,100	1,500	1,0r0	2r0	2
White	.45	.30	.20	.14	900	2,100	1,r00	1,0r0	2r0	2

[1] Specific gravity based on weight and volume when oven-dry.

[2] Basic stresses to be used in determining allowable bolt-bearing stresses.

[3] Driven perpendicular to the grain of the wood and loaded either parallel or perpendicular to the grain.

[4] Inserted perpendicular to the grain of the wood and loaded parallel to the grain.

[5] Grouping to be used in determining allowable connector loads. Group 1 woods provide the weakest and group 4 woods the strongest connector groups.

[6] When graded for density, these species qualify for group 4 connector loads.

TABLE 28.—*Sizes of bright, common wire nails*

Penny	Gage	Length	Diameter D	$D^{3/2}$	Penny	Gage	Length	Diameter D	$D^{3/2}$
		Inch	Inch	Inch			Inch	Inch	Inch
2d	15	1	0.072	0.0193	16d	8	3½	0.162	0.0652
4d	12½	1½	.098	.0307	20d	6	4	.192	.0841
6d	11½	2	.113	.0380	30d	5	4½	.207	.0942
8d	10¼	2½	131	.0474	40d	4	5	.225	.1068
10d	9	3	.148	.0570	50d	3	5½	.244	.1205
12d	9	3¼	.148	.0570	60d	2	6	.262	.1342

among species as well as within individual species, making it difficult to evaluate their behavior. The withdrawal loads for nails driven into wood that is subjected to changes in moisture content may be as much as 75 percent below the values given by the general formula.

Factors That Affect Withdrawal Resistance

The withdrawal resistance of nails is affected by the surface condition of the nail, the types of head and point on the nail, the composition of the nail, the direction of driving, and whether the nails are clinched or unclinched.

The surface condition of nails is frequently modified during the manufacturing process to improve their withdrawal resistance. Such modification is usually done by one of three methods, namely surface coating, surface roughening, or deformation (*10, 28*) of the shank. Other factors that affect the surface conditions of the nail are the oil film remaining on the shank after manufacture or corrosion resulting from storage under adverse conditions, but these factors are so variable that their influence on nail holding cannot be adequately evaluated.

Surface-Coated Nails

A common surface treatment for nails is the so-called "cement coating," which if properly applied may double the resistance of nails to withdrawal immediately after they are driven into the softer woods. In the denser woods, like hard maple, birch, or oak, however, coated nails have practically no advantage over plain nails, since more of the coating is removed in driving. Some of the coating may also be removed in the cleat or facing member before the nail penetrates the foundation member. Different techniques of applying the cement coating and variations in its ingredients may cause large differences in the relative resistance to withdrawal of different lots of cement-coated nails, so that nails may sometimes be obtained that show only a slight initial advantage over plain nails. The increase in withdrawal resistance of cement-coated nails is not permanent but drops off about one-half after a month or so for the softer woods. Cement-coated nails are used extensively in construction of boxes, which are usually built for rough handling and short-time service.

Nails that have special coatings, such as zinc, are intended primarily for uses where corrosion and staining are important factors in permanence and appearance. If the zinc coating is evenly applied, withdrawal resistance may be increased, but extreme irregularities of the coating may actually reduce it. The advantage that zinc-coated nails with an even, uniform coating may have over a plain nail in resistance to immediate withdrawal is usually reduced under repeated cycles of wetting and drying.

Surface-Roughened Nails

A chemically etched nail developed at the Forest Products Laboratory gives somewhat higher withdrawal resistance than other coated nails and retains much of its superiority under varying moisture conditions (*5, 16*). Under impact loading, however, the withdrawal resistance of the etched nails is little different from that of the plain or cement-coated nails under various moisture conditions.

Sand-blasted nails perform in much the same manner as chemically etched nails.

Form of Shank

Nail shanks are varied from a circular form to give an increase in surface area without an increase in nail weight (11, 28). Special nails with barbed, spirally grooved, annular grooved, and other irregular shanks (fig. 27) are offered commercially.

The form and magnitude of the deformations along the shank influence the performance of the nails in the various species. The withdrawal resistance of these nails, except some types of barbed nails, is in general somewhat greater than that of common wire nails in wood remaining at a uniform moisture content. Under conditions involving changes in the moisture content of the wood, however, some of the special nail forms provide considerably greater withdrawal resistance than the common wire nail (6, 7, 28). This is especially true of nails driven into green wood that subsequently seasons. In general, annular grooved nails sustain larger withdrawal loads, and spirally grooved nails sustain greater impact withdrawal work values than the other nail forms.

Nail Points

A nail with a long, sharp point will usually have a higher withdrawal resistance, particularly in the softer woods, than the common wire nail, which usually has a diamond point (12). Sharp points, however, accentuate splitting of certain species, which may reduce withdrawal resistance (fig. 28). A blunt or flat point without taper reduces splitting, but its destruction of the wood fibers when driven reduces withdrawal resistance to less than that of the common wire nail. A nail tapered at the end and terminating in a blunt point will cause less splitting than the common nail and, in the heavier woods, will provide about the same withdrawal resistance. In the less dense woods, its resistance to withdrawal is less than that of the common nail.

Nailheads

Nailhead classifications include flat, oval, countersunk, deep-countersunk, and brad. In general, nails with all types of heads, except the deep-countersunk, brad, and some of the thin flathead nails, are sufficiently strong to withstand the force required to pull them from most woods in direct withdrawal. The deep-countersunk and brad nails are usually driven below the wood surface and are not intended to carry large withdrawal loads. In general, the thickness and diameter of the heads of the common wire nails increase as the size of the nail increases.

Nail Composition

Nails of copper alloys, aluminum alloys, stainless steel, and other alloys are used mainly where corrosion or staining is an important factor in appearance or permanence. Specially hardened nails are also

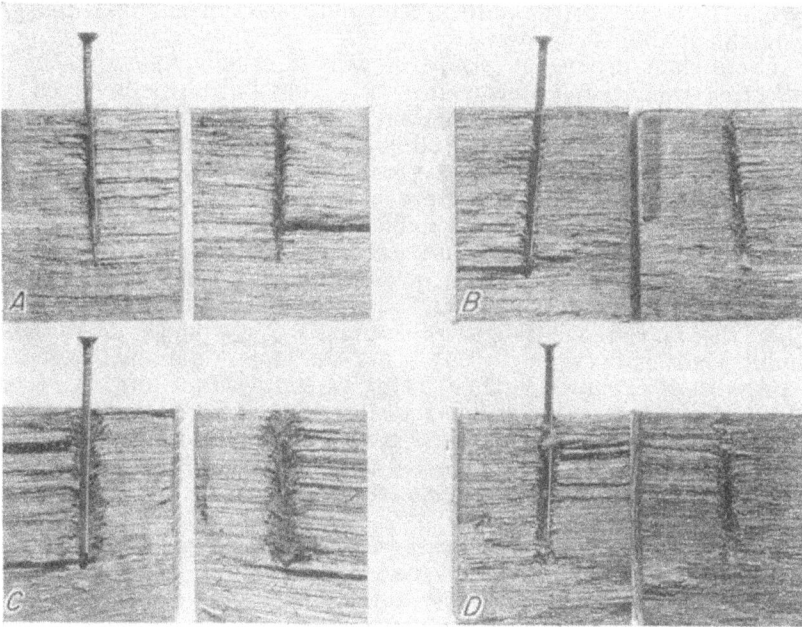

ZM 57967 F

FIGURE 28.—Characteristic fiber distortion caused by nails with four different types of points: A, Sharp; B, common; C, blunt; and D, truncated. The blocks of wood were split after the nails were driven.

frequently used where driving conditions are difficult. In general, the withdrawal resistance of copper nails is somewhat comparable to that of common wire nails when pulled soon after driving.

Direction of Driving

The resistance of nails to withdrawal is generally greatest when they are driven perpendicular to the grain of the wood. When the nail is driven parallel to the wood fibers—that is, into the end of the piece—withdrawal resistance in the softer woods drops to 75 or even 50 percent of the resistance obtained when the nail is driven perpendicular to the grain. The angle at which the nail is driven has less effect on the withdrawal resistance in dense woods than in softer woods. In most species, the ratio between the end-grain and side-grain withdrawal loads of nails pulled after a time interval or after moisture content changes have occurred is usually somewhat higher than that of nails pulled immediately after driving.

The results of withdrawal tests at the Forest Products Laboratory on slant and straight driving (14), when the piece attached is pulled directly away from the main member, show that slant driving is usually superior when nails are driven into dry wood and pulled immediately and decidedly superior when nails are driven into green or partially dry wood that is allowed to season for a month or more.

When nails are driven into green wood and pulled immediately, straight driving is superior.

Cross slant driving of groups of nails is usually somewhat more effective than parallel slant driving. The difficulties involved in starting and driving either parallel- or cross-slanted nails, however, and the loss in depth of penetration, the destruction and mutilation of the wood fibers at the surface in starting the nail, and the breaking down of the wood under the hammer offset to a considerable extent any advantages of slant nailing.

Toenailing, which is a common method of joining wood framework, consists of slant driving a nail or group of nails through the end or edge of an attached member and into a main member. Tests show that the maximum strength of toenailed joints under lateral and uplift loads is obtained by (1) using the largest nail that will not cause excessive splitting; (2) allowing an end distance (distance from the end of the attached member to the point of initial nail entry) of approximately one-third the length of the nail; (3) driving the nail at a slope of 30° with the attached member; and (4) burying the full shank of the nail but avoiding excessive mutilation of the wood from hammer blows (*26*).

Toenailing requires greater skill in assembly but provides joints of greater strength and stability than does ordinary end nailing. In tests of stud-to-sill assemblies with the number and size of nails frequently used in toenailed and end-nailed joints, a joint toenailed with 4 eightpenny common nails was superior to a joint end nailed with 2 sixteenpenny common nails. In such woods as Douglas-fir, toenailing with tenpenny common nails gave greater joint strength than the commonly used eightpenny nails. The allowable withdrawal load per nail in toenailed joints for all conditions of seasoning is equivalent to two-thirds of that calculated by the general formula for withdrawal resistance of nails (*17*).

Prebored Lead Holes

Nails driven into lead holes with a diameter slightly smaller than the nail have somewhat higher withdrawal resistance than nails driven without lead holes. Lead holes also prevent or reduce splitting of the wood.

Clinched Nails

The withdrawal resistance of smooth-shank clinched nails is considerably higher than that of unclinched nails. The ratio between the loads for clinched and unclinched nails varies enormously, depending upon the moisture content of the wood when the nail is driven and withdrawn, the species of wood, the size of nail, and the direction of clinch with respect to the grain of the wood (*2*).

In dry or green wood, a clinched nail provides from 45 to 170 percent more withdrawal resistance than an unclinched nail when withdrawn soon after driving. In green wood that seasons after a nail is driven, a clinched nail gives from 250 to 460 percent greater withdrawal resistance than an unclinched nail. However, this improved strength of the clinched- over the unclinched-nail joint does not justify the use of green lumber, because the joints may loosen as the lumber seasons. Furthermore, the laboratory tests were made with single nails and the effects of drying, such as warping, twisting, and splitting, may reduce the efficiency of a joint that has more than one nail. In such construction as boxes and crates, it is generally necessary to use both clinched and unclinched nails.

Nails clinched across the grain have approximately 20 percent more resistance to withdrawal than nails clinched along the grain.

Nails in Plywood

The nailing characteristics of plywood are not greatly different from those of solid wood except that plywood's greater resistance to splitting when nails are driven near an edge is a definite advantage. The nail-withdrawal resistance of plywood is from 15 to 30 percent less than that of solid wood of the same thickness. The reason is that fiber distortion is less uniform in plywood than in solid wood. For plywood less than one-half inch thick, the high splitting resistance tends to offset the lower withdrawal resistance as compared to solid wood. The withdrawal resistance per inch of penetration decreases with increase in the number of plies. The direction of the grain of the face ply has little influence on the withdrawal resistance along the end or edge of a piece of plywood.

Lateral Resistance

The allowable load for a bright, common wire nail in lateral resistance when driven into the side grain (perpendicular to the wood fibers) of seasoned wood is expressed by the following general formula:

$$p = KD^{3/2}$$

in which p represents the allowable lateral load in pounds per nail; K, a constant; and D, the diameter of the nail in inches. Values of K for a number of the common hardwoods and softwoods are listed in table 27. Values for converting the size of common wire nails from penny to $D^{3/2}$ are listed in table 28.

The constants K listed in table 27 were obtained by dividing the constants in general formulas for the proportional limit loads of nailed joints in different species by a reduction factor of 1.6. The ultimate lateral nail loads in softwoods will be about 6 times the allowable loads expressed by the formula, and in hardwoods about 11 times. These loads apply only for conditions where the side member and the member holding the nail point are of approximately the same density and where the depth of penetration of the nail in the member holding the point is not less than 10 times the diameter of the nail for dense woods and 14 times the diameter for lightweight woods. When metal is held to wood, an increase of about 25 percent can be applied to the allowable lateral nail load.

The allowable lateral load for side-grain nailing given by the general formula applies whether the load is in a direction parallel to the grain of the pieces joined or at right angles to it. When nails are driven into the end grain (parallel with the wood fibers), limited data on softwood species indicate that their maximum resistance to lateral displacement is about two-thirds that for nails driven into the side grain.   Although the average proportional limit loads appear to be about the same for end- and side-grain nailing, the individual results are more erratic for end-grain nailing, and the minimum loads approach only 75 percent of corresponding values for side-grain nailing.

It is therefore recommended that the safe lateral loads for end-grain nailing should be about 60 percent of those computed for side-grain nailing in lightweight species and slightly higher in denser species. If the type of construction, method of loading, and hazard involved are such that it appears safe to the designer to use higher loads, the coefficient preceding $D^{3/2}$ in the equation may be raised accordingly. It is not recommended, however, that smaller factors of safety be used for permanent concealed joints.

Nails driven into the side grain of unseasoned wood give maximum lateral resistance loads approximately equal to those obtained in seasoned wood, but the proportional limit lateral resistance loads are somewhat less.   Since proportional limit loads are of prime importance in establishing allowable loads, it is recommended that the allowable lateral loads obtained by the general formula for seasoned wood be reduced 25 percent for unseasoned wood that will remain wet or be loaded before seasoning takes place.

When nails are driven into green wood, their lateral proportional limit loads after the wood has seasoned are also less than when they are driven into seasoned wood and loaded.   The erratic behavior of a nailed joint that has undergone one or more moisture content changes makes it difficult to recommend an allowable lateral load for a nailed joint under these conditions.   Structural joints should be inspected at intervals and, if it is apparent that a loss in joint strength occurred during drying, the joint should be reinforced with additional nails.   Simply setting the original nails is not sufficient.

Nails in lateral resistance should be driven with sufficient end distance, edge margin, and nail spacing to avoid any unusual splitting of the wood.

Deformed-shank nails carry somewhat higher maximum lateral loads than common wire nails, but both types perform similarly at small distortions in the joint.   In nailed joints subjected to vibration or reversal of stress, as in boxes and crates that are handled roughly, the deformed-shank nails with reduced sections, such as the spirally and annular grooved nails, tend to break more easily than the smooth-shank nails.   The lateral loads for aluminum alloy nails are slightly lower at small distortions in the joint than they are for uniform-shank steel nails, but are somewhat higher at large distortions.   Tests have also indicated that an aluminum nail with a shank diameter 3 to 10 percent larger than that of a steel nail sustains lateral loads at small distortions that are comparable to those for a steel nail.

SPIKES

Common wire spikes are manufactured in the same manner as common wire nails.   They have either a chisel point or a diamond point and are made in lengths of 3 to 12 inches.   For corresponding lengths (3 to 6 inches), they have larger diameters than the common wire nails, and beyond the sixtypenny size they are usually designated by inches of length.

The allowable withdrawal and lateral resistance formulas and limitations given for common wire nails are also applicable to spikes, except that, in calculating the withdrawal load for spikes, the depth of penetration should be reduced by two-thirds the length of the point.

DRIFT BOLTS

The ultimate withdrawal load of a round drift bolt or pin from the side grain of seasoned wood is given by the formula:

$$p = 6,000 \ G^2 D$$

in which p is the ultimate withdrawal load per lineal inch of penetration; G, the specific gravity based on the oven-dry weight and volume of the wood; and D, the diameter of the drift bolt in inches.

This equation provides an average relationship for all species, and the withdrawal load of some species may be above or below the equation values. It also presumes that the bolts are driven into prebored holes having a diameter one-eighth inch less than that of the bolt diameter.

The allowable withdrawal load per lineal inch of penetration should be determined by reducing the ultimate withdrawal load given by the general formula by a factor consistent with the character of the work. For general use, it is suggested that the ultimate withdrawal load be divided by the factor 5.

In lateral resistance, the allowable load for a drift bolt driven into the side grain of wood should not exceed and ordinarily should be taken as less than that for a common bolt of the same diameter (p. 182). The drift bolt should normally be of greater length than the common bolt to compensate for the lack of washers and nut.

WOOD SCREWS

The common types of wood screws have flat, oval, or round heads. The flathead screw is most commonly used if a flush surface is desired. Ovalhead and roundhead screws are used for appearance or when countersinking is objectionable. Besides the head, the principal parts of a screw are the shank, thread, and core (fig. 29).

Withdrawal Resistance

The resistance of wood screws to withdrawal from the side grain of seasoned wood varies directly with the square of the specific gravity of the wood. Tests indicate that, within certain limits, the withdrawal load varies directly with the depth of penetration and the diameter of the screw (4). The effective length of a screw is limited by the length at which the screw fails in tension; this limiting length decreases as the density of the wood increases. The longer lengths of standard screws should therefore be avoided in dense hardwoods.

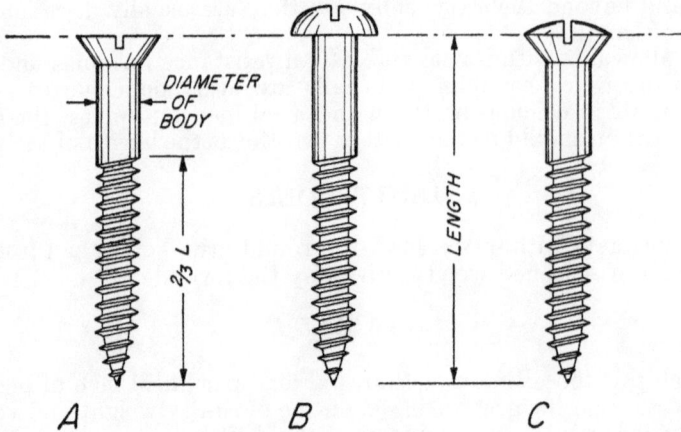

ZM 86795 F

FIGURE 29.—Common types of wood screws: *A*, Flathead; *B*, roundhead; and *C*, ovalhead.

The allowable withdrawal load of common wood screws inserted into the side grain of seasoned wood may be expressed as follows:

$$p = 2{,}370 \ G^2D$$

in which p is the allowable withdrawal load (one-sixth of the ultimate load) per lineal inch of penetration of the threaded part of the screw; G, the specific gravity based on weight and volume when oven-dry; and D, the shank diameter of the screw in inches. Specific gravity values raised to the second power are given for a number of widely used wood species in table 27.

This equation is applicable when screw lead holes in softwoods have a diameter of about 70 percent of the root diameter of the threads, and in hardwoods, about 90 percent. The root diameter for most sizes of screws averages about two-thirds of the shank diameter. The equation may be applied to all species, but inherent characteristics may cause some species to give values 10 to 15 percent above or below the equation values, and an occasional species may vary somewhat more.

The equation values are applicable to the following sizes of screws:

Screw length (inches):	*Gage limits*	Screw length (inches)—Con.	*Gage limits*
¼	1–6	2	7–16
¾	2–11	2½	9–18
1	3–12	3	12–20
1½	5–14		

For lengths and gages outside of these limits, the actual values are likely to be less than the equation values. The withdrawal loads of screws inserted into the end grain of a piece are more or less erratic, but, when splitting is avoided, they should average 75 percent of the load sustained by screws inserted into the side grain.

Lubricating the surface of a screw is recommended to facilitate insertion, especially in the dense woods, because it will have little effect on ultimate withdrawal resistance.

Lateral Resistance

The allowable load in lateral resistance for wood screws in the side grain of seasoned wood is given by the formula (*3, 9*)

$$p = KD^2$$

in which p is the allowable lateral load in pounds; D, the diameter of the screw shank in inches; and K, a constant depending on the inherent characteristics of the wood species. Values for converting screw gage into D^2 are given in the following tabulation:

Screw number or gage:	D	D^2	Screw number or gage—Con.	D	D^2
0	0.060	0.0036	9	0.177	0.0313
1	.073	.0053	10	.190	.0361
2	.086	.0074	11	.203	.0412
3	.099	.0098	12	.216	.0467
4	.112	.0125	14	.242	.0586
5	.125	.0156	16	.268	.0718
6	.138	.0190	18	.294	.0864
7	.151	.0228	20	.320	.1024
8	.164	.0269	24	.372	.1384

Values for K for a number of the common species are given in table 27. These values for K were obtained by dividing the constants in general equations for proportional limit loads by a reduction factor of 1.6. They apply to wood at about 15 percent moisture content. Loads computed by substituting these constants in the equation are expected to allow a slip of from 0.007 to 0.01 inch, depending somewhat on the species and quality of the wood.

The formula applies when the depth of penetration of the screw into the block receiving the point is not less than seven times the shank diameter and when the cleat and the block holding the point are approximately of the same density. This depth of penetration gives an ultimate load of about six times the recommended allowable load. For a depth of penetration of less than seven times the shank diameter, the ultimate load is reduced about in proportion to the reduction in penetration, and the load at the proportional limit is reduced somewhat less rapidly. When the depth of penetration of the screw in the holding block is 4 times the shank diameter, the maximum load will be less than 4 times the load expressed by the formula, and the proportional-limit load will be approximately equal to that given by the formula. When the screw holds metal to wood, the allowable load can be increased by about 25 percent.

In tests at the Forest Products Laboratory to determine the lateral resistance of screws, the part of the lead hole receiving the shank, for hardwoods such as oak, had the same diameter as the shank, and that receiving the threaded part had the same diameter as the root of the thread. For softwoods such as southern yellow pine and Douglas-fir, the lead hole receiving the shank was about seven-eighths the diameter of the shank and that for the threaded parts was about seven-eighths the diameter of the screw at the root of the thread.

A larger coefficient K can be used in the formula for the allowable lateral loads if it appears safe to allow larger loads and greater slips in the design.

Screws should always be turned in.   They should never be started or driven with a hammer, because this practice tears the wood fibers and injures the screw threads, seriously reducing the load-carrying capacity of the screw.

LAG SCREWS

Lag screws are commonly used because of their convenience, particularly where it would be difficult to fasten a bolt or where a nut on the surface would be objectionable. Lag screws range from about 0.2 to 1 inch in diameter and from 1 to 16 inches in length. The threaded part varies with the length and ranges from ¾ inch with the 1- and 1¼-inch screws to half the length for all lengths greater than 10 inches. The equations given herein for withdrawal and lateral loads are based on lag screws having an average yield point of about 45,000 pounds per square inch and an average tensile strength of 77,000 pounds per square inch.   For metal lag screws having greater or lower yield points and tensile strength, the withdrawal loads should be adjusted in proportion to the tensile strength and the lateral loads in proportion to the square root of the yield-point stresses.

Withdrawal Resistance

The allowable load in direct withdrawal of lag screws from seasoned wood may be computed from the equation:

$$p = 1,500 \ G^{3/2} \ D^{3/4}$$

in which p is the allowable withdrawal load (one-fifth of the ultimate load) in pounds per inch of penetration of the threaded part; D, the shank diameter in inches, and G, the specific gravity of the wood based on weight and volume when oven-dry (19).

Lag screws, like wood screws, require prebored holes of the proper size (fig. 30). The lead hole for the shank should be of the same diameter as that of the shank. The diameter of the lead hole for the threaded part varies with the density of the wood: For lightweight softwoods, such as the cedars and white pines, 40 to 70 percent of the shank diameter; for Douglas-fir and southern yellow pine, 60 to 75 percent; and for dense hardwoods, such as the oaks, 65 to 85 percent. The smaller percentage in each range applies to lag screws of the smaller diameters, and the larger percentage to lag screws of larger diameters.   Soap or similar lubricants should be used on the screws to facilitate turning, and lead holes slightly larger than those recommended for maximum efficiency should be used with lag screws of excessive length.

In determining the withdrawal resistance, the allowable tensile strength of the lag screw at the net (root) section should not be exceeded.   Penetration of the threaded part to a distance about 7 times

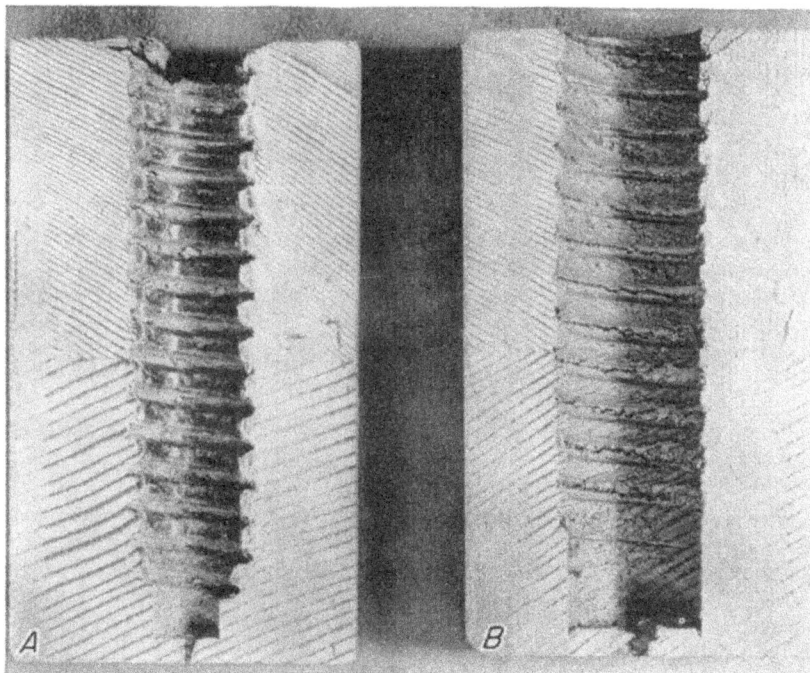

FIGURE 30.—*A*, Clean-cut, deep penetration of thread made by lag screw turned into a lead hole of proper size; *B*, rough shallow penetration of thread made by lag screw turned into oversized lead hole.

the shank diameter in the denser species and 10 to 12 times the shank diameter in the less dense species will develop approximately the ultimate tensile strength of the lag screw.

The resistance to withdrawal of a lag screw from the end-grain surface of a piece of wood is about three-fourths as great as its resistance to withdrawal from the side-grain surface of the same piece.

Lateral Resistance

The allowable lateral loads for lag screws inserted in the side grain and loaded parallel to the grain of a piece of seasoned wood can be computed from the equation:

$$p = KD^2$$

in which p is the allowable lateral load in pounds parallel to the grain; K, a constant depending on the species; and D, the shank diameter of the lag screw in inches (*19*). Values for K for a number of species can be found in table 27. The values given by this formula apply when the thickness of the attached member is 3.5 times the shank diameter of the lag screw, and the depth of penetration in the main member is 7 times the diameter in the harder woods and 11 times the diameter in the softer woods. For other thicknesses, the computed allowable loads should be multiplied by the following factors:

Ratio of thickness of member to shank diameter of lag screw	Factor	Ratio of thickness of member to shank diameter of lag screw	Factor
2	0. 62	4½	1. 13
2½	. 77	5	1. 18
3	. 93	5½	1. 21
3½	1. 00	6	1. 22
4	1. 07	6½	1. 22

When the lag screw is inserted into the side grain of wood and the load is applied perpendicular to the grain, the allowable load given by the lateral resistance formula should be multiplied by the following appropriate factors:

Shank diameter of lag screw (inches)	Factor	Shank diameter of lag screw (inches)	Factor
³⁄₁₆	1. 00	½	0. 65
¼	. 97	⅝	. 60
⁵⁄₁₆	. 85	¾	. 55
⅜	. 76	⅞	. 52
⁷⁄₁₆	. 70	1	. 50

For other angles of loading, the allowable loads may be computed from the parallel and perpendicular values by the use of the Scholten nomograph for determining the bearing strength of wood at various angles to the grain (fig. 31) (18). The nomograph provides values comparable to those given by the Hankinson formula:

$$N = \frac{PQ}{P \sin^2 \theta + Q \cos^2 \theta}$$

where P represents the allowable load or stress parallel to the grain; Q, the allowable load or stress perpendicular to the grain; and N, the allowable load or stress at an inclination θ with the direction of the grain.

Example: P, the allowable load parallel to grain is 6,000 pounds, and Q, the allowable load perpendicular to the grain is 2,000 pounds. N, the allowable load at an angle of 40° to grain is found as follows: Connect with a straight line 6,000 pounds (a) on line OX of the nomograph with the intersection (b) on line OY of a vertical line through 2,000 pounds. The point where this line (ab) intersects the line representing the given angle 40° is directly above the allowable load, 3,285 pounds.

Working values for lateral resistance as computed by the preceding methods are based on complete penetration of the shank into the attached member but not into the foundation member. When the shank penetrates the foundation member, the following increases in allowable loads are permitted:

Ratio of penetration of shank into foundation member to shank diameter	Increase in load (percent)	Ratio of penetration of shank into foundation member to shank diameter	Increase in load (percent)
1	8	5	36
2	17	6	38
3	26	7	39
4	33		

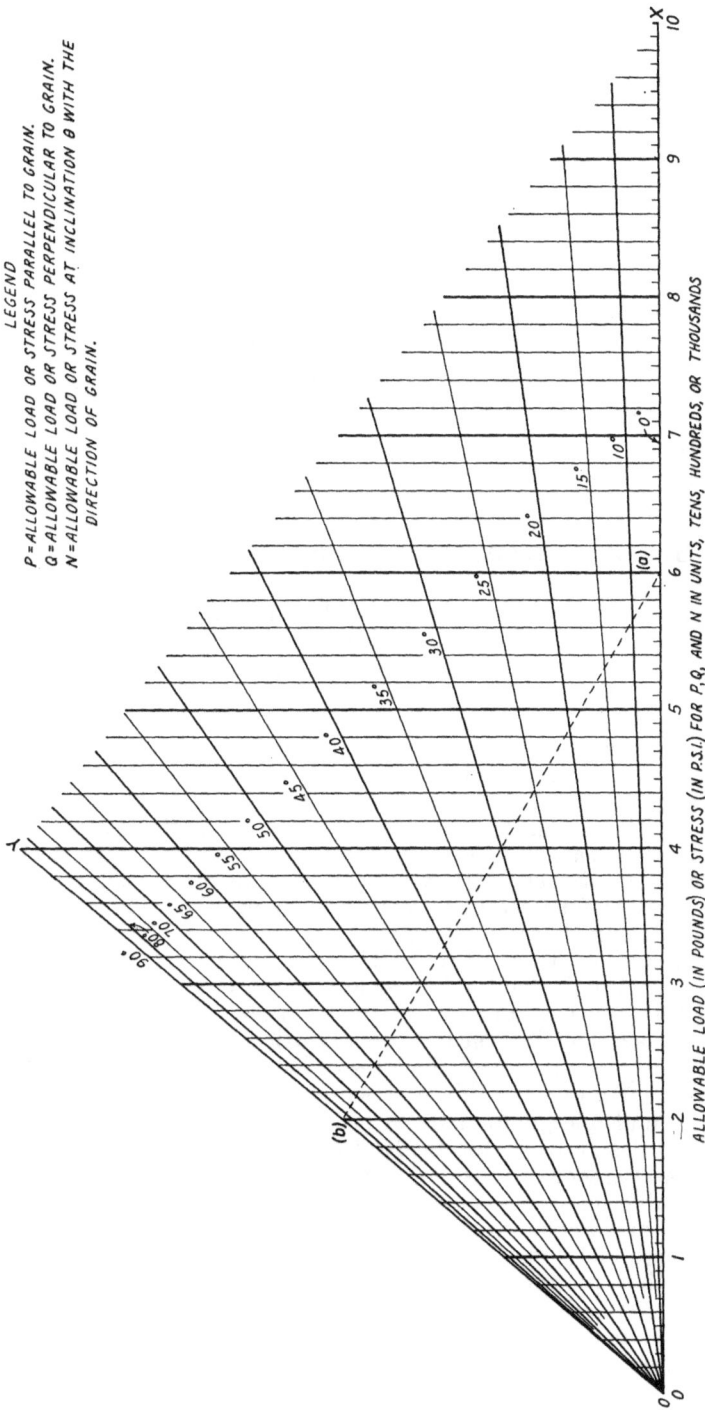

LEGEND

P=ALLOWABLE LOAD OR STRESS PARALLEL TO GRAIN.
Q=ALLOWABLE LOAD OR STRESS PERPENDICULAR TO GRAIN.
N=ALLOWABLE LOAD OR STRESS AT INCLINATION θ WITH THE
DIRECTION OF GRAIN.

ZM 92352 F

FIGURE 31.—Scholten nomograph for determining the bearing strength of wood at various angles to the grain. The dotted line *ab* refers to the example given in the text.

When lag screws are used with metal plates, the allowable lateral loads parallel to the grain may be increased 25 percent, but no increase should be made in the allowable loads when the applied load is perpendicular to the grain.

Lag screws should preferably not be driven into end grain, because splitting may develop under lateral load. If lag screws are so used, however, the allowable loads should be taken as two-thirds those for lateral resistance when lag screws are inserted into side grain and the loads act perpendicular to the grain.

The spacings, end and edge distances, and net section for lag screw joints should be the same as those for joints with bolts of a diameter equal to the shank diameter of the lag screw.

Lag screws should always be inserted by turning with a wrench, not by driving. Soap, beeswax, or other lubricants applied to the screw, particularly with the denser species, will facilitate insertion and prevent damage to the threads, but will not affect the lag screw's holding power.

BOLTS

Bearing Strength of Wood Under Bolts

Basic bolt-bearing stresses (table 27) for calculating the allowable loads for bolted joints acting parallel and perpendicular to the grain were developed from general strength data published by the Forest Products Laboratory (*24, 30*). The average ultimate crushing strength parallel to the grain of small, clear, green specimens was modified for softwoods and hardwoods for duration of loading, increase in strength with drying, variability of the wood, and a reduction factor for safety. The average proportional limit stress perpendicular to the grain of small, clear, green specimens was modified for increase in strength with seasoning, variability of the wood, and a reduction factor for safety. Special attention was given to the development of the basic stresses for certain species that show abnormal strength, splitting, or drying characteristics.

Basic stresses are listed in table 27 for a number of the common species and are applicable for bolted joints in any grade of seasoned lumber that is to be used in a dry, inside location. In locations where the timber is occasionally wet but quickly dried and in locations that are continuously damp or wet, the stresses listed in table 27 should be reduced one-fourth and one-third, respectively. The limited test data show that in multiple joints made with unseasoned members joined at right angles to each other, the loads after the members had seasoned were about 40 percent of those for joints made of seasoned members.

The allowable loads for correctly spaced and alined bolts bearing parallel or perpendicular to the grain of the wood and with the load applied through metal plates to both ends of the bolt are calculated by multiplying the appropriate basic stress (table 27) by a percentage factor (table 29) based on the ratio of the length of bolt in the main member to the bolt diameter (L/D). For loads acting perpendicular to the grain, an additional factor, referred to as a diameter factor, is applied to the basic stress values in accordance with the following bolt diameters:

Diameter of bolt (inches)	Diameter factor	Diameter of bolt (inches)	Diameter factor
1/4	2.50	1 1/4	1.19
3/8	1.95	1 1/2	1.14
1/2	1.68	1 3/4	1.10
5/8	1.52	2	1.07
3/4	1.41	2 1/2	1.03
7/8	1.33	3 or over	1
1	1.27		

TABLE 29.—*Percentage of basic stress used in calculating allowable bearing stresses for common bolts when load is applied through metal splice plates*

Ratio of bolt length to diameter (L/D)	Bolts bearing parallel to grain,[1] when basic stress is—			Bolts bearing perpendicular to grain,[2] when basic stress is—			
	750 to 950 p. s. i.	1,000 to 1,200 p. s. i.	1,300 to 2,000 p. s. i.	150 to 180 p. s. i.	200 to 280 p. s. i.	300 to 350 p. s. i.	400 to 600 p. s. i.
	Percent	*Percent*	*Percent*	*Percent*	*Percent*	*Percent*	*Percent*
1	100.0	100.0	100.0	100.0	100.0	100.0	100.0
2	100.0	100.0	100.0	100.0	100.0	100.0	100.0
3	100.0	100.0	99.0	100.0	100.0	100.0	100.0
4	99.5	97.4	92.5	100.0	100.0	100.0	100.0
5	95.4	88.3	80.0	100.0	100.0	100.0	100.0
6	85.6	75.8	67.2	100.0	100.0	100.0	96.3
7	73.4	65.0	57.6	100.0	100.0	97.3	86.9
8	64.2	56.9	50.4	100.0	96.1	88.1	75.0
9	57.1	50.6	44.8	94.6	86.3	76.7	64.6
10	51.4	45.5	40.3	85.0	76.2	67.2	55.4
11	46.7	41.4	36.6	76.1	67.6	59.3	48.4
12	42.8	37.9	33.6	68.6	61.0	52.0	42.5
13	39.5	35.0	31.0	62.2	55.3	45.9	37.5

[1] For wood splice plates, each of which is 1/2 the thickness of the main member, the allowable loads should be taken as 4/5 those computed for metal splice plates.

[2] No reduction need be made when wood splice plates are used except that the allowable load perpendicular to the grain should never exceed the allowable load parallel to the grain for any given size and quality of bolt and timber.

Allowable Bolt Loads Parallel to the Grain

The allowable load on a bolted timber joint acting parallel to the grain of the wood through metal plates at both ends of the bolt is calculated as follows:

(1) Select from table 27 the basic stress in compression parallel to the grain for the particular species of wood and reduce in accordance with the service condition if the joint is to be used in other than a dry, inside location. Call this stress S_1.

(2) Calculate the ratio of the length of the bolt in the main member to the bolt diameter (L/D). Select the percentage r of basic stress from table 29 for the calculated L/D.

(3) Multiply the basic stress S_1 by the percentage r to obtain the allowable unit stress, S_2, which is assumed to be uniformly distributed.

(4) Multiply S_2 by the projected area of the bolt to obtain the allowable load P_1 for one bolt.

When the load is applied through wood splice plates, each of which is one-half the thickness of the center member, the allowable load is obtained by multiplying P_1 by 0.8.

When the bolt holes are properly centered and alined, the allowable load on a number of bolts of the same or different diameters may be taken as the sum of the individual load capacities.

Allowable Bolt Loads Perpendicular to the Grain

The allowable load on a bolted timber joint acting perpendicular to the grain of the wood and at both ends of the bolt through metal plates or wood side plates, each of which is one-half the thickness of the main member, is calculated as follows:

(1) Select from table 27 the basic stress in compression perpendicular to the grain for the particular species of wood and reduce in accordance with the service condition if the joint is to be used in other than a dry, inside location. Call this stress S_1.

(2) Calculate the ratio of the length of bolt in the main member to the bolt diameter (L/D). Select the percentage r of basic stress from table 29 for the appropriate L/D.

(3) Select the diameter factor v for the appropriate size of bolt (p. 183).

(4) Multiply the basic stress S_1 by the percentage factor r and the diameter factor v to obtain the allowable average unit stress S_2 for the particular L/D.

(5) Multiply the allowable unit stress S_2 by the projected area of the bolt to obtain the allowable load for a single bolt loaded through either wood or metal splice plates.

When the bolt holes are properly centered and alined, the allowable load on a number of bolts of the same or different diameters may be taken as the sum of the individual load capacities.

The allowable bolt loads, both parallel and perpendicular to the grain, determined by the preceding procedure are of such magnitude as to prevent undue permanent distortion or other evident injury in the joint under long-term loading or repeated application of the load.

For loads acting at an angle between the limits of parallel to the grain and perpendicular to the grain, the allowable bolt load may be obtained from the nomograph in figure 31 (18).

Effect of Quality of Bolt on Joint Strength

Both the quality of the wood and the quality of the bolt are factors in determining the proportional limit strength of a bolted joint. The percentages of basic stresses given for calculating safe bearing stresses apply to the common commercial bolts used in building construction. For high-strength bolts, such as aircraft bolts, the reduction values listed in table 29 would be somewhat larger for the larger L/D ratios.

Details of Design

The details of design required in the application of the allowable loads for bolts may be summarized as follows:

(1) A load applied to only one end of a bolt, perpendicular to its axis, may be taken as one-half the symmetrical two-end load.

(2) The center-to-center distance along the grain between bolts acting parallel to the grain should be at least four times the bolt diameter. When a joint is in tension, the bolt nearest the end of a timber should be at a distance from the end of at least 7 times the bolt diameter for softwoods and 5 times for hardwoods. When the joint is in compression, the end margin may be four times the bolt diameter for both softwoods and hardwoods. Any decrease in these spacings and margins will decrease the load in about the same ratio.

(3) For bolts bearing parallel to the grain, the distance from the edge of a timber to the center of a bolt should be at least 1.5 times the bolt diameter. This margin, however, will usually be controlled by (a) the common practice of having an edge margin equal to one-half the distance between bolt rows, and (b) the area requirements at the critical section. (The critical section is that section of the member, taken at right angles to the direction of load, which gives the maximum stress in the member, based on the net area remaining after reductions are made for bolt holes at that section.) For parallel-to-grain loading in softwoods, the net area remaining at the critical section should be at least 80 percent of the total area in bearing under all the bolts in the particular joint under consideration; in hardwoods it should be 100 percent.

(4) For bolts bearing perpendicular to the grain, the margin between the edge toward which the bolt pressure is acting and the center of the bolt or bolts nearest this edge should be at least four times the bolt diameter. The margin at the opposite edge is relatively unimportant. The minimum center-to-center spacing of bolts in the across-the-grain direction for loads acting through metal side plates need only be sufficient to permit the tightening of the nuts. For wood side plates, the spacing is controlled by the rules applying to loads acting parallel to grain if the design load approaches the bolt-bearing capacity of the side plates. When the design load is less than the bolt-bearing capacity of the side plates, the spacing may be reduced below that required to develop their maximum capacity.

Effect of Bolt Holes

The bearing strength of wood under bolts is affected considerably by the size and type of bolt hole into which the bolts are inserted. A bolt hole that is too large causes nonuniform bearing of the bolt, and, if the bolt hole is too small, the wood will split when the bolt is driven. Normally, bolts should fit neatly, so that they can be inserted by tapping lightly with a wood mallet. In general, the smoother the hole, the higher the bearing values will be (fig. 32). Deformations accompanying the load also increase with increase in the unevenness of the bolt-hole surface (fig. 33).

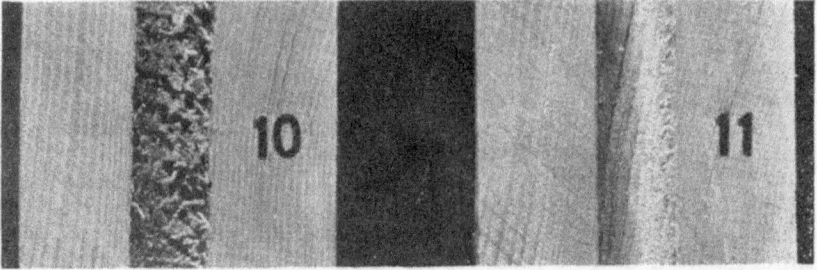

ZM 92351 F

FIGURE 32.—Effect of rate of feed and drill speed on the surface condition of bolt holes drilled in Sitka spruce. The hole on the left was bored with a twist drill rotating at a speed of 200 r. p. m.; the feed rate was 60 inches per minute. The hole on the right was bored with the same drill rotating at 800 r. p. m.; the feed rate was 2 inches per minute.

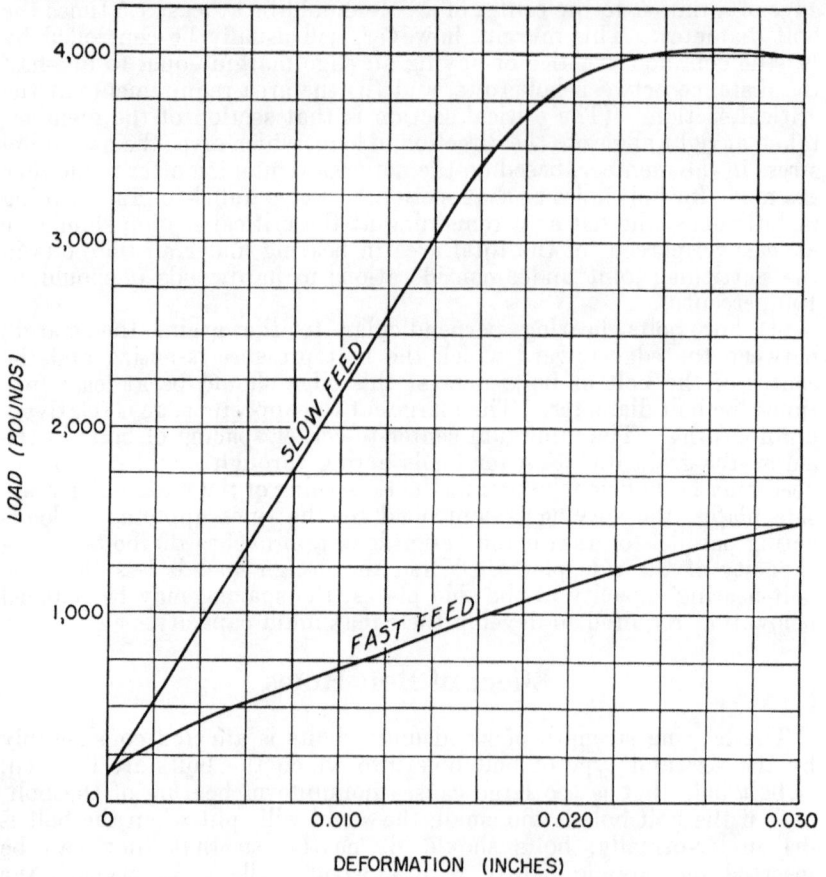

ZM 56698 F

FIGURE 33.—Typical load-deformation curves showing the effects of surface condition of bolt holes, resulting from a slow feed rate and a fast feed rate, on the deformation in a joint when subjected to loading under bolts. The surface conditions of the bolt holes were similar to those illustrated in figure 32.

Rough holes result from the use of dull bits and improper rates of feed and drill speed. A twist drill, preferably machine sharpened, at a 118° angle and operated at a recommended peripheral speed that corresponds to a speed of rotation of approximately 1,350 r. p. m. divided by the diameter of the drill in inches produces uniform, smooth holes. For carbon steel drills, the speed of rotation should be approximately one-half that used for high-speed drills. The rate of feed depends upon the diameter of the drill and the speed of rotation but should enable the drill to cut rather than tear the wood. The drill should produce shavings, not chips (8).

TIMBER-CONNECTOR JOINTS

When timbers are joined with metal or other types of connectors, the strength of the connector joint depends on the type and size of the connector, the species of wood, the thickness and width of the member, the distance of the connector from the end of the timber, the spacing of the connectors, the direction of application of the load with respect to the direction of the grain of the wood, and other factors (17, 21, 22, 23, 27). Safe working long-time loads for metal connectors in common use—split-ring (fig. 34), toothed (fig. 35), and shear-plate (fig. 36)—and for a wood connector are discussed in this section.

Derivation of Safe Working Loads for Long-Continued Loading

The working loads for the split-ring, shear-plate, and toothed connectors provided in tables 30, 31, and 32 were derived from the results of tests of full-scale joints made from the more important commercial species. The species were classified into four groups in accordance with their strength in connector joints. In establishing the working loads, particular consideration was given to (1) the effect of long-continued loading as against the brief loading period involved in the test of joints (see p. 160 for effect of duration of load on working stresses) and (2) allowance for variability in timber quality (1).

Since adequate data are not available on the effect of duration of stress on the strength of connector joints, insofar as the wood is considered, the relation between the load at failure in a standard bending test of a few minutes' duration and the load that will cause failure under long-time loading was assumed to apply. Under constant load a beam will fail at a load only about nine-sixteenths as great as the breaking load found in the standard bending test (1, 15).

Tests have demonstrated that the density or quality of the wood is often the controlling factor in determining the strength of a joint. Consequently, the load carried by a connector in the Laboratory test employing wood of average quality for a species was adjusted to allow for the lower than average material likely to be used in service.

The working loads listed in tables 30, 31, and 32 for connectors acting parallel to the grain were derived by applying a reduction fac-

M 93396 F

FIGURE 34.—Split-ring connector assembly—connector, precut groove, bolt, washer, and nut.

M 32890 F

FIGURE 35.—Toothed connector assembly.

FIGURE 36.—*A*, Shear-plate connector assembly with wood side plates; *B*, shear-plate connector assembly with steel side plates.

tor to the ultimate load, as found in tests. For split-ring and shear-plate connectors, a reduction factor of 4 gave values that met the provision that working loads for these connectors should not exceed five-eighths of the proportional limit test load. Because load-slip curves for the toothed connector do not exhibit a well-defined propor-

TABLE 30.—Safe working loads¹ for 1 split-ring connector and bolt (1 connector unit)

Connector unit	Minimum thickness of member		Minimum width all members	Group 1 woods²		Group 2 woods²		Group 3 woods²		Group 4 woods²	
	With 1 connector only	With 2 connectors in opposite faces 1 bolt³		Load at 0° angle to grain	Load at 90° angle to grain	Load at 0° angle to grain	Load at 90° angle to grain	Load at 0° angle to grain	Load at 90° angle to grain	Load at 0° angle to grain	Load at 90° angle to grain
(1)	(2)	(3)	(4)	(5)	(6)	(7)	(8)	(9)	(10)	(11)	(12)
	In.	*In.*	*In.*	*Lb.*	*Lb.*	*Lb.*	*Lb.*	*Lb.*	*Lb.*	*Lb.*	*Lb.*
2½-inch connector and ½-inch bolt	1	1⅝	3⅝	1,490	880	1,735	1,025	2,065	1,230	2,395	1,435
	1⅛+	2+	3⅝	1,785	1,055	2,085	1,230	2,480	1,475	2,875	1,725
	1⅜	1⅝	5½	2,415	1,400	2,795	1,620	3,355	1,945	3,915	2,270
4-inch connector and ¾-inch bolt	1¼	2	5½	2,775	1,610	3,215	1,865	3,860	2,240	4,500	2,610
	1½	2⅝	5½	3,380	1,960	3,915	2,270	4,695	2,725	5,480	3,175
	1⅝+	3+	5½	3,445	1,995	3,985	2,310	4,780	2,775	5,580	3,235

¹ The safe working loads apply to seasoned timbers in dry, inside locations for a long-continued load. It is assumed also that the joints are properly designed with respect to such features as centering of connectors, adequate end distance, and suitable spacing.

² Group 1 woods provide the weakest connector joints, and group 4 woods the strongest; species included in the groups are given in table 27.

³ A 3-member assembly with 2 connector units would therefore take double the safe working loads indicated in columns 5–12.

TABLE 31.—*Safe working loads* [1] *for 1 toothed connector and bolt (1 connector unit)*

Connector unit	Minimum thickness of member		Minimum width all members	Group 1 woods [2]		Group 2 woods [2]		Group 3 woods [2]		Group 4 woods [2]	
	With 1 connector only	With 2 connectors in opposite faces 1 bolt [3]		Load at 0° angle to grain	Load at 45° to 90° angle to grain	Load at 0° angle to grain	Load at 45° to 90° angle to grain	Load at 0° angle to grain	Load at 45° to 90° angle to grain	Load at 0° angle to grain	Load at 45° to 90° angle to grain
(1)	(2)	(3)	(4)	(5)	(6)	(7)	(8)	(9)	(10)	(11)	(12)
	In.	*In.*	*In.*	*Lb.*	*Lb.*	*Lb.*	*Lb.*	*Lb.*	*Lb.*	*Lb.*	*Lb.*
2-inch connector with ½-inch bolt	1	1⅝	2⅝	780	520	900	600	1,000	665	1,100	735
	1⅛+	2+	2⅝	860	570	990	660	1,100	735	1,210	805
2⅝-inch connector with ⅝-inch bolt	1	1⅝	3⅝	1,170	780	1,350	900	1,500	1,000	1,650	1,100
	1⅛	2	3⅝	1,295	865	1,495	995	1,660	1,105	1,825	1,220
	1⅜+	2½+	3⅝	1,460	975	1,690	1,125	1,875	1,250	2,060	1,375
3⅜-inch connector with ¾-inch bolt	1	1⅝	4⅝	1,520	1,015	1,755	1,170	1,950	1,300	2,145	1,430
	1⅛	2	4⅝	1,665	1,110	1,920	1,280	2,135	1,425	2,350	1,565
	1½	2⅝	4⅝	1,910	1,270	2,200	1,465	2,445	1,630	2,690	1,795
	1⅝+	3+	4⅝	2,055	1,370	2,370	1,580	2,630	1,755	2,895	1,930
4-inch connector with ¾-inch bolt	1	1⅝	5½	1,835	1,220	2,115	1,410	2,350	1,565	2,585	1,725
	1⅛	2	5½	1,985	1,320	2,290	1,525	2,540	1,695	2,795	1,865
	1½	2⅝	5½	2,235	1,490	2,575	1,720	2,865	1,910	3,150	2,100
	1⅝+	3+	5½	2,385	1,590	2,750	1,835	3,055	2,035	3,360	2,240

[1] The safe working loads apply to seasoned timbers in dry, inside locations for a long-continued load. It is assumed also that the joints are properly designed with respect to such features as centering of connectors, adequate end distance, and suitable spacing.

[2] Group 1 woods provide the weakest connector joints, and group 4 woods the strongest; species included in the groups are given in table 27.

[3] A 3-member assembly with 2 connector units would therefore take double the safe working loads indicated in columns 5–12.

TABLE 32.—*Safe working loads*[1] *for 1 shear-plate connector and bolt (1 connector unit)*

Connector unit	Minimum thickness of member		Minimum width all members	Group 1 woods[2]		Group 2 woods[2]		Group 3 woods[2]		Group 4 woods[2]	
	With 1 connector only	With 2 connectors in opposite faces 1 bolt[3]		Load at 0° angle to grain	Load at 90° angle to grain	Load at 0° angle to grain	Load at 90° angle to grain	Load at 0° angle to grain	Load at 90° angle to grain	Load at 0° angle to grain	Load at 90° angle to grain
(1)	(2)	(3)	(4)	(5)	(6)	(7)	(8)	(9)	(10)	(11)	(12)
	In.	*In.*	*In.*	*Lb.*	*Lb.*	*Lb.*	*Lb.*	*Lb.*	*Lb.*	*Lb.*	*Lb.*
	----------	1⅝	3½	1,470	852	1,704	988	2,045	1,186	2,386	1,384
2⅝-inch connector with ¾-inch bolt	----------	2	3½	1,785	1,035	2,069	1,230	2,483	1,440	2,667	1,680
	1⅝+	2⅝+	3½	1,889	1,096	2,191	1,271	2,629	1,525	2,667	1,779
	----------	1¾	5½	1,901	1,102	2,204	1,278	2,645	1,533	3,086	1,789
	----------	2	5½	2,120	1,229	2,458	1,426	2,950	1,711	3,442	1,996
4-inch connector with ¾-inch bolt	----------	2⅝	5½	2,486	1,442	2,882	1,672	3,458	2,006	4,035	2,340
	----------	1⅝	5½	2,660	1,540	3,080	1,790	3,700	2,150	4,320	2,510
	----------	3	5½	2,705	1,568	3,137	1,819	3,763	2,182	4,391	2,547
	1¾+	3⅝+	5½	2,851	1,653	3,306	1,917	3,967	2,301	4,625	2,684
	----------	1¾	5½	1,901	1,105	2,204	1,278	2,645	1,533	3,086	1,787
	----------	2	5½	2,120	1,229	2,458	1,426	2,950	1,711	3,442	1,996
4-inch connector with ⅞-inch bolt	----------	2⅝	5½	2,486	1,442	2,882	1,672	3,458	2,006	4,035	2,340
	----------	1⅝	5½	2,660	1,540	3,080	1,790	3,700	2,150	4,320	2,510
	----------	3	5½	2,705	1,568	3,137	1,819	3,763	2,182	4,391	2,547
	1¾+	3⅝+	5½	2,851	1,653	3,306	1,917	3,967	2,301	4,625	2,684

[1] The safe working loads apply to seasoned timbers in dry, inside locations for a long-continued load. It is assumed also that the joints are properly designed with respect to such features as centering of connectors, adequate end distance, and suitable spacing. The tabulated loads apply for metal or wood side plates except that, for 4-inch shear plates with metal side plates, the parallel-to-grain loads may be increased 5, 11, and 18 percent for groups 2, 3, and 4 woods, respectively. The allowable loads for all loading except wind shall not exceed 2,900 pounds for 2⅝-inch shear plates or 4,970 pounds and 6,760 pounds for 4-inch shear plates with ¾- and ⅞-inch bolts, respectively; for wind loading, the corresponding allowable loads shall not exceed 3,870 pounds, 6,630 pounds, and 9,020 pounds.

[2] Group 1 woods provide the weakest connector joints, and group 4 woods the strongest; species included in the groups are given in table 27.

[3] A 3-member assembly with 2 connector units would therefore take double the safe working loads indicated in columns 5–12.

tional limit, no factor based on proportional limit loads was considered. For this connector, the larger reduction factor of 4½ was applied to the ultimate load.

Tests of connectors under loads bearing perpendicular to grain, although less extensive and less numerous than those for parallel bearing, have been sufficient to establish a generally applicable relationship between the two directions. This relationship was used in deriving allowable loads for perpendicular bearing. Ultimate load was given less consideration for perpendicular than for parallel bearing, and greater dependence was placed on other factors, such as the load at proportional limit and at given slips of the joint.

The figures quoted as the ratios between working loads and the loads found in test are in no instance true factors of safety. For example, the reduction factor of 4 for split-ring and shear-plate connectors includes allowances for duration of stress and for variability as well as a margin for safety. Thus, after the values from test are multiplied by a factor of nine-sixteenths as an allowance for a long-continued load and by three-fourths to cover variability of the wood, the actual factor of safety for a connector joint is on the order of 1¾ (4 x $\frac{9}{16}$ x ¾=1$\frac{11}{16}$) if the working load acts over a long period. The tests from which working loads were derived were on specimens carefully made from seasoned material, under favorable conditions, and by experienced workmen (23).

In tabulating the safe working load for a connector joint of any number of members, the unit is one connector with a bolt in shear. For any joint assembly in which more than one connector unit is used in the contact faces with the same bolt axis, the total safe working load is the sum of the safe working loads of each connector unit. For example, in tables 30, 31, and 32, minimum actual thickness of the members is given for a joint assembly of 3 members employing 2 connectors in opposite faces with a common bolt; this assembly is equivalent to 2 connector units, and, therefore, the safe working load will be twice the corresponding value shown for a 1-connector assembly (columns 5 to 12). The loads as given apply only when the joints are properly designed with respect to such features as centering of connectors, adequate end distances, and suitable spacing of connectors.

Modification of Working Loads and Factors To Be Considered in Their Use

Some of the factors that affect the safe working loads of connectors were included in deriving the tabular values. Others require modification of the values listed in accordance with the provisions outlined in the following paragraphs.

Wind or Earthquake Loads

In designing for wind or earthquake forces acting alone, or acting in conjunction with dead and live loads, the safe working loads for the various connectors may be increased by the following percentages, provided the number and size of connectors is not less than that required for the combination of dead and live loads alone:

Increase (percent) [1]

Split-ring connector, any size, bearing in any direction	50
Shear-plate connector, any size, bearing parallel to grain	33½
Shear-plate connector, any size, bearing perpendicular to grain	50
Toothed-ring connector, 2-inch, bearing in any direction	50
Toothed-ring connector, 4-inch, bearing in any direction	25

[1] Percentages for shear-plate connectors bearing at intermediate angles and for toothed-ring connectors of other sizes can be obtained by interpolation.

Impact Forces

Impact may be disregarded up to the following percentage of the static effect of the live load producing the impact:

Impact allowances (percent) [1]

Split-ring connector, any size, bearing in any direction	100
Shear-plate connector, any size, bearing parallel to grain	66⅔
Shear-plate connector, any size, bearing perpendicular to grain	100
Toothed-ring connector, 2-inch, bearing in any direction	100
Toothed-ring connector, 4-inch, bearing in any direction	50

[1] Percentages for shear-plate connectors bearing at intermediate angles and for toothed-ring connectors of other sizes may be obtained by interpolation.

One-half of any impact load that remains after disregarding the percentages indicated should be included with the other dead and live loads in obtaining the total force to be considered in designing the joint.

Factor of Safety Not Reduced

The procedures described for increasing the allowable loads on connectors for forces suddenly applied and forces of short duration do not reduce the actual factor of safety of the joint but are recommended because of the favorable behavior of wood under such forces. The differentiation among types and sizes of connector and directions of bearing is due to variations in the extent to which distortion of the metal, as well as the strength of the wood, affects the ultimate strength of the joint.

Special Design Considerations

Conditions of design may be encountered, with respect to the kind of load on a structure and the period of its continuation, that are neither "long continued" nor "suddenly applied." These conditions require or justify special consideration and possible modifications, other than those that have been indicated, of the working loads listed in tables 30 to 32. For such conditions, it may be assumed that 90 and 80 percent of the stress that causes failure in 5 minutes (time usually assumed for wind load) will cause failure in 50 minutes and 10 hours, respectively (see p. 160 for effect of duration of load on working stresses). Special considerations are also involved when connectors are used in plywood (*27*).

Exposure and Moisture Condition of Wood

The loads listed in tables 30, 31, and 32 apply to seasoned timbers used where they will remain dry. If the timbers will be more or less continuously damp or wet in use, two-thirds of the tabulated values should be used. The amount by which the loads should be reduced to adapt them to other conditions of use depends upon the extent to which the exposure favors decay, the required life of the structure or part, the frequency and thoroughness of inspection, the original cost and the cost of replacements, the proportion of sapwood and the durability of the heartwood of the species if untreated, and the character and efficiency of the treatment if treated. These factors should be evaluated for each individual design. Industry recommendations for the use of connectors when the condition of the lumber is other than continuously wet or continsously dry are given in National Design Specification for Stress-Grade Lumber and Its Fastenings (*17*).

Ordinarily, before fabrication of connector joints, timbers should be seasoned to a moisture content corresponding as nearly as practical to that which they will attain in service. This is particularly desirable for timber for roof trusses and other structural units used in dry locations and in which shrinkage is an important factor. Urgent needs for some types of construction sometimes result in the erection of many timber-connector structures and structural units employing green or inadequately seasoned lumber. Since such lumber subsequently dries out in most buildings, causing shrinkage and opening the joints, it is essential that adequate maintenance measures be adopted. The maintenance for connector joints in green lumber should include inspection of the structural units and tightening of all bolts as needed during the time the units are coming to moisture equilibrium, which is normally during the first year. Frequently, the first inspection should be made within 3 to 6 months after erection.

Grade and Quality of Lumber

The timber for which the working loads for connectors are applicable should conform to the general requirements in regard to the quality of structural lumber given in the grading rule books of lumber manufacturers' associations for the various commercial species.

With the recommended safe loads for connectors, it is assumed that the wood at the joints is clear and relatively free from checks, shakes, and splits. The wood should be either free from knots, or, if knots are assumed to be present in the longitudinal projection of the net section within a length from the critical section of one-half the diameter of the connector, the area of the knots should be subtracted from the area of the critical section. It is also assumed that cross grain at the joint does not exceed a slope of 1 in 10.

Loads at an Angle With the Grain of Wood

The safe working loads for the split-ring and shear-plate connectors for intervening angles of 0° to 90° between direction of load and grain may be obtained by the formula given under the heading Compressive

Strength on Surfaces Inclined to the Grain on page 80 or by the nomograph in figure 31 (*18*). With the toothed connectors, the safe working load at an inclination to the grain of 0° to 45° may be obtained with the previously mentioned formula; but from 45° to 90° it is equal to the working load perpendicular to the grain.

Thickness of Member

The relationship between the loads for the different thicknesses of lumber is based on test results for connector joints. The least thickness of member given in tables 30, 31, and 32 for the various sizes of connectors is, in general, the minimum that should be used. The loads listed for the greatest thickness of member in each type and size of connector unit are the maximum loads to be used for all thicker lumber. The loads for wood members of thicknesses intermediate to those listed can be obtained by direct interpolation.

Width of Member

The width of member listed for each type and size of connector is the minimum that should be used. When the connectors are bearing parallel to the grain, no increase in load occurs with an increase in width over the minimum. When they are bearing perpendicular to the grain, the load increases about 10 percent for each 1-inch increase in width of member over the minimum widths required for each type and size of connector, up to twice the diameter of the connectors. When the connector is placed off center and the load is applied continuously in one direction only, the proper working load can be determined by considering the width of member as equal to twice the edge distance (the distance between the center of the connector and the edge of the member toward which the load is acting), but the distance between the center of the connector and the opposite edge should not be less than one-half the permissible minimum width of the member.

End Distance and Spacing

The working load values in tables 30, 31, and 32 apply when the distance of the connector from the end of the member (end distance *e*) and the spacing (*s*) between connectors in multiple joints are not factors in the strength of the joint (fig. 37, *A*). When the end distance or spacing for connectors bearing parallel to the grain is less than that required to develop the full load, the proper reduced working load for design may be obtained by multiplying the working loads in tables 30, 31, and 32 by the appropriate strength ratio given in table 33. For example, the load for a 4-inch split-ring connector bearing parallel to the grain, when placed 7 or more inches from the end of a Douglas-fir tension member that is 1⅝ inches thick, is 4,780 pounds. When the end distance is only 5¼ inches, the strength ratio obtained by direct interpolation from the values given in table 33 is 0.81, and the load equals 0.81 times 4,780 or 3,870 pounds.

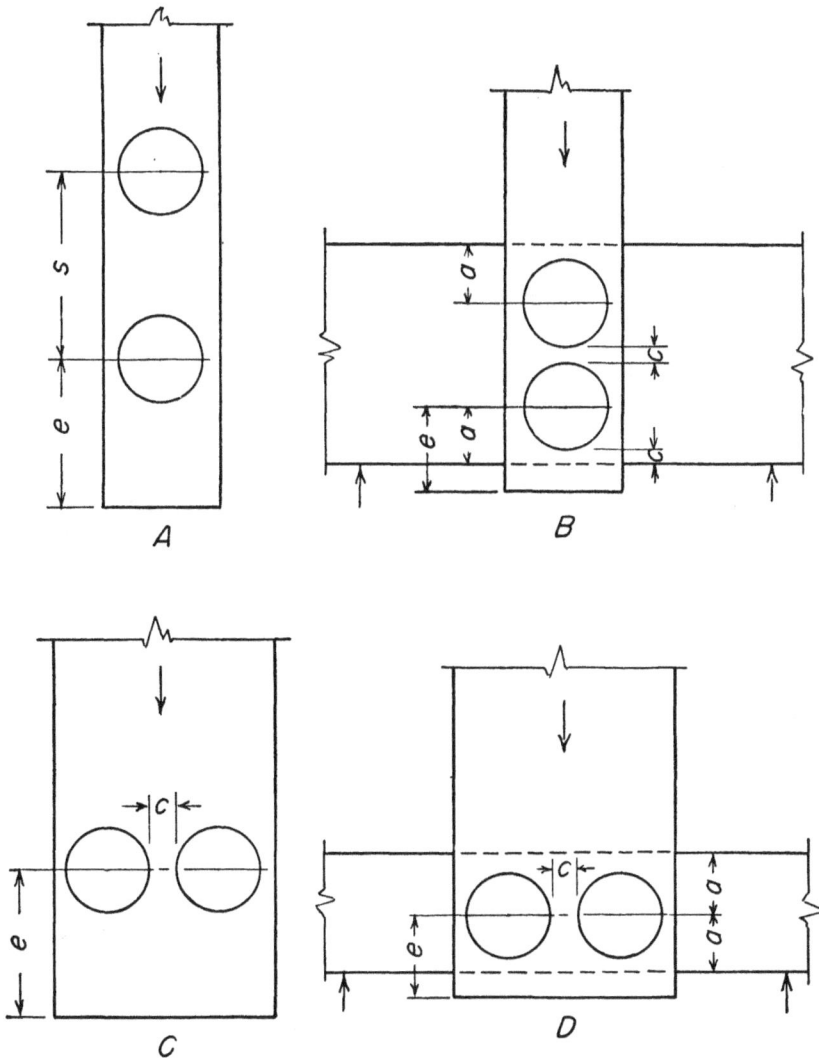

LEGEND:

e - END DISTANCE

s - SPACING PARALLEL TO GRAIN

a - EDGE DISTANCE

c - CLEAR DISTANCE

ZM 39253 F

FIGURE 37.—Types of multiple-connector joints: A, Joint strength dependent upon end distance e and connector spacing s; B, joint strength dependent upon end e, clear c, and edge a distances; C, joint strength dependent upon end e and clear c distances; D, joint strength dependent upon end e, clear c, and edge a distances.

Placement of Multiple Connectors

Preliminary investigations of the placement of connectors in a multiple joint, together with the observed behavior of single connector joints tested with variables that simulate those in a multiple joint, furnish a basis for some suggested design practices.

When two or more connectors in the same face of a member are in a line at right angles to the grain of the member and are bearing parallel to the grain (fig. 37, *C*), the clear distance (*c*) between the connectors should not be less than one-half inch.

When two or more connectors are acting perpendicular to the grain and are spaced on a line at right angles to the length of the member (fig. 37, *B*), the rules for the width of member and edge distances (p. 196) used with one connector are applicable to the edge distances for multiple connectors. The clear distance between the connectors (*c*) should be equal to the clear distance from the edge of the timber toward which the load is acting to the connector nearest this edge (*c*).

In a joint with two or more connectors spaced on a line parallel to the grain and with the load acting perpendicular to the grain (fig. 37, *D*), the available data indicate that the load for multiple connectors is not equal to the sum of the loads for individual connectors and that somewhat more favorable results are obtained in tests if the connectors are staggered so that they do not act along the same line

TABLE 33.—*Strength ratio for connectors for various longitudinal spacings and end distances* [1]

| Connector and diameter (inches) | Spacing [2] | Strength ratio | End distance [3] | | Strength ratio |
			Tension member	Compression member	
	Inches	*Percent*	*Inches*	*Inches*	*Percent*
Split-ring:					
2½	6¾+	100	5½+	4+	100
2½	3⅜	50	2¾	2½	62
4	9+	100	7+	5½+	100
4	4⅞	50	3½	3¼	62
Shear-plate:					
2⅝	6¾+	100	5½+	4+	100
2⅝	3⅜	50	2¾	2½	62
4	9+	100	7+	5½+	100
4	4½	50	3½	3¼	62
Toothed-ring:					
2	4+	100	3½+	2+	100
2	2	50	2	----------	67
2⅝	5¼+	100	4⅝+	2⅝+	100
2⅝	2⅝	50	2⅝	----------	67
3⅜	6¾+	100	5⅞+	3⅜+	100
3⅜	3⅜	50	3⅜	----------	67
4	8+	100	7+	4+	100
4	4	50	4	----------	67

[1] Strength ratio for spacings and end distances intermediate to those listed may be obtained by interpolation, and multiplied by the loads in tables 30, 31, and 32 to obtain design load. The strength ratio applies only to those connector units affected by the respective spacings or end distances. The spacings and end distances should not be less than the minimum shown.

[2] Spacing is distance from center to center of connectors (fig. 37, *A*).

[3] End distance is distance from center of connector to end of member (fig. 37, *A*).

with respect to the grain of the transverse member. Industry recommendations for various angle-to-the-grain loadings and spacings are given in reference (*17*).

The placement of connectors in joints with members at right angles to each other is subject to the minimum limitation of either member. It is virtually impossible to set up general rules regarding the alinement, spacing, and margin of connectors to cover all possible directions of the applied load. The designer must rely upon a sense of propor.ion and fitness in applying the rules set forth to a condition of loading that is within the limits discussed.

Cross Bolts

Cross bolts placed at or near the end of timbers joined with connectors, or at intermediate panel points will provide additional safety. They may also be used to reinforce members that have, through change in moisture content in service, developed checks to an undesirable degree.

Net Section

The stress in the net area (whether in tension or compression) of sawed timbers, which is the area remaining at the critical section after subtracting the projected area of the connectors and bolt from the full cross-sectional area of the member, should not exceed the safe stress of clear wood in compression parallel to the grain. In using this stress, it is assumed that knots do not occur in the longitudinal projection of the net section within a length of one-half the diameter of the connector from it.

In laminated timbers, knots may occur in the inner laminations at the connector location without being apparent from the outside of the member. It is impractical to assure that there are no knots at or near the connector. In laminated construction, therefore, the safe stress at the net section is limited to the safe compressive stress for the member, which takes account of the effect of knots.

Examples of Connector-Joint Design

(1) Calculate the safe working strength of a tension joint of seasoned coast-type Douglas-fir in which 2 pieces 3⅝ inches thick and 5½ inches wide are joined end to end by means of side plates 1⅝ inches thick, 5½ inches wide, and 28 inches long, when four 4-inch split-ring connectors and two ¾-inch bolts are used. In this arrangement, 2 connectors and a concentric bolt are placed symmetrically on either side of the butt joint at a distance of 7 inches from the ends of the members and side plates. This end distance, as shown in table 33, is adequate to develop the full design load.

The working load given in table 30 for one 4-inch split-ring connector, when used in 1 face of a Douglas-fir member 1⅝ inches thick or as 1 of 2 connectors used in opposite faces of a member 3 inches thick, is 4,780 pounds. The safe load of the joint for 2 connectors equals 2 times 4,780 or 9,560 pounds.

(2) Calculate the safe working strength of the joint in example (1) when the side plates are 16 inches instead of 28 inches in length. By

placing the connectors halfway between the ends of the sides plates and the butt joint, the end distance is 4 inches. The strength ratio as interpolated from values given in table 33 for a 4-inch end distance is 0.68, and the safe load accordingly equals 0.68 times 9,560 or 6,500 pounds.

(3) Calculate the safe working strength of a joint of seasoned southern yellow pine in which 2 tension side members 1⅝ inches thick and 5½ inches wide are joined at right angles to opposite faces of a center timber 3⅜ inches thick and 5½ inches wide by means of two 4-inch split-ring connectors and a ¾-inch bolt.

The load for 1 of two 4-inch split-ring connectors used in opposite faces of a member 3 inches thick and 5½ inches wide and bearing perpendicular to the grain is 2,775 pounds (table 30). The load for 1 connector bearing parallel to the grain in 1 face of a side member 1⅝ inches thick and with an end distance of 7 inches is 4,780 pounds (table 30). The safe load of the joint, which is governed by the load in the center member, equals 2 times 2,775 or 5,550 pounds.

(4) Calculate the safe working strength of the joint in example (3) when the distance from the end of the side plates overlapping the center member to the center of the bolt hole is 3½ instead of 7 inches.

The strength ratio for an end distance of 3½ inches is 0.62 (table 33). The load for one 4-inch split-ring connector in the side member, hence, equals 0.62 times 4,780 or 2,964 pounds. This is larger than the working load for one connector in the center member. The strength of the joint, therefore, is still governed by the load in the center member and, as before, is 5,550 pounds.

Other Connector Types

In addition to the connectors discussed, the Forest Products Laboratory made an exploratory investigation of several other types of connectors intended for use in building construction, such as the bulldog, claw-plate, circular-spike, and Kübler wood-dowel connector (20). Many of these connectors are of European origin. Some of the metal connectors are no longer manufactured, and others have a very limited use. A brief presentation is included here on the Kübler wood dowel in order to provide information on the load-carrying capacity of a wood connector.

The Kübler doubly coned dowels were originally made of cast iron. Later, oak was tried and found satisfactory in Europe. Two sizes were tested by the Laboratory, one approximately 2½ inches and the other approximately 4 inches in diameter. Bolts ½ inch in diameter were used with each size of dowel, and the bolt holes in the test specimens were bored ¹⁄₁₆-inch larger than the bolts.

The recommended working load for one Kübler wood dowel is as follows: 2½-inch dowel parallel to grain, 1,650 pounds; perpendicular to grain, 990 pounds; 4-inch dowel parallel to grain, 3,750 pounds; perpendicular to grain, 1,425 pounds. These loads are for oak dowels used in seasoned southern yellow pine or Douglas-fir (coast type). They can be safely used with all the species in groups 3 and 4 listed in table 27 and are based on a factor of 4 on ultimate loads of specimens tested at the Forest Products Laboratory. For group 1 woods use 80 percent and for group 2 use 90 percent of these values.

LITERATURE CITED

(1) AMERICAN SOCIETY FOR TESTING MATERIALS.
    1949. TENTATIVE METHODS FOR ESTABLISHING STRUCTURAL GRADES OF
        LUMBER. Amer. Soc. Testing Mater. Standard D245–49T, 27
        pp.
(2) BORKENHAGEN, E. H., AND KUELLING, H. J.
    1948. CLINCHING OF NAILS IN CONTAINER CONSTRUCTION. U. S. Forest
        Prod. Lab. Rept. R1777, 4 pp., illus. [Processed.]
(3) DEWELL, H. D.
    1917. TIMBER FRAMING. 275 pp., illus., San Francisco.
(4) FAIRCHILD, I. J.
    1926. HOLDING POWER OF WOOD SCREWS. U. S. Natl. Bur. Standards
        Technol. Paper 319, 28 pp., illus.
(5) GAHAGAN, J. M., AND BEGLINGER, E.
    1932. NEW NAIL-TREATING PROCESS INCREASES HOLDING POWER. U. S.
        Forest Prod. Lab. Rpt. R970, 3 pp. [Processed.]
(6) GIESE, H., AND HENDERSON, S. M.
    1947. EFFECTIVENESS OF ROOFING NAILS FOR APPLICATION OF METAL
        BUILDING SHEETS. Iowa Agr. Expt. Sta. Res. Bul. 355; 525–592,
        illus.
(7) ———— BODY, L. L., AND DALE, A. C.
    1950. EFFECT OF MOISTURE CONTENT OF WOOD ON WITHDRAWAL RESISTANCE
        OF ROOFING NAILS. Agr. Engin. 31 (4): 178–181, 183, illus.
(8) GOODELL, H. R., AND PHILIPPS, R. S.
    1944. BOLT-BEARING STRENGTH OF WOOD AND MODIFIED WOOD: EFFECT
        OF DIFFERENT METHODS OF DRILLING BOLT HOLES IN WOOD AND
        PLYWOOD. U. S. Forest Prod. Lab. Rpt. 1523, 10 pp., illus.
        [Processed.]
(9) KOLBERK, A., AND BIRNBAUM, M.
    1913. TRANSVERSE STRENGTH OF SCREWS IN WOOD. Cornell Civ. Engin.
        22: 31–41, illus.
(10) MARKWARDT, L. J.
    1952. HOW SURFACE CONDITION OF NAILS AFFECTS THEIR HOLDING POWER
        IN WOOD. U. S. Forest Prod. Lab. Rpt. D1927, 4 pp., illus.
        [Processed.]
(11) ———— AND GAHAGAN, J. M.
    1929. THE GROOVED NAIL. Packing and Shipping 56 (1): 12–14, illus.
(12) ———— AND GAHAGAN, J. M.
    1930. EFFECT OF NAIL POINTS ON RESISTANCE TO WITHDRAWAL. U. S.
        Forest Prod. Lab. Rpt. R1226, 3 pp., illus. [Processed.]
(13) ———— AND GAHAGAN, J. M.
    1931. MECHANISM OF NAIL HOLDING. Barrel and Box and Packages 36 (8):
        26–27, illus.
(14) ———— AND GAHAGAN, J. M.
    1952. SLANT DRIVING OF NAILS. DOES IT PAY? Packing and Shipping
        56 (10): 7–9, 23, 25, illus.
(15) ———— AND WILSON, T. R. C.
    1935. STRENGTH AND RELATED PROPERTIES OF WOODS GROWN IN THE
        UNITED STATES. U. S. Dept. Agr. Tech. Bul. 479, 99 pp., illus.
(16) MARTIN, T. J., AND VAN KLEECK, A.
    1941. FASTENING. U. S. Patent No. 2,268,323. U. S. Pat. Office, Off.
        Gaz. 533: 1226.
(17) NATIONAL LUMBER MANUFACTURERS ASSOCIATION.
    1953. NATIONAL DESIGN SPECIFICATION FOR STRESS-GRADE LUMBER AND
        ITS FASTENINGS. 66 pp., illus.
(18) NEWLIN, J. A.
    1939. BEARING STRENGTH OF WOOD AT ANGLE TO THE GRAIN. U. S.
        Forest Prod. Lab. Rpt. R1203, 1 p., illus. [Processed.]
(19) ———— AND GAHAGAN, J. M.
    1938. LAG SCREW JOINTS: THEIR BEHAVIOR AND DESIGN. U. S. Dept. Agr.
        Tech. Bul. 597, 26 pp., illus.
(20) PERKINS, N. S., LANDSEM, P., AND TRAYER, G. W.
    1933. MODERN CONNECTORS FOR TIMBER CONSTRUCTION. U. S. Dept.
        Com., Natl. Com. Wood Util., and U. S. Dept. Agr., Forest
        Service. 147 pp., illus.

(21) SCHOLTEN, J. A.
    1938. MODERN CONNECTORS IN WOOD CONSTRUCTION. Agr. Engin. 19
        (5): 201–203, illus.
(22) ———
    1940. CONNECTOR JOINTS IN WOOD CONSTRUCTION. Railway Purchases
        and Stores 33 (9): 431–435, illus.
(23) ———
    1944. TIMBER-CONNECTOR JOINTS, THEIR STRENGTH AND DESIGN. U. S.
        Dept. Agr. Tech. Bul. 865, 106 pp., illus.
(24) ———
    1946. STRENGTH OF BOLTED TIMBER JOINTS. U. S. Forest Prod. Lab.
        Rept. R1202, 2 pp., illus. (Revised.) [Processed.]
(25) ———
    1950. NAIL-HOLDING PROPERTIES OF SOUTHERN HARDWOODS. South.
        Lumberman 181 (2273): 208–210
(26) ———AND MOLANDER, E. G.
    1950. STRENGTH OF NAILED JOINTS IN FRAME WALLS. Agr. Engin.
        31 (11): 551–555.
(27) STERN, E. G.
    1940. A STUDY OF LUMBER AND PLYWOOD JOINTS WITH METAL SPLIT-RING
        CONNECTORS. Pa. Engin. Expt. Sta. Bul. 53, 90 pp., illus.
(28) ———
    1950. IMPROVED NAILS FOR BUILDING CONSTRUCTION. Va. Polytech.
        Inst., Engin. Expt. Sta. Bul. 76, 23 pp.
(29) ———
    1950. NAILS IN END-GRAIN LUMBER. Timber News and Machine Wood-
        worker 58 (2138): 490–492, illus.
(30) TRAYER, G. W.
    1932. BEARING STRENGTH OF WOOD UNDER BOLTS. U. S. Dept. Agr.
        Tech. Bul. 332, 40 pp., illus.
(31) U. S. FOREST PRODUCTS LABORATORY.
    1931. NAIL-HOLDING POWER OF AMERICAN WOODS. U. S. Forest Prod.
        Lab. Tech. Note 236, 4 pp., illus. [Processed.]
(32) ———
    1940. GENERAL OBSERVATIONS ON THE NAILING OF WOOD. U. S. Forest
        Prod. Lab. Tech. Note 243, 2 pp. [Processed.]
(33) ———
    1941. NAILING DENSE HARDWOODS. U. S. Forest Prod. Lab. Tech. Note
        247, 3 pp. [Processed.]
(34) U. S. HOUSING AND HOME FINANCE AGENCY.
    1947. TECHNIQUE OF HOUSE NAILING. 53 pp., illus.

SOLID AND BUILT-UP STRUCTURAL MEMBERS

Wood structures take many forms, and it is beyond the scope of this publication to treat each one in detail. This section deals only with fundamental considerations related to the simpler structural elements, such as beams and columns. Industry recommendations covering the design, fabrication, and erection of timber structures are discussed in Standards for Timber Construction (1).[14]

SOLID BEAMS OF RECTANGULAR CROSS SECTION

Wood beams are usually of rectangular section and of uniform depth throughout their span. They should be so designed that the following stresses do not exceed the allowable unit values: (1) The extreme fiber stress caused by flexural loads; (2) the maximum horizontal shear stress; and (3) the stress in compression across the grain at the end bearings. Beams should also be so designed that their deflection under load will not exceed the limit fixed by the intended use of the structure. If actual and nominal sizes of lumber differ, the allowable unit values must be used with the actual sizes.

Extreme Fiber Stress

Either the ultimate bending strength or the safe load for a wood beam of rectangular cross section may be calculated from the formula

$$M=\frac{Ffbd^2}{6}$$

where M is the bending moment in inch-pounds, F is a form or depth factor, f is the unit flexural stress in the extreme fiber in pounds per square inch, and b is the width and d is the depth of the beam in inches. If the ultimate bending strength is desired, f is the modulus of rupture. If the safe load is desired, f is the safe flexural stress for the kind and grade of material used. In either case, M must be expressed in terms of the load and the span in order to calculate the stress. For example, if a simply supported beam has a total load W uniformly distributed over a span L,

$$M=\frac{WL}{8}$$

If the beam has a concentrated load P at the center of the span,

$$M=\frac{PL}{4}$$

---

[14] Italic numbers in parentheses refer to Literature Cited, p. 230.

Depth Factor

A depth factor is introduced in the basic design formula for rectangular beams to take into account the somewhat lower unit strength developed in deep beams as compared to shallow beams. In arriving at the basic stress values for bending strength in table 25, a depth factor was assumed that corresponds approximately to a 12-inch depth of beam. In most designs of structural lumber beams of rectangular shape, if working stresses derived from the basic stress values of that table are used, no further adjustment for depth is necessary; that is, the basic formula can be used in the form

$$M = \frac{fbd^2}{6}$$

If the depth of a rectangular beam is more than 16 inches, an additional depth factor should be calculated by the formula of page 258. That formula is for use with the basic stress values of table 25 and therefore gives a factor of unity when the depth is 12 inches.

Combined Loading Effects

If flexural stress in the extreme fiber is combined with direct stress resulting from the application of end compression or tension, the combined stress is determined by algebraic addition of the acting stresses. For example, compressive stress is added to bending stress in the upper fiber of a beam and subtracted from bending stress in the lower fiber of the beam. Stresses so combined should not exceed allowable values given by the following formulas.

If a transversely loaded beam is also under end compression and is stayed against lateral buckling, the allowable stress is such that

$$\frac{P/A}{c} + \frac{M/S}{f} = 1$$

where P/A is the unit direct stress induced by axial load in compression parallel to grain (P is the axial compressive load and A is the area of the cross section), M/S is the flexural stress induced in bending (M is the bending moment and S is the section modulus, equal to $bd^2/6$ in a rectangular beam), c is the allowable working stress in compression parallel to grain, and f is the allowable working stress in bending.

As the end compression approaches zero, the term $\frac{P/A}{c}$ also approaches zero and the preceding equation approaches

$$\frac{M/S}{f} = 1$$

or

$$M = fS = \frac{fbd^2}{6}$$

the basic formula for flexural stress.

If the unit stress from end compression is large in comparison with the flexural stress and lateral buckling is possible, the beam should be designed as a column with side loads by the methods given on page 219.

If a transversely loaded beam is also in axial tension, the allowable stress is such that

$$\frac{P/A}{t}+\frac{M/S}{f}=1$$

where P/A is the unit direct stress induced by axial load in tension. t is the allowable working stress in tension parallel to grain, and M/S and f are as previously defined. In many cases t and f have the same value (table 25), and the equation becomes

$$P/A+M/S=t=f$$

Horizontal Shear in Beams

The general formula for horizontal shear in beams is

$$q=\frac{VQ}{It}$$

in which all units are in inches or pounds, q is the maximum horizontal shearing stress, Q is the statical moment of the area either above or below the neutral axis and about that axis (as used here, statical moment is a product of an area and the distance from the center of gravity of that area to an axis), V is the external or vertical shear, I is the moment of inertia of the section about the neutral axis, and t is the width of the beam at the neutral axis.

For a rectangular beam b inches wide and d inches deep, the general formula becomes

$$q=\frac{3V}{2bd}$$

Stress Concentration Effects

Shearing stress is increased at certain locations in a beam because of stress concentration effects. The most common concentration of stress is at the base of shakes, checks, or splits, where there is an abrupt change of cross section. This concentration causes shear failures in beams at relatively low stress values, even when calculated on the net cross section after deduction of the area of actual opening. Since checking occurs to some extent in practically all structural beams, a large factor for stress concentration was introduced in the basic shear stress values of table 25.

Another shear stress concentration may occur immediately adjacent to points of application of a concentrated load or reaction. Such a concentration is not common, because allowable bearing values in compression perpendicular to grain ordinarily require fairly large bearing areas; furthermore, building regulations usually specify minimum lengths of bearing that do not result in critical stress concentrations. Special cases involving applications of load or reaction to areas measuring one-half inch or less in length should be examined for stress concentration effects (*16*).

Shear in Checked Beams

Because the upper and lower parts of a beam checked near the neutral axis act partly as two beams and partly as a unit, a part of the end reaction is resisted internally by each half of the beam acting independently and consequently is not associated with shearing stress at the neutral plane (*12*). In using the basic formula for horizontal shear, in which *b* is the full width of the beam and *q* the allowable unit shear stress based on the assumption that checks are present, allowance for two-beam action should be made in calculating the allowable vertical shear *V*.

The following procedure is recommended for calculating the horizontal shear on the neutral plane in checked beams:

(1) Use the basic shear formula as given.

(2) Use for the allowable shearing stress a working stress value derived from the appropriate basic stress in table 25.

(3) In calculating the reactions represented by the vertical shear value *V* in the formula, (*a*) take into account any relief to the beam under consideration resulting from the load being distributed to adjacent parallel beams by flooring (p. 209) or other members of the construction, (*b*) neglect all loads within a distance from both supports equal to the height of the beam, (*c*) if there are any moving loads, place the largest one at a distance from the support equal to three times the height of the beam, and (*d*) treat all other loads in the usual manner. If a timber does not qualify for shear resistance under the foregoing procedure, which under certain conditions may be overconservative, the reactions for the concentrated loads should be more accurately determined by the following equation:

$$r' = \frac{10P' (L-x') \left(\frac{x'}{h}\right)^2}{9L \left[2 + \left(\frac{x'}{h}\right)^2\right]}$$

in which r' is the reaction to be used as due to a load P'; L, the span in inches; x', the distance in inches from the reaction to the load P'; and h, the height of the beam in inches.

Compression Across the Grain

The area of end bearing should be such that the unit stress in compression perpendicular to grain does not exceed the allowable value given in table 25. The stresses in compression perpendicular to grain given in that table apply to bearings of any length at the ends of a beam and to bearings 6 inches or more in length at any other place along the beam. They may be increased when applied to small plates and washers as explained on page 163. In calculating the bearing area at the ends of beams, no allowance need be made for the fact that, as the beam bends, the pressure upon the inside edge of the bearing is greater than that at the opposite edge. The wood yields enough so that the pressures become equalized before serious damage occurs.

Deflections of Beams

When service conditions require that beams be designed for stiffness, the dimensions are determined by applying the usual deflection formulas, the deflection being kept within a prescribed fraction of the span. The deflection is often limited to one three-hundred-sixtieth of the span of framing over plastered ceilings and over unplastered ceilings to one two-hundred-fortieth of the span. A ratio of one two-hundreth of the span has been used for highway bridges, but some engineers advocate a more severe limitation. A commonly recorded limitation for stringers in railroad bridges and trestles is one three-hundreth of the span. For floors that support shafting, deflection limitations much more severe than the preceding are sometimes required.

Wood beams usually sag in time; that is, the deflection increases beyond what it was immediately after the load was first applied. Green timbers, especially, will sag if allowed to season under load, although partially seasoned material will also sag to some extent. In thoroughly seasoned beams, there are small changes in deflection with changes in moisture content but little permanent increase in deflection.

Calculations from the values for modulus of elasticity in table 25 will give the initial deflection of a beam. If deflection under long-time load is to be limited, it is customary to design for an initial deflection of about one-half the value permitted for long-time deflection. This can be done by doubling any long-time loads when calculating deflection, by using one-half the values given in table 25 for modulus of elasticity, or by any equivalent method.

Notched Beams

Beams are often notched at the ends to improve clearance or to bring the top surfaces level with adjacent beams or girders (fig. 38). Occasionally they are also notched at intermediate points, either bottom or top, in order to clear other parts of a structure or to receive other members.

When notched on the lower side at the ends (fig. 38, A), the strength of a short, relatively deep beam is decreased by an amount depending on the shape of the notch and on the relation of the depth of the notch to the depth of the beam. It is recommended that in designing beams with square-cornered notches at the ends of the beams the desired bending load be checked against the load obtained by the equation

$$V = \frac{2}{3}\left(\frac{bd_e^2 q}{d}\right)$$

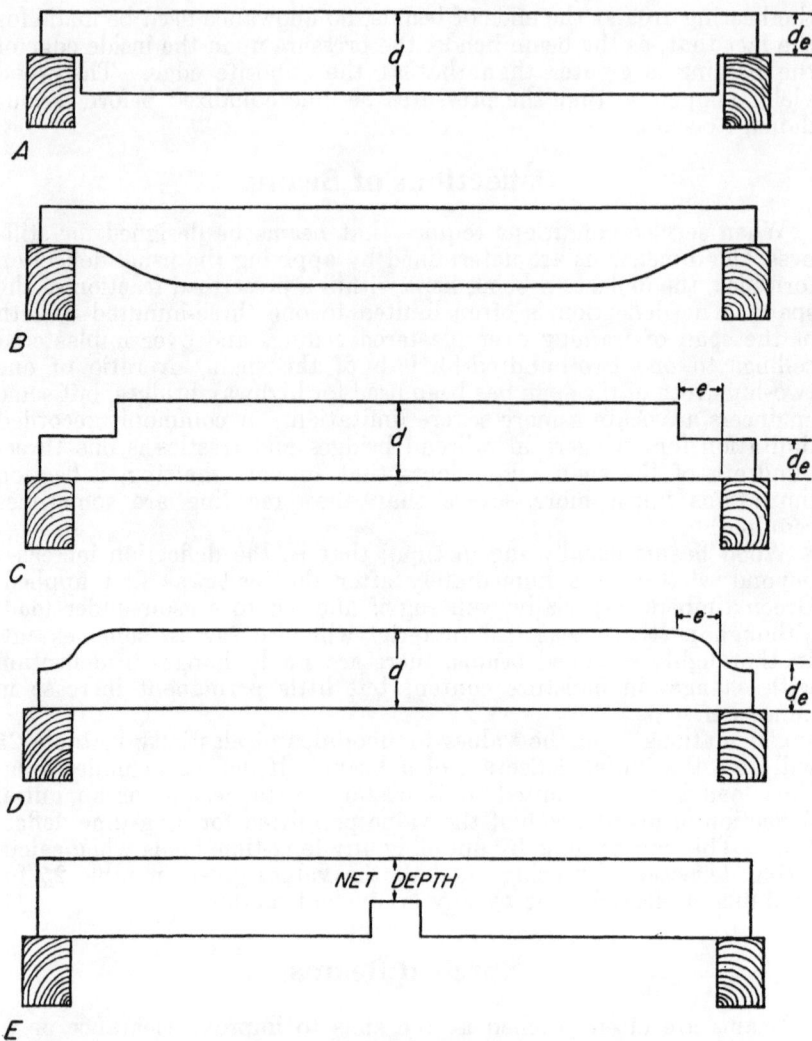

ZM 91876 F

FIGURE 38.—Notched beams: *A*, Notched at ends to one-half depth of beam; *B*, beam with gradual change in cross section at ends to eliminate concentration of stress; *C*, notched on the upper side; *D*, beveled on the upper side; *E*, notched to one-half depth of beam at center.

where V is the vertical shear; b, the width of the beam; d_e, the actual end depth above the notch; d, the total depth of the beam; and q, the working stress in horizontal shear.

In setting up this equation the actual end depth is used as the effective depth to resist shear, and the safe shear stress is reduced by multiplying it by the ratio of the effective depth to the total depth. Experiments made at the Forest Products Laboratory on short deep beams with relatively high shearing stresses substantiate this recommendation (*18*). The experiments also showed that by cutting away the lower corner at the end notches, in order to obtain a gradual change in cross section, the breaking loads were raised very markedly (fig. 38, *B*).

Beams are occasionally notched or beveled on the upper side (figs. 38, *C*, 38, *D*), and these beams present a less severe condition from the standpoint of stress concentration effects than beams notched or beveled on the lower side. Experiments with notches on the upper side indicate that shear stress should be checked with the formula

$$V = \frac{2}{3} qb \left[d - \left(\frac{d-d_e}{d_e} \right) e \right]$$

where V is the vertical or external shear; q is the working stress in shear; b is the width and d is the depth of the beam; d_e is the depth below the notch; and e is the distance that the notch extends inside the inner edge of the support. If e exceeds d, this formula is not used; rather, the shear strength is evaluated on the basis of the depth of the beam below the notch, d_e. For a beam with a bevel on the upper side (fig. 38, *D*) instead of a square notch, d_e is taken as the height of the beam at the inner edge of the support and e as the distance from the support to the start of the bevel. The depth of a notch on the upper side should in no case exceed 40 percent of the depth d of the beam.

When notches are at or near the middle of the length of a beam (fig. 38, *E*), the net depth should be used in determining the bending strength. Tests have shown that, as far as breaking loads are concerned, this rule is sufficiently conservative. The general tendency, however, of a notch on the top or bottom of a beam and near the point of maximum moment is to lower the proportional limit load and to start compression or tension failure at lower loads than would be expected of an unnotched beam of a depth equal to the net depth of the notched beam.

The stiffness of a beam is practically unaffected by notches.

Distribution of Concentrated Load Between Joists

Tests at the Forest Products Laboratory indicate that loads up to 400 pounds concentrated on an area of ⅓ square foot of a conventional dwelling floor system are distributed laterally for a considerable distance with only 20 to 30 percent carried by the joist directly under the load. The floor system in the tests consisted of joists, bridging, a diagonally laid subfloor, and a 1-inch finish floor laid across the joists. The load was applied at midspan.

The general problem of load distribution has been extensively investigated, particularly with reference to moving loads on bridge floor systems (*3*, *10*). Recommendations from those investigations indicate that about ¼ to ½ of a concentrated load on a joisted floor may be carried by the joist under the load. The proportion is higher when the loaded joist is near the side of the floor system than when it is at the center. Plank floors on beams that are spaced 5 or more feet apart show little or no distribution of load to beams other than the one directly under the load.

SOLID BEAMS OF CIRCULAR SECTION AND OF SQUARE SECTION WITH A DIAGONAL VERTICAL

Such wood members as poles and masts usually have circular cross sections, and sometimes ordinary beams and columns are of circular section. For architectural effects, a wood beam of square cross section is sometimes placed with a diagonal of the section vertical. In calculating the bending strength of such members, it is necessary to include a form factor in the beam formula.

The design formula for a beam of circular section is

$$M = \frac{Ff\pi r^3}{4}$$

where M is the bending moment, F is the form factor, f is the unit flexural stress in the outer fiber, and r is the radius of the cross section. The form factor F for a circular wood beam has a value of 1.18, and the formula then becomes

$$M = \frac{1.18\, f\pi r^3}{4} = 0.927\, fr^3$$

If circular wood beams have an appreciable taper, the maximum value for stress in the outer fiber under uniformly distributed load, as in a joist or rafter, occurs away from the center of the span toward the small end of the beam. The location of maximum stress varies with the taper, being at midspan with no taper and at a distance from the small end of 0.27 times the span if the small end has one-half the diameter of the large end. With a single concentrated load at midspan, the maximum stress occurs at midspan unless the small end has less than one-half the diameter of the large end.

Wood poles set in the ground are designed as vertical tapered cantilever beams. It can be shown that, if a horizontal force is applied at the top end of the pole, the maximum stress occurs where the diameter is 1½ times the top diameter. With forces uniformly distributed along the pole, the maximum stress occurs where the diameter is 3 times the top diameter. If application of either of these rules indicates that the maximum stress is below the ground line, the maximum stress will occur at the ground line.

The design formula for a beam of square cross section placed so that a diagonal of the section is vertical is

$$M = \frac{Ff a^3}{6\sqrt{2}}$$

where M is the bending moment, F is the form factor, f is the unit flexural stress in the outer fiber, and a is the side of the square.

The form factor F for a square beam with a diagonal vertical has a value of 1.414, and the formula then becomes

$$M = \frac{1.414\, f a^3}{6\sqrt{2}} = \frac{f a^3}{6}$$

It can be shown from the preceding formulas that a beam of square cross section has the same bending strength whether placed in the usual manner or with a diagonal of the section vertical and that a circular beam (without taper) has the same strength as a square beam of the same cross-sectional area.

BUILT-UP BEAMS WITH MECHANICAL FASTENINGS

Beams are sometimes built up from lumber of relatively small dimensions in order to (1) utilize small pieces of lumber, (2) reduce seasoning degrade, and (3) eliminate knots or other strength-reducing features. Two general methods are employed: (1) Several pieces of the same depth, but relatively thin, may be placed side by side and spiked or bolted together, either with or without intervening spaces; (2) two or more pieces of the same width may be placed on top of each other and bolted or otherwise fastened together with keys to transmit the shear stress, or fastened together by continuous sheathing on the sides, or glued.

The following discussion is restricted to built-up beams with mechanical fastenings. Glued laminated beams are discussed on pages 247–263.

Vertically Laminated Beams

Laminated beams with the laminations vertical and bolted together, in tests at the Forest Products Laboratory (4), were as strong and stiff as solid beams of the same external dimensions. The laminated beams tested consisted of 5 nominal 2- by 12-inch planks each 16 feet long fastened together side by side with bolts spaced as indicated in figure 39. They were practically free from such features as knots, but about one-half of the pieces contained seasoning checks. The solid beams were dense and of a high structural grade.

As a rule, the stiffer planks in a vertically laminated beam take a larger percentage of the load than those less stiff. First failure, however, usually occurs in the piece with the largest knot or other strength-reducing feature, regardless of its stiffness.

The planks must be well fastened together in order to prevent buckling of individual planks. If spikes are employed, through bolts or bolts and connectors should be provided. Spikes are seldom used for beams more than 10 inches in depth without through bolts. It is

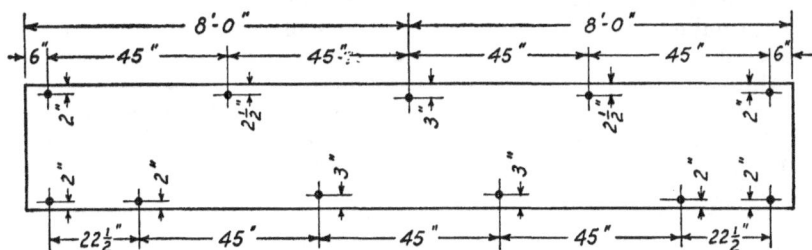

ZM 22052 F

FIGURE 39.—Bolt spacing of beams with vertical laminations used in tests at the Forest Products Laboratory.

good practice also to place two bolts at each end of the beam, as shown in figure 39.

A girder is sometimes built of one or more full-length members with narrow pieces nailed or spiked between or on the side of the members in a diagonal position, forming a flat inverted V (fig. 40, A). Tests show that narrow pieces so nailed add little or nothing to the strength or stiffness of the girder (5). If a girder consists of two full-length outside laminations with a full-width center lamination butt-jointed at the center of the span (fig. 40, B), the butt-jointed lamination adds to the girder about one-half as much strength and stiffness as a full-length lamination would add.

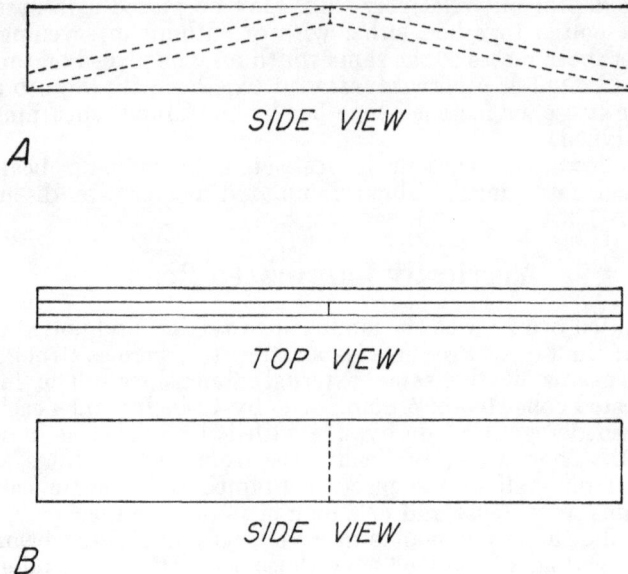

ZM 91877 F

FIGURE 40.—A, girders with narrow pieces nailed diagonally; B, a butt-jointed center lamination.

Sheathed Beams

A sheathed compound beam may be made of diagonal boards or planks nailed to each side of two timbers placed one on top of the other (fig. 41). The diagonal sheathing runs in opposite directions on the two sides. In tests (7), the efficiency of beams of this type for a span-depth ratio of 12 was about 70 percent and for a ratio of 24 about 80 percent as compared to the strength of solid beams of the same size. Deflections of the sheathed beams, however, were about double those of the solid beams.

Long before a beam broke in these tests, the diagonals split open or the nails were partly drawn out or bent over the wood, thereby permitting the two timbers to slide on each other. Beams of this type should not be used where the increased deflection would be objectionable.

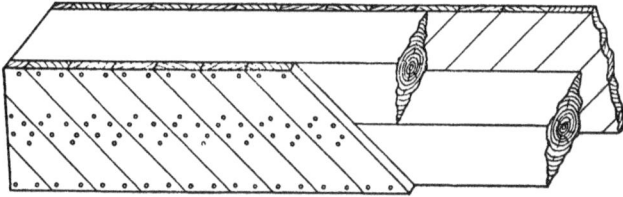

ZM 22050 F

FIGURE 41.—A sheathed compound beam composed of diagonal boards or planks nailed to each side of two timbers.

Keyed Beams

In a keyed compound beam the shear between adjacent timbers is taken by wood or metal keys, and the timbers are fastened together with bolts, as in figure 42. Kidwell (7) tested several types of keyed beams and recommended that for ordinary purposes an efficiency of 75 percent be allowed when oak keys are used and 80 percent when cast iron keys are used. He gave formulas for the number and spacing of keys. This information may also be found in books on timber framing (6). The keys are sometimes tapered like wedges, so that a driving fit may be obtained (fig. 42), but they are also made of flat plates or blocks without taper, sometimes known as tables.

Connectors of the toothed-plate, toothed-ring, or split-ring type (p. 187) are suitable for keys of compound beams. Tests at the Forest Products Laboratory indicated efficiencies up to 85 percent in strength for keyed beams with split-ring connectors spaced 6 inches along the beam length. Deflections were appreciably greater than for one-piece beams.

Well-seasoned timber should be used in the construction of keyed beams and great care exercised in framing.

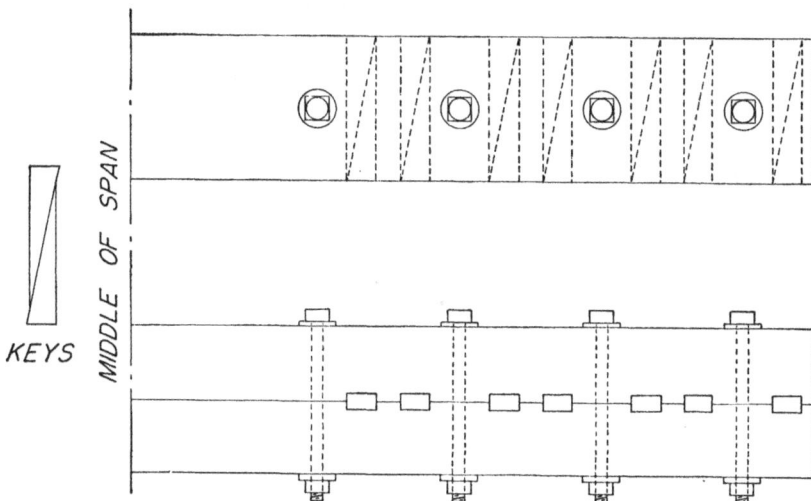

ZM 22075 F

FIGURE 42.—A keyed compound beam in which the shear between adjacent timbers is taken by wood or metal keys.

Composite Beams or Slabs

Composite beams or slabs can be made of lumber in combination with other structural materials (15). Bridge decks have been made with a base of laminated lumber covered with a concrete mat several inches thick. The lumber base is composed of alternating wide and narrow longitudinal planks, spiked, bolted, or doweled together and sometimes prefabricated in panels 2 to 4 feet wide. The wide planks are grooved and milled, with concrete filling the recesses to form keys that transmit horizontal shear and prevent uplift of the concrete from the wood. The concrete is lightly reinforced against temperature and shrinkage stresses, or, in continuous spans, may be reinforced for tensile stresses. The lumber base must be designed to carry the load of green concrete until the concrete has set and can contribute to the strength of the deck. The planks should have preservative treatment, since moisture may enter the joints and provide conditions favorable to decay. Figure 43 shows a typical cross section of such a composite deck.

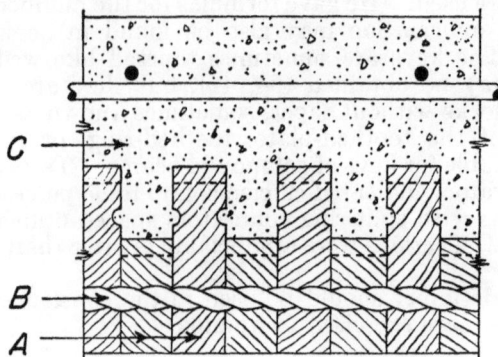

TRANSVERSE SECTION

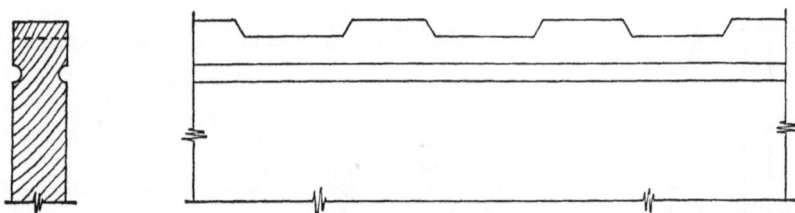

DETAIL OF WIDE PLANK

M 93470 F

FIGURE 43.—Construction of a composite wood-concrete deck. The transverse section shows: A, Laminated lumber base of wide and narrow planks; B, twisted transverse dowel for holding planks together; C, concrete mat with reinforcing bars. The upper portions of the wide planks are milled and longitudinally grooved, as shown in the detail, to provide keys that transmit horizontal shear and prevent separation of the concrete from the wood.

Trussed Beams

Trussed beams can be economically used for long spans and large loads when headroom permits. In the simplest form (fig. 44, A) the wood beam is supported at the center by a single wood strut or king-post, which in turn is supported by metal tie rods passing up to a thick steel plate at each end of the beam. To avoid secondary stresses the centerlines of beam, rod, and end support should intersect in a point. If the available headroom is above the beam the trussed beam illustrated in figure 44, B may be used.

For the greater spans, or where headroom is small or loads are concentrated at two intermediate points, queen trusses (figs. 44, C, 44, D) are preferable to king trusses.

Sometimes 2 tie rods are used with a single beam, 1 being placed on each side. Also, 2 beams may have a single rod between them or 3 beams may have 2 rods between them. It is desirable to give the assembly as much depth as conditions permit in order to reduce the stresses. Greater stability is obtained with beams continuous for the full span than with beams jointed over the struts.

Stresses in the beam itself and in the reinforcing struts or rods (fig. 44) can be calculated in terms of loads or reactions by the principles of stress analysis as set forth in textbooks on structural design.

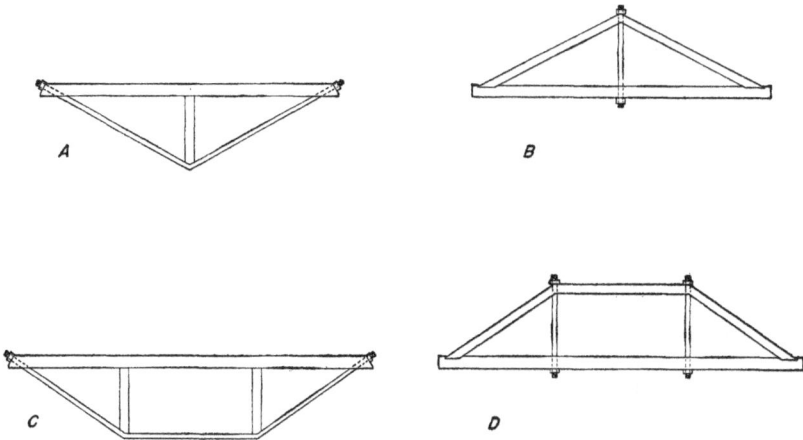

ZM 22073 F

FIGURE 44.—A and B, King types of trussed beams; C and D, queen types.

LATERAL SUPPORT OF BEAMS

Deep narrow beams are laterally unstable and require restraint against sidewise buckling. In floor joists, buckling is prevented by diagonal or other bridging between joists or by the flooring nailed to the joists. The following approximate rules are recommended:

(1) If the ratio of depth to breadth does not exceed 2, no lateral support is needed.

(2) If the ratio is 3, the ends should be held in position.

(3) If the ratio is 4, the piece should be held in line by spiking or bolting as in a vertically laminated beam.

(4) If the ratio is 5, one edge should be held in line, as by flooring nailed to the edge of a joist.

(5) If the ratio is 6, diagonal bridging as in floor joists should be used.

(6) If the ratio is 7, both edges should be held in line.

(7) If a beam is subject to both flexure and compression parallel to grain, the ratio of depth to breadth may be as much as 5, if one edge is held firmly in line, for example, by rafters (or by roof joists) and diagonal sheathing. If the dead load is sufficient to induce tension on the under side of rafters, the ratio may be as high as 6.

Additional rules apply to the design of roof truss members that are subjected to combined flexure and compression (10).

Special conditions of lateral instability may require a more refined method of analysis (17) than is afforded by the foregoing approximate rules.

SOLID WOOD COLUMNS OF RECTANGULAR CROSS SECTION

Length Classes

Solid wood columns fall into three length classes characterized by the type of failure. When the length does not exceed 11 times the least dimension, failure is by crushing. At lengths between 11 and K times the least dimension, failure is a combination of crushing and lateral buckling (11). The significance of K is explained in the following paragraphs. Beyond this range, wood columns generally fail by lateral deflection or buckling, behaving essentially as Euler columns.

Since most wood columns are rectangular or square in section, the ratio of unsupported length to least dimension, or the slenderness ratio, is commonly expressed as L/d, in which L is the unsupported length in inches and d the least side in inches.

For short columns or posts, that is, the first class of columns with L/d ratios of 11 or less, the same unit stress is recommended for all lengths. For these columns

$$\frac{P}{A} = c$$

in which P/A is the working stress for the column and c is the safe unit compressive stress parallel to the grain for the grade of material and the loading condition under consideration. Basic stresses from which safe working stresses (c) are derived are given in table 25.

For columns of the intermediate class, that is, with L/d ratios between 11 and K, the following formula (11) is recommended:

$$\frac{P}{A} = c\left[1 - \frac{1}{3}\left(\frac{L}{Kd}\right)^4\right]$$

in which P/A and c have the same meaning as before and K is a constant depending upon c and the modulus of elasticity. The value K is the minimum value of L/d at which the column will behave as a Euler column. This value is obtained when

$$\frac{P}{A} = \frac{2c}{3}$$

In the Euler formula for long columns (see following paragraph), P/A has the value $2c/3$ when

$$\frac{L}{d} = \frac{\pi}{2} \sqrt{\frac{3E}{6c}}$$

where E is the modulus of elasticity. If this value of L/d is designated as K and a reduction factor of 3 is introduced for E, then

$$K = \frac{\pi}{2} \sqrt{\frac{E}{6c}} = 0.64 \sqrt{\frac{E}{c}}$$

Values of E are given in table 25.

For columns with L/d ratios equal to K or greater, the Euler formula with a reduction factor of 3 is:

$$\frac{P}{A} = \frac{0.274E}{\left(\dfrac{L}{d}\right)^2}$$

The values of c (table 25) used in the formulas for short and for intermediate columns have safety factors about equivalent to that in the Euler formula.

The reduction factor of 3 is applied to the Euler maximum load and in the formula for determining K to convert from laboratory test conditions to the conditions of use (including long-time loading) and to provide a factor of safety. Conditions of ordinary hazard to columns as primary structural members are assumed. Conditions of extraordinary hazard or danger to human life may require that a larger safety factor be obtained through a reduced working stress value. Conversely, in situations of more limited hazard, or where property damage only may be involved, the designer may economize with a smaller factor of safety by increasing the working stress. Columns to be used continuously dry or subjected to short-time periods of loading may be designed with somewhat less reduction from the Euler breaking load (see pp. 158–161) without encroaching on the safety factor.

The slenderness ratio (L/d) of a solid wood column should in no case exceed 50.

Examples: Calculate the safe load for a 10- by 10-inch column if its grade and loading condition are such that c and E are 1,300 and 1,600,000 pounds per square inch, respectively, and the column is (1) 8 feet long; (2) 14 feet long; (3) 20 feet long. The actual dimensions of the 10- by 10-inch timber are 9½ by 9½ inches.

For a length of 8 feet the column has a slenderness ratio of 10 and is therefore within the short-column class. The unit stress c is multiplied by the area to give the safe load:

$$9\tfrac{1}{2}\times9\tfrac{1}{2}\times1,300=117,300 \text{ pounds}$$

In calculating the safe load for the 14-foot column, K must first be determined and then the L/d ratio, which if less than K places the column in the intermediate-length class. Since E is 1,600,000 pounds per square inch,

$$K=0.64\sqrt{\frac{1,600,000}{1,300}}=22.4$$

For a length of 14 feet the L/d ratio is

$$\frac{14\times12}{9.5}=17.7$$

which is between 11 and 22.4. The column is therefore in the intermediate class and

$$\frac{P}{A}=1,300\left[1-\frac{1}{3}\left(\frac{17.7}{22.4}\right)^{4}\right]$$

$$=1,130 \text{ pounds per square inch.}$$

The safe load is:

$$9\tfrac{1}{2}\times9\tfrac{1}{2}\times1,130=102,000 \text{ pounds}$$

With a 20-foot length L/d is 25.3, which is greater than K, and the column is in the Euler class. Then

$$\frac{P}{A}=\frac{0.274\times1,600,000}{(25.3)^{2}}$$

$$=685 \text{ pounds per square inch}$$

and the safe load is:

$$9\tfrac{1}{2}\times9\tfrac{1}{2}\times685=61,800 \text{ pounds}$$

Alternate Design Method

While ultimate strength values in wood columns show that the three length classes exist, design at the working stress level is sometimes simplified by omitting the intermediate class and using the long-column formula for all lengths that give values not exceeding the safe stress on short columns (10). Working stresses by this method range from 0 to 15 percent higher than those calculated by the preceding formulas for intermediate columns.

Side Loads and Eccentricity

The general formulas for safe eccentric or combined side and end loadings of wood columns as developed by Newlin (*21*) are

$$\frac{P/A\left(\frac{6e}{d}\right)+M/S+zP/A}{f}+\frac{P/A}{c}=1$$

for columns with slenderness ratios of 11 or less, and

$$\frac{P/A\left(\frac{15e}{2d}\right)+M/S+zP/A}{f-P/A}+\frac{P/A}{c}=1$$

for columns with slenderness ratios of 20 or more, where P/A is the direct compressive stress induced by axial load; M/S, the flexural stress induced by side loads that are independent of end load; c, the allowable unit working stress in compression parallel to grain for a column of the slenderness ratio under consideration, but with centrally applied axial load and no side load; f, the allowable unit working stress in flexure that is permitted where only flexural stress exists; e, the eccentricity; d, the width of column, measured in the direction of side loads or eccentricity (equivalent to the depth when the column is considered as a beam subject to flexural forces); and z, the ratio of flexural to direct compressive stress when both result from the same loading (remains constant while load varies).

Stresses for columns with slenderness ratios between 11 and 20 are determined by straight-line interpolation between the formula for a slenderness ratio of 11 and the formula for a slenderness ratio of 20.

These formulas may be simplified for some conditions of loading by dropping out certain terms; for example, if there are side loads and a concentrically applied end load, e becomes zero, and the first term in the numerator of the equations disappears. The formulas can be solved for P/A or for M/S if this will facilitate their use. Solution of the second formula for P/A leads to a quadratic equation with two roots; the root giving the smaller numerical value to P/A should be used.

These formulas are applicable only to columns of square or rectangular cross section.

If the side loads are such that maximum deflection and maximum flexural stress do not occur at midlength of the column, it is generally satisfactory to consider M/S as the maximum flexural stress due to the load or loads, regardless of its position in the length of the column. When the point of this maximum stress is near the end of the column, a slight error on the side of overload will occur.

Columns With Side Brackets

Formulas for eccentric loading assume that the eccentric load is applied at the end of the column. If an eccentric load is applied by a bracket at some point below the upper end, conditions of stress and deflection are somewhat more favorable. An exact solution of the problem is difficult, but the following simple rule is safe and, for brackets in the upper quarter of the length of the column, is quite accurate.

Assume that a bracket load P at a distance a from the center of a column (fig. 45) is replaced by the same load P centrally applied at the top of the column plus a side load P_1 applied at midheight.

Calculate P_1 from the formula

$$P_1 = \frac{3aL_1P}{L^2}$$

where L is the length of the column and L_1 is the length from the bottom of the column to the top of the bracket. Take care that a, L, and L_1 are expressed in the same units of measurement.

The assumed centrally applied load P can be added to other column loads, and the assumed side load P_1 is used to calculate flexural stress M/S in the formula for side loads.

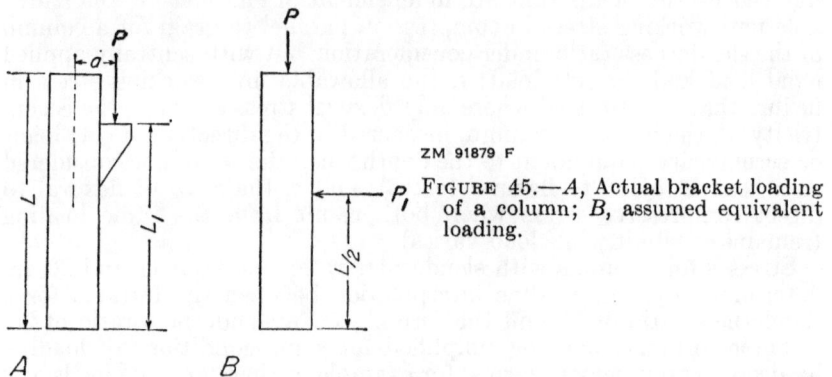

ZM 91879 F

FIGURE 45.—A, Actual bracket loading of a column; B, assumed equivalent loading.

BUILT-UP COLUMNS WITH MECHANICAL FASTENINGS

This section refers to built-up columns with pieces joined by nails, bolts, or other mechanical fastenings. Glued laminated columns are discussed on pages 252, 256, 261, and 263.

Design of Rectangular Sections

No arrangement of pieces with any kind of mechanical fastenings will make a built-up column fully equal in strength to a one-piece column of comparable material and dimensions (14). Arrangements with parallel planks and cover plates (fig. 46, A) or with planks boxed around a solid core (fig. 46, B) give the following percentages of the strength of a one-piece column of the same dimensions and quality:

L/d ratio:	Strength of 1-piece column (percent)	L/d ratio:—Continued	Strength of 1-piece column (percent)
6	82	18	65
10	77	22	74
14	71	26	82

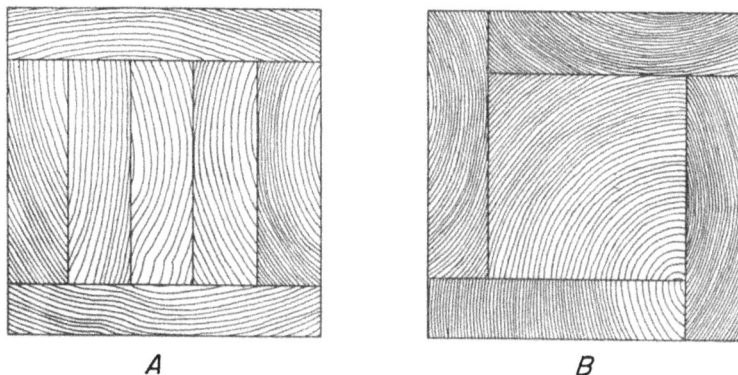

FIGURE 46.—Two types of built-up columns: *A*, Edges tied together with cover plates; *B*, boxed solid core.

For L/d ratios of 10 or greater, these percentages apply to both built-up columns in which the individual pieces are full length and columns in which butt-jointed pieces are used.

Compression Joints

Butt joints in compression members should be perpendicular to the direction of stress and should be accurately cut and closely fitted. In short posts (L/d less than 10), pieces butted end to end fail at 75 to 80 percent of the crushing strength of full-length pieces. If the joint in a short column has full lateral support (such as may occur in the compression members of trusses) and a metal plate not thinner than 20 gage is tightly fitted between the butted pieces, no reduction of compressive strength need be assumed (*10*).

Wrinkling or Twisting of Compression Members With Thin Outstanding Flanges

Columns whose cross section includes thin outstanding flanges or parts (fig. 47) are elastically unstable (*17*). They can fail by wrinkling of the outstanding part or by torsion of the whole section about its own axis at loads less than those indicated by the formulas on pages 216–218 for crushing or bending failure.

If an outstanding flange under a compressive load projects from a column section that is high in torsional stiffness, the wrinkling stress may be critical; that is, the outstanding part may wrinkle and make the remaining section unsafe. The stress at which wrinkling may occur is given by

$$p = 0.07 \, E \, \frac{h^2}{b^2}$$

where p is the critical wrinkling stress, E is the modulus of elasticity, and h is the thickness and b the width of the outstanding flange.

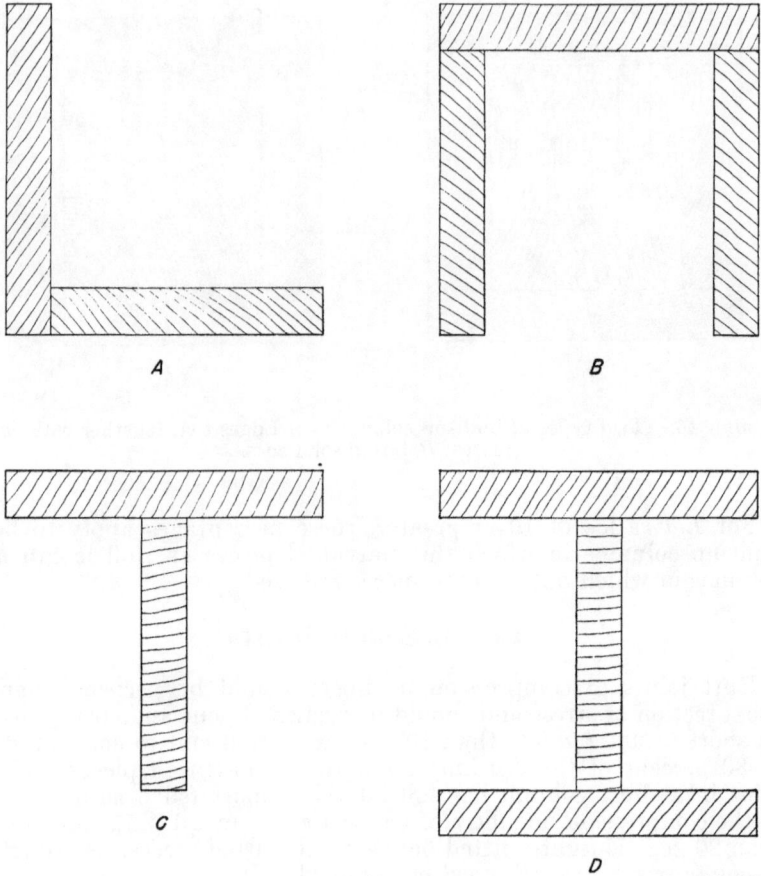

ZM 23108 F

FIGURE 47.—Typical cross sections of compression members that may fail through elastic instability: A, L-section: B, U-section; C, T-section; D, I-section.

If one or more relatively wide and thin parts project from a compression member that does not have great torsional stiffness, the member may twist about its own longitudinal axis. Such twisting is likely to occur with sections like A of figure 47. The critical stress is given by

$$p = 0.044 \, E \, \frac{h^2}{b^2}$$

with notation as in the preceding equation. Resistance to twisting can be increased above the value indicated in this equation by using generous fillets in the internal angles of the section.

The coefficients given in the equation for wrinkling and twisting were obtained from tests of Sitka spruce, but they are applicable to other species of wood without appreciable error. The rigidity of the section may, in some instances, be such that the critical stress occurs at a value between the values given by the wrinkling and twisting formulas. For circumstances requiring precise design, reference should be made to a more comprehensive discussion of the subject (17).

Stresses given by the formulas for wrinkling or twisting are critical or failing stresses, and the unit stress on the compression member must be kept below them. The amount of reduction or the factor of safety required on the critical values is left for the structural designer to determine on the basis of the particular conditions under which the member is to be used.

Box Columns of Wood

Tests for strength of box columns of wood made at the Virginia Polytechnic Institute (8) indicate that the design formulas for short, intermediate, and long columns, as on pages 216–218, are applicable to box columns of square cross section. Where L/d appears in the formulas for intermediate or long solid columns, it is replaced by

$$\frac{L}{\sqrt{d_1^2 + d_2^2}}$$

for square box columns, d_1 being the outside dimension and d_2 the inside dimension of the cross section of the box column. The formulas were adapted, by means of suitable reduction coefficients related to the arrangement and fastening of members and the condition of end fixity, to allow for load capacities somewhat lower than those of solid columns of equal cross section. It was recommended that the ratio of width to thickness of the side members of the box should not exceed 10. The work included an analysis of economics in the use of lumber made possible by box-column design.

Spaced Columns

Spaced columns (fig. 48) consist essentially of two parallel compression members separated by spacer blocks and joined through those blocks by means of timber connectors. The compression members and the spacer blocks are often of the same thickness. The spaced column may be used for the direct support of vertical loads, but often appears as a chord or other member in a wood truss.

The compression members in spaced columns are characterized by their slenderness ratio L/d, the ratio of the length, L, to the thickness of one member, d (fig. 48). This ratio should not exceed 80. The length L is the full length of a free-standing column, or the length between panel points or other points of lateral support in a truss. A load may be applied directly to the ends of the compression members, in which case the end spacer blocks do not extend beyond the ends of the column. In trusses, however, the end spacer blocks may extend beyond the ends of the column, and the load may be applied through them and their connectors to the compression members.

Connectors should be of a size and number to restrain differential movement between the compression members. Since the forces causing differential movements are less in short than in long spaced

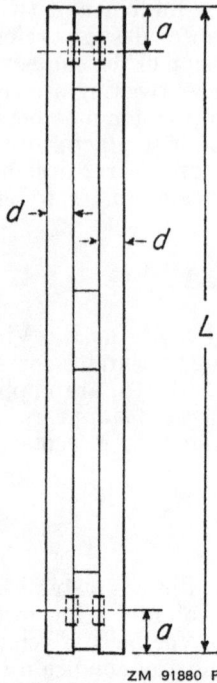

FIGURE 48.—A spaced column made from two compression members joined through spacer blocks by means of timber connectors.

ZM 91880 F

columns, the connector requirements are correspondingly less.   The connector or connectors on each face of each end spacer block should bear safely a load (p. 187) calculated by multiplying the cross-sectional area of one compression member (without deduction for cuts made to receive connectors) by the following unit stress values:

(1) For slenderness ratios of 11 or less, use 0 unit stress.

(2) For slenderness ratios of 60 to 80, use one-fourth of the basic stress value in compression parallel to grain for the species (table 25).

(3) For slenderness ratios between 11 and 60, use a unit stress obtained by straight-line interpolation between 0 for a slenderness ratio of 11 and the value for a slenderness ratio of 60.

Spacer blocks joined to the compression members by means of connectors are required at both ends of the spaced column.   A spacer block at midlength or, in long columns, two blocks at the one-third points of length are also required.   All spacer blocks must be of equal thickness and at least as thick as the compression members.   They should, if possible, have the grain of wood running parallel to the length of the compression members.   The minimum length of the end spacer blocks is determined by the end distances required by the connectors (p. 196), and their width should be equal to or greater than the width of the compression members.   Middle blocks should be of such size and so fastened as to insure that the compression members will maintain their spacing under load.   Spiking or bolting is generally adequate for joining the middle blocks to the compression members. Truss joint members may take the place of end or middle spacer blocks if they meet the requirements for size, location, and number and size of connectors.

Spaced columns are designed from the ordinary column formulas of pages 216–218, considering each compression member as an unsupported simple column and taking the sum of the safe loads for the two compression members. Certain modifications of the formulas are made to convert to spaced-column conditions. Test values show that loads supported by compression members in the short-column class are limited by the net section of the members after deduction of the cuts made to receive the connectors and bolts. Loads on long columns are not so limited. Further, there is a fixity factor that increases the strength of members in the long-column but not in the short-column class. The fixity factor is 2.5 if connectors in the end spacer blocks are at a distance of $L/20$ or less from the end of the column and 3.0 if the connectors are between $L/20$ and $L/10$ from the end of the column and the end blocks extend to the end of the column.

The fixity factor is most conveniently recognized in design by applying it to the value for E in the column formulas of pages 216–218. The column formulas become

$$P/A = \frac{0.274 \ FE}{(L/d)^2}$$

for long columns, and

$$P/A = c \left[1 - \frac{1}{3} \left(\frac{L}{Kd} \right)^4 \right]$$

for intermediate columns, where

$$K = \frac{\pi}{2} \sqrt{\frac{FE}{6c}}$$

where F is the fixity factor.

The load P calculated by either of these formulas must not exceed the safe load on a short column of the same grade, and of net cross section as reduced by connector or bolt holes. Where it is practical to exclude knots from the connector areas, somewhat higher values of P can be realized (10).

The compression members in spaced columns are usually of such proportions that critical loads for long columns are governed by the stiffness in a direction perpendicular to the width of those members. If the compression members are not much wider than they are thick, the stiffness in a direction parallel to their width may govern. It can be shown that the strength of a long spaced column is the same for both directions of bending when the ratio of width to thickness of the compression members equals the square root of the fixity factor.

The required cross-sectional area and basic stress value for the compression members are used as previously described to determine the required number and size of connectors in the end spacer blocks. These connectors in turn determine the minimum length of the end spacer blocks.

The formulas for spaced columns give safe load values for long-time loading and are subject to increase for shorter durations of load (p. 160). Increases for drying (p. 158) are reflected in the values chosen for E and c.

COLUMNS OF CIRCULAR CROSS SECTION

Round and square wood members of the same cross-sectional area will carry the same loads in both bending and compression and will have approximately the same stiffness. To compute the required size of a round column, design first for a square column and then use a round column of a diameter that will give an area equivalent to the area of the square. If the column is tapered, the diameter for use as d and for computing A in the Euler formula should be taken as the diameter at a distance of one-third of the length from the smaller end. This will give a diameter of a round column necessary to prevent failure from buckling. The unit compressive stress at the small end of the column that will result from the calculated load should also be computed, since it must not exceed the allowable stress for a short column.

BUILT-UP WOOD ARCHES WITH MECHANICAL FASTENINGS

Horizontally Laminated Arches

Arches with mechanically joined laminations whose width is horizontal generally lack efficiency against thrust and flexural loads. Glued construction (p. 247) is best for horizontally laminated arches. If mechanical fastenings are used, horizontally laminated arches have greater lateral stability than vertically laminated arches and are occasionally alternated with the latter where this stability is an important factor.

Vertically Laminated and Segmented Arches

Curved segmented rafters or roof arches can be made by joining with bolts or spikes relatively short pieces of lumber with their width vertical and upper edges bandsawed to a curved outline. A common construction for rafters of this type is of 3 thicknesses of nominal 1-inch lumber (fig. 49).

Transverse joints at the ends of the segments of vertically laminated and segmented arches transmit no bending stress and only limited compressive stress, so that design should be based on the net unjointed section of the arch. Working stresses suitable for the design of one-piece members of the same species and grade may be used with that net section. For bending strength and stiffness, all segments should extend through the full effective depth of the cross section of the arch. Owing to the curvature of the member, stresses at the ends of the segments are not parallel to the direction of grain; if the radius of curvature is short, this fact should be taken into account (p. 253) in setting the working stresses.

A segmented, laminated arch does not have a high degree of lateral stability. If 1 of its edges is held firmly in line by the roof deck, the ratio of the depth to the thickness of its cross section should not exceed 5.

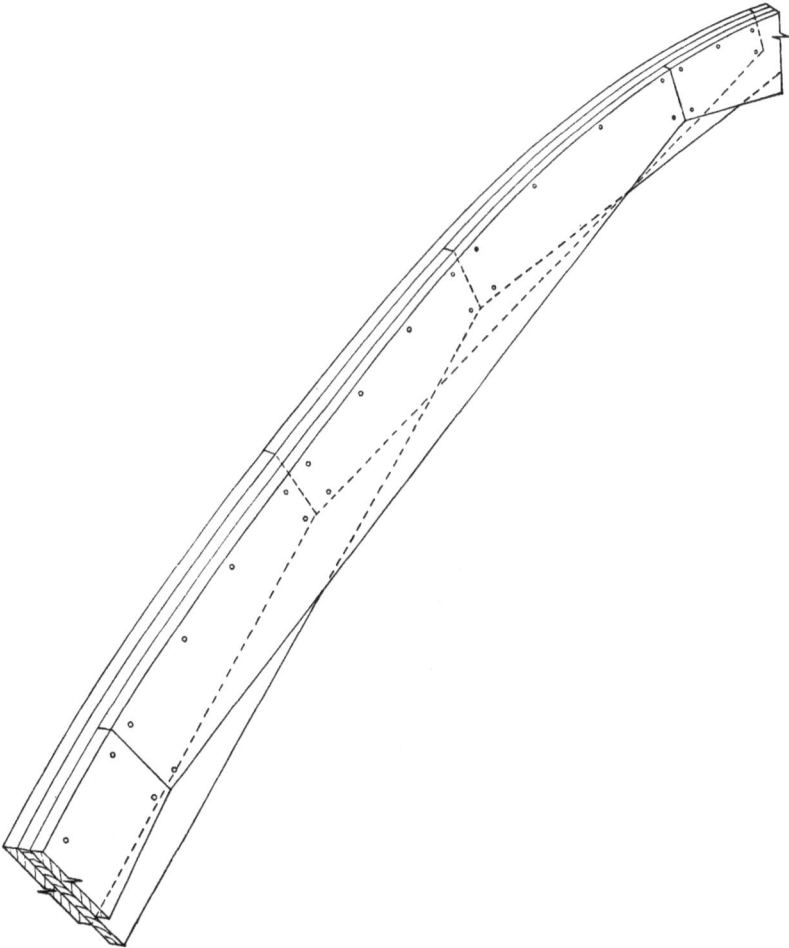

ZM 91881 F

FIGURE 49.—Perspective view of a part of a vertically laminated segmented arch.

Tests of curved rafters at the Forest Products Laboratory (20) indicate that the same amount of lumber when bent and glued in horizontal laminations is somewhat stiffer and considerably stronger than when nailed in vertical laminations. Gluing of the nailed vertically laminated rafters improved both their strength and their stiffness.

LAMELLA ROOFS

The lamella roof (fig. 50) is a curved roof framed by a system of intersecting skewed arches made up of short segments of lumber bolted at their intersections. The intersection of arches in two directions adds to the strength and stability against horizontal forces. The multiplicity of joints in a lamella roof system permits considerable vertical deformation under load, which should be taken into account in the design (*19*). Since the arches are skewed, their thrust has components in both the transverse and longitudinal directions of the building. Transverse components are resisted by tie rods and braces, as with conventional arches. Longitudinal components are resisted by the longitudinally placed roof decking and by sill planks extending lengthwise of the building.

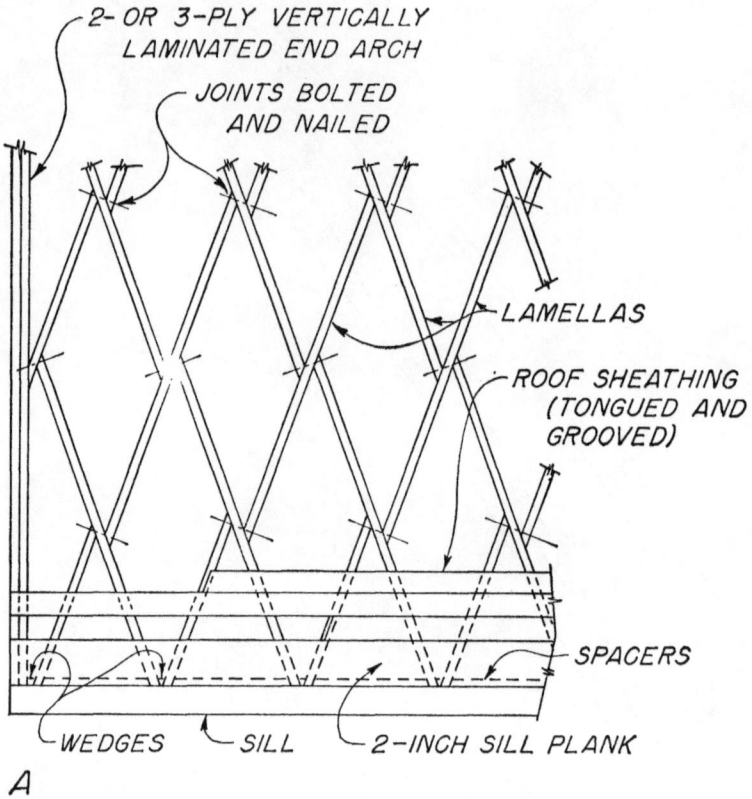

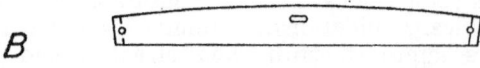

M 93471 F

FIGURE 50.—*A*, Lamella roof plan; *B*, side view of a lamella.

To minimize deformations in a lamella system, all joints should be carefully fitted and well tightened. If lamellas are of unseasoned wood, shrinkage in drying will permit slippage in each joint, which has a cumulative effect of shortening the arch system. Such shortening can be compensated by an additional camber or crown in the system as first constructed. Use of well-seasoned lumber minimizes such joint slippage and consequent change in shape.

Since lamella arches are commonly rather flat, it is important that they be built to accurate line and contour if design assumptions are to be valid (*19*). Falsework under the full length of span is required for support of the system during construction, as with other types of built-in-place arches.

TENSION MEMBERS

Unit design stresses for tension members are derived from the basic stress values in column 2 of table 25. Care should be taken that the net cross section of tension members at points cut or bored to receive fastenings is adequate for the loads imposed. Tension members, unless very long, ordinarily do not require lateral support.

POLE FRAMING

Pole framing (fig. 51) is one of the simplest types of building frames used to give shape and stability to a building. In general, it can be adapted to any type of building (*2, 9, 13, 22*). It was used extensively by the pioneers of this country, but due to the short service life of the poles and the crude construction methods, it bowed to other types of construction as sawed lumber and other materials became available. In recent years, because of rising building costs and the availability of methods for making wood more durable, the use of pole

M 90379 F

FIGURE 51.—A modern pole-frame barn under construction. (Photo courtesy University of Wisconsin.)

framing has been revived. Because no foundation is needed and the amount of bracing required is small, both labor and material costs are less for pole framing than for other types of framing.

Pole-frame buildings are generally 1-story buildings with the poles spaced from 10 to 13 feet apart and set from 4 to 5½ feet in the ground, depending on the type of soil and the height of building. Plates and girts are secured and braced to the poles to provide a skeleton frame for the attachment of roof rafters and end and side wall sheathing.

Proper preservative treatment of the poles to resist decay and termites (see p. 435) is essential if they are to serve as long as the other parts of the structure. Firm soil or concrete footings are needed to support the poles against settlement, and firm packing of the soil about the poles is required for anchorage and stability. In general, mechanical fastenings, such as nails and bolts, are used to secure the plates, girts, and braces to the poles.

Poles in building frames act as columns in carrying the building loads and as vertical beams in resisting wind or other horizontal forces. Engineering principles of design for round columns (p. 226) and round beams (p. 210) can be used where loads and stresses are known.

Literature Cited

(1) American Institute of Timber Construction.
    1954. timber construction standards   58 pp., illus.
        Washington, D. C.
(2) Bigalow, I. W.
    1950. pole barn construction.   N. Y. Agr. Col. (Cornell) Bul. 401,
        9 pp., illus. [Processed.]
(3) Boomsliter, G. P., Cather, C. M., and Worrell, D. T.
    1951. distribution of wheel loads on a timber bridge floor.   W.
        Va. Engin. Expt. Sta. Res. Bul. 24, 31 pp., illus.
(4) Heck, G. E.
    1921. built-up southern yellow pine timbers tested for strength.
        Natl. Lumber Mfrs. Assoc. Wood Construct. Inform. Serv. Ser.
        E–2B, 9 pp., illus.
(5) ———
    1950. comparative tests of nailed built-up beams of various con-
        structions.   South. Lumberman 180 (2260): 58, 60.
(6) Jacoby, H. S., and Davis, R. P.
    1930. timber design and construction.   Ed. 2, 234 pp., illus. New
        York.
(7) Kidwell, E.
    1898. efficiency of built-up wooden beams.   Amer. Inst. Mining
        Engin. 27: 732–818.
(8) Kinzey, B. Y., Jr.
    1950. wood box columns and their design.   Va. Polytech. Inst.,
        Wood Res. Lab., 15 pp., illus.
(9) Miller, T. A. H.
    1931. use of logs and poles in farm construction.   U. S. Dept. Agr.
        Farmers' Bul. 1660, 25 pp., illus. (Slightly rev. 1948.)
(10) National Lumber Manufacturers Association.
    1953. national design specification for stress-grade lumber and
        its fastenings.   66 pp., illus.
(11) Newlin, J. A., and Gahagan, J. M.
    1930. tests of large timber columns and presentation of the
        forest products laboratory column formula.   U. S. Dept.
        Agr. Tech. Bul. 167, 44 pp., illus.

(12) NEWLIN, J. A., HECK, G. E., AND MARCH, H. W.
    1934. NEW METHOD FOR CALCULATING LONGITUDINAL SHEAR IN CHECKED
        WOODEN BEAMS. Amer. Soc. Mech. Engin. Trans. 56 (10):
        739–744.
(13) ROBINSON, D. D., AND ROSS, C. R.
    1947. POLES AND FENCE POSTS FOR OREGON FARMS. Oreg. Agr. Col.
        Ext. Bul. 681, 24 pp., illus.
(14) SCHOLTEN, J. A.
    1931. BUILT-UP WOOD COLUMNS CONSERVE LUMBER. Engin. News-Rec.
        107 (9), 11 pp., illus.
(15) SCOFIELD, W. F. ,AND O'BRIEN, W. H.
    1949. MODERN TIMBER ENGINEERING. Ed. 3, 147 pp., illus. South. Pine
        Assoc.
(16) SMITH C. B., AND VOSS, A. W.
    1948. STRESS DISTRIBUTION IN A BEAM OF ORTHOTROPIC MATERIAL SUB-
        JECTED TO A CONCENTRATED LOAD. Natl. Advisory Comm.
        Aeronaut. Tech. Note 1486, 37 pp., illus.
(17) TRAYER, G. W., AND MARCH, H. W.
    1931. ELASTIC INSTABILITY OF MEMBERS HAVING SECTIONS COMMON IN
        AIRCRAFT CONSTRUCTION. Natl. Advisory Comm. Aeronaut.
        Rpt. 382, 42 pp., illus.
(18) U. S. FOREST PRODUCTS LABORATORY.
    1935. ROUNDING NOTCHES MAKES STRONGER JOISTS. Amer. Lumberman
        3039: 46, illus.
(19) VON KARMAN, T.
    ANALYSIS OF THE LAMELLA ROOF: (1) VERTICAL FORCES: (2) HORI-
        ZONTAL FORCES. 29 pp., illus. Guggenheim Aeronaut. Lab.,
        Calif. Inst. Technol., Pasadena.
(20) WILSON, T. R. C.
    1939. GLUED LAMINATED WOODEN ARCH. U. S. Dept. Agr. Tech. Bul.
        691, 122 pp., illus.
(21) WOOD, L. W.
    1950. FORMULAS FOR COLUMNS WITH SIDE LOADS AND ECCENTRICITY.
        U. S. Forest Prod. Lab. Rpt. R1782, 17 pp., illus. (Processed.]
(22) WOOLEY, J. C.
    1952. REPAIRING AND CONSTRUCTING FARM BUILDINGS. 261 pp., illus.
        New York, N. Y.

GLUING OF WOOD

Gluing is done extensively in woodworking and in the production of laminated wood (p. 247), plywood (p. 275), and sandwich materials (p. 291). Modern glues, processes, and techniques vary as widely as the products made, and developments have been many in recent years. In general, however, it remains true that the quality of a glued joint depends upon (1) the kind of wood and its preparation for use, (2) the kind and quality of the glue and its preparation for use, (3) the details of the gluing process, (4) the types of joints, and (5) the conditioning of the joints (7).[15] Depending on the glue used, service conditions also affect the performance of the joint to a greater or lesser extent.

GLUING PROPERTIES OF DIFFERENT WOODS

Table 34 gives the gluing properties of the woods most widely used for glued products. The classifications are based on the average quality of side-grain joints of wood that is approximately average in density for the species, when glued with animal, casein, starch, urea resin, and resorcinol resin glues. A species is considered to be glued satisfactorily when the strength of the joint is approximately equal to the strength of the wood (7).

Whether it will be easy or difficult to obtain a satisfactory joint depends upon the density of the wood, the structure of the wood, the presence of extractives or infiltrated materials in the wood, and the kind of glue. In general, heavy woods are more difficult to glue than lightweight woods, hardwoods are more difficult to glue than softwoods, and heartwood is more difficult than sapwood. Several species vary considerably in their gluing characteristics with different glues (table 34).

GLUES USED IN WOODWORKING

Table 35 describes briefly the characteristics, preparation, and uses of the types of glue most commonly used to make joints in wood.

Animal glues (10) have long been used extensively in woodworking; starch glues came into general use, especially for veneering, early in this century; casein glue (11) and vegetable protein glues, of which soybean is the most important, gained commercial importance during and immediately following World War I for gluing lumber and veneer into products that required moderate water resistance. Synthetic resin glues were developed more recently but now surpass many of the older glues in importance as woodworking glues. Phenol resin glues (9) are widely used to produce plywood for severe service conditions. Urea resin glues (12) are used extensively in producing plywood for furniture and interior paneling. Resorcinol (9) and phenol-resorcinol resin glues are useful for gluing lumber into products that will withstand exposure to the weather. Polyvinyl resin emulsion glues are used in assembly joints of furniture.

---

[15] Italic numbers in parentheses refer to Literature Cited, p. 245.

TABLE 34.—*Classification of various hardwood and softwood species according to gluing properties*

HARDWOODS

Group 1	Group 2	Group 3	Group 4
(Glue very easily with different glues under wide range of gluing conditions)	(Glue well with different glues under a moderately wide range of gluing conditions)	(Glue satisfactorily under well-controlled gluing conditions)	(Require very close control of gluing conditions, or special treatment to obtain best results)
Aspen.	Alder, red.	Ash, white.[2]	Beech, American.
Chestnut, American.	Basswood.[1]	Cherry, black.[1][2]	Birch, sweet and yellow.[2]
Cottonwood.	Butternut.[1][2]	Dogwood.[2]	Hickory.[2]
Willow, black.	Elm:	Maple, soft.[1][2]	Maple, hard.
Yellow-poplar.	American.[2]	Oak:	Osage-orange.
	Rock.[1][2]	Red.[2]	Persimmon.
	Hackberry.	White.	
	Magnolia.[1][2]	Pecan.	
	Mahogany.[2]	Sycamore.[1][2]	
	Sweetgum.[1]	Tupelo:	
		Black.[1]	
		Water.[1][2]	
		Walnut, black.	

SOFTWOODS

Baldcypress.	Cedar, eastern red-.[2]	Cedar, Alaska-.[2]	
Cedar, western red-.[3]	Douglas-fir.		
Fir, white.	Hemlock, western.[3]		
Larch, western.	Pine:		
Redwood.	Eastern white.[3]		
Spruce, Sitka.	Southern yellow.[1]		
	Ponderosa.		

[1] Species is more subject to starved joints, particularly with animal glue, than the classification would otherwise indicate.
[2] Glued more easily with resin glues than with nonresin glues.
[3] Glued more easily with nonresin glues than with resin glues.

Broadly, synthetic resin glues are of two types—thermosetting and thermoplastic. Thermosetting resins, once cured, are not softened by heat. Thermoplastic resins will soften when reheated.

In addition to the major woodworking glues listed in table 35, glues made from a base of blood albumin have had very limited use in the United States but are more commonly used in Europe in the production of plywood. They are mixed at the time of use and are applied at room temperatures, but most of them require hot pressing. They are highly water resistant but not so durable as some of the synthetic resin glues. Defibrinated blood is sometimes used as a filler or extender for phenol resin glues.

Many brands of glues made from fish, animal, or vegetable derivatives are sold in liquid form, ready for application. Their principal use in woodworking is for small jobs and repair work. They are variable in quality and low in water resistance and durability under damp conditions. The better brands are moderate in dry strength and set fairly quickly. They are applied cold, usually by brush, and are pressed cold. They stain wood only slightly, if at all.

GLUES FOR BONDING WOOD TO METAL

Glues capable of producing bonds of high strength and durability between wood and metal are comparatively new in woodworking. Many of them are combinations of a thermosetting resin, often of the phenolic type, and a thermoplastic resin or elastomer, such as a polyvinyl resin or synthetic rubber (3). The vehicle may be water, alcohol, ethyl acetate, or some other solvent.

Some of these wood-metal adhesives are spread on one or both of the surfaces to be bonded, the solvent is evaporated, and the bonding operation is completed under heat and pressure. With other wood-metal adhesives, the bonding is done in 2 steps, usually with 2 different adhesives. In the first step, one adhesive is spread as a primer on the metal and cured at elevated temperatures. In the second step, the primed metal is bonded to the wood with a room-temperature-setting resin adhesive of the woodworking type. The metal surface must be thoroughly cleaned, and generally more care is required in removing solvents before bonding than is required in wood-wood bonds.

Adhesives formulated with casein and rubber latex, both natural and synthetic, have been used for several years to bond wood to metal. These formulations have the advantage of curing at room temperature, and the bonds are satisfactory where highest strength and durability are not essential.

DRYING AND CONDITIONING WOOD FOR GLUING

The moisture content of wood at the time of gluing has much to do with the final strength of joints, the development of checks in the wood, and the warping of the glued members. Glues will adhere to wood with any moisture content up to 15 percent; water-resistant glues will adhere to wood with even higher moisture content. Large changes in the moisture content of the wood after gluing, however, cause shrinking or swelling stresses that may seriously weaken both the wood and the joints.

The most satisfactory moisture content of wood at the time of gluing is that which, when increased by the moisture of the glue, approximately equals the average moisture content that the glued member will have in service. In gluing 1-inch boards or thicker pieces, this relation can be attained by proper seasoning. In veneer or other thin pieces, however, the moisture added by the glue frequently exceeds the moisture content of the wood in service (7). Under these conditions the wood cannot be dried enough before gluing to avoid redrying after gluing. The amount of moisture added to wood in gluing varies from less than 1 percent in lumber to 45 percent or more in thin plywood. The thickness of the wood, the number of plies, the density of the wood, the glue mixture, the quantity of glue spread, and the gluing procedure (hot pressing or cold pressing) all affect the increase in moisture content of the wood.

In general practice, adjustments cannot be made for all these widely varying factors, and it is seldom necessary to dry lumber to a moisture content below 5 percent or higher than 12 percent. Lumber with a

TABLE 35.—*Characteristics, preparation, and uses of the most commonly used woodworking glues*

Class	Form and testing	Properties	Preparation and application	Uses
Animal	Many grades sold in dry form; liquid glues available; quality determined by tests on solutions of the glue.[1]	High dry strength; low resistance to moisture and damp conditions; stain wood very slightly, if at all.	Dry form mixed with water, soaked, and melted; solution kept warm during application; liquid forms applied as received; both pressed at room temperatures.	Used extensively in furniture assembly joints, cabinetmaking, and millwork.
Casein and vegetable protein.	Several brands sold in dry powder form; may also be prepared from raw materials by user; quality determined by tests on glue and by tests on wood joints.[2]	High to low dry strength; moderate to low water resistance and moderately durable under damp conditions; pronounced dulling effect on tools; stain some woods badly.	Mixed with cold water, applied and pressed cold.	Used for gluing lumber and veneer for purposes requiring moderate moisture resistance.
Starch	Different grades sold in dry form; also available in liquid form ready to use; quality determined chiefly by tests on wood joints.	High in dry strength; low resistance to moisture and damp conditions; stain some woods moderately.	Dry forms mixed with water, usually with addition of caustic soda, and heated; applied cold; liquid form applied as received; both pressed cold.	Used primarily in gluing veneer.
Urea resin	Several brands sold as dry powders, others as liquids; may be blended with melamine or other resins; quality determined by tests on glue and by tests on wood joints.[3]	High in both wet and dry strength; moderately durable under damp conditions; moderate to low resistance to temperatures in excess of 150° F.; white or tan in color; stain wood very slightly if at all.	Dry form mixed with water; hardeners, fillers, and extenders may be added by user to either dry or liquid form; applied at room temperatures, some formulas cure at room temperatures, others require hot pressing at about 250° F.	Used extensively in gluing veneer for furniture and other interior uses and to some extent in gluing lumber and assembly joints.
Melamine resin	Comparatively few brands available; usually marketed as a powder with or without catalyst; quality determined by tests on glue and by tests on wood joints.[4]	High in both wet and dry strength; very resistant to moisture and damp conditions; stain wood very slightly, if at all; white to tan in color.	Mixed with water and applied at room temperatures; heat required to cure (250° to 300° F.).	Used to a limited extent in bag molding and in gluing lumber and veneer for purposes requiring colorless and highly resistant glue lines.
Phenol resin	Several brands available, some dry powders, others as liquids, and one as dry film; quality determined by tests on glue and by tests on wood joints.[4]	High in both wet and dry strength; very resistant to moisture and damp conditions, more resistant than wood to high temperatures; stain wood very slightly; dark red in color.	Film form used as received; powder form mixed with solvent, often alcohol and water, at room temperature; hardeners and fillers often added by users; most common types require hot pressing at about 300° F.	Used primarily for production of highly resistant plywood.

Resorcinol resin----	Several brands available in liquid form; catalyst supplied separately; some brands are combinations of phenol and resorcinal resins; relatively high priced; quality determined by tests on glue and by tests on wood joints.[4]	High in both wet and dry strength; very resistant to moisture and damp conditions; more resistant than wood to high temperatures; stain wood very slightly; dark red in color.	Mixed with catalyst and applied at room temperatures; resorcinal glues cure at room temperatures; some resorcinol-phenol blends cure best under moderate heat (100° to 200° F.).	Used primarily for gluing lumber or assembly joints that must withstand severe service conditions.
Polyvinyl resin emulsion.	Several brands are available, varying to some extent in properties; marketed in liquid form ready to use; quality determined largely by tests on glue joints.	Generally high in dry strength; low resistance to moisture and elevated temperatures; joints tend to yield under continued stress; white in color; stain wood little if at all.	Marketed as a liquid ready to use; applied and pressed at room temperatures.	Used in assembly joints in furniture, cabinetmaking, and millwork.

[1] Federal Specification C-G-451. Glues; animal. 1931.
[2] Federal Specification C-G-456. Glue; casein. 1941.
[3] Federal Specification C-G-496. Glue; Resin-type (Liquid and Powder), 1944.
[4] Military Specification MIL-A-397A. Adhesive, room-temperature and intermediate-temperature setting, resin (phenol, resorcinol, and melamine base). 1951.

moisture content of 5 to 6 percent is generally satisfactory for gluing into furniture, interior millwork, and similar items.   Lumber for outside use should generally contain 10 to 12 percent of moisture before gluing.   A moisture content of 3 to 5 percent in veneer at the time of gluing is satisfactory even for thin plywood in furniture, interior millwork, and similar products.   For certain uses, such as plywood for boxes, veneer at a moisture content of 10 to 12 percent or even higher is acceptable and may be desirable from a manufacturing standpoint.

Lumber that has been dried to the approximate average moisture content desired for gluing may still show differences between various boards and between the interior and the surfaces of individual pieces. Large differences in the moisture content of pieces that are glued together result eventually in considerable stress on glue joints and tend to produce warping of the product.   For this reason, it is desirable for many purposes to condition wood to a relatively uniform moisture content after drying and before gluing.   Variations of 1 percent or less between boards of the same species and size may be disregarded, since they may occur even after a long conditioning period.   Lumber that is to be glued should also be free from drying degrade, such as casehardening, warp, checks, and splits, although, if quality is of secondary importance, some types of degrade may be permitted.

MACHINING LUMBER FOR GLUING

Wood surfaces that are to be glued should be smooth and true, free from machine marks, chipped or loosened grain, and other surface irregularities.   Preferably, machining should be done just before gluing, so that the surfaces do not become distorted from subsequent moisture changes.   For uniform distribution of gluing pressure, each lamination or ply should be uniform in thickness.   A small variation in thickness in each lamination or ply may cause a considerable variation in the thickness of the pile when a number of similar pieces are piled in the same order as they come from the surfacer.

Surfaces made by saws are usually rougher than those made by planers, jointers, and other machines equipped with cutterheads. Saws perfected recently, however, produce a smoother cut than older types.   By making it possible to glue sawed joints more extensively, these new saws save both labor and material.   Joints can be made between smoothly sawed surfaces that are approximately equal in strength to those between planed surfaces.   Unless the saws are very well maintained, however, joints between sawed surfaces are generally weaker than those between well-planed or jointed surfaces.

In cabinet work and other constructions where joints are exposed to view, the sawed joint is often more conspicuous than the planed joint.

Wood surfaces are intentionally roughened by some operators by tooth planing, scratching, or sanding with coarse sandpaper in the belief that rough surfaces are better for gluing.   Tests of joints made using good gluing practices, however, show no benefit from roughening the surfaces.

Preparing Veneer for Gluing

Veneer is cut by sawing, slicing, or rotary processes (5). Sawed veneer is produced in long narrow strips, usually from flitches selected for figure and grain. The two sides of the sheet are equally firm and strong, and either side may be glued or exposed to view with the same results.

Sliced veneer is also cut in the form of long strips by moving a flitch or block against a heavy knife. Because the veneer is forced abruptly away from the flitch by the knife, it tends to have fine checks or breaks on the knife side. The checked side is called the open or loose side, the other side the closed or tight side. The open side is likely to show imperfections in finishing and therefore should be the glue side whenever possible. For matching face stock where the checked side of part of the sheets must be the finish side, the veneer must be well cut.

The rotary-cut process produces continuous sheets of flat-grained veneer by revolving a log against a knife. The half-round process, the back-cut process, and other modifications of straight rotary-cutting are used to produce highly figured veneer from stumps, burls, and other irregular parts of logs. In these processes a part of a log, stump, or burl is placed off center in a lathe and is rotary-cut into small sheets of veneer. All rotary-cut veneer has an open and a closed side (4), although it may be difficult to distinguish one from the other if the veneer is well cut. When rotary-cut veneer is used for faces, the checked or open side should if possible, be the glue side.

Because veneer usually is not resurfaced before it is glued, it must be carefully cut. Provided the veneer is well cut, there is no appreciable difference in any property except appearance of plywood made from veneer produced by any of the three processes. Veneer selected to be glued should be (1) uniform in thickness, (2) smooth and flat, (3) free from large checks, decay, or other quality-reducing features, and (4) straight grained. For plywood of the lower grades, however, some of these requirements may be modified.

Proper Gluing Conditions

To insure a strong joint in wood, a film of glue unbroken by air bubbles or foreign particles should be in contact with the wood surfaces over the entire joint area. This result is obtained by controlling the details of the gluing operation (2, 7, 8).

Making strong glue joints with glues applied in liquid condition depends primarily upon a proper correlation between gluing pressure and glue consistency during pressing. The consistency of the glue mixture, once it is spread on the wood, is extremely variable. It depends upon such factors as the kind of glue, glue-water proportion of the mixture, quantity of glue spread, moisture content of the wood, temperature of the glue, room, and wood, the time between spreading and pressing, and the extent to which the glue-coated surfaces are exposed to the air. Room-temperature-setting glues usually thicken and harden steadily after spreading until they are cured. Hot-press glues often thin out during the initial heating period and then thicken and harden as curing progresses.

Pressure is used to squeeze the glue out into a thin continuous film between the wood layers, to force air from the joint, to bring the wood surfaces into intimate contact with the glue, and to hold them in this position during the setting or curing of the glue. A light pressure should be used with a thin glue or one that becomes thin during curing, a heavy pressure with a thick glue, and corresponding variations in pressure should be made with glues of intermediate consistency. The strongest joints usually result when the consistency of the glue permits the use of moderately high pressures—100 to 250 pounds per square inch (7).

Joints should be kept under pressure at least until they have enough strength to withstand the interior stresses tending to separate the wood pieces. In cold-pressing operations, under favorable gluing conditions, this stage will be reached in 2 to 7 hours, depending upon the curing characteristics of the glue and upon the thickness, density, and absorptive characteristics of the wood. A longer pressing period is advisable as a precautionary measure when operating conditions permit (13).

In hot-pressing operations the time required varies with the temperature of the platens, the thickness and kind of material being pressed, and the kind of glue. The variation in time in actual practice is from about 2 minutes to as much as 30 minutes. Depending somewhat upon the size of the joint and the type of glue, the time under pressure may be reduced to a few seconds by heating the glue joint with high-frequency electric current.

Gluing with the dry forms of glues requires special conditions that vary somewhat with the particular glue and the product.

TYPES OF GLUED JOINTS

Side-Grain Surfaces

With most species of wood, straight, plain joints between side-grain surfaces (fig. 52, A) can be made substantially as strong as the wood itself in shear parallel to the grain, tension across the grain, and cleavage. The tongued-and-grooved joint (fig. 52, B) and other shaped joints have the theoretical advantage of larger gluing surfaces than the straight joints, but in practice they do not give higher strength with most woods. Furthermore, the theoretical advantage is often lost, wholly or partly, because the shaped joints are more difficult to machine than straight, plain joints so as to obtain a perfect fit of the parts. Because of poor contact, the effective holding area and strength may actually be less on a shaped joint than on a flat surface. The principal advantage of the tongued-and-grooved and other shaped joints is that the parts can be more quickly alined in the clamps or press. A shallow tongue-and-groove is usually as useful in this respect as a deeper cut and is less wasteful of wood.

End-Grain Surfaces

It is practically impossible to make end-butt joints (fig. 53, A) sufficiently strong or permanent to meet the requirements of ordinary service. With the most careful gluing possible, not more than about

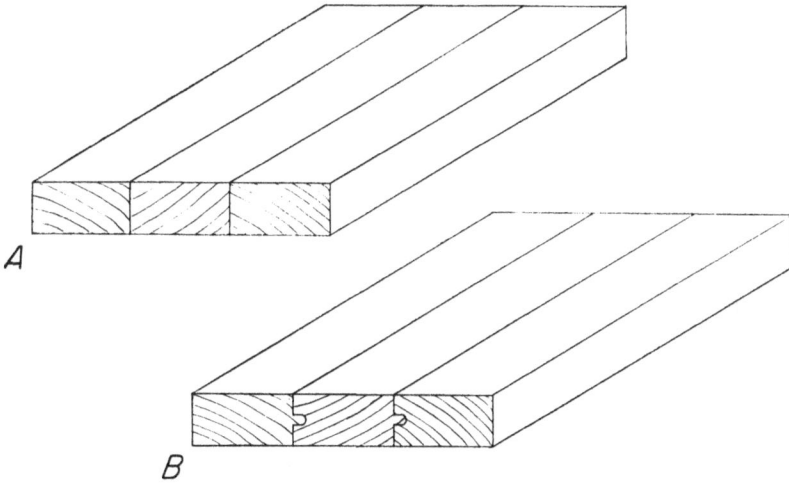

M 93137 F

FIGURE 52.—Side-to-side-grain joints: *A*, Plain; *B*, tongued-and-grooved.

25 percent of the tensile strength of the wood parallel with the grain can be obtained in butt joints. In order to approximate the tensile strength of various species, a scarf, serrated, or other form of joint that approaches a side-grain surface must be used (fig. 53). The plain scarf is perhaps the easiest to glue and entails fewer machining difficulties than the many-angle forms. The efficiencies of scarf joints of different slopes are discussed in the section on strength of glued structural members (p. 255).

End-to-Side-Grain Surfaces

End-to-side-grain joints (fig. 54, *A*) are also difficult to glue properly and, further, are subjected in service to unusually severe stresses as a result of unequal dimensional changes in the two members of the joint as their moisture content changes. It is therefore necessary to use irregular shapes of joints, dowels, tenons, or other devices to reinforce such a joint in order to bring side grain into contact with side grain or to secure larger gluing surfaces (fig. 54). All end-to-side-grain joints should be carefully protected from changes in moisture content.

CONDITIONING GLUED JOINTS

When boards are glued edge to edge, the wood at the joint absorbs water from the glue and swells. If the glued assembly is surfaced before this excess moisture is dried out or distributed, more wood is removed along the swollen joints than elsewhere. Later, when the joints dry and shrink, permanent depressions are formed that may be very conspicuous in a finished panel (6).

When pieces of lumber are glued edge to edge or face to face, the glue moisture need not be dried out but simply allowed to distribute

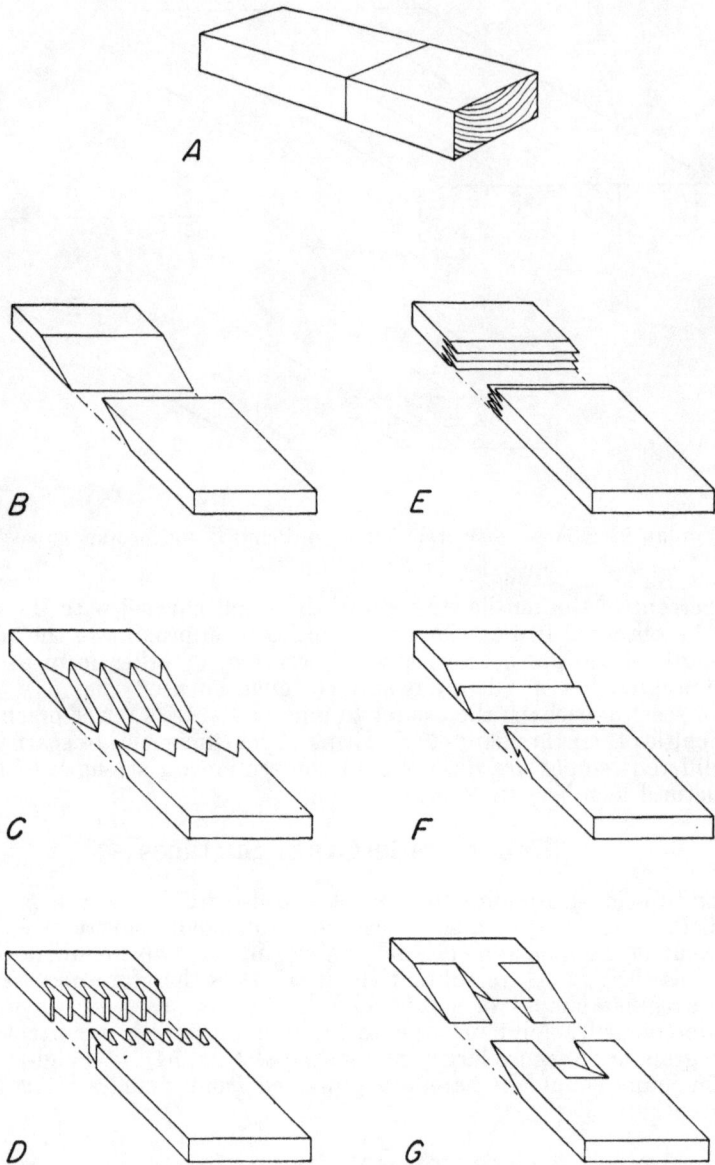

FIGURE 53.—End-to-end-grain joints: *A*, End butt; *B*, plain scarf; *C*, serrated scarf; *D*, finger; *E*, Onsrud; *F*, hooked scarf; *G*, double-slope scarf.

itself uniformly throughout the wood.   Approximately uniform distribution of glue moisture can usually be obtained by conditioning the stock after gluing for 24 hours at 160° F., 4 days at 120° F., or at least 7 days at room temperature with the relative humidity, in each case, adjusted to prevent significant drying.

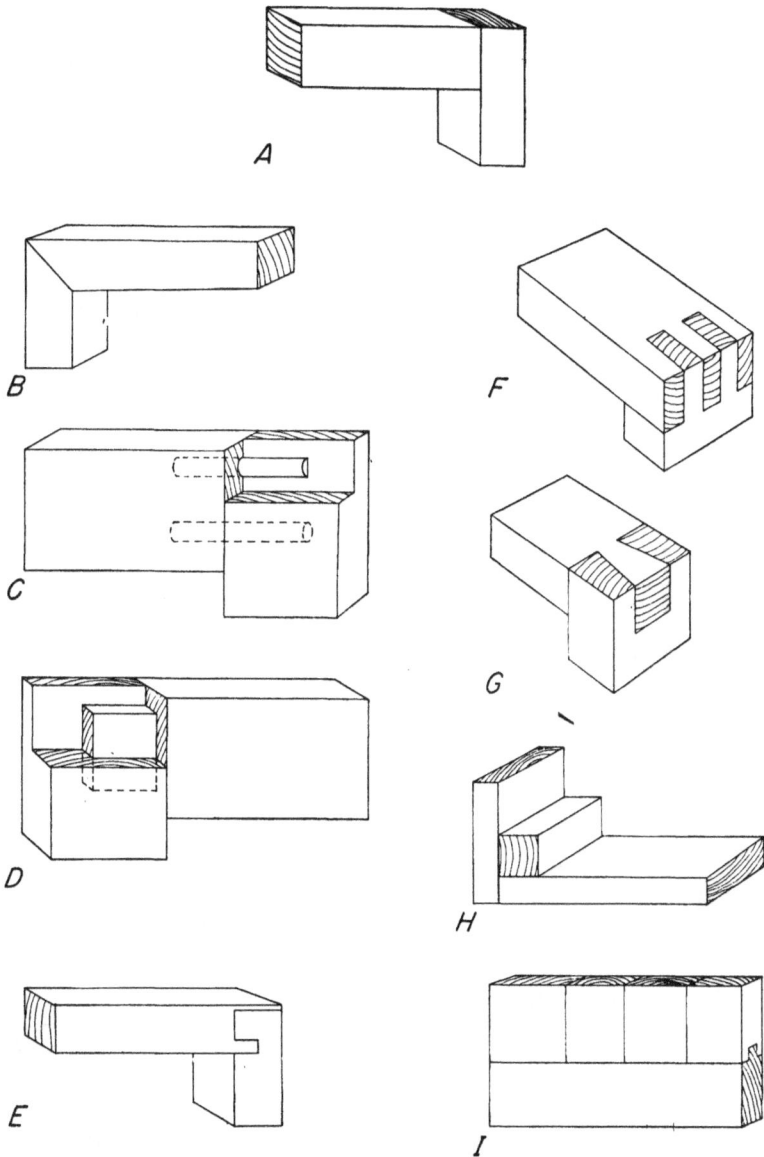

FIGURE 54.—End-to-side-grain joints: *A*, Plain; *B*, miter; *C*, dowel; *D*, mortise and tenon; *E*, dado tongue and rabbet; *F*, slip or lock corner; *G*, dovetail; *H*, blocked; *I*, tongued-and-grooved.

In plywood, veneered panels, and other constructions made by gluing together thin layers of wood, it is advisable to condition the panels to average service moisture content. In cold-gluing operations, it is frequently necessary to dry out at least a part of the moisture added in gluing. The drying is most advantageously done under controlled conditions and time schedules (7).

Drying such glued products to excessively low moisture content materially increases warping, opening of joints, and checking. Following hot-press operations, the panels will often be very dry, and it may be desirable to recondition them under circumstances that will cause them to regain moisture.

DURABILITY OF GLUED MEMBERS

The durability of glue joints in wood members depends upon the type of glue, the service conditions, the gluing technique, and the design or construction of the joints. Moisture conditions are particularly important not only because of the effect of moisture on the glue itself but because of the effect of moisture content on the internal stresses developed on the glue joint. These internal stresses, and consequently the behavior of the joints in service, are also dependent upon the design of the joint, the thickness of the plies or laminations, and the density and shrinkage characteristics of the species used.

Available evidence indicates that joints well designed and well made with any of the commonly used woodworking glues will retain their strength indefinitely under conditions such that the moisture content of the wood does not exceed approximately 15 percent and the temperature remains within the range of human comfort.

Low temperatures seem to have no significant effect on the strength of glue joints, but some glues have shown evidence of deterioration when exposed to temperatures of about 158° F. Joints that were well made with phenol resin, resorcinol resin, or melamine resin glues have proved more durable than the wood when exposed to water, to warmth and dampness, to alternative wetting and drying, and to temperatures sufficiently high to char the wood (14). These glues are sufficiently durable for use in products that are exposed to the weather.

Tests have shown that joints made with urea resin glues are highly resistant to water and to wetting and drying but that they tend to weaken when subjected to temperatures of about 158° F. or higher. Data considered adequate for predicting the retention of strength of urea resin glue bonds exposed to temperatures between 158° F. and the human-comfort zone are not available. Casein and soybean glue joints will withstand temporary exposure to dampness or water without permanent loss in strength, but if the moisture content of the wood continuously or repeatedly exceeds about 18 percent, they will eventually lose strength; the rate of loss of strength may vary, however, depending upon such factors as species and construction.

Polyvinyl resin emulsion glues have moderate resistance to dampness, but low resistance to water; the tendency of the joints to yield under stress generally increases as the temperature increases. Joints made with animal glue or with starch glue are not suited to damp service conditions. If toxic materials are added to animal glues, serviceability of the glued products in warm, damp environments is increased slightly.

Treatments that can be used to increase the durability of glued members include: (1) Coatings that reduce the moisture content changes in the wood and (2) impregnation of the wood with preservatives. Moisture-excluding coatings reduce the rate of moisture content changes in wood and lessen the swelling stresses that occur during temporary periods of exposure to damp conditions; they do not protect wood effectively during prolonged exposure to damp conditions. By impregnating glued members with preservatives, the deteriorating effects of prolonged exposure to damp conditions can be reduced. The useful life of casein-bonded plywood may be doubled or tripled by impregnating it with preservative chemicals, such as coal-tar creosote, that protect both glue and wood (1).

LITERATURE CITED

(1) BROUSE, D.
1932. METHODS OF INCREASING DURABILITY OF PLYWOOD. Woodworking Indus. 11 (2): 30–31, 33, illus.
(2) DOUGLASS, W. D., and PETTIFOR, C. B.
1932. THIRD AND FINAL REPORT OF THE ADHESIVES RESEARCH COMMITTEE (GT. BRIT.). Dept. Sci. Indus. Res., 109 pp., illus., London.
(3) EICKNER, H. W., and BLOMQUIST, R. F.
1950. ADHESIVES FOR BONDING WOOD TO METAL. U. S. Forest Prod. Lab. Rpt. 1768, 15 pp., illus. [Processed.]
(4) FLEISCHER, H. O.
1949. EXPERIMENTS IN ROTARY VENEER CUTTING. Forest Prod. Res. Soc. Proc. 3: 20 pp., illus.
(5) KNIGHT, E. V., and WULPI, M.
1927. VENEERS AND PLYWOOD, THEIR CRAFTSMANSHIP AND ARTISTRY. 372 pp., illus., New York.
(6) SELBO, M. L.
1952. EFFECTIVENESS OF DIFFERENT CONDITIONING SCHEDULES IN REDUCING SUNKEN JOINTS IN EDGE GLUED LUMBER PANELS. Forest Prod. Res. Soc. Jour. 2 (1): 8 pp., illus.
(7) TRUAX, T. R.
1929. GLUING OF WOOD. U. S. Dept. Agr. Bul. 1500, 78 pp., illus.
(8) ————
1930. GLUING WOOD IN AIRCRAFT MANUFACTURE. U. S. Dept. Agr. Tech. Bul. 205, 58 pp., illus.
(9) U. S. DEPARTMENT OF DEFENSE.
1951. ADHESIVE, ROOM-TEMPERATURE AND INTERMEDIATE-TEMPERATURE SETTING, RESIN (PHENOL, RESORCINOL, AND MELAMINE BASE). Mil. Spec. MIL–A–397A, 15 pp., illus.
(10) U. S. FEDERAL SPECIFICATIONS BOARD.
1931. GLUE; ANIMAL. Fed. Spec. C–G–451, 7 pp., illus.
(11) ————.
1941. GLUE; CASEIN. Fed. Spec. C–G–456, 7 pp., illus.
(12) ————.
1946. GLUE; UREA-RESIN. Fed. Spec. C–G–496, 10 pp., illus.
(13) U. S. FOREST PRODUCTS LABORATORY.
1951. WOOD AIRCRAFT INSPECTION AND FABRICATION. ANC Bul. 19, 335 pp., illus. Issued by Subcommittee on Air Force-Navy-Civil Aircraft Design Criteria, Aircraft Committee, Munitions Board.)
(14) WANGAARD, F. F.
1946. SUMMARY OF INFORMATION ON THE DURABILITY OF WOODWORKING GLUES. U. S. Forest Products Laboratory Rpt. 1530, 36 pp., illus. [Processed.]

GLUED STRUCTURAL MEMBERS

Glued structural members are of two types—glued laminated members and wood-plywood members of built-up cross section. Both types offer certain advantages.

GENERAL CHARACTERISTICS OF GLUED LAMINATED CONSTRUCTION

Parallel-grain or laminated construction (fig. 55) refers to two or more layers of wood glued together with the grain of all layers approximately parallel. The laminations may vary as to species, number, size, shape, and thickness.

Parallel-grain or laminated construction has long been used for furniture parts, for cores of veneered panels, and for sporting goods. The first use of glued laminated structural members was in Europe, where it is reported that as early as 1907 laminated arches were made of dry softwood lumber and casein glue. Improvements in casein glue during World War I aroused further interest in the fabrication of glued laminated structural members, at first for aircraft and later extending to the framing members of buildings. One of the early examples of the use of glued laminated arches in the United States was in a building erected in 1934 by the Forest Products Laboratory. This installation was followed by many others in gymnasiums, churches, halls, factories, hangars, and barns. The development of the highly moisture-resistant synthetic resin glues permitted expansion of glued laminated construction into such uses as bridges, trucks, and marine construction where a high degree of resistance to severe service conditions is required (4)[16]. With increased public acceptance of glued laminated construction, laminating has increased steadily until it now forms an important segment of the woodworking industry.

Advantages of Glued Laminated Members

If glued laminated members are made of dry lumber, they will remain relatively constant in dimensions under normally dry use conditions. Hence, the development of splits, checks, and loosened fastenings and connections is greatly reduced.

Another advantage of laminated members is that they can be produced in a thoroughly dry condition in a relatively short time.

Laminating also permits more nearly complete utilization of available supplies, because larger members can be built up of short-length and narrow lumber. Moreover, selective placement of the laminations to disperse knots and other natural characteristics results in improved strength.

---

[16] Italic numbers in parentheses refer to Literature Cited. p. 273.

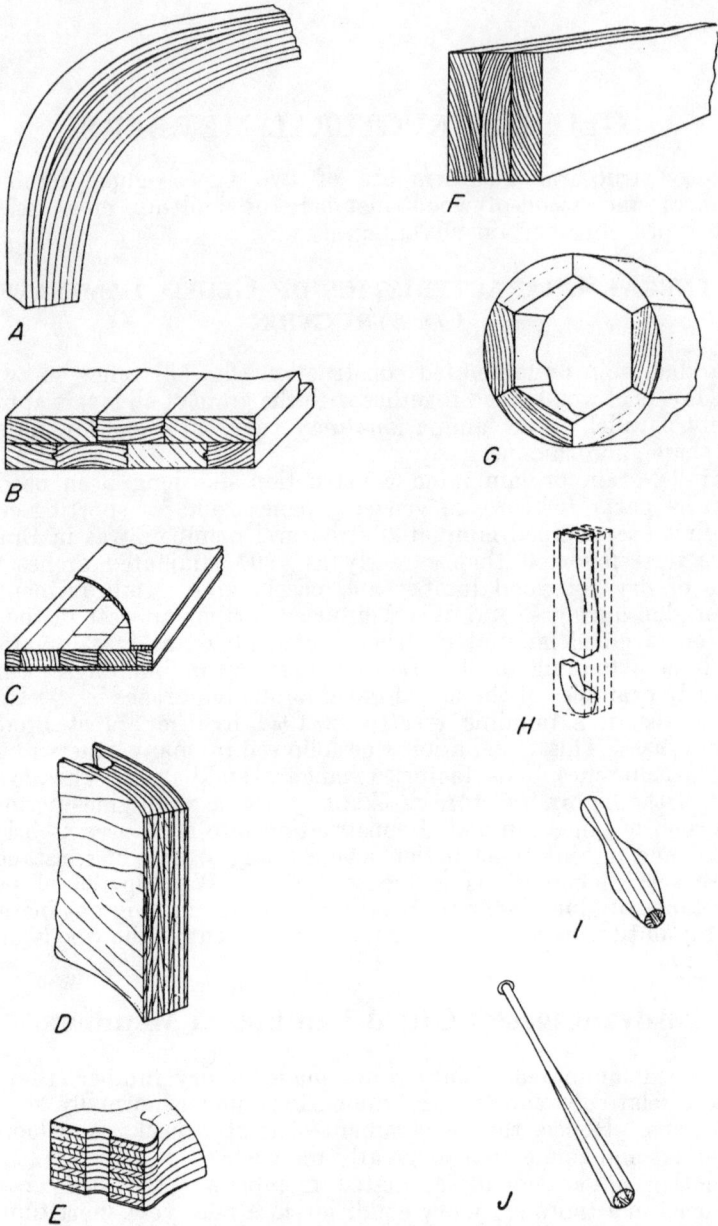

ZM 92379 F

FIGURE 55.—Types of parallel-grain (laminated) construction: A, Section of arch; B, section of table top; C, section of door stile or rail; D, section of grand piano; E, section of airplane propeller hub; F, section of beam; G, section of a column; H, chair leg; I, bowling pin; J, baseball bat.

Laminating structural members by gluing, however, involves certain economic considerations and fabricating techniques not encountered in producing solid timbers. The cost of preparing lumber for gluing and of the laminating operation, for example, usually raises the final cost of the product above that of solid green timbers. For constructions in which green timbers are satisfactory, therefore, more time and higher costs are required to cut and season lumber and to laminate than to cut solid green timbers. In addition, the laminating process itself requires equipment, plant facilities, and fabricating skills not required for producing solid green timbers.

Avoiding Internal Stresses

For best results in making glued laminated members, it is important to avoid the development of internal stresses when the member is exposed to conditions that change its moisture content. Differences in shrinking and swelling are the fundamental causes of internal stresses, and laminations should be of such character that they shrink or swell similar amounts in the same direction. If laminations are of the same species or of species with similar shrinkage characteristics, if they are all flat-grained or all edge-grained material, and if they are of the same moisture content, the assembly will be reasonably free from stresses on the glued joints and have little tendency to change shape or to check or open at the joints. Laminations that have an abnormal tendency to shrink endwise because they have excessive cross grain or compression wood (p. 97) should not be included in constructions that must remain flat, because serious bowing or cupping may result.

While observance of these principles is desirable, practical considerations may prevent exact conformance. In softwood structural members intended for interior use, for example, segregation of flat-grained from edge-grained material is generally unnecessary, and a range in moisture content among the laminations in the same assembly no greater than 5 percent may be permitted without significant effect on serviceableness.

Preservative Treatment of Laminated Members

If the laminated member is to be used under conditions that raise its moisture content to more than about 20 percent, either the heartwood of durable species should be used or the wood should be treated with preservative chemicals and only highly moisture-resistant glues should be used. If the size and shape of the members permit, laminated members can be treated with preservatives after gluing, but penetration perpendicular to the planes of the glue joints will be distinctly retarded at the first glue line. Providing suitable precautions are observed, laminations may be treated and then glued. The treated laminations should be conditioned and must be resurfaced just before gluing. Not all preservative-treated wood can be glued with all glues, but, if suitable glues and treatments are selected and if the gluing is carefully done, laminated members can be produced that are entirely serviceable under moist, warm conditions favoring decay.

Species for Laminating

Softwoods, principally Douglas-fir and southern yellow pine, are most commonly used for laminating such structural members as arches and beams. Boat timbers, on the other hand, are often made of white oak because it is moderately durable under wet conditions. Other species can also be used, of course, when their mechanical and physical properties are suited for the purpose.

Quality of Glue Joints

The quality of glue joints in laminated members intended for service under dry conditions is usually evaluated by means of the block shear test (1). Acceptable criteria are often based on unit shear strength and percentage of wood failure considered satisfactory for the species being used. For laminated members that must withstand severe service conditions, however, the block shear test alone does not provide an adequate evaluation. To serve satisfactorily under severe conditions, the glue joints should be capable of withstanding, without significant delamination, high internal stresses that develop as a result of rapid wetting and drying (2). Industry standards (13, 14, 20) covering the design and fabrication of glued structural laminated wood have been issued, and at least two military specifications cover structural glued laminated items for specific uses (16, 17).

STRENGTH OF GLUED LAMINATED MEMBERS

From the standpoint of strength and design, glued laminated members have certain definite advantages. For example, a laminated member made of thoroughly seasoned wood for use under continuously dry conditions may be designed on the basis of the strength of the dry wood, regardless of the size of the finished member, while large, one-piece sawed members must be designed on the basis of green strength, regardless of service conditions. In a laminated member designed to resist bending loads, high-grade laminations may be placed in the outer portions of the beam, where their high strength may be effectively used, and lower grade laminations in the inner portion, where their low strength will not greatly affect the overall strength of the member. By selective placement of the laminations, the knots can be scattered and improved strength can be obtained. Even with random assembly of laminations, studies have indicated that knots are unlikely to occur one above another in several adjacent laminations.

Aside from the beneficial effect of dispersion of imperfections that results from laminating, however, available test data do not indicate that laminating improves strength properties over those of a comparable solid piece. That is, the simple act of gluing together pieces of wood does not, of itself, improve strength properties, unless the laminations are so thin that the glue bonds significantly affect the strength of the member. For most laminating, where material in lumber thicknesses is used, that would not occur.

The various factors affecting strength are discussed in the following paragraphs. The principal determinants of strength are knots, cross grain, and end-joint efficiency. It should be noted that the effects of these 3 factors are not cumulative; that is, the lowest of the 3 controls the strength. Where other effects, such as those of curvature, are applicable, they should be applied in addition to those for knots, cross grain, and end joints.

Effect of Knots on Bending Strength and Stiffness

The effect of knots on the bending strength and stiffness of laminated members depends upon the number, size, and position, with respect to the neutral axis of the member, of the knots close to the critical section. Specifically, the bending properties depend upon the sum of the moments of inertia, about the gravity axis of the full cross section, of the areas occupied by all knots within 6 inches to either side of the critical section. This sum may be represented by the symbol I_K. The moment of inertia of the full (or gross) cross section of the member about its own gravity axis is represented by I_G. The relations between bending strength and stiffness and the ratio I_K/I_G are shown in figures 56 and 57.

The curves shown in figures 56 and 57 were empirically derived from tests of laminated beams containing knots in various concentrations (22).

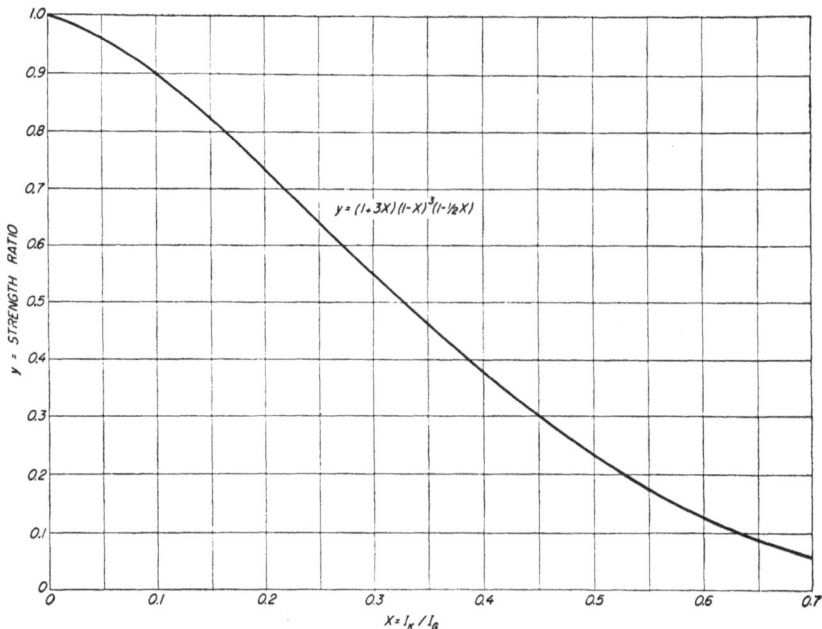

$$y = (1 + 3X)(1-X)^3(1-\tfrac{1}{2}X)$$

x axis: $X = I_K / I_G$

y axis: $y =$ STRENGTH RATIO

ZM 75226 F

FIGURE 56.—Design curve relating allowable flexural stress to moment of inertia of areas occupied by knots in laminations of laminated beams.

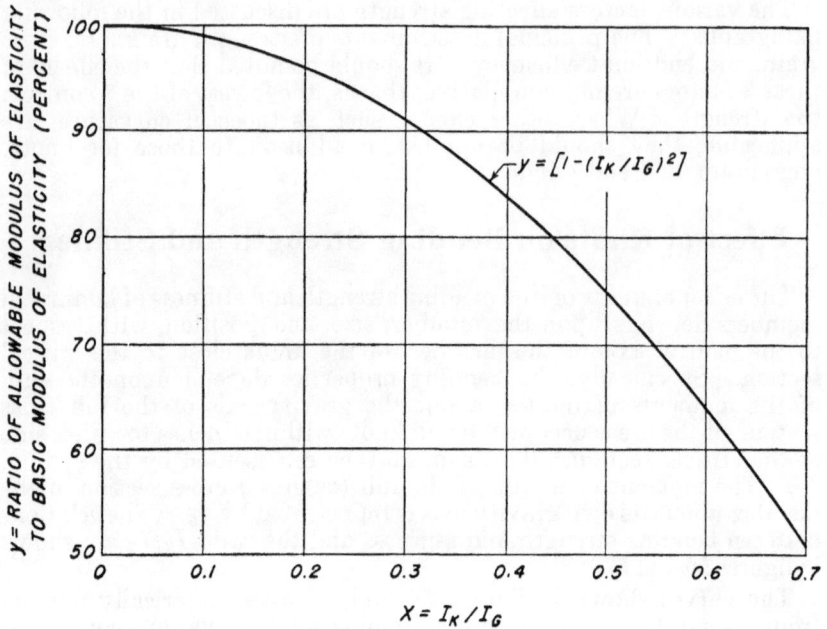

ZM 81023 F

FIGURE 57.—Design curve relating allowable modulus of elasticity to moment of inertia of areas occupied by knots in laminations of laminated beams.

Effect of Knots on Compressive Strength

The compressive strength of laminated members depends upon the proportion of the cross-sectional area of each lamination occupied by the largest knot in the lamination. Figure 58 shows an empirically derived relationship between compressive strength and K/b, where b is the lamination width and K is the average of the largest knot sizes in each of the laminations (22).

Effect of Knots on Tensile Strength

Test data relating knot size to tensile strength are not available. Figure 59, however, represents a relation between tensile strength and K/b derived from figure 58.

Effect of Cross Grain on Strength

The effect of cross grain on strength is given in table 15. For laminated beams, it is possible to vary the cross-grain requirements at different points in the depth of the beam in accordance with the stress requirements. That is, steeper cross grain may be permitted in laminations in the interior of the beam than in the laminations at and near the outside. The permitted variation should be based on the assumption of linear variation of stress across the depth.

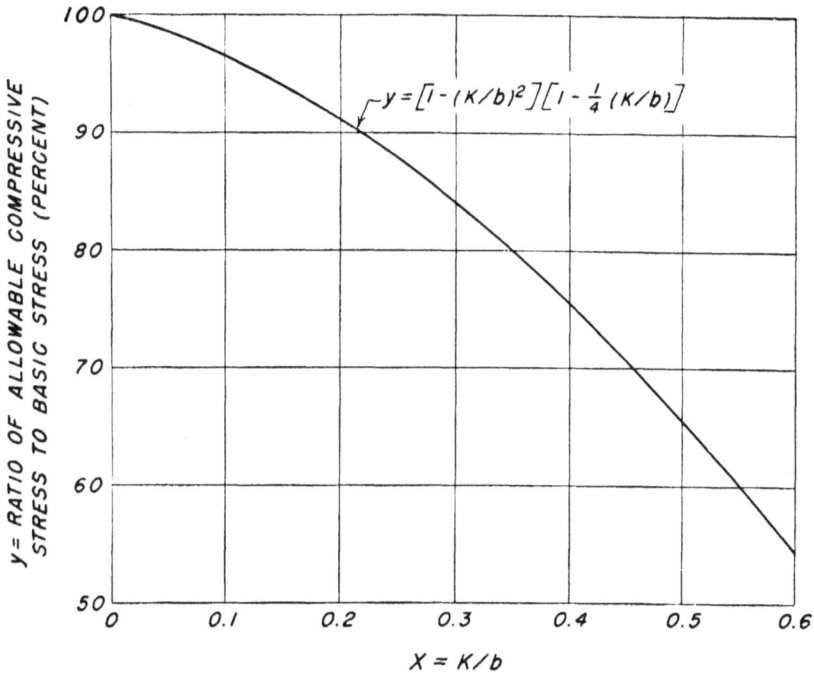

$$y = \left[1 - (K/b)^2\right]\left[1 - \tfrac{1}{4}(K/b)\right]$$

FIGURE 58.—Design curve relating allowable compressive stress to size of knots in laminations of laminated short columns.

Effect of Curvature of Laminations on Bending Strength

Stress is induced when laminations are bent to curved forms, such as arches and curved rafters. While much of this stress is quickly relieved, some remains and tends to reduce the strength of a curved member. The ratio of the allowable design stress in curved members to that in straight members is

$$1.00 - \frac{2,000}{(R/t)^2}$$

where t is the thickness of a lamination and R is the radius of the curve to which it is bent, both R and t being in the same units (21).

Effect of Lamination Thickness on Strength

Tests (22) have indicated that lamination thickness has no effect on the strength of straight laminated members. Lamination thickness does, however, affect the strength of curved members, depending upon the radius to which the lamination is bent (see preceding section).

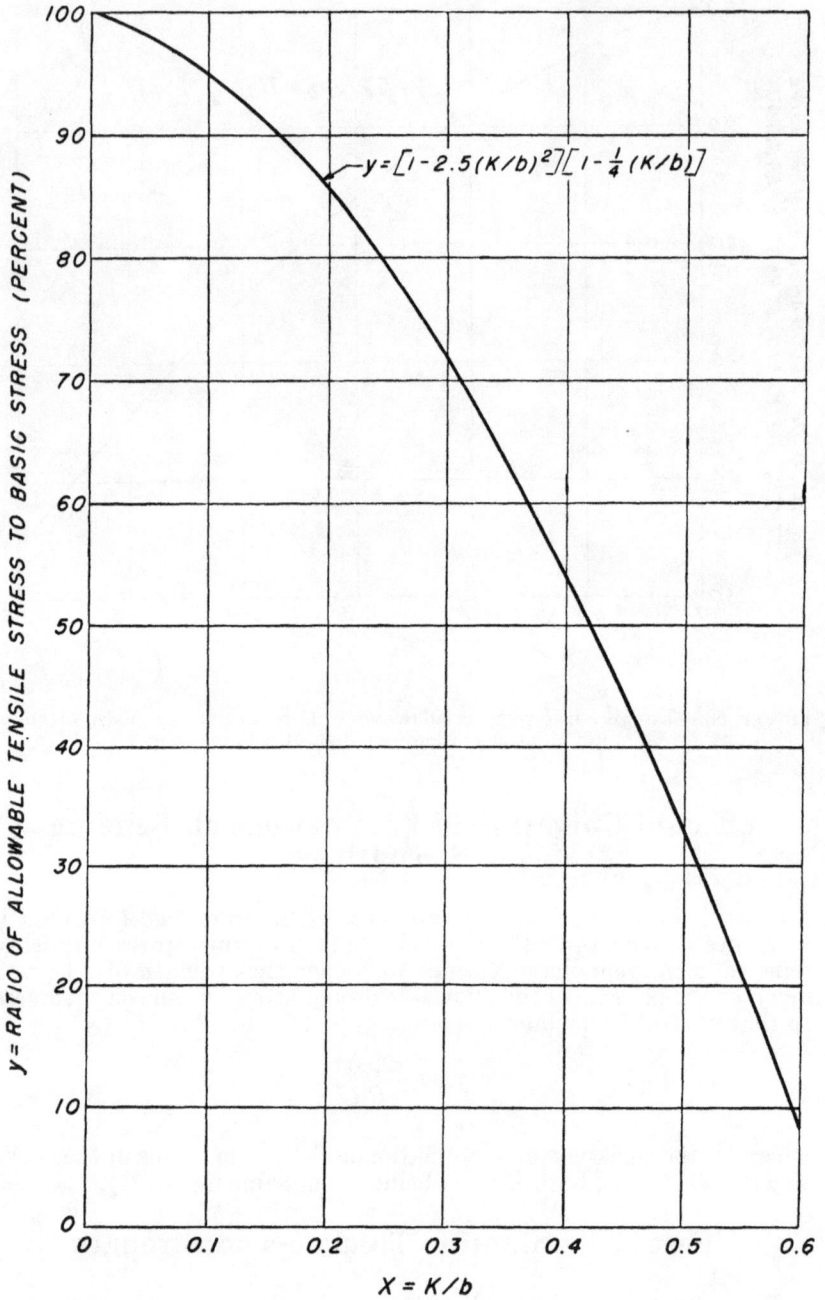

FIGURE 59.—Design curve relating allowable tensile stress to size of knots in laminations of laminated tension members.

Effect of End Joints on Strength

For a large proportion of laminated members, because of their size, pieces of wood will need to be jointed end-to-end to provide laminations of sufficient length. In most cases, the strength of members is reduced by the presence of end joints (*4*, *21*, *22*).

Scarf Joints

Laminations containing scarf joints (fig. 53) and located in the tension portion of bending members or in tension members should not be stressed to more than the following percentages of the stress permitted in an unjointed member:

Scarf slope:	Joint factor (percent)
1 in 12 or flatter	90
1 in 10	85
1 in 8	80
1 in 5	65

These limitations apply to interior laminations as well as to the lamination at the tension face. In a beam with 40 equal laminations, for example, the stress at the outer face of the second lamination is 95 percent of that at the outer face of the bottom lamination. Hence, if the bottom lamination is continuous, and the second lamination contains a scarf joint sloping at 1 in 12 at the critical section, the working stress (stress in extreme fiber) should be restricted to about 95 percent of the basic value, in order not to exceed the 90 percent permissible at the outer face of the second lamination.

Working stresses need not be reduced if laminations containing scarf joints are located in the compression portion of bending members or in compression members.

Stepped or hooked scarf joints (fig. 53) may be assumed to have the same strength as a plain scarf joint of the same slope, except that the area occupied by the step or hook should be considered as ineffective in transmitting stress. The slope of a stepped scarf should be taken as the slope of the plane portion of the scarf, not as the distance between tips divided by the lamination thickness.

A double-slope scarf joint (patent applied for), as illustrated in figure 53, has been used to some extent for end-jointing laminations. If made by comparable fabricating techniques, a joint of this type should have the same strength as a plain scarf joint of the same slope.

Scarf joints in tension and bending members should be well scattered. It is suggested that, in highly stressed areas, scarf joints in adjacent laminations be spaced, center to center, at intervals of not less than 24 times the lamination thickness. In bending members spacing of scarf joints may vary linearly from $24t$ in areas of maximum allowable stress to $0t$ in areas of zero stress.

Where it is not feasible to maintain a specific pattern of scarf-joint distribution, joints should be well scattered—that is, a conscious effort should be made to prevent concentration of joints—and somewhat lower strength should be assumed for scarf joints stressed in tension than was given previously. Under such conditions, the following joint factors are suggested in lieu of those given previously:

Scarf slope: Joint factor
                                                               (percent)
  1 in 12 or flatter_____ 85
  1 in 10_____ 80
  1 in 8_____ 75
  1 in 5_____ 60

The slope of scarf joints in compression members or in the compression portion of bending members should not be steeper than 1 in 5; a similar limitation of 1 in 10 is suggested for tension members and the tension portion of bending members. Because there is some question as to the durability of steep scarf joints for exterior use or other severe exposure, scarf joints with a slope steeper than 1 in 10 should not be used.

Butt Joints

Butt joints can transmit no tensile stress and can transmit compressive stress only after considerable deformation or if a metal bearing plate is tightly fitted between the abutting ends. In normal assembly operations, such fitting would not be done, and it is therefore necessary to assume that butt joints are ineffective in transmitting both tensile and compressive stresses. Because of this ineffectiveness, and because butt joints cause concentration of both shear stress and longitudinal stress, they are not recommended for use in structural members.

If butt joints are used in the compression portion of bending members, all laminations at a particular cross section containing butt joints should be disregarded in computing the moment of inertia of that cross section. If butt joints occur in the tension portion of a bending member, the effective moment of inertia to be used in design is the net moment of inertia (computed about the mid-depth of the beam and considering laminations with butt joints ineffective) further reduced by multiplying it by a factor of 0.8.

If butt joints are used in compression members, the effective cross-sectional area should be computed by subtracting from the gross cross-sectional area the area of all laminations containing butt joints at a single cross section. In addition, laminations that contain butt joints and that are adjacent to others containing butt joints should be considered only partially effective if the spacing of the joints in adjacent laminations is less than 50 times the lamination thickness. The effective area of such adjacent laminations should be computed by multiplying their gross area by the following percentages:

Butt-joint spacing   Effectiveness factor
(t=lamination thickness)   (percent)
      $30t$ 90
      $20t$ 80
      $10t$ 60

Butt joints are not recommended for use in tension members. If they are used, the effective cross-sectional area may be computed as indicated for compression members, except that an additional reduction of 20 percent should be made.

Other Types of End Joints

In addition to scarf joints and butt joints, other types of end joints may be used in laminated members. In general, however, few data are available on their effect on strength. When data are not available for establishing spacing requirements, strength, and like factors for these joints, they may be treated as butt joints.

Effect of Edge Joints on Strength

It is sometimes necessary to joint pieces edge-to-edge to provide laminations of sufficient width. For tension members, compression members, and horizontally laminated bending members, the strength of such joints is of little importance to the overall strength of the member. Therefore, from the standpoint of strength alone, it is unnecessary that edge joints be glued. Other considerations, however, such as the appearance of face laminations or the possibility that water will enter the unglued joints and promote decay, may dictate that edge joints should be glued.

In vertically laminated beams, sufficient laminations must be edge glued to provide adequate shear resistance in the beam. Not only is adequate initial strength required of such joints, but they must be durable enough to retain that strength under the conditions to which the beam is exposed in service.

Effect of Shakes, Checks, and Splits on Shear Strength

In general, shakes, checks, and splits have little effect on the shear strength of laminated beams. Shakes generally occur infrequently but should be excluded from material for laminations or placed only in positions of low shear. Most laminated members are made from laminations that are thin enough to season readily without developing checks and splits. Since checks and splits lie in a radial plane, and since the majority of laminations are flat grained, checks and splits will be so positioned in horizontally laminated beams that they will not affect shear strength.

In cases where shakes, checks, and splits affect the shear strength of a beam, as in vertically laminated beams, their effect may be taken into account in the manner described for one-piece sawed members (p. 206).

Vertically Laminated Beams

The effect of knots on the strength of vertically laminated beams may be taken into account by determining the strength ratio of each lamination by the method described in the section on stress grades and working stresses and averaging these ratios to determine the strength ratio for the beam (22). The effects of cross grain on strength and limitations on end joints are as described on pages 252–257.

Effect of Beam Depth on Bending Strength

Unit strength values, computed by the usual engineering methods from test data on wood beams, decrease as the depth of the beam increases. Basic stresses for extreme fiber in bending have been taken to correspond to beams of 12-inch depth, so that the effect of depth on strength is not a serious problem for one-piece sawed members. Laminated members, however, can be of considerable depth, and the effect of this factor on strength should be considered.

An empirical formula for the depth-effect factor, F_h, is

$$F_h = 0.81 \frac{H^2 + 143}{H^2 + 88}$$

where H is the depth of the beam.

The form of the beam has a similar effect, as discussed under structural members of lumber (p. 210).

DESIGN OF GLUED LAMINATED MEMBERS

Engineering formulas applicable to solid wood structures are applicable also to structures of laminated wood. The fact that laminated members can be made in curved form, however, introduces some circumstances not ordinarily met in the design of wood structures.

The application of ordinary engineering formulas to deep, sharply curved bending members may introduce appreciable error in the calculated stresses. For such cases, the methods applicable to curved beams, as described in standard text books on mechanics, should be used.

When bending moments are applied to curved members, stresses are set up in a direction parallel to the radius of curvature (perpendicular to grain). Stresses so induced should be appropriately limited.

Procedures for establishing design stresses and factors affecting the design of glued laminated structural members are discussed in detail by Freas and Selbo (4).

Basic Stresses

Basic stresses for use in determining working stresses for the design of laminated members to be used under dry conditions, as in most covered structures, are given in table 36. For structures to be used under wet conditions of service, as in most exterior service, the basic stresses are the same as those given for solid sawed timbers (table 25).

Estimation of I_K/I_G for Bending Members

Unless the location of knots is controlled when laminations are assembled in beams and arches, an assumption must be made as to the concentration of knots near the critical cross section, in order to estimate the factor I_K/I_G for the determination of the appropriate strength ratio. A conservative assumption would place a knot of maximum permitted size at the critical cross section in every lamination. Experience with lumber grades indicates that such a concentration of knots is unlikely to occur.

When data on the frequency of occurrence of the sums of knot sizes in 1-foot lengths for a specific grade are available, it is possible to apply statistical procedures to these data and to arrive at a value of I_K/I_G suitable as a basis for design. The same procedure may be used for members in which two grades of lumber are used. Values of I_K/I_G and of the strength ratio, based on a statistical approach, have been developed for a limited number of grades of Douglas-fir and southern yellow pine (14, 20).

When a statistical analysis is not available for a grade or species, it is suggested that either definite limitations on knot concentrations be imposed in assembly of laminations or, if random assembly of laminations is used, a conservative assumption as to knot concentration be used as a basis for design. If specific limitations on knot

TABLE 36.—*Basic stresses for structural members laminated from clear material and under long-time service at maximum design load and under dry conditions (as in most covered structures)—for use in determining working stresses according to grade of laminations and other applicable factors*

SOFTWOODS

Species [1]	Extreme fiber in bending or tension parallel to grain	Maximum longitudinal shear	Compression perpendicular to grain	Compression parallel to grain	Modulus of elasticity in bending
	P. s. i.	*P. s. i.*	*P. s. i.*	*P. s. i.*	*1,000 p. s. i.*
Baldcypress (cypress)	2,400	170	330	2,000	1,300
Cedar:					
Alaska-	2,000	150	275	1,450	1,300
Atlantic (southern whitecedar) and northern white-	1,400	115	195	1,050	900
Port-Orford-	2,000	150	275	1,650	1,600
Western redcedar	1,600	135	220	1,300	1,100
Douglas-firs:					
Coast type	2,750	150	350	2,000	1,800
Coast type, close-grained	2,950	150	375	2,150	1,800
Rocky Mountain type	2,000	135	310	1,450	1,300
All types, dense	3,200	150	410	2,350	1,800
Fir:					
Balsam	1,600	115	165	1,300	1,100
California red, grand, noble, and white	2,000	115	330	1,300	1,200
Hemlock:					
Eastern	2,000	115	330	1,300	1,200
Western (west coast hemlock)	2,400	125	330	1,650	1,500
Larch, western	2,750	150	350	2,000	1,600
Pine:					
Eastern white (northern white), ponderosa, sugar, and western white (Idaho white)	1,600	135	275	1,400	1,100
Jack	2,000	135	240	1,450	1,200
Lodgepole	1,600	100	240	1,300	1,100
Red (Norway pine)	2,000	135	240	1,450	1,300
Southern yellow	2,750	180	350	2,000	1,800
Dense	3,200	180	410	2,350	1,800
Redwood	2,200	115	275	1,850	1,300
Close-grained	2,400	115	295	2,000	1,300
Spruce:					
Engelmann	1,400	115	195	1,100	900
Red, white, and Sitka	2,000	135	275	1,450	1,300
Tamarack	2,200	160	330	1,850	1,400

See footnote at end of table.

TABLE 36.—*Basic stresses for structural members laminated from clear material and under long-time service at maximum design load and under dry conditions (as in most covered structures)—for use in determining working stresses according to grade of laminations and other applicable factors*—Continued

HARDWOODS

Species [1]	Extreme fiber in bending or tension parallel to grain	Maximum longitudinal shear	Compression perpendicular to grain	Compression parallel to grain	Modulus of elasticity in bending
	P. s. i.	*P. s. i.*	*P. s. i.*	*P. s. i.*	*1,000 P. s. i.*
Ash:					
Black	1,800	150	330	1,150	1,200
Commercial white	2,550	210	550	2,000	1,600
Beech, American	2,750	210	550	2,200	1,800
Birch, sweet and yellow	2,750	210	550	2,200	1,800
Cottonwood, eastern	1,400	100	165	1,100	1,100
Elm:					
American and slippery (soft elm)	2,000	170	275	1,450	1,300
Rock	2,750	210	550	2,200	1,400
Hickory, true and pecan	3,500	235	660	2,750	2,000
Maple, black and sugar (hard maple)	2,750	210	550	2,200	1,800
Oak, commercial red and white	2,550	210	550	1,850	1,600
Sweetgum (gum, red gum, sap gum)	2,000	170	330	1,450	1,300
Tupelo, black (black gum) and water	2,000	170	330	1,450	1,300
Yellow-poplar (poplar)	1,800	150	240	1,450	1,300

[1] Species names from approved check list, U. S. Forest Service. Commercial designations are shown in parentheses.

concentration are imposed, the limitations adopted may be used as a basis for calculation of I_K/I_G.

Under a conservative assumption, each lamination would contain, at the critical cross section, a knot of the maximum size permitted in the grade. Then, for members containing laminations of a single grade, the value of I_K/I_G will be numerically equal to the ratio K/b, where K is the maximum permitted knot size and b is the finished width of the lamination. Where two grades are combined—for example, in a member with an outer group of laminations on each side of higher grade than the central group of laminations—the value of K/b for the member, in computing I_K/I_G, depends upon the contribution of all the groups to the moment of inertia of the member. For example, assume a member having 15 laminations, of which the central 9 have a K/b value of 0.50 and the outer 3 on each side have a K/b value of 0.25. It may be assumed that for all laminations K/b equals 0.25 and that the central 9 laminations have an additional K/b of 0.25. The central group, however, contributes to the moment of inertia only in the proportion $(9/15)^3$. Therefore, the value of I_K/I_G will be:

$$0.25 + 0.25 \ (9/15)^3 = 0.304$$

Where it is feasible to establish a specific pattern of knot locations during assembly of the member, I_K/I_G may be evaluated from the knot locations specified by the pattern.

Estimation of K/b for Axially Stressed Members

Unless the location of knots is controlled when axially stressed members are assembled, some assumption must be made as to the concentration of large knots in a member. A conservative assumption would be that a knot of maximum permissible size occurs in each 3-foot length of each lamination.

A statistical method based on the frequency of occurrence of the largest knots in 3-foot lengths of lumber has been developed for estimating the average of the largest knot sizes in a member. This method has been applied to a limited number of Douglas-fir and southern yellow pine grades (14, 20).

If grades or species are to be used for which a statistical analysis has not been made, it is suggested that the conservative assumption for concentration of large knots be used as a basis for design stresses. Under this assumption, the factor K/b, for use with figure 58 or figure 59, will involve a value of K equal to the size of the largest knot permitted in the grade. For members composed of laminations of more than one grade, the stress applicable to the lower grade should be used.

Bending Members

In designing bending members, a number of factors are involved in choosing an appropriate working stress—the effects of knots, cross grain, end joints, height or depth of the beam, and curvature of the laminations. The first four factors apply to either straight or curved members and are not to be combined. The basic stress is multiplied by the lowest of these four factors to obtain the working stress of straight members.[17] For curved members, the resulting value is multiplied also by the curvature factor.

Moment of Inertia

If scarf joints with slopes not steeper than 1 in 5 are present, the moment of inertia of the full cross section of the beam may be considered in design. If butt joints are present, the moment of inertia should be reduced as described for them.

Deflections

Experience with figure 57 and statistically derived values of I_K/I_G indicates that, for the majority of cases, the reduction of modulus of elasticity below the basic value will be less than 5 percent. In most instances, therefore, it will be sufficiently accurate to base calculations of deflection on basic values of modulus of elasticity. Use of either basic values of modulus of elasticity or values derived from figure 57 will give only immediate values of deflection. Where deflection is critical, consideration must be given to the added deflection that occurs under long-time loading (see p. 161).

---

[17] This may be unconservative for members having span-depth ratios of less than about 14. That is, for such members, it may be necessary to modify the lowest of the other three strength ratios by multiplying by the height factor.

Radial Stresses

When curved members are subjected to bending moments, stresses are set up in a direction parallel to the radius of curvature (perpendicular to grain). If the moment increases the radius (makes the member straighter), the stress is tension; if it decreases the radius (makes the member more sharply curved), the stress is compression (*21*). The stress is a maximum at the neutral axis and is approximately

$$S_R = \frac{3}{2} \frac{M}{Rbh}$$

where M is the bending moment, R is the radius of curvature of the centerline, and b and h are, respectively, the width and height of the cross section.

For compression perpendicular to grain, values of S_R should be limited to those shown in tables 25 and 36. For tension perpendicular to grain, the stress should be limited, for softwoods, to about one-third and, for hardwoods, to about three-eighths of the working stress in shear.

Deep, Curved Members

It is known from the principles of mechanics that the stresses in sharply curved members subjected to bending are in error when computed by the ordinary formulas for straight beams. The amount of the error depends upon the relation of the depth of the member to the centerline radius. In solid timber construction, of course, this is not a problem. In laminated construction, however, deep, sharply curved members, such as arches with abrupt knees, may be used. It is probable that limitations on the sharpness to which laminations can be bent will limit the curvature of deep beams or arches. The analytical methods applicable to curved beams should be used, however, to determine whether or not the usual engineering formulas may be applied without significant error to deep, sharply curved members of glued laminated wood.

Axially Stressed Members

Three factors must be considered in choosing working stresses for axially stressed members—the effects of knots, cross grain, and end joints. The factor giving the lowest strength ratio is the one that determines the working stress; the three factors are not to be combined. For compression members, the effect of plain scarf joints need not be considered, since they are assumed to be 100 percent efficient in transmitting compression stress.

Effective Cross-Sectional Area

If plain scarf joints with slopes not steeper than 1 in 5 are present, the full cross section may be considered effective. If butt joints are present, the cross-sectional area should be reduced as was described for these joints.

Columns of Different Classes

The usual formulas for determining the load-carrying capacities of wood columns may be used for laminated columns. For short and intermediate columns, the proper value of the compressive strength should be calculated. For all classes of columns, the effect of joints should be considered in arriving at the proper value of the effective area.

Fastenings

Design loads or stresses for fastenings that are applicable to solid wood members (p. 165) are applicable also to laminated members. Since greater depths are possible with laminated members than with solid members, however, design of fastenings for deep laminated members requires special consideration. In these members, considerable shrinkage may occur between widely separated bolts, and, if the bolts are held in position by metal shoes or angles, large splitting forces may be set up. It is desirable, therefore, to have the moisture content of the member as near as possible to that which it will attain in service. Slotting of the bolt holes in the metal fitting will tend to relieve the splitting stresses. Cross bolts will assist in preventing separation if splitting does occur; they will not, however, prevent splits.

DESIGN OF WOOD-PLYWOOD BEAMS AND GIRDERS

Glued structural members of built-up cross section, with plywood webs and either solid or laminated flanges, may be designed to carry the same loads as members of rectangular cross section. The cross sections of such wood-plywood beams and girders may have the form of a box, an I, or a double I, as shown in figure 60.

Because the design of members with built-up cross sections places material where it may be stressed to the best advantage, considerable

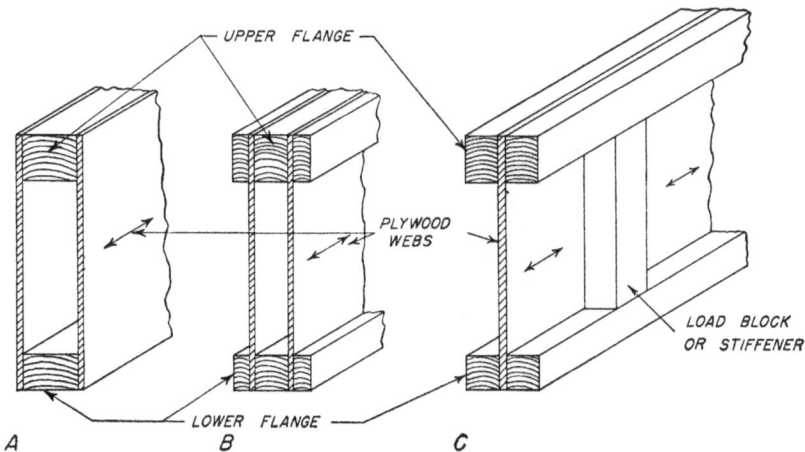

ZM 89772 F

FIGURE 60.—Typical cross sections for wood-plywood beams: A, Box beams; B, double I-beam; C, I-beam.

savings of wood and weight are possible. Fabrication costs, however, are higher than for members of rectangular cross section and may offset the savings in material.

The design of a wood-plywood beam or girder is similar to that of a steel-plate girder: (1) The flanges are designed to withstand bending moments; (2) the plywood webs are designed to withstand shear; (3) the joints between the webs and the flanges are designed to transfer stresses from the webs to the flanges; and (4) stiffeners and load blocks are provided to distribute concentrated loads and to limit the lengths of the unsupported panels in the webs. The design of such a girder is a "cut and try" procedure, but, once a trial section is selected and loads are known, stresses can be calculated by using conventional formulas.

Limiting Considerations in the Design of Wood-Plywood Beams

The overall height of a wood-plywood beam or similar structural member is governed by one of the following considerations: (1) Clearance or other headroom limitations. (2) Width of the plywood economically available for use in a beam. This is usually 4 feet or some even fraction, such as one-half, one-third, or one-fourth, of 4 feet, so that waste will be minimized. (3) Ratio of height to width and span. The member must not be laterally unstable under design loads when it is not restrained.

The choice of the type of cross section that should be used in a given application depends upon several factors. If plywood of sufficient thickness is not available to permit utilization of single-web members (I-beam), members with multiple webs may be required. For some architectural appearances, the smooth surface of the box beam may be desired. In certain exposures, however, because of the danger that moisture may condense on or otherwise enter the interior of a box beam, it may be desirable to select an I-beam. The I-beam is easier to inspect for deterioration while in service than multiple-web beams, because all parts of the I-beam are exposed.

Design of Flanges

The flanges of a wood-plywood member carrying bending moments are designed so that the compressive stresses in compression flanges at design load do not exceed the compressive working stress of the wood and so that the tensile stresses in the tension flange do not exceed the tensile working stress of the wood. For flanges that are of laminated construction, the working stresses and limitations for scarf joints given in the section on glued laminated members (p. 255) should be used as the limiting values of stress.

Calculation of Stress in Tension and Compression

The stress in the extreme fiber in compression is calculated by the following formula:

$$S_c = \frac{MC_c}{I}$$

where S_c is the fiber stress in compression, in pounds per square inch; M, the bending moment at the section under investigation, in inch-pounds; C_c, the distance from the neutral axis to the extreme fiber in the compression flange, in inches; and I, the moment of inertia of the section about the neutral axis, in inches[4].

The stress in the extreme fiber in tension is calculated by the following formula:

$$S_t = \frac{MC_t}{I}$$

where S_t is the fiber stress in tension, in pounds per square inch; M the bending moment at the section under investigation, in inch-pounds; C_t, the distance from the neutral axis to the extreme fiber in the tension flange, in inches; and I the moment of inertia of the section about the neutral axis, in inches[4].

The neutral axis of a box beam or an I-beam is at midheight if the tension and compression flanges are of the same size, but if flanges of unequal size are used, it is necessary to locate the neutral axis by a summation of the moments of the areas of the cross section. Unequal flanges are more economical of material than those of equal size, because the tensile strength of wood is much higher than the compressive strength. The location of the neutral axis of a beam with unequal flanges may be computed by the formula:

$$e = \frac{b \left[t_c (d - t_c) - t_t (d - t_t) \right]}{2 (bt_c + bt_t + nkdt_w)}$$

where e is the distance from midheight of the beam to the neutral axis; n the number of webs in the cross section of the beam; and k the modular ratio between the webs and flanges. For plywood with the grain of some plies parallel and the grain of other plies perpendicular to the grain of the flanges, the modular ratio k is equal to the ratio of the combined thickness of the plies with grain parallel to the grain of the flanges to the thickness of the plywood. For plywood with the grain oriented at 45° to the grain of the wood in the flanges, the modular ratio k is approximately $\frac{1}{4}$. The symbols b, t_c, t_t, t_w, and d are the dimensions shown in figure 61.

Customarily, for preliminary calculation, the webs are ignored both in locating the neutral axis and in computing the moment of inertia. When this is done, the formula simplifies to

$$e = \frac{t_c (d - t_c) - t_t (d - t_t)}{2 (t_c + t_t)}$$

The moment of inertia for a member with a built-up cross section is computed by the summation of the moments of inertia of the elements of the cross section about their respective gravity axes and summation of the areas of the elements multiplied by the square of the distances from the neutral axis to the respective gravity axes. The two summations are added to give the general formula

$$I = \sum I_0 + \sum A x^2$$

where I represents the moment of inertia of the entire cross section; I_0 the moment of inertia of each element about its gravity axis; A the area of each element; and x, the distance from the neutral axis of the entire cross section to the gravity axis of an element.

The specific formula for computing the moment of inertia I of the cross section of a wood-plywood member may be written as follows:

$$I = b\left[\frac{t_c^3 + t_t^3}{12} + t_c\left(C_c - \frac{t_c}{2}\right)^2 + t_t\left(C_t - \frac{t_t}{2}\right)^2\right] + \frac{nkt_w}{12}(d^3 + 12de^2)$$

The symbols are the same as used in previous formulas or are shown in figure 61.

Again, for preliminary computations, the contribution of the webs to the moment of inertia is often neglected. Then the specific formula is simplified to

$$I = b\left[\frac{t_c^3 + t_t^3}{12} + t_c\left(C_c - \frac{t_c}{2}\right)^2 + t_t\left(C_t - \frac{t_t}{2}\right)^2\right]$$

When there are holes in the flanges for attaching other parts of the structure or for other purposes, the moments of inertia should be calculated on the net cross section (gross cross section minus area occu-

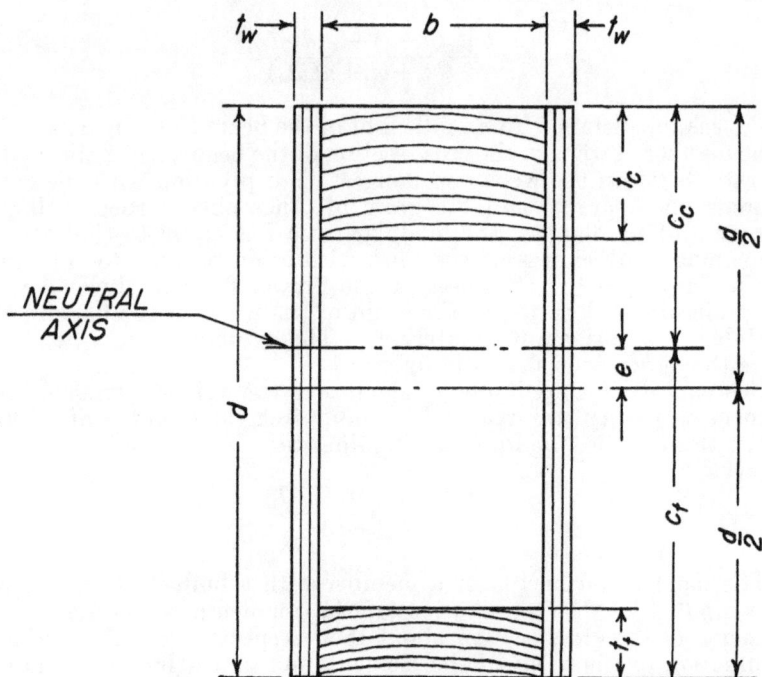

ZM 89773 F

FIGURE 61.—Component parts of a wood-plywood beam and symbols for use in formulas when the component parts have essentially equal moduli of elasticity.

pied by holes) for computing stresses, but the location of the neutral axis should be calculated on the gross cross section. Additional information on the design of flanges may be found in Forest Products Laboratory Report 1513 (7).

The preceding formulas for calculating the position of the neutral axis and the moment of inertia for built-up beams are based on the assumption that the moduli of elasticity for the component parts are essentially equal. If these moduli differ materially, the formulas will give values that are in error, and the magnitude of the error will vary with the difference in moduli of elasticity. The following calculations are more exact, since they take into account differences in moduli of elasticity.

The product of the modulus of elasticity and the moment of inertia, EI, is calculated by the following formula:

$$EI = E_1 \frac{1}{12} bt_c{}^3 + E_2 \frac{1}{12} bt_t{}^3 + E_3 \frac{1}{12} nt_w d^3 + E_1 bt_c \left(a - \frac{t_c}{2}\right)^2$$

$$+ E_2 bt_t \left(a - d + \frac{t_t}{2}\right)^2 + E_3 nt_w d \left(a - \frac{d}{2}\right)^2$$

where terms previously used have the same meanings and apply to the cross section shown in figure 62, E_1, E_2, and E_3 are the compressive moduli of elasticity of the compression flange, tension flange, and web, respectively, in pounds per square inch, and a is the distance from the outside of the compression flange to the axis about which statical moments are taken.

When the axis in question coincides with the neutral axis of the built-up beam, the sum of the statical moments will equal zero. To find the distance a at this point, the expression for EI is differentiated with respect to a and set equal to zero, giving the following equation:

$$E_1 bt_c (2a - t_c) + E_2 bt_t (2a - 2d + t_t) + E_3 nt_w d (2a - d) = 0$$

Solving this equation for a gives:

$$a = \frac{E_1 bt_c{}^2 + 2E_2 bt_t d - E_2 bt_t{}^2 + E_3 nt_w d^2}{2 (E_1 bt_c + E_2 bt_t + E_3 nt_w d)}$$

This value of a can then be substituted in the formula for determination of EI, and the EI value for the built-up beam can be computed.

To compute stresses in the component parts of the beam, the proper moment of inertia value I can be obtained by dividing the EI value for the beam by the modulus of elasticity for the part in question. For example, the formula for the stress in the extreme fiber in compression is:

$$S_c = \frac{MC_c E_1}{EI}$$

where the symbols have the meanings previously indicated.

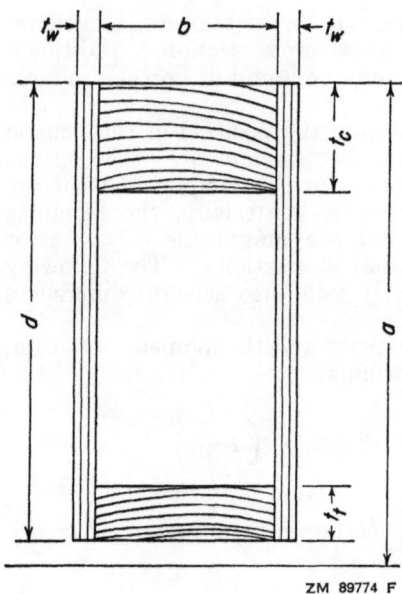

FIGURE 62.—The component parts of a wood-plywood beam and the formula symbols for computing stress when the component parts have different moduli of elasticity.

ZM 89774 F

Design of Webs

The webs of a wood-plywood member are designed so that shear stresses in the webs will not exceed design allowables when expected maximum loads occur in service. Shear webs with the grain of the veneers in the plywood oriented at 45° to the principal axis of the member have considerably higher shear strength for a given thickness than do webs with the grain of the veneers in the plywood oriented at 0° and 90° to the axis of the member. The manufacture of 45° plywood is wasteful of material, however, and plywood with this grain orientation is used only in structures where weight is of paramount importance, such as wing beams in aircraft.

The allowable working stresses in shear for the plywood webs should be determined as given for plywood elsewhere in this handbook (p. 281).

Calculation of Shear Stress in Webs

The stress in shear in the webs is calculated by the following formula:

$$S_s = \frac{VQ}{nIt_w}$$

where S_s is the shear stress in the plywood webs, in pounds per square inch; V, the total shear at the location under investigation, in pounds; Q, the statical moment of the area external to the axis of maximum shear stress, in $\overline{inches}^3$; n, the number of webs; I, the moment of inertia of the cross section about the neutral axis, in $\overline{inches}^4$, based on the gross section and computed as outlined in the section on calculation of stress in tension and compression; and t_w the thickness of one web, as shown in figure 61, in inches.

The statical moment of the area external to the neutral axis is calculated by a summation of the moments of the areas of the elements external to the neutral axis. Either those areas above or those below the neutral axis may be summed. The two statical moments are equal and may be represented by the following formulas:

$$Q=\frac{1}{2}[bt_c\,(2C_c-t_c)+nkt_w\,C_c^2]$$

or

$$Q=\frac{1}{2}[bt_t\,(2C_t-t_t)+nkt_w\,C_t^2]$$

where Q is the statical moment of the area external to the neutral axis in $\overline{\text{inches}}^3$; n, the number of webs; and K, the modular ratio between webs and flanges (see section on calculation of stress in tension and compression). The symbols b, t_c, C_c, t_w, t_t, and C_t are the dimensions shown in figure 61. More detailed information on the design of webs may be found in material published earlier (5, 6, 9, 10, 18, 19).

The statical moments should be calculated on the gross section. No corrections should be made for holes or similar reductions in the area of the flanges.

Design of Joints Between Flanges and Webs

Stresses in the flanges must be transferred to the webs when a wood-plywood member is under load. Usually, the wood-plywood member is glued, and stresses are transferred in shear along the glue line. Members should be designed so that these stresses do not exceed design allowables under expected loads.

Mechanical fastenings (such as split-ring connectors) may be used in large members to transfer the shear stress. When this is done, the total shear force between flanges and webs should be distributed among the fastenings according to design allowables given in the section on timber fastenings.

The allowable stress at the glue lines between flanges and webs is limited by the strength of the plywood in shear acting in the plane of the plywood. This type of shear is designated as rolling shear, and the allowable shear stress is approximately three-eighths of the allowable unit stress for horizontal shear of the species of wood in the plywood if the grain of the wood in the plies is parallel or perpendicular to the principal axis of the member. If the grain in the plywood is oriented at 45° to the principal axis of the member, the allowable shear stress is approximately one-half of the allowable unit stress for horizontal shear of the species of wood used in the plywood. Further information on these allowable stresses is given on page 275 and in Forest Products Laboratory Report 1630 (11).

Calculation of Shear Stress Along the Line Between Flanges and Webs

The shear stress along the line between the flanges and webs is calculated by the following formula:

$$S_g = \frac{VQ_f}{nIt_t}$$

where S_g is the shear stress at the lines between the webs and the flanges, in pounds per square inch; V, the total vertical shear, in pounds; Q_f, the statical moment of the area external to the lines between the webs and the flanges, about the neutral axis, in $\overline{\text{inches}}^{3}$; n, the number of glue lines between the webs and the flanges; I, the moment of inertia of the gross cross section of the member about the neutral axis, in $\overline{\text{inches}}^{4}$; and t_t, the thickness of the tension flange, in inches.

The statical moment of the area external to the lines between the webs and flanges, about the neutral axis, may be expressed by the following formula, since the tension flange is the area external to the lines between the webs and component parts of the tension flange:

$$Q_f = bt_t \left(C_t - \frac{t_t}{2} \right)$$

These formulas for shear stress and statical moment are applicable to wood-plywood members of box or I cross section and to double-I members when the total amount of tension-flange material on the outsides of the webs is equal to the total amount between the webs. When the amounts are not equal, the shear stresses are greater along the lines adjacent to the greater flange areas, and stresses must be investigated for these portions separately.

If the compression flanges are larger than the tension flanges, the shear stresses along the lines between the webs and the compression flanges are less than those along the lines between the webs and the tension flanges, and it is not necessary to investigate these stresses. Examples of this type of calculation are given in Forest Products Laboratory Report 1551 (8).

Design of Stiffeners and Load Blocks

The design of stiffeners and load blocks as to number and sizes does not lend itself to a rational procedure (5), but certain general rules can be given that will help the designer of a wood-plywood structure obtain a satisfactory structural member.

Stiffeners serve a dual purpose in a structural member of this type. One function is to limit the size of the unsupported panel in the plywood web, and the other is to restrain the flanges from moving toward each other as the beam is stressed.

Stiffeners should be glued to the webs and should be in contact with both flanges, as shown in figure 60. A rational way of determining how thick the stiffener in contact with the web should be is not available, but it appears, from tests of box beams made at the Forest Products Laboratory, that a thickness of at least six times the thickness of the plywood web is sufficient. Because stiffeners must also resist the tendency of the flanges to move towards each other, the stiffeners should be as wide as the flanges.

The spacing of the stiffeners is relatively unimportant for the shear stresses that are allowed in plywood webs in which the grain of the wood in some plies is parallel and the grain of the wood in other plies is perpendicular to the axis of the member. Maximum allowed stresses are below those which will produce buckling. Stiffeners placed with a clear distance between stiffeners equal to or less than two times the clear distance between flanges are adequate.

Load blocks are special stiffeners placed along a wood-plywood structural member at points of concentrated load. Load blocks should be designed so that stresses caused by a load that bears against the side-grain material in the flanges do not exceed the design allowables for the flange material in compression perpendicular to grain.

Deflection of Wood-Plywood Members When Used as Beams

The deflection of a wood-plywood beam is produced by bending stresses (compression stresses in the compression flange and tension stresses in the tension flange) and by shear stresses in the plywood webs. For solid wood beams, the deflection due to shear is small, as compared to the deflection due to bending, for the spans usually used.

The deflections due to bending stresses in a wood-plywood beam can be calculated from the conventional formula for flexure:

$$y = \frac{kWl^3}{EI}$$

where y is the center deflection, in inches; k, a constant depending on the type of loading and support; W, the total load, in pounds; l, the span, in inches; E, the modulus of elasticity in compression, in pounds per square inch; and I, the moment of inertia of the section about the neutral axis, in inches^4.

The deflection due to shear in a beam is more difficult to estimate than the deflection due to bending. Tests of a large number of wood-plywood beams loaded at the third points indicate that the actual stiffness of beams with plywood webs in which the grain was oriented at 45° was about 80 percent of the stiffness calculated for bending alone. For beams with plywood webs in which the grain was oriented at 0° and 90° to the axis of the beam, the actual stiffness of the beams averaged about 40 percent of the stiffness calculated for bending alone.

Third-point loading does not produce a relation between bending and shear in a beam that is too greatly different from that usually encountered in service. Therefore, to estimate the total deflection of a wood-plywood beam, it is usually satisfactory to calculate the deflection due to bending and increase this deflection by 25 percent if 45° plywood webs are used and by 150 percent if 0° or 90° webs are used. Other means of estimating the deflections due to shear can be used (3, 12).

If several beams or girders of the same size are used in a structure, and deflections are important, it is desirable to check the deflection characteristics of actual members by test loading them and observing the deflections. If mechanical fastenings are used to connect the webs and flanges, it is necessary to test load typical members, because it is not possible to calculate the accumulative slip in the mechanical fastenings.

Lateral Stability of Beams and Girders

The resistance of a box or I member to buckling in a lateral direction when loaded is important if the member is not restrained and forced to deflect in a plane by other parts of a structure. A deep, narrow beam may fail by buckling (in much the same manner as a long column) before the ultimate bending strength is reached if it is not restrained and forced to deflect in the plane of the load. The calculation of critical loads for any beam with a built-up cross section, like a box beam or I-beam, is complex and involved. Tests at the Forest Products Laboratory indicate that box beams and I-beams of usual proportions are not critical in this respect at stresses allowed by design, if the spans are not excessive, and the ratios of moments of inertia about the neutral axis to moments of inertia about an axis perpendicular to the neutral axis do not exceed 30 to 1.

The following rules for lateral deflection are conservative and can be used to check a design for lateral stability:

(1) If the ratio of the moment of inertia of the cross section about the neutral axis to the moment of inertia about the axis perpendicular to the neutral axis does not exceed 5 to 1, no lateral support is required.

(2) If the ratio of the moments of inertia is between 5 to 1 and 10 to 1, the ends of the beam should be held in position at the bottom flange at supports.

(3) If the ratios of the moments of inertia are between 10 to 1 and 20 to 1, the beams should be held in line at the ends. (Both the top and bottom flanges should be restrained from horizontal movement in planes perpendicular to the axis of the beam.)

(4) If the ratio of the moments of inertia is between 20 to 1 and 30 to 1, one edge should be held in line.

(5) If the ratio of the moments of inertia is between 30 to 1 and 40 to 1, the beam should be restrained by bridging or other bracing at intervals of not more than 8 feet.

(6) If the ratio of the moments of inertia is greater than 40 to 1, the compression flanges should be fully restrained and forced to deflect in a plane.

A more exact analysis of lateral stability which may be used for design if desired, is given by Trayer and March (15).

LITERATURE CITED

(1) AMERICAN SOCIETY FOR TESTING MATERIALS.
    1947. METHODS OF TESTING PLYWOOD, VENEER, AND OTHER WOOD-BASE MATERIALS. Amer. Soc. Testing Mater. Standard D805–47. Pt. 4: 691–713, illus.

(2) ———
    1950. METHOD OF TEST FOR INTEGRITY OF GLUE JOINTS IN LAMINATED WOOD PRODUCTS FOR EXTERIOR SERVICE (TENTATIVE). Amer. Soc. Testing Mater. Standard D1101–50T. Pt. 4: 847–848.

(3) DOUGLAS-FIR PLYWOOD ASSOCIATION.
    1948. TECHNICAL DATA ON PLYWOOD. Sec. 7: 1–4, illus.; Sec. 9: 1–15, illus. Tacoma, Wash.

(4) FREAS, A. D., and SELBO, M. L.
    1953. FABRICATION AND DESIGN OF GLUED LAMINATED WOOD STRUCTURAL MEMBERS. U. S. Dept. Agr. Tech. Bul. 1069, 220 pp., illus.

(5) LEWIS, W. C., and DAWLEY, E. R.
    1943. STIFFENERS IN BOX BEAMS AND DETAILS OF DESIGN. U. S. Forest Prod. Lab. Rpt. 1318–A, 4 pp., illus. [Processed.]

(6) ——— HEEBINK, T. B., and COTTINGHAM, W. S.
    1944. BUCKLING AND ULTIMATE STRENGTHS OF SHEAR WEBS OF BOX BEAMS HAVING PLYWOOD FACE GRAIN DIRECTION PARALLEL OR PERPENDICULAR TO THE AXIS OF THE BEAMS. U. S. Forest Prod. Lab. Rpt. 1318–D 16 pp., illus. [Processed.]

(7) ——— HEEBINK, T. B., and COTTINGHAM, W. S.
    1944. EFFECTS OF CERTAIN DEFECTS AND STRESS CONCENTRATING FACTORS ON THE STRENGTH OF TENSION FLANGES OF BOX BEAMS. U. S. Forest Prod. Lab. Rpt. 1513, 10 pp., illus. [Processed.]

(8) ——— HEEBINK, T. B., and COTTINGHAM, W. S.
    1945. EFFECT OF INCREASED MOISTURE CONTENT ON THE SHEAR STRENGTH AT GLUE LINES OF BOX BEAMS AND ON THE GLUE-SHEAR AND GLUE-TENSION STRENGTHS OF SMALL SPECIMENS. U. S. Forest Prod. Lab. Rpt. 1551, 11 pp., illus. [Processed.]

(9) ——— HEEBINK, T. B., COTTINGHAM, W. S., and DAWLEY, E. R.
    1943. BUCKLING IN SHEAR WEBS OF BOX AND I-BEAMS AND THE EFFECT UPON DESIGN CRITERIA. U. S. Forest Prod. Lab. Rpt. 1318–B, 7 pp., illus. [Processed.]

(10) ——— HEEBINK, T. B., COTTINGHAM, W. S., and DAWLEY, E. R.
    1944. ADDITIONAL TESTS OF BOX AND I-BEAMS TO SUBSTANTIATE FURTHER THE DESIGN CURVES FOR PLYWOOD WEBS IN BOX BEAMS. U. S. Forest Prod. Lab. Rpt. 1318–C, 9 pp., illus. [Processed.]

(11) MARKWARDT, L. J., and FREAS, A. D.
    1950. APPROXIMATE METHODS OF CALCULATING THE STRENGTH OF PLYWOOD. U. S. Forest Prod. Lab. Rpt. 1630, 5 pp., illus. (Rev.) [Processed.]

(12) NEWLIN, J. A., and TRAYER, G. W.
    1941. DEFLECTION OF BEAMS WITH SPECIAL REFERENCE TO SHEAR DEFORMATIONS. U. S. Forest Prod. Lab. Rpt. 1309, 19 pp., illus.; reprint from Natl. Advisory Com. Aeronaut. Rpt. 180. [Processed.]

(13) SOUTHERN HARDWOOD PRODUCERS INC.; APPALACHIAN HARDWOOD MANUFACTURERS, INC.; AND NORTHERN HEMLOCK AND HARDWOOD MANUFACTURERS ASSOCIATION.
    1952. STANDARD SPECIFICATIONS FOR THE DESIGN AND FABRICATION OF HARDWOOD GLUED LAMINATED LUMBER. 12 pp., illus. Memphis, Tenn.

(14) SOUTHERN PINE ASSOCIATION.
    1951. STANDARD SPECIFICATIONS FOR DESIGN AND FABRICATION OF STRUCTURAL GLUED LAMINATED SOUTHERN PINE, 12 pp., illus. New Orleans, La.

(15) TRAYER, G. W., and MARCH, H. W.
    1931. ELASTIC INSTABILITY OF MEMBERS HAVING SECTIONS COMMON IN AIRCRAFT CONSTRUCTION. Natl. Advisory Com. Aeronaut. Rpt. 382, 42 pp., illus.

(16) U. S. DEPARTMENT OF DEFENSE.
            1950. LAMINATED WHITE OAK (FOR SHIP AND BOAT USE). Mil. Specif.
                    MIL-0-15154, 10 pp., illus.
(17) ————
            1952. LAMINATED DOUGLAS FIR (FOR SHIP USE). Mil. Specif. MIL-F-
                    2038, 11 pp., illus.
(18) U. S. FOREST PRODUCTS LABORATORY.
            1943. DESIGN OF PLYWOOD WEBS FOR BOX BEAMS. U. S. Forest Prod. Lab.
                    Rpt. 1318, 7 pp., illus. [Processed.]
(19) ———— and ANC-23 PANEL ON SANDWICH CONSTRUCTION FOR AIRCRAFT.
            1951. DESIGN OF WOOD AIRCRAFT STRUCTURES. ANC Bul. 18, 234 pp.,
                    illus.   (Issued by Subcommittee on Air Force-Navy-Civil Aircraft
                    Design Criteria, Aircraft Committee, Munitions Board.)
(20) WEST COAST LUMBERMEN'S ASSOCIATION.
            1951. STANDARD SPECIFICATIONS FOR THE DESIGN AND FABRICATION OF
                    STRUCTURAL GLUED (SOFTWOOD) LAMINATED LUMBER.   12 pp.,
                    illus. Portland, Oreg.
(21) WILSON, T. R. C.
            1939. GLUED LAMINATED WOODEN ARCH. U. S. Dept. Agr. Tech.   Bul. 691,
                    123 pp., illus.
(22) ——— and COTTINGHAM, W. S.
            1947. TESTS OF GLUED LAMINATED WOOD BEAMS AND COLUMNS AND DEVEL-
                    OPMENT OF PRINCIPLES OF DESIGN.   U. S. Forest Prod. Lab. Rpt.
                    R1687, 36 pp., illus. [Processed.]

PLYWOOD AND OTHER CROSSBANDED PRODUCTS

Plywood is a term generally used to designate glued wood panels that are made up of layers, or plies, with the grain of 1 or more layers at an angle, usually 90°, with the grain of the others (*1, 4, 5, 6, 7, 8*).[18] The outside plies are called faces or face and back, the center ply or plies are called the core, and the plies immediately below the face and back, laid at right angles to them, are called the crossbands (fig. 63). The essential features of plywood are embodied in other glued constructions with many variations of details (*3*). The core may be veneer, lumber, or various combinations of veneer and lumber; the total thickness may be less than 1/16 inch or more than 3 inches; the different plies may vary as to number, thickness, and kinds of wood; and the shape of the members may also vary. The crossbands and their arrangement largely govern the properties, particularly the warping characteristics, and the uses of all such constructions (*2*).

As compared with solid wood, the chief advantages of plywood are its approach to equalization of strength properties along the length and width of the panel, greater resistance to checking and splitting, and less change in dimensions with changes in moisture content. The greater the number of plies for a given thickness, the more nearly equal are the strength and shrinkage properties along and across the panel and the greater the resistance to splitting.

ARRANGEMENT OF PLIES

The tendency of crossbanded products to warp as the result of stresses caused by shrinking and swelling is largely eliminated by balanced construction. This construction consists of arranging the plies in pairs about the core, so that for each ply there is an opposite, similar, and parallel ply. Matching the plies involves a consideration of (1) thickness, (2) kind of wood with particular reference to shrinkage and density, (3) moisture content at the time of gluing, and (4) angle or relative direction of the grain.

The use of an odd number of plies permits an arrangement that gives a substantially balanced effect; that is, when 3 plies are glued together with the grain of the outer 2 plies at right angles to the grain of the center ply, the stresses are balanced and the panel tends to remain flat with moisture content changes. With 5, 7, or some other uneven number of plies, the forces may be similarly balanced. If only two plies are glued together with the grain of one at right angles to the grain of the other, each ply tends to distort the other when moisture content changes occur, and cupping usually results. Similar results are likely when any even number of plies are used, unless the 2 center plies are parallel and act essentially as 1 ply.

The use of balanced construction is highly important in thin panels that must remain flat. In thicker members some deviation from balanced construction is possible without serious consequences. For example, with lumber cores that are properly crossbanded the face

---

[18] Italic numbers in parentheses refer to Literature Cited, p. 289.

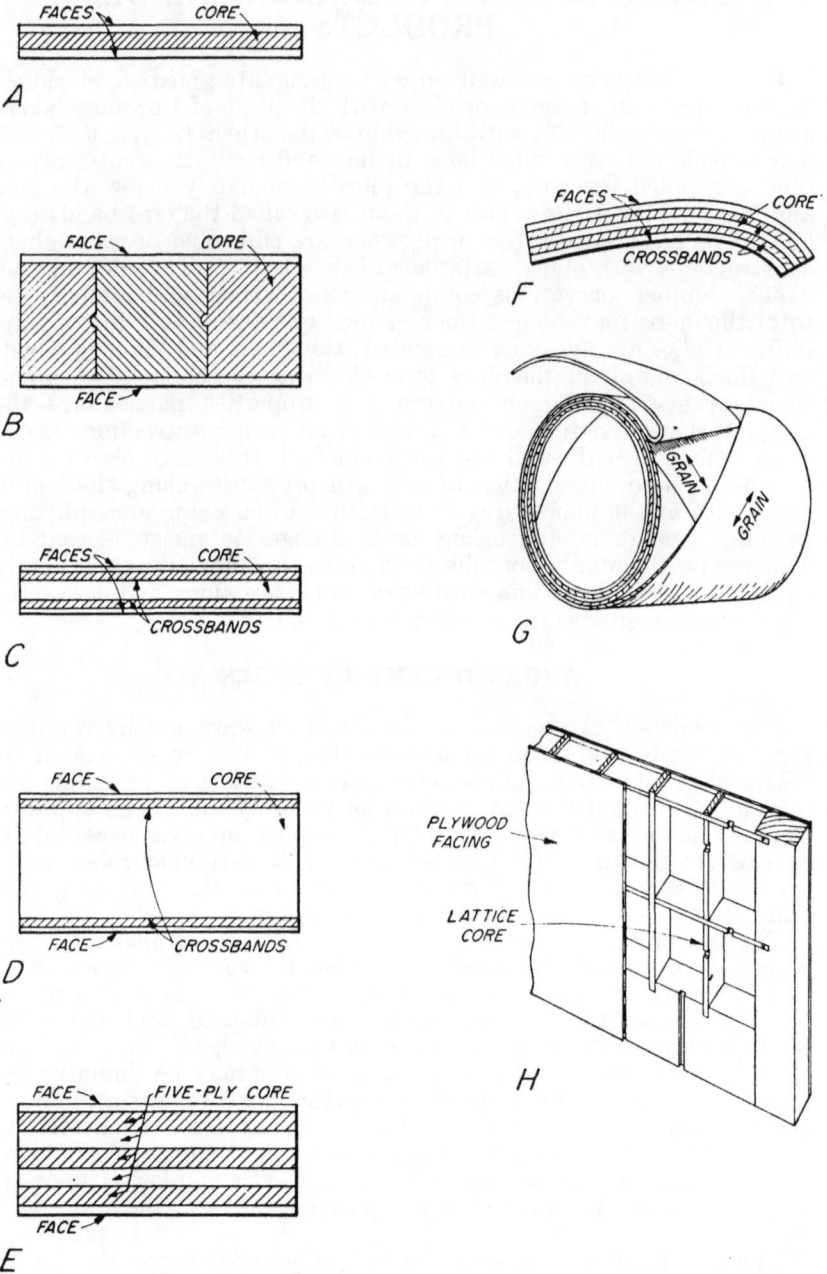

ZM 92378 F

FIGURE 63.—Types of plywood and crossbanded construction: *A*, 3-ply (all veneer); *B*, 3-ply (lumber core); *C*, 5-ply (all veneer); *D*, 5-ply (lumber core); *E*, 7-ply (all veneer); *F*, 5-ply bent work (all veneer); *G*, 5-ply, spirally wrapped (all veneer); *H*, section of hollow-core door.

and back plies may be dissimilar without any noticeable effect; whereas if dissimilar face and back plies were used in thin three-ply panels, the warping might be very objectionable. In certain curved members, the natural cupping tendency of an even number of plies may even be used to advantage.

Since the outer or face plies of a crossbanded construction are restrained on only one side, changes in moisture content induce relatively large stresses on the outer glue joints. The magnitude of stresses depends upon such factors as thickness of plies, density and shrinkage of the wood involved, and the amount of change in moisture content. In general, one-eighth inch is about the maximum thickness of face plies that can be held securely in place when dense woods are used and large changes in moisture content occur, and thinner faces are advisable under such conditions of service.

QUALITY OF PLIES

In thin plywood the quality of all the plies affects the shape and permanence of form of the panel. All plies should be straight grained, smoothly cut, and of sound wood.

In thick, five-ply lumber-core panels the crossbands, in particular, affect the quality and stability of the panel. Imperfections in the crossbands, such as marked differences in the texture of the wood or irregularities in the surface, are easily seen in the panel through thin surface veneers. Cross grain that runs sharply through the crossband veneer from one face to the other causes the panels to cup. Cross grain that runs diagonally across the face of the crossband veneer causes the panel to twist unless the two crossbands are laid with their grain parallel. Lack of observance of this simple precaution accounts for much warping in crossbanded construction (2).

The best woods for cores of high-grade panels are those of low density and shrinkage, of slight contrast between springwood and summerwood, and of species that are glued easily. Edge-grained cores are better than flat-grained cores because they shrink less in width. In softwoods with pronounced summerwood, moreover, edge-grained cores are better because the hard bands of summerwood are less likely to show through thin veneer and the panels show fewer irregularities in the surfaces. In most species, a core made entirely of either quartersawed or plainsawed material remains more uniform in thickness with moisture content changes than one in which the two types of material are combined.

Distinct distortion of surfaces has been noted, particularly in softwoods, when the core boards were neither distinctly flat grained nor edge grained.

GRADES AND TYPES OF PLYWOOD

Broadly speaking, two classes of plywood are available—hardwood and softwood. Most softwood plywood is made of Douglas-fir, but western hemlock, white fir, ponderosa pine, redwood, and other species are used. Hardwood plywood is made of many species.

Various grades and types are manufactured. "Grade" is determined by the quality of veneer, and "type" by the moisture resistance of the glue joints. The types and grades in general use are listed in commercial standards established by the industry with the assistance of the Department of Commerce. Separate commercial standards cover Douglas-fir plywood, western softwood plywood, ponderosa pine and sugar pine plywood, and hardwood plywood (*4, 5, 6, 7*).

Douglas-fir plywood.—Two types of Douglas-fir plywood are listed in Commercial Standard CS45–48—Interior and Exterior. The Interior type is expected to retain its form and practically all of its strength when occasionally subjected to a thorough wetting and subsequent normal drying. It is commonly bonded with soybean glue or with an extended resin glue of the phenol type. Exterior-type plywood is expected to retain its form and strength when repeatedly wet and dried and otherwise subjected to the elements and to be suitable for permanent exterior use. It is commonly bonded with hot-press, phenol resin glue.

Within each type, several grades are established by the quality of the veneer on the two faces of a panel. In descending order of quality, the veneer is designated as A, B, C, or D. Grade A–A plywood, for example, has Grade A veneer on both faces, and Grade A–D plywood has Grade A veneer on one face and Grade D on the other. As a general rule, Grade C is used for the inner plies of the Exterior type and Grade D for the inner plies of the Interior type.

Softwood plywood other than Douglas-fir.—The requirements of Commercial Standard CS122–49 for western softwood plywood resemble those of the commercial standard for Douglas-fir. The same two types—Interior and Exterior—are provided for, and the grade is determined by the quality of veneer used on the two faces of each panel. The grades of veneer are designated A, B, C, and D, as with Douglas-fir veneer, but one higher grade, A–1, is provided.

Commercial Standard CS157–49 provides for only an Interior type of ponderosa pine and sugar pine plywood. The grades of veneer are designated in descending order of quality as Good, Sound, Solid, Sheathing, and Backs. Provisions are made for eight grades of plywood, as for example "Good 2 Sides" and "Solid 1 Side," depending upon the quality of veneer used on the faces and backs.

Hardwood plywood.—Four types of hardwood plywood—Technical, Type I, Type II, and Type III—are described in Commercial Standard CS35–49. For the most part the difference between the types is based on the resistance of the glue bond to severe service conditions, although Technical differs from Type I in the permissible thickness and arrangement of plies. Glue bonds conforming to the requirements of both Technical and Type I plywood are high in durability and correspond closely to those required for Exterior-type Douglas-fir plywood. In general, resistance of the glue bonds of Type II hardwood plywood resembles that of Interior-type Douglas-fir plywood. Good dry strength but no water resistance is required of the glue bonds in Type III hardwood plywood.

The grade of the plywood is determined by the quality of the veneers used on the faces and backs of the panels. The veneer is graded 1, 2, 3, and 4 in order of descending quality. The requirements are the same for all hardwood species for Grades 2, 3, and 4; Grade 1 requirements are described separately for each species. The plywood grade is designated by combining the designation of the grade of the veneer on the face and back, as Grade 1–2 or Grade 2–3. Veneers of Grade 2 or 3 are generally used for inner plies.

Plywood for special purposes.—Most of the plywood produced in this country conforms in general to the commercial standards. Some plywood, however, is produced to meet specific demands or to conform to specifications that were drawn to secure a product suitable for a special purpose.

A considerable quantity of plywood is produced with overlays of paper or fabric impregnated with synthetic resins. Frequently, an overlay is applied to one face to provide a desired decorative effect combined with good wearing characteristics. This product is widely used for counter and table tops. If a decorative type of overlay is applied on one face, a plain overlay is usually applied on the back to reduce the tendency to warp that would otherwise result. Often the overlay is applied primarily to improve general serviceability, to improve painting characteristics, or to reduce face checking. In such cases, the decorative layer is omitted, and the same overlay of plain color is used for both face and back. As a general rule, decorative overlays are laid in a separate operation after the plywood panel has been fabricated, while plain overlays are laid at the time of gluing the veneer into plywood.

Specifications for aircraft structural plywood sharply define glue-line quality, species, thickness of plies, and mechanical properties of the veneer in order to insure a product suitable for use in the structural elements of aircraft (8). Large corporations likewise may develop their own specifications for plywood for special purposes.

Unless very large quantities are involved, procurement of special-purpose plywood is difficult, costly, and often wasteful. For most purposes, plywood conforming to commercial standards is adequate.

DETERMINATION OF STRENGTH OF PLYWOOD

Determination of the strength properties of plywood by testing all of the many possible combinations of ply thickness, species, and number of plies is impractical. For this reason, formulas relating the strength properties of plywood to the construction of the plywood and to the properties of the component plies were developed mathematically and checked by tests to verify their applicability. Plywood strength properties are therefore not given directly but are given as formulas that permit calculations for any specific case.

Plywood may be used under loading conditions such that stiffeners will be required to prevent it from buckling. It may also be used in the form of cylinders or curved plates. Such uses are beyond the scope of this handbook, but they are discussed in ANC Bulletin 18 (9).

It is obvious from its construction that a strip of plywood cannot be so strong in tension, compression, or bending as a strip of solid wood of the same size. Those plies having their grain direction oriented at 90° to the direction of stress can contribute only a fraction of the strength contributed by the corresponding areas of a solid strip, since they are stressed perpendicular to the grain. Strength properties in the directions parallel and perpendicular to the face grain tend to be equalized in plywood, since in some interior plies the grain direction is parallel to the face grain and in others it is perpendicular.

The formulas given in this handbook may be used, in general, for calculating the strength properties of plywood at proportional limit or ultimate or for estimating working stresses, depending upon the strength property that is substituted in the formulas for the property of the veneer. Procedures for evaluating the effect of such features as knots and cross grain on the strength of veneer are not available. Factors by which basic stresses must be reduced to obtain safe working stresses must therefore be a matter of judgment based on each individual case. Strength properties at proportional limit and ultimate that are suitable for use in the formulas are given in table 12. Basic stresses are given in tables 25 and 36.

Properties in Edgewise Compression

Modulus of Elasticity

The modulus of elasticity of plywood in compression parallel to or perpendicular to the face-grain direction is equal to the weighted average of the moduli of elasticity of all plies parallel to the applied load. That is,

$$E_w \text{ or } E_x = \frac{1}{h} \sum_{i=1}^{i=n} E_i h_i$$

where E_w is the modulus of elasticity of plywood in compression parallel to the face grain; E_x, the modulus of elasticity of plywood in compression perpendicular to the face grain; E_i, the modulus of elasticity parallel to the applied load of the veneer in ply i; h_i, the thickness of the veneer in ply i; h, the thickness of the plywood; and n, the number of plies.

Those plies whose grain direction is perpendicular to the applied load contribute so little to the value of E_w and E_x that they may be omitted from the calculation with little error.

The modulus of elasticity in compression at angles to the face-grain direction other than 0° or 90° is given approximately by:

$$\frac{1}{E_\theta} = \frac{1}{E_w} \cos^4 \theta + \frac{1}{E_x} \sin^4 \theta + \frac{1}{G_{wx}} \sin^2 \theta \cos^2 \theta$$

where E_θ is the modulus of elasticity of a plywood strip in compression at an angle θ to the face grain; G_{wx} is the modulus of rigidity in shear for the conditions in which all 4 of the shear stress components are directed parallel to the plane of the plywood (1 pair of components parallel to and 1 pair perpendicular to the face grain); and the other terms are as defined in the previous formula. Formulas for computing values of G_{wx} are elsewhere in this section.

When compressive loads are applied at an angle to the face grain of plywood elements, shear strains (angular displacements) result in addition to the direct compressive shortening. ANC Bulletin 18 (9) gives methods for calculating these shear strains as well as the dimensional changes resulting from the effect of Poisson's ratio.

Strength Properties

The compressive strength of plywood subjected to edgewise forces is given by:

$$F_{cw} = \frac{E_w}{E_{cL}} F_{cL}$$

$$F_{cx} = \frac{E_x}{E_{cL}} F_{cL}$$

where F_{cw} is the compressive strength of plywood parallel to the face grain; F_{cx}, the compressive strength of plywood perpendicular to the face grain; F_{cL}, the compressive strength of the veneer parallel to the grain; and E_{cL}, the modulus of elasticity of the veneer parallel to the grain.

Values of modulus of elasticity in bending for solid wood are given in table 12. These values, increased by 10 percent, may be used for E_{cL}. If more than one species is used in the longitudinal plies, values for the species having the lowest ratio of F_{cL}/E_{cL} should be used in the formulas given.

When plywood is loaded at an angle to the face grain, its ultimate compressive strength or working stress in compression may be computed from:

$$F_{c\theta} = \frac{1}{\sqrt{\dfrac{\cos^4 \theta}{F_{cw}^2} + \dfrac{\sin^4 \theta}{F_{cx}^2} + \dfrac{\sin^2 \theta \cos^2 \theta}{F_{swx}^2}}}$$

where $F_{c\theta}$ is the compressive strength of plywood at an angle θ to the face grain, and F_{swx} is the shear strength of plywood for the condition with all 4 of the shear stress components directed parallel to the plane of the plywood (1 pair of components parallel and 1 pair perpendicular to the face grain); and the other terms are as previously defined.

Properties in Edgewise Tension

Modulus of Elasticity

Values of modulus of elasticity in tension will be the same as those in compression.

Strength Properties

The strength of a plywood strip in tension parallel or perpendicular to the face grain may be taken as the sum of the strength values of the plies having their grain direction parallel to the applied load. For this purpose, the tensile strength may be taken as equal to the modulus of rupture.

The tensile strength parallel to the face grain will be designated as F_{tw} and the tensile strength perpendicular to the face grain as F_{tx}.

The tensile strength at an angle to the face grain may be computed from:

$$F_{t\theta} = \cfrac{1}{\sqrt{\cfrac{\cos^4\theta}{F_{tw}{}^2} + \cfrac{\sin^4\theta}{F_{tx}{}^2} + \cfrac{\sin^2\theta\cos^2\theta}{F_{swx}{}^2}}}$$

where $F'_{t\theta}$ is the tensile strength of plywood at an angle θ to the face grain.

Properties in Edgewise Shear

Modulus of Rigidity

The modulus of rigidity of plywood may be calculated from:

$$G_{wx} = \frac{1}{h} \sum_{i=1}^{i=n} G_i h_i$$

$$G_{fwx} = \frac{1}{I} \sum_{i=1}^{i=n} G_i I_i$$

where G_{wx} is the modulus of rigidity of plywood subjected to shear but not to bending; G_{fwx}, the modulus of rigidity of plywood subjected to bending as well as shear; and G_i, the modulus of rigidity of the ith ply in the plane of the sheet; I_i, the moment of inertia of the ith ply about the neutral plane of the plywood; and I, the moment of inertia of the total cross section about its centerline.

When the plywood is made of a single species of wood:

$$G_{fwx} = G_{wx} = G_{LT} \text{ for rotary-cut veneers;}$$

$$G_{fwx} = G_{wx} = G_{LR} \text{ for quarter-sliced veneer.}$$

Values of G_{LT} and G_{LR} are given in terms of the modulus of elasticity parallel to grain (E_L) in table 13.

The modulus of rigidity at an angle to the face grain may be computed from:

$$\frac{1}{G_\theta} = \frac{1}{G_{wx}} \cos^2 2\theta + \left[\frac{1}{E_{cw}} + \frac{1}{E_{cx}}\right] \sin^2 2\theta$$

Strength Properties

The ultimate strength of plywood elements in shear, with the shearing forces parallel and perpendicular to the face-grain direction, is given by the empirical formulas

$$F_{swx} = 55\,\frac{n-1}{h} + \frac{9}{16h} \sum_{i=1}^{i=n} F_{swxi}\, h_i$$

where n is the number of plies and F_{swxi} is the shear strength of the ith ply.

In using this formula, the factor $(n-1)/h$ should not be assigned a value greater than 35.

To convert F'_{swx} to a basic stress, divide it by 4, thus taking into account variability, long-time loading, and factor of safety.

In some commercial grades of plywood, gaps in the core or cross-bands are permitted. These gaps reduce the shear strength of ply-wood, and the formula just given should be corrected to account for this effect. This may be done by subtracting from the number of plies (n) in the first term twice the number of plies containing openings at any one section and omitting from the summation in the second term all plies containing openings at any one section. Since the first term represents the contribution of the glue layers to shear, twice the number of plies containing openings at any one section is subtracted to account for the lack of glue on each side of the opening. The modification for the effect of core gaps just outlined represents a logically derived procedure not confirmed by test. When test data become available, this procedure may be modified.

When the plywood is stressed in shear at an angle to the face grain, ultimate shear strength with face grain in tension or compression is given by the following formulas:

$$F_{s\theta t} = \frac{1}{\sqrt{\left(\dfrac{1}{F_{tw}{}^2} + \dfrac{1}{F_{cx}{}^2}\right)\sin^2 2\theta + \dfrac{\cos^2 2\theta}{F_{swx}{}^2}}}$$

$$F_{s\theta c} = \frac{1}{\sqrt{\left(\dfrac{1}{F_{cw}{}^2} + \dfrac{1}{F_{tx}{}^2}\right)\sin^2 2\theta + \dfrac{\cos^2 2\theta}{F_{swx}{}^2}}}$$

For the special case in which the shear stresses are applied at 45° to the face grain:

$$F_{s45t} = \frac{F_{tw}}{\sqrt{1 + \left(\dfrac{F_{tw}}{F_{cx}}\right)^2}}$$

$$F_{s45c} = \frac{F_{tx}}{\sqrt{1 + \left(\dfrac{F_{tx}}{F_{cw}}\right)^2}}$$

Properties in Flexure

Modulus of Elasticity

The modulus of elasticity in bending is equal to the average of the moduli of elasticity parallel to the span of the various plies weighted according to their moment of inertia about the neutral plane. That is,

$$E_{fw} \text{ or } E_{fx} = \frac{1}{I}\sum_{i=1}^{i=n} E_i I_i$$

where E_{fw} is the modulus of elasticity of plywood in bending when the face grain is parallel to the span; E_{fx}, the modulus of elasticity of plywood in bending when the face grain is perpendicular to the span; E_i, the modulus of elasticity of the ith ply in the span direction; I_i, the moment of inertia of the ith ply about the neutral plane of the plywood; and I, the moment of inertia of the total cross section about its centerline.

For symmetrically constructed plywood, the neutral plane will be at the centerline. For unsymmetrically constructed plywood, the neutral plane is usually not at the centerline of the geometrical section. For such cases, the distance from the neutral plane to the extreme compression (or tension) fiber, c, is given by:

$$c = \frac{\sum\limits_{i=1}^{i=n} h_i E_i c_i}{\sum\limits_{i=1}^{i=n} h_i E_i}$$

where c_i is the distance from the extreme compression (or tension) fiber to the centerline of the ith ply.

When all plies are of the same thickness and species, these general formulas reduce, for plywood made from rotary-cut veneer, to:

$$\text{3-ply, } E_{fw} = \frac{E_L}{27} \left(\frac{E_T}{E_L} + 26 \right)$$

$$E_{fx} = \frac{E_L}{27} \left(1 + 26 \frac{E_T}{E_L} \right)$$

$$\text{5-ply, } E_{fw} = \frac{E_L}{125} \left(26 \frac{E_T}{E_L} + 99 \right)$$

$$E_{fx} = \frac{E_L}{125} \left(26 + 99 \frac{E_T}{E_L} \right)$$

$$\text{7-ply, } E_{fw} = \frac{E_L}{343} \left(99 \frac{E_T}{E_L} + 244 \right)$$

$$E_{fx} = \frac{E_L}{343} \left(99 + 244 \frac{E_T}{E_L} \right)$$

$$\text{9-ply, } E_{fw} = \frac{E_L}{729} \left(244 \frac{E_T}{E_L} + 485 \right)$$

$$E_{fx} = \frac{E_L}{729} \left(244 + 485 \frac{E_T}{E_L} \right)$$

For quarter-sliced veneer, the formulas will be the same, except that E_T/E_L will be replaced by E_R/E_L. The symbols E_T and E_R denote the moduli of elasticity in the tangential and radial directions, respectively. Values of E_T/E_L and E_R/E_L for some species are given in table 13.

For species for which values of the ratio are not available, it is suggested that E_T/E_L be taken as 0.05 and E_R/E_L as 0.10.

Strength Properties

The resisting moment of plywood strips having face grain parallel to the span is given by:

$$M = 0.85 \frac{E_{fw}}{E_L} \frac{F_b I}{c}$$

For face grain perpendicular to the span,

$$M = 1.10 \frac{E'_{fx}}{E_L} \frac{F_b I}{c} \text{ for 3-ply plywood}$$

$$M = 0.90 \frac{E'_{fx}}{E_L} \frac{F_b I}{c} \text{ for plywood having 5 or more plies}$$

where M is the resisting moment of the plywood; F_b, the strength property of the outermost longitudinal ply; c, the distance from the neutral plane to the outer fiber of the outermost longitudinal ply; and E'_{fx} is the same as E_{fx}, except that the outermost ply in tension is neglected, and the other terms are as defined previously.

For plywood having five or more plies, the use of E_{fx} in place of E'_{fx} in calculating the resisting moment will result in negligible error. It should be noted that E'_{fx} is used only in strength calculations and is not to be used in deflection calculations.

Formulas are not available for calculating bending strength at an angle to the face grain.

Approximate Methods

Table 37 gives approximate methods of calculating the strength and stiffness of plywood. The methods given therein will be useful in estimating properties and, in some cases, may be expected to give values sufficiently accurate for general use. The methods are not applicable to cases where the plywood may be expected to buckle.

Splitting Resistance

Plywood permits fastening with nails or screws close to the edges because it offers much greater resistance to splitting than ordinary wood. Also, because of the equalization of strength properties along and across a sheet and the resistance to splitting resulting from the crossbanded construction, plywood panels covering relatively large areas are less liable to damage from concentrated or impact loads than similar panels made of ordinary lumber.

In removing nails from plywood, some care must be used to pull them straight out, or nearly so, because splintering of the outside ply may result if the nails are pulled or pried out at an angle.

SHRINKAGE OF PLYWOOD

The shrinkage of plywood varies with the species, the ratios of ply thicknesses, the number of plies, and the combination of species. The average shrinkage values obtained at the Forest Products Laboratory in drying, from a soaked to an oven-dry condition, 3-ply panels having all plies in any one panel of the same thickness and species was about 0.45 percent parallel to the face grain and 0.67 percent perpendicular to the face grain, with ranges of from 0.2 to 1 percent and 0.3 to 1.2 percent, respectively. The panels tested ranged in thickness from $\frac{1}{10}$ to $\frac{1}{2}$ inch. For all practical purposes, shrinkage of plywood in thickness does not differ from that of solid wood.

TABLE 37.—*Design method and allowable stresses for calculating the strength and stiffness of plywood* [1]

Property, and direction of stress with respect to direction of face grain	Area to be considered	Unit stress to be used
Tension:		
Parallel or perpendicular.	Parallel plies [2] only	Unit stress for extreme fiber in bending.
±45°	Full cross-sectional area	$\frac{1}{6}$ unit stress for extreme fiber in bending.
Compression:		
Parallel or perpendicular.	Parallel plies [2] only	Unit stress in compression parallel to grain.
±45°	Full cross-sectional area	$\frac{1}{3}$ unit stress in compression parallel to grain.
Bearing at right angles to plane of plywood.	Loaded area	Unit stress in compression perpendicular to grain.
Load in bending, parallel or perpendicular.	Bending moment $M=KSI/c$ where $S=$unit stress for extreme fiber in bending; $I=$moment of inertia computed on basis of parallel plies only; $c=$distance from neutral axis to outer fiber of outermost ply having its grain in the direction of the span; $K=1.50$ for 3-ply plywood having the grain of the outer plies perpendicular to the span; $K=0.85$ for all other plywood.	Unit stress for extreme fiber in bending.
Deflection in bending, parallel or perpendicular.	Deflection may be calculated by the usual formulas, taking as the moment of inertia that of the parallel plies plus $\frac{1}{20}$ that of the perpendicular plies. (When face plies are parallel, the calculation may be simplified, with but little error, by taking the moment of inertia as that of the parallel plies only.)	Unit value for modulus of elasticity.
Deformation in tension or compression, parallel or perpendicular.	Parallel plies [2] only	Unit value for modulus of elasticity.
Shear through thickness:		
Parallel or perpendicular.	Full cross-sectional area	Double unit stress in horizontal shear.[3]
±45°	do	4 times unit stress in horizontal shear.

See footnotes at end of table.

TABLE 37.—*Design method and allowable stresses for calculating the strength and stiffness of plywood* [1]—Continued

Property, and direction of stress with respect to direction of face grain	Area to be considered	Unit stress to be used

Shear in plane of plies:
  Parallel or perpendicular
	Full shear area: Plywood beams. Horizontal shear.
	
	Area of contact between plywood and flange or framing member:
	  I- or box-beams with plywood webs_____
	  Shear between plies of web or between web and flange:
	
	Panels having plywood covers stressed in compression or tension, or both. Shear between plies or between cover and framing members when depth of member exceeds twice its width and end headers are used or when depth is not more than twice the width and no headers are used.
	  Interior framing members_____
	
	Framing members at edge of panel__
	
		$\frac{3}{4}$ unit stress in horizontal shear.
		
		$\frac{3}{8}$ unit stress in horizontal shear.
		
		$\frac{3}{4}$ unit stress in horizontal shear.
		
		$\frac{3}{8}$ unit stress in horizontal shear.

±45°_____	Area of contact between plywood and flange or framing member:
	  I- or box-beams with plywood webs_____
	  Shear between plies of web or between web and flange:
	
		$\frac{1}{2}$ unit stress in horizontal shear.

See footnote at end of table.

TABLE 37.—*Design method and allowable stresses for calculating the strength and stiffness of plywood* [1]—Continued.

Property, and direction of stress with respect to direction of face grain	Area to be considered	Unit stress to be used

	Area of contact between plywood and flange or framing member—Continued
	  Shear between plies of web or between web and flange—Continued
	Panels having plywood covers stressed in compression, or tension, or both. Shear between plies or between cover and framing members when depth of member exceeds twice its width and end headers are used or when depth is not more than twice the width and no headers are used.
	  Interior framing members..........
		Unit stress in horizontal shear.

	Framing members at edge of panel	½ unit stress in horizontal shear.

[1] The suggested simplified methods of calculation apply reasonably well with usual plywood types under ordinary conditions of service. It is recognized, however, that they are not entirely valid for all types of plywood and plywood constructions, or for all spans and span-depth ratios. Also the methods given are not applicable to structures so proportioned that the plywood is in the buckling range, in which event the results will be too high.

[2] By "parallel plies" is meant those plies whose grain direction is parallel to the direction of principal stress.

[3] This value should be reduced if gaps occur at edge joints (see p. 283).

HOLLOW-CORE CONSTRUCTION

The term "hollow-core" refers to a type of crossbanded construction, somewhat resembling lumber-core plywood, in which the center ply or core is not solid but built up of wood strips, wood lattice, sections of paper cylinders, or other materials offering only partial support to the facings. Perhaps the most common application of the construction is the hollow-core, flush door (fig. 63, *H*) in which the space between the stiles and rails contains a lattice and the facings are three-ply plywood. The lattice or other supporting material used in the core space must be spaced so as to provide sufficient mchanical support to prevent buckling or distortion of the facings, and, as in conventional plywood, the construction as a whole must be balanced to prevent warping. When properly fabricated, the construction is lightweight in relation to its strength.

Literature Cited

(1) American Society for Testing Materials.
    1949. definitions of terms relating to veneer and plywood (tentative). Amer. Soc. Testing Mater. Standard D1038–49T. Pt. 4, 733–738.

(2) Brouse, D.
    1940. some causes of warping in plywood and veneered products. U. S. Forest Prod. Lab. Rpt. R1252, 8 pp., illus. [Processed.]

(3) Perry, T. D.
    1942. modern plywood. 366 pp., illus. New York.

(4) U. S. Department of Commerce.
    1948. douglas-fir plywood. Com. Standard CS45–48, 25 pp., illus.

(5) ———
    1949. hardwood plywood. Com. Standard CS35–49, 27 pp., illus.

(6) ———
    1949. western softwood plywood. Com. Standard CS122–49, 22 pp., illus.

(7) ———
    1949. ponderosa pine and sugar pine plywood. Com. Standard CS157–49, 13 pp.

(8) U. S. Department of Defense.
    1950. plywood and veneer: aircraft flat panel. Mil. Specif. MIL–P–6070, 12 pp., illus.

(9) U. S. Forest Products Laboratory.
    1951. design of wood aircraft structures. ANC Bul. 18, 234 pp., illus. (Issued by Subcommittee on Air Force-Navy-Civil Aircraft Design Criteria, Aircraft Committee, Munitions Board.)

STRUCTURAL SANDWICH CONSTRUCTION

Structural sandwich construction is a layered construction formed by bonding two thin facings to a thick core (fig. 64). The thin facings are usually of a strong, dense material, since they are the principal load-carrying parts of the construction. The core, which is of a weaker, lightweight material, separates and stabilizes the thin facings and carries shearing loads. The entire assembly provides a structural element of high strength and stiffness in proportion to its weight. Sandwich construction is also economical, since only small amounts of the relatively expensive facing material are used and the core materials are usually inexpensive. The materials are positioned so that each is used to its best advantage.

M 93157 F

FIGURE 64.—A cutaway section of sandwich construction with plywood facings and a paper honeycomb core.

Specific nonstructural advantages can be incorporated in a sandwich construction by proper selection of facing and core materials. An impermeable facing can be employed to act as a moisture barrier for a wall or roof panel in a house; an abrasion-resistant facing can be used for the top facing of a floor panel; and decorative effects can be obtained by using panels with plywood or plastic facings for walls, doors, tables, and other furnishings. Core material can be chosen to provide thermal insulation, fire resistance, and decay resistance.

The component parts of the sandwich construction should be compatible with service requirements. Moisture-resistant facings, cores, and adhesives should be employed if the construction is to be exposed to adverse moisture conditions. Similarly, heat-resistant or decay-resistant facings, cores, and adhesives should be used if exposure to elevated temperatures or decay organisms is expected.

FABRICATION OF SANDWICH PANELS

Facing Materials

One of the advantages of sandwich construction is the great latitude it provides in choice of facings and the opportunity to use thin sheet materials because of the nearly continuous support by the core. The stiffness, stability, and, to a large extent, the strength of the sandwich are determined by the characteristics of the facings. Some of the different facing materials used include plywood, single veneers or plywood overlaid with a resin-treated paper; hardboard; asbestos board; metals, such as aluminum, enameled steel, stainless steel, or magnesium sheet; wallboard; fiber-reinforced plastics or laminates; and veneer bonded to metal.

Core Materials

Many lightweight materials, such as balsa wood, rubber or plastic foams, and formed sheets of cloth, metal, or paper, have been used as core for sandwich construction. From the standpoint of economy, availability, and other advantages for mass production, the type of core made from paper treated with synthetic resin (5, 9) [19] appears to be promising. The paper core is made by forming flat sheets of dense paper in various ways to yield a much larger volume of low-density honeycomblike material (fig. 64).

Since sandwich panels are likely to be subjected to damp or wet conditions, the presence of resin in the paper core is necessary to yield a product that is permanently strong and stiff in the wet condition. Treatment with 15 percent of resin provides paper of good strength when wet, decay resistance, and handling characteristics during fabrication (10). Resin amounts in excess of about 15 percent do not seem to produce a gain in strength commensurate with the increased quantity of resin required.

Manufacturing Operations

The principal operation in the manufacture of sandwich panels is the bonding of the facings to the core, usually in hot-press equipment. Special presses are indicated for sandwich panel manufacture, because the pressures required are usually lower than can be obtained in the range of good pressure control on ordinary plywood or plastic presses. Since pressure requirements are low, however, simple and perhaps less costly presses could be used. Continuous roller presses or bag-molding equipment may also be suitable. Certain special problems

---

[19] Italic numbers in parentheses refer to Literature Cited, p. 297.

arise in the pressing of sandwich panels, but their manufacture is basically not complicated (9).

The facing materials may need to be cleaned and primed before the adhesive is applied, especially if they are metallic.

In certain sandwich panels, loading rails or edgings are placed between the facings at the time of assembly. Special fittings or equipment, such as heating coils, plumbing, or electrical wiring conduit, can be placed more easily in the panel during manufacture than after it is completed.

Some of the most persistent difficulties in the use of sandwich panels are caused by the necessity for edges, inserts, and connectors for panels. In some cases, the problem involves tying together thin facing materials without causing severe stress concentrations, and in other cases, such as furniture, the problem is caused by "show-through" of core or inserts through decorative facings. These difficulties are minimized by the choice of materials in which the rate and degree of differential dimensional movement between core and insert are at a minimum.

Further information on fabrication of sandwich panels is given in other Laboratory papers (4, 9, 12, 14).

STRUCTURAL DESIGN OF SANDWICH CONSTRUCTION

The structural design of sandwich construction may be compared to the design of an I-beam; the facings of the sandwich represent the flanges of the I-beam, and the sandwich core represents the I-beam web. The core of the sandwich serves, through the bonding adhesive, to carry shearing loads and to support the thin facings against lateral wrinkling caused by compressive loads in the facings.

In general, the procedure is to provide facings thick enough to carry the compression and tension stresses and then to space the facings with a core thick enough to impart stiffness and bending strength to the construction. The core should be strong enough to carry the required shearing loads. The construction should be checked for possible buckling, as for a column or panel in compression, and for possible wrinkling of the facings.

The core material itself is assumed to contribute nothing to the stiffness of the sandwich construction, because it usually has a low modulus of elasticity. The facing moduli of elasticity are usually at least 100 times as great as the core modulus of elasticity. The core material may also have a small shear modulus. This small shear modulus causes increased deflections of sandwich constructions subjected to bending and decreased buckling loads of columns and edge-loaded panels, compared to constructions in which the core shear modulus is large. The effect of this low shear modulus is greater for short beams and columns and small panels than it is for long beams and columns and large panels.

The stiffness of sandwich construction having facings of equal or unequal thickness is given by:

$$D = \frac{f_1 f_2 E_1 E_2 \, (h+c)^2}{4 \, (f_1 E_1 + f_2 E_2)}$$

where D is the stiffness per unit width of sandwich construction (product of modulus of elasticity and moment of inertia of the cross section); f_1, f_2 are the facing thicknesses; E_1, E_2 are the moduli of elasticity of the facings; h is the total sandwich thickness; and c is the core thickness. The stiffness is used to compute the deflections and the buckling loads of sandwich panels.

The midspan deflection of a panel of sandwich construction, with simply supported ends and free edges, under a uniform transverse load is given by:

$$w = \frac{5Pa^3}{384Db}\left[1 + \frac{192\, cD}{5a^2\, G_c\, (h+c)^2}\right]$$

where w represents the midspan deflection; P, the total load on sandwich panel; a, the span length; G_c, the shear modulus of core material; and b, the width of the sandwich panel.

In a strip of sandwich construction subjected to both bending moments and shear loads the mean facing stresses are given by:

$$S_{1,\,2} = \frac{2M}{f_{1,\,2}\,(h+c)\,b}$$

where $S_{1,\,2}$ is the mean compression or tension stress in facing 1 or 2; $f_{1,\,2}$, the thickness of facing 1 or 2; and M, the bending moment.

Under the same conditions, the shear stress in the core is given by:

$$q = \frac{2V}{(h+c)\,b}$$

where q is the shear stress in the core and V, the shear load on the sandwich.

The buckling load of a sandwich panel at least twice as wide as it is thick and loaded as a simply supported column is given by:

$$P = \frac{\pi^2 D b}{a^2 \left[1 + \dfrac{4\pi^2\, cD}{a^2\, (h+c)^2\, G_c}\right]} \quad \text{for } \frac{4\pi^2\, cD}{a^2\, (h+c)^2\, G_c} < 1.0$$

and

$$P = bh\, G_c \quad \text{for } \frac{4\pi^2\, cD}{a^2\, (h+c)^2\, G_c} \geq 1.0$$

where a is the column length and b, the panel width.

The preceding formulas are basically those needed for the design of sandwich constructions. Formulas have been derived for various loading and edge conditions for constructions of orthotropic or isotropic materials. The behavior of thin facings in regard to wrinkling and core shear failure has also been investigated. Analyses of these more specific problems of design may be found in other laboratory papers (1, 2, 3, 6, 7, 8, 13).

DIMENSIONAL STABILITY AND BOWING OF SANDWICH PANELS

In a sandwich panel any dimensional movement of one facing with respect to the other due to changes in moisture content and temperature causes bowing of an unrestrained panel (9). Thus, although the use of dissimilar facings is often desirable from an economic or decorative standpoint, the dimensional instability of the facings during panel manufacture or exposure may rule out possible benefits. If dimensional change of both facings is equal, the length and width dimensions of the panel will increase or decrease but bowing will not result.

The problem of dimensional stability is chiefly related to the facings, because the core does not have enough stiffness to cause bowing of the panel or to cause it to remain flat. The magnitude of the bowing effect, however, depends on the thickness of the core.

It is possible to calculate mathematically the bowing of a sandwich construction if the percent expansion of each facing is known. The maximum deflection due to bowing caused by the expansion of one facing resulting from temperature or moisture differential is given approximately by

$$\Delta = \frac{ka^2}{800h}$$

where k is the percent expansion of one facing as compared to the opposite facing; a, the length of the panel; and h, the total sandwich thickness.

In conventional construction, vapor barriers, usually of asphalt-impregnated paper or metal foil, are often installed to block migration of vapor to the cold side of a wall. Various methods have been tried or suggested for reducing vapor movement through sandwich panels, which causes a moisture differential with resultant bowing of the panels. These include bonding of metal foil within the sandwich construction; blending aluminum flakes with the resin bonding adhesives; and use of plastic vapor barriers between veneers, overlay papers, special finishes, or metal or plastic facings. Because added cost is likely, some of these should not be resorted to unless their need has been demonstrated.

A large test unit (14) simulating use of sandwich panels of various kinds under housing conditions has been constructed at the Laboratory for the purpose of observing bowing of panels and general performance. The experimental assembly shown in figure 65 represents the type of construction used in the test unit.

As a generalization, the bowing of sandwich panels is probably neither greater nor less than that of stressed-skin prefabricated housing panels having similar facings.

THERMAL INSULATION OF SANDWICH PANELS

Satisfactory thermal insulation of sandwich panels can best be obtained by using cores having low thermal conductivity, although the use of reflective layers on the facings is of some value. Paper cores

M 76939 F

FIGURE 65.—Experimental assembly used to investigate the performance of sandwich panels for house construction.

of the types previously mentioned have thermal conductivity values (k values) ranging from 0.30 to 0.65 British thermal units per hour per 1° F. per square foot per inch of thickness, depending on the particular core construction (5, 9, 11).   The k value is also affected by the cell size, density, and resin content of the core.

An improvement in the insulation value can be realized by filling the honeycomb core with insulation or a foamed-in-place resin. A reduction in the k values of a corrugated core from 0.46 to 0.40 British thermal units per hour per 1° F. per square foot per inch of thickness was obtained when a phenolic resin was foamed into the core. A slightly lower value was obtained through the use of fill insulation.

FIRE RESISTANCE OF SANDWICH PANELS

In tests at the Laboratory, the fire resistance of wood-faced sandwich panels was appreciably higher than that of hollow panels faced with the same thickness of plywood (9). Fire resistance was greatly increased when coatings that intumesce on exposure to heat were applied to the core material.

The spread of fire through the honeycomb core depended to a large extent on the alinement of the flutes in the core. In panels having flutes perpendicular to the facings, only slight spread of flame occurred (9). In corrugated core in which one-half of the flutes were parallel to the length of the panel, the spread of flame occurred in the vertical direction along open channels. This could no doubt be improved by placing a barrier sheet at the top of the panel or at intervals in the panel height, or, if strength requirements permit, by simply turning the length of the core blocks at 90° to the vertical direction.

LITERATURE CITED

(1) BOLLER, K. H., and NORRIS, C. B.
      1950. EFFECT OF SHEAR STRENGTH ON MAXIMUM LOADS OF SANDWICH
            COLUMNS. U. S. Forest Prod. Lab. Rpt. 1815, 12 pp., illus.
            [Processed.]
(2) ERICKSEN, W. S.
      1950. DEFLECTION UNDER UNIFORM LOAD OF SANDWICH PANELS HAVING
            FACINGS OF UNEQUAL THICKNESS. U. S. Forest Prod. Lab. Rpt.
            1583-C, 32 pp., illus. [Processed.]
(3) ———— and MARCH, H. W.
      1950. COMPRESSIVE BUCKLING OF SANDWICH PANELS HAVING FACINGS OF
            UNEQUAL THICKNESS. U. S. Forest Prod. Lab. Rpt. 1583-B,
            30 pp., illus.
(4) HEEBINK, B. G., MOHAUPT, A. A., and KUNZWEILER, J. J.
      1947. FABRICATION OF LIGHTWEIGHT SANDWICH PANELS OF THE AIRCRAFT
            TYPE. U. S. Forest Prod. Lab. Rpt. 1574, 30 pp., illus. [Processed.]
(5) KUENZI, E. W.
      1951. PAPER HONEYCOMB AS A CORE FOR STRUCTURAL SANDWICH CON-
            STRUCTION. Amer. Soc. Testing Mater. Symposium on Struc-
            tural Sandwich Constructions, Spec. Tech. Pub. 118, illus.
(6) ———— and ERICKSEN, W. S.
      1951. SHEAR STABILITY OF FLAT PANELS OF SANDWICH CONSTRUCTION.
            U. S. Forest Prod. Lab. Rpt. 1560, 42 pp., illus. [Processed.]
(7) MARCH, H. W., and SMITH, C. B.
      1949. FLEXURAL RIGIDITY OF A RECTANGULAR STRIP OF SANDWICH CON-
            STRUCTION. U. S. Forest Prod. Lab. Rpt. 1505, 19 pp., illus.
            [Processed.]
(8) NORRIS, C. B., ERICKSEN, W. S., MARCH, H. W., SMITH, C. B., and BOLLER,
      K. H.
      1949. WRINKLING OF THE FACINGS OF SANDWICH CONSTRUCTIONS SUB-
            JECTED TO EDGEWISE COMPRESSION. U. S. Forest Prod. Lab.
            Rpt. 1810, 68 pp., illus. [Processed.]

(9) SEIDL, R. J.
    1952. PAPER-HONEYCOMB CORES FOR STRUCTURAL SANDWICH PANELS.
        U. S. Forest Prod. Lab. Rpt. R1918, 21 pp., illus. [Processed.]
(10) ——— KUENZI, E. W., FAHEY, D. J., and MOSES, C. S.
    1951. PAPER-HONEYCOMB CORES FOR STRUCTURAL BUILDING PANELS:
        EFFECT OF RESINS, ADHESIVES, FUNGICIDES, AND WEIGHT OF
        PAPER ON STRENGTH AND RESISTANCE TO DECAY. U. S. Forest
        Prod. Lab. Rpt. R1796, 16 pp., illus. [Processed.]
(11) TEESDALE, L. V.
    1949. THERMAL INSULATION MADE OF WOOD-BASE MATERIALS: ITS AP-
        PLICATION AND USE IN HOUSES. U. S. Forest Prod. Lab. Rpt.
        R1740, 40 pp., illus. [Processed.]
(12) U. S. FOREST PRODUCTS LABORATORY.
    1951. SANDWICH CONSTRUCTION FOR AIRCRAFT, PART I: FABRICATION,
        INSPECTION, DURABILITY, AND REPAIR. ANC Bulletin 23, 120
        pp., illus. (Issued by Subcommittee on Air Force-Navy-Civil
        Aircraft Design Criteria, Aircraft Committee, Munitions Board.)
(13) ———
    1951. SANDWICH CONSTRUCTION FOR AIRCRAFT, PART II: DESIGN CRITERIA
        FOR SANDWICH CONSTRUCTION. ANC Bulletin 23, 101 pp., illus.
        (Issued by Subcommittee on Air Force-Navy-Civil Aircraft
        Design Criteria, Aircraft Committee, Munitions Board.)
(14) U. S. FOREST PRODUCTS LABORATORY.
    1948. PHYSICAL PROPERTIES AND FABRICATION DETAILS OF EXPERI-
        MENTAL HONEYCOMB-CORE SANDWICH HOUSE PANELS. U. S.
        Housing and Home Finance Agency Tech. Paper 7, 23 p., illus.

BENT WOOD MEMBERS

Curved members of wood are produced by bandsawing or by bending. Bandsawing often wastes considerable wood and produces curved members that have cross grain, or grain that runs out along the curved surfaces. Bending produces curved members that are free from this type of cross grain and, if well done with properly selected lumber, is also less wasteful of wood. Bending is used particularly for producing members of sharp curvature, such as parts of furniture and ship and boat frames.

The bending may be done with or without a plasticizing treatment, such as steaming. The bent member may be solid or laminated. The grain of all laminations may be parallel, or that of adjacent laminations may be perpendicular. The laminations may be individually bent and then glued together in the curved shape, or glued together and then bent.

STEAM BENDING OF SOLID PIECES

Principles of Bending

Wood of certain species steamed or soaked in boiling water can be compressed as much as 25 to 30 percent parallel to the grain. The same wood can be stretched only 1 to 2 percent. Because of the relation between attainable tensile and compressive deformations, it is necessary, if bending involves severe deformation, that most of the deformation be forced to take place as compression. The inner or concave side must assume the maximum amount of compression and the outer or convex side, zero strain or a slight tension. To accomplish this, a metal strap equipped with end fittings is used. The strap makes contact with the outer or convex side and, acting through the end fittings, places the whole piece of wood in compression. The tensile stress that would normally develop in the outer side of the piece of wood during bending is borne by the metal strap (6).[20]

Selection of Stock

In general, hardwoods possess better bending quality than softwoods, and certain hardwoods surpass others in this quality. The species commonly used to produce bent members are: White oak, red oak, elm, hickory, ash, beech, birch, maple, walnut, sweetgum, and mahogany. Softwoods have poor bending quality, as a rule, and are not often used in bending operations. Yew and yellow-cedars are probable exceptions to this rule. Douglas-fir, southern yellow pine, the cedars, and redwood are used for ship and boat planking, for which purpose they are often bent to moderate curvature after being steamed or soaked.

Bending stock should be free from serious cross grain and distorted grain, such as may occur near knots. The slope of cross grain should not be greater than about 1 to 15. Decay, knots, shake, pith, surface checks, and exceptionally light or brashy wood should be avoided.

[20] Italic numbers in parentheses refer to Literature Cited, p. 308.

Such irregularities tend to cause the pieces of wood to fail during bending, particularly if they are on the face that is to be the inner or concave side at the time of bending and if the piece is to be bent to a sharp curvature.

Seasoning Bending Stock

Although green wood can be bent to produce most curved members, difficulties are introduced in the drying and fixing of the bent piece and in reducing the moisture content of the bent piece to a figure suited for the end use. Another disadvantage with green stock is that hydrostatic pressure may be developed during bending. Hydrostatic pressure may cause compression failures on the concave side if the wood is compressed by an amount greater than the air space in the cells of the green wood.

Bending stock that has been dried to a low moisture content requires a long steaming or soaking process to increase its moisture content to the point where it can be made sufficiently plastic for successful bending. For most chair and furniture parts, the moisture content of the bending stock should be 12 to 20 percent before it is steamed. The preferred moisture content varies with the severity of the curvature to which the wood is bent and the method used in drying and fixing the bent member. For example, chair-back slats, which have a slight curvature and are subjected to severe drying conditions between steam-heated platens, can be produced successfully from stock at a moisture content of 12 percent. The moisture content values desired for the bending of chair parts fall within the range of those for partially or thoroughly air-dried stock.

Plasticizing of Bending Stock

Heat and moisture make certain species of wood sufficiently plastic for bending operations. Steaming at atmospheric or a low gage pressure (fig. 66) or soaking in boiling or nearly boiling water are satisfactory methods of plasticizing wood. Wood at 20- to 25-percent moisture content needs to be heated without losing moisture; at lower moisture content heat and moisture must be added. As a consequence, the recommended plasticizing processes are steaming or boiling for about ½ hour per inch of thickness for wood at 20- to 25-percent moisture content and steaming or boiling for about 1 hour per inch of thickness for wood at lower moisture content values. Steaming at high pressures causes wood to become plastic, but wood treated with high-pressure steam generally does not bend so successfully as wood treated at atmospheric or low pressure. Certain chemicals plasticize wood, but wood so plasticized seldom bends as successfully as wood plasticized with steam at atmospheric pressure.

Bending Operation and Apparatus

After being plasticized, the stock should be quickly placed in the bending apparatus and bent to shape. The bending apparatus consists essentially of a form (or forms) and a means of forcing the piece

M 59799 F

FIGURE 66.—Removing a stick from a steaming retort.

of steamed wood against the form. If the curvature to be obtained demands a difference of much more than 3 percent between lengths of the outer and inner surfaces of the pieces, then the apparatus should include a device for applying end pressure. This generally takes the form of a metal strap or pan provided with end blocks, end bars, or clamps (figs. 67, 68).

Fixing the Bend

After being bent, the piece should be cooled and dried while held in its curved shape. One method is to dry the piece in the bending machine between the plates of a hot-plate press. Another method is to secure the bent piece to the form and place both the piece and the form in a drying room. Still another is to keep the bent piece in a minor strap with tie rods or stays, so that it can be removed from the form and placed in a drying room. When the bent member has dried to a moisture content suited for its intended use, the restraining devices can be removed and the piece will hold its curved shape.

Characteristics of Bent Wood

After a bent piece of wood is dried, the curvature will be maintained unless the wood undergoes changes in moisture content. An increase in moisture content causes the piece to lose some of its curvature. A decrease in moisture content causes the curve to become sharper,

M 4064 F

FIGURE 67.—A pan loaded with steamed stock, between the plates of a hot-plate press, ready for bending into chair-back posts.

M 39136 F

FIGURE 68.—Apparatus for bending heavy members. The end blocks of the metal strap are equipped with reversed levers made from I-beams, which prevent the end blocks from overturning and sliding off, help to regulate end pressure, and reduce the tendency of the stick to assume reverse or back bends near the ends.

although repeated changes in moisture content bring about a gradual straightening. These changes are caused primarily by lengthwise swelling or shrinking of the inner (concave) face, the fibers of which were wrinkled or folded during the bending operation.

A bent piece of wood possesses less strength than a similar but unbent piece. The reduction in strength brought about by bending, however, is seldom serious enough to affect the utility value of the member.

STEAM BENDING OF LAMINATED MEMBERS

Straight laminated members can be steamed and bent after they are glued by operations similar to those used for solid members (5). The glue used for laminating should be one that will not be affected by the steaming or boiling treatment. In assembling the laminations for a member that is to be bent, the best pieces should preferably be selected for the outer laminations. It is especially important to place a high-quality piece on the side that is to be concave after bending. Butt joints within the member should be designed to resist longitudinal compression.

PRODUCTION OF BENT LAMINATED MEMBERS WITHOUT STEAM TREATMENT

Laminated curved members are produced from dry stock by bending and gluing together in one operation several comparatively thin pieces without softening them by steam or hot water. This process has the following advantages over the bending of single-piece members: (1) The laminations can be made so thin that bending to the required radius involves only moderate stress and deformation of the wood fibers; consequently, the use of steam or hot water is unnecessary and less subsequent drying and conditioning is required; (2) because of the moderate stress induced in bending, stronger parts are produced; (3) the tendency of laminated members to change shape with changes in moisture content resulting from changes in relative humidity is less than that of single-piece bent members; (4) ratios of thickness of member to radius of curvature that are impossible in bending single pieces can readily be obtained by laminating; also members having reversed curvature are more readily made by laminating thin plies, and curved parts of any desired length can be produced by staggering the joints in the laminations.

Both softwoods and hardwoods can be used for laminated bent structural members, and thin material of any species can be satisfactorily bent for such purposes. The choice of species and glue is dependent primarily upon the cost, required strength, and demands of the application.

Laminating of curved pieces in the United States was originally applied chiefly to comparatively small parts, such as those in furniture and pianos. The principle has been extended rapidly, however, to the construction of wood arches for use as roof supports in farm, industrial, and public-assembly buildings, and to the construction of ship timbers (4, 12).

VENEERED CURVED MEMBERS

Veneered curved members are usually produced by gluing veneer to one or both faces of a curved solid wood base (*7, 8, 9, 10*). The bases are ordinarily bandsawed to the desired shape or bent from a piece grooved with saw kerfs on the concave side at right angles to the directions of bend. Sometimes series of curved pieces are bandsawed contiguously from the same block of wood and reassembled in a gluing operation in the same relative position with glue-coated pieces of veneer between, thus producing a number of similar veneered curved pieces in one operation. Pressure is then applied to the whole series. In other assemblies single members are bandsawed to the desired curvature and then faced with a layer of veneer, and pressure is applied by means of flexible bands and clamps while the glue sets. Pieces bent by making saw kerfs on the concave side are commonly reinforced and kept to the required curvature by gluing splines, veneer, or other pieces to the curved base.

Veneering over curved solid wood finds use mainly in furniture. The grain of the veneer is commonly laid in the same general direction as the grain of the curved wood base. The use of crossband veneers, that is, veneers laid with the grain at right angles to the grain of the base and face veneer, reduces the tendency of the member to split.

PLYWOOD CURVED MEMBERS

Curved plywood is produced (1) by bending and gluing the plies in one operation, or (2) by bending previously glued flat plywood. Curved plywood made by method (1) is more stable in curvature than plywood curved by method (2).

Plywood Bent and Glued Simultaneously

In bending and gluing plywood in a single operation, glue-coated pieces of veneer are assembled and pressed over or between curved forms, and pressure and sometimes heat applied until the glue sets and holds the assembly to the desired curvature (*3, 11*). Some of the laminations are at an angle, usually 90°, to other laminations, as in the manufacture of flat plywood. The grain direction of the thicker laminations is normally parallel to the axis of the bend in order to facilitate bending.

The advantages of bending and gluing plywood simultaneously to form a curved shape are similar to those for curved laminated members. and in addition the cross plies give the curved members properties characteristic of cross-banded plywood.

Molded Plywood

Although any piece of curved plywood may properly be considered to be molded plywood, the term "molded" is usually reserved for plywood that is glued to the desired shape, either between curved forms, as described previously, or, more commonly, by means of fluid pressure. The molding of plywood by means of fluid pressure applied with flexible bags or blankets of some impermeable material (bag molding) is used in making plywood parts of various degrees of compound curvature (*1, 2*). These parts vary in size from small aircraft fairings to complete boat hulls.

Misnomers, such as "plastic plywood," "plastic planes," or just "plastics" have been applied to structures of molded plywood that are actually made from wood veneers bonded with synthetic resin adhesives. By weight, these structures contain about 80 percent of wood and 20 percent of resin adhesive. Except for variations in shape, the product is essentially the same as flat-pressed plywood.

Plywood Bent After Gluing

After the plies are glued together, flat plywood is often bent by methods that are somewhat similar to those used in bending solid wood. To bend plywood properly to shape, it must be plasticized by some means, usually moisture or heat, or a combination of both. The amount of curvature that can be introduced into a flat piece of plywood depends on numerous variables, such as moisture content, direction of grain, thickness and number of plies, species and quality of veneer, and the technique applied introducing the bend. Plywood is normally bent over a form or a bending mandrel.

Limited data are available on the relation between approximate breaking radius and thickness of plywood, veneer, and lumber under several conditions. Some of these were derived from actual factory practice and some from tests at the Forest Products Laboratory. An approximate relation between thickness and breaking radius of plywood of aircraft quality is presented in figure 69, *A* and *B*. These graphs are based primarily on tests of plywood between 0.035 and 0.375 inch in thickness, but it is believed that the curve for the 90° angle and 10 percent moisture content may also be applied in estimating the breaking radius of veneer or air-dry lumber between 0.01 and 1 inch in thickness. The angle referred to on each graph is the angle between the grain direction of the face plies and the axis of the bend.

Most of the tests were made at the 0° or 90° angle, but a few tests on wider strips of plywood at 45° indicated that there was little or no difference between the results at 0° and 45° conditions. The term "hot soaked" means that the plywood is thoroughly soaked in hot or boiling water until it sinks, after which it is bent over a mandrel heated to approximately 300° F.; the term "10-percent moisture content" means plywood of this moisture content bent over a cold mandrel. The tests were made by slowly bending 1-inch-wide strips of aircraft plywood around a series of mandrels of decreasing size until fracture occurred. Soaking in water at room temperature would probably produce a degree of flexibility or a breaking radius intermediate between that shown for dry plywood and that for hot-water-soaked plywood. Bending plywood at 10-percent moisture content over a hot mandrel also might be considered an intermediate treatment, and a breaking radius between those of dry and hot-soaked plywood anticipated.

In applying these data, it is desirable to multiply the approximate breaking radius obtained from figure 69 by a factor of safety in order (1) to obtain a working radius that will provide against overstressing in bending, (2) to allow for inapplicability of figure 69 to the case

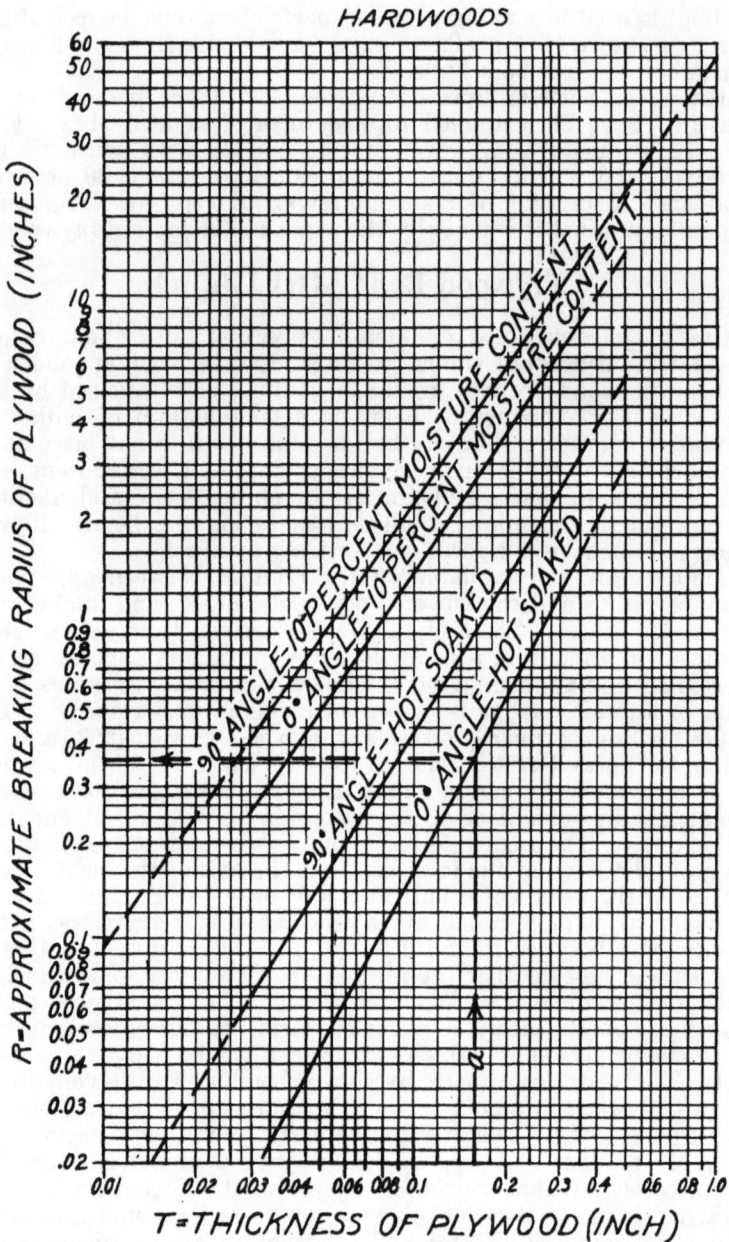

FIGURE 69, A.—Hardwoods.   Approximate relation between thickness and breaking radius of plywood between 0.035 and 0.375 inch in thickness, of aircraft
construction and quality.   Tests made by slowly bending plywood strips 1
inch wide over mandrels with no support on the tension side.   Angle referred
to on the graph is the angle between the grain direction of the face plies and the
axis of the bend.   Broken line a refers to the example given in the text.

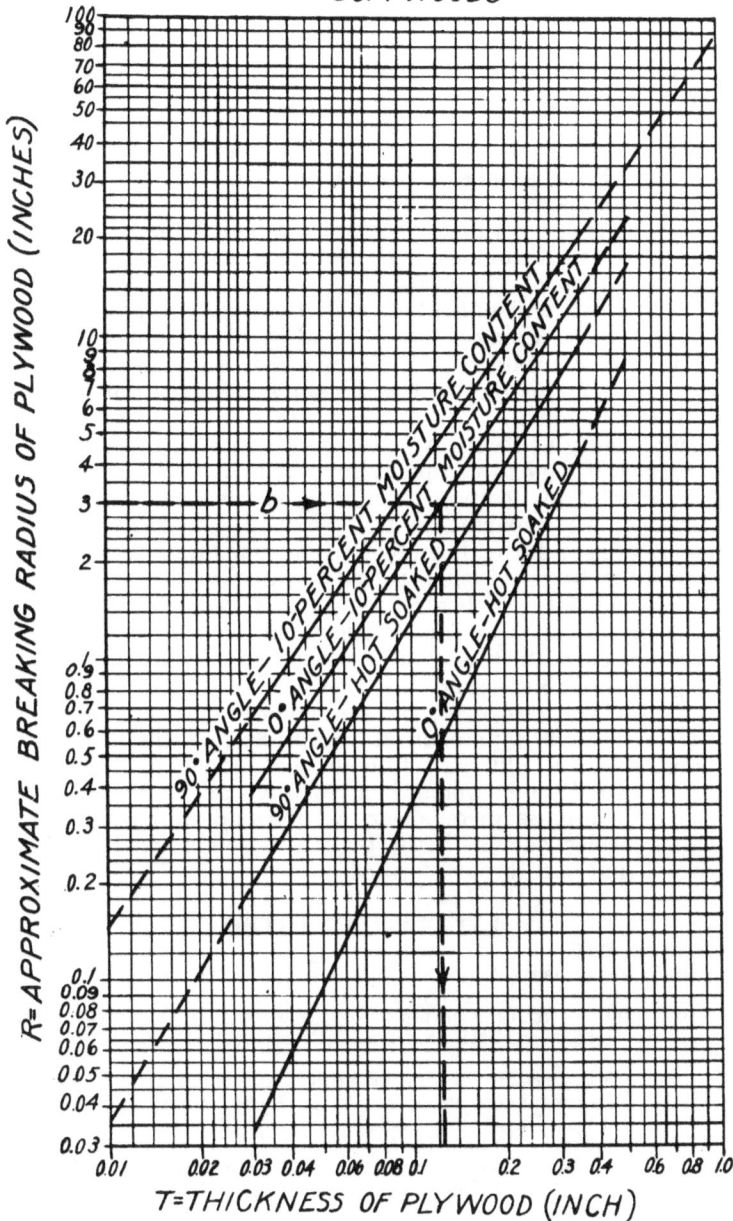

SOFTWOODS

R= APPROXIMATE BREAKING RADIUS OF PLYWOOD (INCHES)

T=THICKNESS OF PLYWOOD (INCH)

ZM 48523 F

FIGURE 69, B.—Softwoods. Approximate relation between thickness and breaking radius of plywood between 0.035 and 0.375 inch in thickness, of aircraft construction and quality. Tests made by slowly bending plywood strips 1 inch wide over mandrels with no support on the tension side. Angle referred to on the graph is the angle between the grain direction of the face plies and the axis of the bend. Broken line *b* refers to the example given in the text.

in hand, (3) to avoid face checking of plywood, and (4) to reduce the forces required to form the bend and hold the bent part in position. Values of 2 to 3 are suggested for the factor of safety against breakage applied to the radius in adapting these data to the bending of commercial plywood.

Examples showing the use of figure 69.—After soaking in hot water, to what radius can 0.160-inch birch plywood be bent on a hot mandrel, with the face grain parallel to the axis of bend and allowing a factor of 3 against breakage? Starting at $T=0.160$ inch on the graph for hardwoods, R is found (line a on graph) to be 0.37 inch, which, multiplied by 3, gives 1.11 inches as the safe bending radius.

What thickness of Douglas-fir plywood can be safely bent to a 9-inch radius dry and unheated, with the face grain parallel to the axis of the bend? Using a factor of safety of 3, the breaking radius is read on the softwood curve as 3, which is 9 divided by the factor of safety. From this radius, the thickness is found (line b on graph) to be about 0.125 inch.

Flat plywood glued with a waterproof adhesive can be bent after gluing or postformed to compound curvatures. No simple criterion, however, is available for predetermining whether a specific compound curvature can be imparted to flat plywood. Soaking the plywood and the use of heat during forming are aids in manipulation. Normally the plywood to be postformed is first thoroughly soaked in hot water and then dried between heated male and female dies attached to a hydraulic press. If the use of postforming for bending flat plywood to compound curvatures is contemplated, exploratory trials to determine the practicability and the best procedure are recommended. Plywood has been postformed to compound curvatures that require an upset of about 8 percent in compression without evidence of wrinkling or shear failures. It should be remembered that in postforming plywood to compound curvatures, all of the deformation must be by compression or shear, as plywood cannot be stretched. Hardwood species, such as birch, poplar, and gum, are normally used in plywood that is to be postformed.

LITERATURE CITED

(1) HEEBINK, B. G.
   1945. METHODS OF BAG-MOLDING PLYWOOD. Indus. Plastics 1 (4).
(2) ———
   1946. FLUID-PRESSURE MOLDING OF PLYWOOD. U. S. Forest Prod. Lab. Rpt. R1624, 19 pp., illus. [Processed.]
(3) ———
   1950. SOME METHODS OF GLUING LIGHT LAMINATED OR PLYWOOD CURVED SHAPES FROM VENEER. U. S. Forest Prod. Lab. Rpt. R1485, 3 pp., illus. [Processed.]
(4) KNAUSS, A. C., and SELBO, M. L.
   1948. LAMINATING OF STRUCTURAL WOOD PRODUCTS BY GLUING. U. S. Forest Prod. Lab. Rpt. D1635, 63 pp., illus. (Revised.) [Processed.]
(5) MCKEAN, H. B., BLUMENSTEIN, R. R., and FINNORN, W. F.
   1952. LAMINATING AND STEAM-BENDING OF TREATED AND UNTREATED OAK FOR SHIP TIMBERS. South. Lumberman 185 (2321).
(6) PECK, E. C.
   1950. BENDING SOLID WOOD TO FORM. U. S. Forest Prod. Lab. Rpt. D1764, 29 pp., illus. [Processed.]
(7) PERRY, T. D.
   1930. METHODS USED IN VENEERING CURVED WOOD FORMS. Veneers 24 (9).

(8) PERRY, T. D.
    1930. INSTRUCTIONS FOR BUILDING CURVED PLYWOOD. Veneers 24 (10).
(9) ——
    1930. PRINCIPLES INVOLVED IN BENDING PLYWOOD. Veneers 24 (11).
(10) ——
    1951. CURVES FROM FLAT PLYWOOD. Wood Prod. 56 (4).
(11) TRUAX, T. R.
    1929. GLUING OF WOOD. U. S. Dept. Agr. Bul. 1500, 78 pp., illus.
(12) U. S. FOREST PRODUCTS LABORATORY.
    1945. WOOD—A MANUAL FOR ITS USE IN WOODEN VESSELS. Issued by
    Navy Bureau of Ships, 235 pp., illus.

CONTROL OF MOISTURE CONTENT AND SHRINKAGE OF WOOD

Equilibrium Moisture Content

Any piece of wood will give off or take on moisture from the surrounding atmosphere until the amount of moisture in the wood balances that in the atmosphere. The moisture content of the wood at the point of balance is called the equilibrium moisture content and is expressed as a percentage of the oven-dry weight of the wood.

Assuming constant temperature, the ultimate moisture content that a given piece of wood will attain depends entirely upon the relative humidity of the atmosphere surrounding it. The relationship between equilibrium moisture content and relative humidity at different temperatures is shown in table 38. For practical use in control of dry-kiln or storage-room conditions, comparative values are shown for different dry-bulb temperatures and for different wet-bulb depressions. For example, at a dry-bulb temperature of 70° F. and a wet-bulb depression of 9° below dry-bulb temperature, which corresponds to a relative humidity of 59 percent, the equilibirum moisture content of the wood is 10.9 percent. Changes in relative humidity in the higher ranges cause greater changes in equilibirum moisture content than do corresponding changes in the lower ranges of relative humidity. Although different species exhibit some differences in their reactions to relative humdity, for practical purposes table 38 applies to the wood of any species.

Wood in service is exposed to daily and seasonal changes in relative humidity. Thus, wood is virtually always undergoing at least slight changes in moisture content because of its tendency to come to a balance with the relative humidity of the surrounding air. The changes are gradual and may be further retarded by protective coatings, such as varnish, lacquer, or paint. The practical object of all correct seasoning, handling, and storing methods is to minimize moisture content variations in wood in service by fabricating or installing the wood at a moisture content corresponding to the average atmospheric conditions to which it will be exposed.

Shrinkage of Wood

Wood, like many other materials, shrinks as it loses moisture and swells as it absorbs moisture.

Wood from the tree may contain from 30 to 300 percent of water (*3, 11*),[21] based on the weight of the oven-dry wood. This water may be separated roughly into two parts, that contained as free water in the cell cavities and intercellular spaces of the wood and that held as absorbed water in the capillaries of the walls of such wood elements as fibers and ray cells. The absorbed water is of primary interest in the consideration of shrinkage. When all of the free water is removed but all of the absorbed water remains, wood is said to have reached

---

[21] Italic numbers in parentheses refer to Literature Cited, p. 336.

TABLE 38.—*Relative humidity [1] and equilibrium moisture content [2] table for use with dry-bulb temperatures and wet-bulb depressions*

Wet-bulb depression (°F.)

Each cell gives relative humidity (roman) over equilibrium moisture content (*italic*). A dash (—) indicates no value is tabulated.

Temperature dry-bulb (°F.)	1	2	3	4	5	6	7	8	9	10	11	12	13	14	15	16	17	18	19	20	21	22	23	24	25	26	27	28	29	30	32	34	36	38	40	45	50
30	89	78 *15.9*	67 *12.9*	57	46 *9.0*	36 *7.4*	27 *5.7*	17 *3.9*	6 *1.6*																												
35	90	81 *16.8*	72 *13.9*	63 *11.9*	54 *10.3*	45 *8.8*	37 *7.4*	28 *6.0*	19 *4.5*	11 *2.9*	3 *0.8*																										
40	92	83 *16.8*	75	68	60	52	45 *8.6*	37 *7.4*	29 *6.2*	22 *5.0*	15 *3.5*	8 *1.9*																									
45	93	85 *17.6*	78 *14.8*	72	64 *11.2*	58 *9.9*	51 *9.5*	44 *8.5*	37 *7.4*	31 *6.5*	25 *5.3*	19 *4.2*	12 *2.9*	6 *1.5*																							
50	93	85 *18.3*	80 *15.6*	76	68 *12.0*	62 *11.5*	56 *10.3*	50 *9.5*	44 *8.5*	38 *7.6*	32 *6.8*	27 *5.7*	21 *4.8*	16 *3.9*	10 *2.8*	5 *1.5*																					
55	93	86 *19.0*	82 *16.3*	76 *15.1*	70 *12.7*	65 *11.5*	60 *11.0*	54 *10.1*	49 *9.3*	44 *8.4*	39 *7.6*	34 *6.8*	28 *6.0*	24 *5.3*	19 *4.5*	14 *3.6*	9 *2.5*	5 *1.3*																			
60	94	88 *19.5*	82 *16.9*	80 *15.4*	68 *13.4*	62 *12.2*	56 *11.0*	51 *10.3*	44 *9.3*	38 *8.4*	34 *7.6*	32 *6.8*	27 *6.0*	21 *5.3*	16 *4.5*	13 *3.6*	13 *2.5*	13 *1.5*	9 *2.3*	5 *1.3*	1 *0.2*																
65	94	89 *19.9*	83 *17.4*	76 *15.6*	70 *12.7*	65 *11.6*	60 *10.3*	54 *9.4*	49 *8.6*	44 *7.6*	39 *6.7*	34 *5.7*	28 *4.8*	24 *3.9*	19 *2.8*	14 *1.5*	9	5			8 *2.3*	6 *1.4*	2 *0.4*														
70	95	90 *19.9*	84 *17.8*	80 *15.9*	75 *12.7*	70 *11.5*	66 *11.6*	61 *11.6*	56 *10.9*	52 *10.1*	48 *9.4*	44 *8.8*	39 *8.3*	36 *7.7*	32 *7.2*	29 *6.6*	24 *6.0*	20 *5.5*	16 *4.9*	13 *4.3*	12 *3.7*	12 *2.9*	15 *2.3*	6 *1.5*	3 *0.7*												
75	95	90.5 *20.5*	86 *18.2*	81 *16.1*	77 *14.9*	74 *13.5*	70 *13.0*	66 *12.1*	62 *11.2*	58 *10.5*	54 *9.8*	51 *9.3*	47 *8.7*	44 *8.2*	41 *7.7*	37 *7.2*	34 *6.7*	31 *6.2*	28 *5.6*	24 *5.1*	21 *4.7*	18 *4.1*	15 *3.5*	12 *2.9*	10 *2.3*	7 *1.7*	4 *0.9*	1									
80	95	90.6 *20.6*	87 *18.5*	83 *16.8*	79 *15.5*	76 *14.0*	73 *13.5*	70 *12.9*	66 *12.0*	63 *11.2*	59 *10.5*	56 *10.0*	53 *9.6*	50 *9.0*	47 *8.5*	44 *8.1*	41 *7.6*	38 *7.1*	36 *6.7*	33 *6.3*	30 *5.8*	28 *5.5*	25 *5.2*	23 *4.8*	20 *4.4*	18 *4.0*	15 *3.4*	13 *3.0*	11 *2.4*	9 *1.9*	4 *0.9*						
85	96	91 *21.0*	88 *18.8*	85 *17.2*	80 *15.7*	78 *14.5*	74 *13.6*	71 *12.8*	68 *11.8*	65 *11.1*	61 *10.5*	58 *10.1*	55 *9.6*	52 *9.1*	49 *8.5*	47 *8.1*	44 *7.6*	41 *7.2*	39 *6.7*	36 *6.3*	34 *5.6*	31 *5.2*	29 *4.8*	27 *4.8*	24 *4.6*	22 *3.9*	19 *3.4*	17 *3.0*	15 *2.4*	13 *2.0*	9 *0.9*	4					
90	96	92 *21.1*	89 *18.9*	85 *17.3*	81 *15.9*	78 *14.7*	75 *13.7*	72 *12.8*	69 *12.0*	66 *11.4*	63 *10.7*	60 *10.2*	57 *9.7*	55 *9.3*	52 *8.8*	49 *8.4*	47 *8.0*	44 *7.6*	42 *7.2*	39 *6.8*	37 *6.5*	34 *6.1*	32 *5.7*	30 *5.3*	28 *4.9*	26 *4.6*	23 *4.2*	22 *3.8*	20 *3.3*	17 *2.8*	14 *2.1*	10 *1.3*	6 *0.4*	2			
95	96	92 *21.3*	89 *19.0*	86 *17.4*	83 *16.1*	80 *14.9*	77 *13.9*	73 *12.9*	70 *12.1*	68 *11.6*	65 *11.0*	62 *10.2*	59 *9.7*	56 *9.6*	54 *9.1*	51 *8.8*	49 *8.2*	46 *7.9*	44 *7.5*	41 *7.1*	39 *6.8*	37 *6.4*	35 *6.0*	33 *5.7*	30 *5.3*	28 *5.1*	26 *4.8*	24 *4.4*	22 *4.0*	21 *3.3*	17 *2.3*	14 *1.5*	10 *0.6*	7			
100	96	93 *21.3*	89 *19.0*	86 *17.5*	83 *16.1*	80 *15.0*	77 *13.9*	73 *13.1*	70 *12.4*	68 *11.8*	65 *11.2*	62 *10.6*	59 *10.1*	56 *9.6*	54 *9.2*	51 *8.7*	49 *8.5*	46 *8.1*	44 *7.8*	41 *7.4*	39 *7.0*	37 *6.7*	35 *6.4*	33 *6.1*	30 *5.7*	28 *5.4*	26 *5.2*	24 *4.9*	22 *4.6*	21 *4.2*	17 *3.6*	13 *3.1*	10 *2.4*	7 *1.6*	4 *0.7*		

Equilibrium moisture content table. Relative humidity values in roman type; equilibrium moisture content values in italic type. Rows are temperature (°F); each cell gives relative humidity (roman) / equilibrium moisture content (italic).

Temp																																			
105	8 *1.8*	11 *2.4*	14 *3.1*	17 *3.6*	20 *4.2*	23 *4.6*	26 *4.8*	28 *5.2*	30 *5.4*	32 *5.7*	34 *6.1*	35 *6.4*	37 *6.7*	40 *6.9*	42 *7.3*	44 *7.6*	45 *7.9*	48 *8.3*	50 *8.7*	53 *9.0*	55 *9.4*	58 *9.8*	60 *10.3*	63 *10.8*	66 *11.3*	69 *11.9*	71 *12.5*	74 *13.2*	77 *14.0*	80 *15.1*	83 *16.2*	87 *17.5*	90 *19.0*	93 *21.4*	96 —
110	11 *2.4*	14 *3.1*	17 *3.5*	20 *4.0*	23 *4.5*	26 *4.8*	28 *5.2*	30 *5.4*	32 *5.6*	34 *5.9*	36 *6.2*	36 *6.5*	38 *6.7*	42 *7.0*	44 *7.2*	46 *7.5*	48 *7.9*	50 *8.2*	52 *8.6*	54 *8.9*	57 *9.3*	60 *9.6*	62 *10.1*	65 *10.5*	67 *11.1*	70 *11.6*	73 *12.2*	76 *12.9*	79 *13.7*	82 *15.1*	84 *16.2*	87 *17.5*	90 *19.0*	93 *21.4*	93 —
115	14 *3.0*	17 *3.5*	20 *4.0*	23 *4.3*	25 *4.6*	29 *4.8*	30 *5.2*	33 *5.4*	34 *5.6*	36 *6.0*	38 *6.3*	40 *6.6*	43 *6.8*	43 *7.2*	45 *7.6*	48 *7.8*	50 *8.2*	52 *8.5*	54 *8.9*	56 *9.3*	58 *9.7*	61 *10.0*	63 *10.4*	66 *11.0*	68 *11.4*	71 *12.0*	74 *12.7*	76 *13.4*	79 *14.1*	82 *15.1*	85 *16.2*	88 *17.5*	90 *19.0*	93 *21.4*	—
120	2 *0.4* 17 *2.9*	20 *3.9*	22 *3.9*	25 *4.3*	27 *4.7*	29 *5.0*	31 *5.2*	33 *5.6*	35 *5.8*	38 *6.1*	40 *6.3*	41 *6.6*	43 *7.0*	45 *7.2*	47 *7.6*	49 *8.0*	51 *8.2*	53 *8.6*	54 *9.0*	58 *9.4*	61 *9.7*	63 *10.1*	65 *10.5*	67 *11.0*	69 *11.5*	72 *12.1*	74 *12.7*	77 *13.4*	80 *14.2*	82 *15.2*	85 *16.2*	88 *17.4*	91 *19.0*	93 *21.3*	—
125	5 *1.1* 19 *3.3*	22 *3.7*	24 *4.2*	26 *4.6*	28 *5.0*	30 *5.2*	33 *5.6*	35 *5.8*	36 *6.1*	38 *6.3*	40 *6.7*	41 *7.0*	45 *7.2*	47 *7.5*	48 *7.9*	51 *8.2*	53 *8.6*	55 *9.0*	58 *9.4*	60 *9.7*	63 *10.1*	65 *10.5*	67 *11.0*	69 *11.5*	72 *12.1*	73 *12.7*	77 *13.4*	80 *14.1*	82 *15.2*	85 *16.2*	88 *17.4*	91 *18.9*	94 *21.2*	—
130	8 *1.6* 19 *3.7*	22 *3.7*	24 *4.8*	27 *5.0*	29 *5.2*	31 *5.4*	33 *5.8*	35 *6.0*	37 *6.1*	40 *6.5*	41 *6.8*	43 *7.1*	45 *7.4*	47 *7.6*	48 *8.1*	51 *8.7*	53 *9.0*	55 *9.4*	57 *9.7*	60 *10.0*	63 *10.5*	65 *11.0*	67 *11.5*	69 *12.1*	72 *12.7*	73 *13.4*	76 *14.1*	80 *15.0*	83 *16.1*	86 *17.3*	88 *18.9*	91 *21.2*	94 —	—
140	10 *2.0* 22 *4.1*	24 *4.2*	26 *4.6*	29 *5.0*	31 *5.2*	33 *5.6*	36 *5.8*	38 *6.0*	40 *6.4*	41 *6.6*	43 *6.9*	45 *7.2*	47 *7.3*	48 *7.6*	51 *8.0*	53 *8.4*	55 *8.7*	58 *9.1*	60 *9.4*	62 *10.0*	65 *10.5*	68 *11.0*	70 *11.6*	73 *12.1*	75 *12.7*	77 *13.4*	80 *14.0*	83 *15.0*	86 *16.0*	89 *17.4*	91 *18.9*	94 *21.3*	—
150	14 *2.6* 24 *4.4*	26 *4.5*	30 *4.8*	32 *5.2*	35 *5.4*	38 *5.6*	40 *6.1*	41 *6.2*	43 *6.4*	46 *6.7*	48 *6.9*	49 *7.2*	51 *7.5*	53 *7.8*	54 *8.0*	56 *8.3*	58 *8.6*	60 *9.0*	62 *9.4*	64 *9.9*	67 *10.4*	69 *11.0*	72 *11.6*	73 *12.2*	75 *12.9*	78 *13.7*	81 *14.5*	84 *15.4*	86 *16.6*	90 *18.4*	92 *20.2*	94 —	—
160	18 *2.9* 25 *4.5*	27 *4.7*	30 *5.1*	32 *5.3*	35 *5.5*	38 *5.8*	41 *6.0*	43 *6.4*	45 *6.7*	46 *6.9*	48 *7.1*	50 *7.4*	52 *7.6*	53 *8.0*	55 *8.3*	57 *8.5*	60 *8.9*	62 *9.1*	64 *9.6*	66 *10.1*	69 *10.7*	72 *11.3*	74 *11.8*	75 *12.4*	78 *13.2*	81 *13.9*	84 *14.8*	86 *16.2*	88 *18.1*	92 *19.8*	94 —	—
170	21 26 *5.0*	28 *5.2*	31 *5.4*	33 *5.8*	35 *6.0*	38 *6.1*	40 *6.4*	43 *6.4*	45 *6.8*	47 *7.0*	49 *7.2*	51 *7.4*	53 *7.5*	55 *7.8*	57 *8.0*	58 *8.4*	60 *8.7*	62 *9.0*	64 *9.6*	65 *10.1*	67 *10.6*	72 *11.2*	73 *11.8*	75 *12.4*	77 *13.1*	79 *14.0*	81 *14.9*	84 *16.4*	86 *17.7*	92 *19.4*	95 —	—
180	24 *3.2* 27 *5.2*	30 *5.2*	32 *5.6*	35 *6.0*	38 *6.2*	41 *6.4*	43 *6.6*	45 *6.9*	47 *7.1*	49 *7.2*	51 *7.5*	53 *7.8*	55 *7.9*	57 *8.0*	58 *8.3*	60 *8.6*	62 *8.8*	65 *9.1*	67 *9.4*	69 *10.1*	73 *10.8*	75 *11.4*	76 *12.1*	78 *12.9*	79 *13.7*	82 *14.5*	85 *15.2*	89 *16.5*	91 *18.1*	94 *19.4*	96 —	—
190	26 *3.3* 28 *5.4*	30 *5.4*	33 *5.9*	36 *6.1*	38 *6.2*	40 *6.4*	43 *6.7*	45 *6.8*	46 *7.1*	48 *7.2*	51 *7.4*	52 *7.6*	54 *7.7*	55 *8.0*	57 *8.2*	60 *8.6*	63 *8.8*	65 *9.2*	66 *9.6*	69 *10.0*	75 *10.5*	76 *11.4*	77 *12.4*	79 *13.4*	82 *14.4*	84 *15.4*	88 *16.9*	90 *18.4*	92 *20.3*	94 *21.4*	96 —	—
200	28 *3.3* 30 *5.7*	34 *5.8*	36 *6.3*	38 *6.4*	40 *6.6*	43 *7.0*	45 *7.2*	48 *7.4*	49 *7.6*	51 *7.7*	53 *7.9*	54 *8.1*	56 *8.4*	58 *8.5*	60 *8.7*	63 *9.1*	65 *9.6*	66 *9.8*	69 *9.8*	72 *10.5*	73 *11.1*	74 *11.8*	77 *12.4*	79 *13.2*	82 *14.0*	84 *14.9*	86 *16.4*	90 *18.1*	92 *19.8*	94 *21.7*	96 —	—
210	30 *3.3* 32 *5.8*	36 *6.1*	38 *6.1*	40 *6.3*	43 *6.4*	44 *6.7*	47 *7.1*	49 *7.2*	51 *7.4*	53 *7.5*	55 *5.8*	57 *6.0*	59 *6.1*	60 *6.3*	61 *6.5*	63 *6.8*	64 *6.9*	65 *7.1*	67 *7.4*	68 *7.6*	70 *8.3*	71 *8.7*	73 *9.0*	76 *9.7*	78 *10.6*	79 *11.1*	81 *11.7*	83 *13.0*	85 *13.8*	86 *14.6*	90 *16.0*	92 *17.7*	94 —	—

1 Relative humidity values in roman type.
2 Equilibrium moisture content values in italic type.

the fiber saturation point. The fiber saturation point is approximately 30-percent moisture content for all species.

Shrinkage occurs if the moisture content is reduced to a value below that of the fiber saturation point and is proportional to the amount of moisture lost below 30-percent moisture content (*7, 15*). Wood dried to 15-percent moisture content has attained about one-half of the total shrinkage possible. For each 1-percent loss in moisture content below the fiber saturation point, the wood shrinks about one-thirtieth of the total. Likewise, for each 1-percent increase in moisture content the piece swells about one-thirtieth of the total swelling possible. The total swelling is equal numerically to the total shrinkage. Shrinking and swelling are expressed as percentages based on the green dimensions of the wood.

As a piece of wood dries, the outer parts are reduced to a moisture content below the fiber saturation point considerably sooner than the inner parts. Thus, the whole piece may show some shrinkage before the average moisture content reaches the fiber saturation point.

Wood shrinks most in the direction of the annual growth rings (tangentially), somewhat less across these rings (radially), and very little, as a rule, along the grain (longitudinally) (*2*). The combined effects of radial and tangential shrinkage on the shape of various sections in drying from the green conditions are illustrated in figure 70. Wood that contains cross grain or irregular grain will shrink more longitudinally than straight-grained wood. Local shrinkage in such pieces may cause distortion.

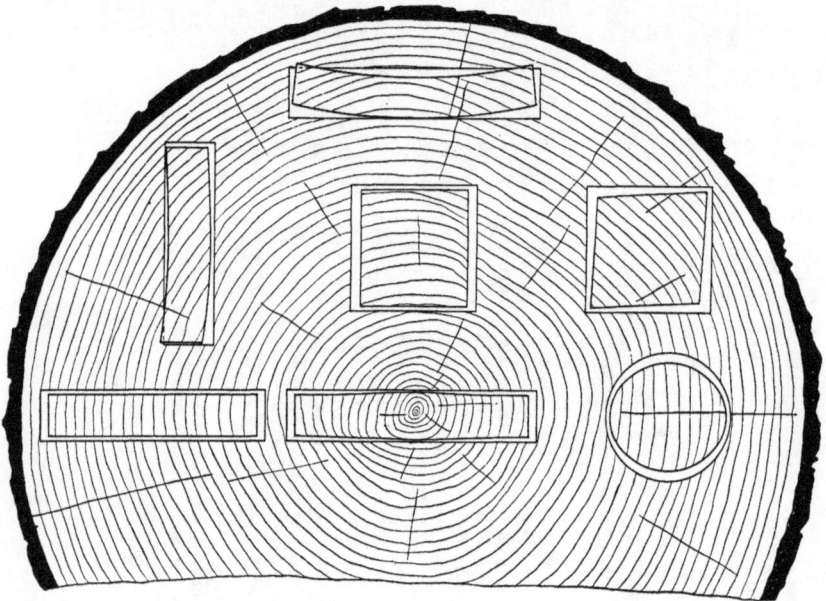

ZM 12494 F

FIGURE 70.—Characteristic shrinkage and distortion of flats, squares, and rounds as affected by the direction of the annual rings. Tangential shrinkage is about twice as great as radial.

Table 39 gives the average tangential, radial, and volumetric shrinkage for numerous species during drying from the green condition to 20-, 6-, and 0-percent moisture content. The ratio of total radial to total tangential shrinkage ranges from 1:1.1 to 1:3.7.

In general, the heavier species of wood shrink more across the grain than lighter ones. Heavier pieces also shrink more than lighter pieces of the same species. Hardwoods generally shrink more than softwoods. Species, however, do not always conform to the general shrinkage pattern. For example, basswood is a light wood but shrinks considerably more than black locust, a heavy wood.

TABLE 39.—*Shrinkage values of wood based on its dimensions when green*

Species	Shrinkage								
	Dried to 20-percent moisture content [1]			Dried to 6-percent moisture content [2]			Dried to 0-percent moisture content		
	Radial	Tangential	Volumetric	Radial	Tangential	Volumetric	Radial	Tangential	Volumetric
SOFTWOODS	*Percent*	*Percent*	*Percent*	*Percent*	*Percent*	*Percent*	*Percent*	*Percent*	*Percent*
Baldcypress	1.3	2.1	3.5	3.0	5.0	8.4	3.8	6.2	10.5
Cedar:									
Alaska	.9	2.0	3.1	2.2	4.8	7.4	2.8	6.0	9.2
Atlantic white	1.0	1.8	2.9	2.3	4.3	7.0	2.9	5.4	8.8
Eastern redcedar	1.0	1.6	2.6	2.5	3.8	6.2	3.1	4.7	7.8
Incense	1.1	1.7	2.5	2.6	4.2	6.1	3.3	5.2	7.6
Northern white	.7	1.6	2.4	1.8	3.9	5.8	2.2	4.9	7.2
Port-Orford	1.5	2.3	3.4	3.7	5.5	8.1	4.6	6.9	10.1
Western redcedar	.8	1.7	2.3	1.9	4.0	5.4	2.4	5.0	6.8
Douglas-fir:									
Coast type	1.7	2.6	3.9	4.0	6.2	9.4	5.0	7.8	11.8
Intermediate type	1.4	2.5	3.6	3.3	6.1	8.7	4.1	7.6	10.9
Rocky Mountain type	1.2	2.1	3.5	2.9	5.0	8.5	3.6	6.2	10.6
Fir:									
Alpine	.9	2.5	3.1	2.1	5.9	7.5	2.6	7.4	9.4
Balsam	1.0	2.3	3.7	2.3	5.5	9.0	2.9	6.9	11.2
California red	1.3	2.4	4.1	3.2	5.8	9.8	4.0	7.2	12.2
Corkbark	.9	2.5	3.0	2.2	5.9	7.2	2.8	7.4	9.0
Grand	1.1	2.5	3.7	2.7	6.0	8.8	3.4	7.5	11.0
Noble	1.5	2.7	4.6	3.6	6.6	11.0	4.5	8.2	13.8
Pacific silver	1.5	3.3	4.6	3.7	7.8	11.0	4.6	9.8	13.8
White	1.1	2.4	3.3	2.6	5.7	7.8	3.2	7.1	9.8
Hemlock:									
Eastern	1.0	2.3	3.2	2.4	5.4	7.8	3.0	6.8	9.7
Western	1.4	2.6	4.0	3.4	6.3	9.5	4.3	7.9	11.9
Larch, western	1.4	2.7	4.4	3.4	6.5	10.6	4.2	8.1	13.2
Pine:									
Eastern white	.8	2.0	2.7	1.8	4.8	6.6	2.3	6.0	8.2
Jeffrey	1.5	2.2	3.3	3.5	5.4	7.9	4.4	6.7	9.9
Limber	.8	1.7	2.7	1.9	4.1	6.6	2.4	5.1	8.2
Lodgepole	1.5	2.2	3.8	3.6	5.4	9.2	4.5	6.7	11.5
Pitch	1.3	2.4	3.6	3.2	5.7	8.7	4.0	7.1	10.9
Pond	1.7	2.4	3.7	4.1	5.7	9.0	5.1	7.1	11.2
Ponderosa	1.3	2.1	3.2	3.1	5.0	7.7	3.9	6.3	9.6
Red	1.5	2.4	3.8	3.7	5.8	9.2	4.6	7.2	11.5
Southern yellow:									
Loblolly	1.6	2.5	4.1	3.8	5.9	9.8	4.8	7.4	12.3
Longleaf	1.7	2.5	4.1	4.1	6.0	9.8	5.1	7.5	12.2
Shortleaf	1.5	2.6	4.1	3.5	6.2	9.8	4.4	7.7	12.3
Slash	1.8	2.6	4.1	4.4	6.2	9.8	5.5	7.8	12.2

See footnotes at end of table.

TABLE 39.—*Shrinkage values of wood based on its dimensions when green*—Con.

Species	Shrinkage								
	Dried to 20-percent moisture content [1]			Dried to 6-percent moisture content [2]			Dried to 0-percent moisture content		
	Radial	Tangential	Volumetric	Radial	Tangential	Volumetric	Radial	Tangential	Volumetric
SOFTWOODS—con.	*Percent*	*Percent*	*Percent*	*Percent*	*Percent*	*Percent*	*Percent*	*Percent*	*Percent*
Pine—Continued									
Sugar	1.0	1.9	2.6	2.3	4.5	6.3	2.9	5.6	7.9
Table-mountain	1.1	2.3	3.6	2.7	5.4	8.7	3.4	6.8	10.9
Western white	1.4	2.5	3.9	3.3	5.9	9.4	4.1	7.4	11.8
Pinyon	1.5	1.7	3.3	3.7	4.2	7.9	4.6	5.2	9.9
Redwood:									
Old-growth	.9	1.5	2.3	2.1	3.5	5.4	2.6	4.4	6.8
Second-growth	.7	1.6	2.4	1.8	3.9	5.7	2.2	4.9	7.1
Spruce:									
Black	1.4	2.3	3.8	3.3	5.4	9.0	4.1	6.8	11.3
Engelmann	1.1	2.2	3.5	2.7	5.3	8.3	3.4	6.6	10.4
Red	1.3	2.6	3.9	3.0	6.2	9.4	3.8	7.8	11.8
Sitka	1.4	2.5	3.8	3.4	6.0	9.2	4.3	7.5	11.5
Tamarack	1.2	2.5	4.5	3.0	5.9	10.9	3.7	7.4	13.6
Yew, Pacific	1.3	1.8	3.2	3.2	4.3	7.8	4.0	5.4	9.7
HARDWOODS									
Alder, red	1.5	2.4	4.2	3.5	5.8	10.1	4.4	7.3	12.6
Apple	2.0	3.5	6.1	4.7	8.4	14.7	5.9	10.5	18.4
Ash:									
Black	1.7	2.6	5.1	4.0	6.2	12.2	5.0	7.8	15.2
Blue	1.3	2.2	3.9	3.1	5.2	9.4	3.9	6.5	11.7
Green	1.5	2.4	4.2	3.7	5.7	10.0	4.6	7.1	12.5
Oregon	1.4	2.7	4.4	3.3	6.5	10.6	4.1	8.1	13.2
Pumpkin	1.2	2.1	4.0	3.0	5.0	9.6	3.7	6.3	12.0
White	1.6	2.6	4.5	3.8	6.2	10.7	4.8	7.8	13.4
Aspen:									
Bigtooth	1.1	2.6	3.9	2.6	6.3	9.4	3.3	7.9	11.8
Quaking	1.2	2.2	3.8	2.8	5.4	9.2	3.5	6.7	11.5
Basswood, American	2.2	3.1	5.3	5.3	7.4	12.6	6.6	9.3	15.8
Beech, American	1.7	3.7	5.4	4.1	8.8	13.0	5.1	11.0	16.3
Birch:									
Alaska paper	2.2	3.3	5.6	5.2	7.9	13.4	6.5	9.9	16.7
Gray	1.7	--------	4.9	4.2	--------	11.8	5.2	--------	14.7
Paper	2.1	2.9	5.4	5.0	6.9	13.0	6.3	8.6	16.2
Sweet	2.2	2.8	5.2	5.2	6.8	12.5	6.5	8.5	15.6
Yellow	2.4	3.1	5.6	5.8	7.4	13.4	7.2	9.2	16.7
Buckeye, yellow	1.2	2.7	4.2	2.9	6.5	10.0	3.6	8.1	12.5
Butternut	1.1	2.1	3.5	2.7	5.1	8.5	3.4	6.4	10.6
California-laurel	1.0	2.8	4.1	2.3	6.8	9.9	2.9	8.5	12.4
Catalpa, northern	.8	1.6	2.4	2.0	3.9	5.8	2.5	4.9	7.3
Cherry:									
Black	1.2	2.4	3.8	3.0	5.7	9.2	3.7	7.1	11.5
Pin	.9	3.4	4.3	2.2	8.2	10.2	2.8	10.3	12.8
Chestnut, American	1.1	2.2	3.9	2.7	5.4	9.3	3.4	6.7	11.6
Cottonwood:									
Eastern and southern	1.3	3.1	4.7	3.1	7.4	11.3	3.9	9.2	14.1
Black	1.2	2.9	4.1	2.9	6.9	9.9	3.6	8.6	12.4
Cucumbertree	1.7	2.9	4.5	4.2	7.0	10.9	5.2	8.8	13.6
Dogwood, flowering	2.5	3.9	6.9	5.9	9.4	16.6	7.4	11.8	20.8

See footnotes at end of table.

TABLE 39.—*Shrinkage values of wood based on its dimensions when green*—Con.

Species	Shrinkage								
	Dried to 20-percent moisture content [1]			Dried to 6-percent moisture content [2]			Dried to 0-percent moisture content		
	Radial	Tangential	Volumetric	Radial	Tangential	Volumetric	Radial	Tangential	Volumetric
HARDWOODS—con.									
	Percent	*Percent*	*Percent*	*Percent*	*Percent*	*Percent*	*Percent*	*Percent*	*Percent*
Elm:									
American	1.4	3.2	4.9	3.4	7.6	11.7	4.2	9.5	14.6
Rock	1.6	2.7	4.7	3.8	6.5	11.3	4.8	8.1	14.1
Slippery	1.6	3.0	4.6	3.9	7.1	11.0	4.9	8.9	13.8
Greenheart	1.1	1.4	2.7	2.7	3.4	6.4	3.4	4.2	8.0
Hackberry	1.6	3.0	5.6	3.8	7.1	13.5	4.8	8.9	16.9
Hickory:									
Pecan [3]	1.6	3.0	4.5	3.9	7.1	10.9	4.9	8.9	13.6
True:									
Mockernut	2.6	3.7	6.0	6.2	8.8	14.3	7.8	11.0	17.9
Pignut	2.4	3.8	6.0	5.8	9.2	14.3	7.2	11.5	17.9
Shagbark	2.3	3.3	5.6	5.6	8.0	13.4	7.0	10.0	16.7
Shellbark	2.5	4.2	6.4	6.1	10.1	15.4	7.6	12.6	19.2
Holly, American	1.6	3.3	5.6	3.8	7.9	13.5	4.8	9.9	16.9
Honeylocust	1.4	2.2	3.6	3.4	5.3	8.6	4.2	6.6	10.8
Hophornbeam, eastern	2.8	3.3	6.5	6.8	8.0	15.5	8.5	10.0	19.4
Iroko	1.1	1.6	2.8	2.7	3.8	6.8	3.4	4.8	8.5
Ironbark, gray	1.9	2.8	4.7	4.5	6.7	11.2	5.6	8.4	14.0
Khaya	1.4	1.9	2.9	3.3	4.6	7.0	4.1	5.8	8.8
Lauan, red	1.1	2.7	3.9	2.6	6.4	9.4	3.3	8.0	11.7
Locust, black	1.5	2.4	3.4	3.7	5.8	8.2	4.6	7.2	10.2
Madrone, Pacific	1.9	4.1	6.0	4.5	9.9	14.5	5.6	12.4	18.1
Magnolia, southern	1.8	2.2	4.1	4.3	5.3	9.8	5.4	6.6	12.3
Mahogany	1.2	1.7	2.7	2.9	4.0	6.4	3.6	5.0	8.0
Mangrove	1.8	--------	5.3	4.3	--------	12.6	5.4	--------	15.8
Maple:									
Bigleaf	1.2	2.4	3.9	3.0	5.7	9.3	3.7	7.1	11.6
Black	1.6	3.1	4.7	3.8	7.4	11.2	4.8	9.2	14.0
Red	1.3	2.7	4.4	3.2	6.6	10.5	4.0	8.2	13.1
Silver	1.0	2.4	4.0	2.4	5.8	9.6	3.0	7.2	12.0
Sugar	1.6	3.2	5.0	3.9	7.6	11.9	4.9	9.5	14.9
Oak:									
Black	1.5	3.2	4.7	3.6	7.8	11.4	4.5	9.7	14.2
Bur	1.5	2.9	4.2	3.5	7.0	10.2	4.4	8.8	12.7
California black	1.2	2.2	4.0	2.9	5.3	9.7	3.6	6.6	12.1
Canyon live	1.8	3.2	5.4	4.3	7.6	13.0	5.4	9.5	16.2
Chestnut	1.8	3.2	5.6	4.4	7.8	13.4	5.5	9.7	16.7
Laurel	1.3	3.3	6.3	3.2	7.9	15.2	4.0	9.9	19.0
Live	2.2	3.2	4.9	5.3	7.6	11.8	6.6	9.5	14.7
Oregon white	1.4	3.0	4.5	3.4	7.2	10.7	4.2	9.0	13.4
Pin	1.4	3.2	4.8	3.4	7.6	11.6	4.3	9.5	14.5
Post	1.8	3.3	5.4	4.3	7.8	13.0	5.4	9.8	16.2
Northern red	1.3	2.7	4.5	3.2	6.6	10.8	4.0	8.2	13.5
Rocky Mountain white	1.4	2.4	4.2	3.3	5.8	10.0	4.1	7.2	12.5
Scarlet	1.5	3.2	4.6	3.7	7.8	11.0	4.6	9.7	13.8
Southern red	1.5	2.9	5.4	3.6	7.0	13.0	4.5	8.7	16.3
Swamp chestnut	1.7	3.6	5.5	4.2	8.6	13.1	5.2	10.8	16.4
Swamp red	1.8	3.5	5.9	4.4	8.5	14.2	5.5	10.6	17.7
Water	1.4	3.1	5.5	3.4	7.4	13.1	4.2	9.3	16.4
White	1.8	3.0	5.3	4.2	7.2	12.6	5.3	9.0	15.8
Willow	1.7	3.2	6.3	4.0	7.7	15.1	5.0	9.6	18.9

See footnotes at end of table.

TABLE 39.—*Shrinkage values of wood based on its dimensions when green*—Con.

Species	Shrinkage								
	Dried to 20-percent moisture content [1]			Dried to 6-percent moisture content [2]			Dried to 0-percent moisture content		
	Radial	Tangential	Volumetric	Radial	Tangential	Volumetric	Radial	Tangential	Volumetric
HARDWOODS—con.	Percent	Percent	Percent	Percent	Percent	Percent	Percent	Percent	Percent
Osage-orange			3.1			7.4			9.2
Persimmon, common	2.6	3.7	6.4	6.3	9.0	15.3	7.9	11.2	19.1
Sassafras	1.3	2.1	3.4	3.2	5.0	8.2	4.0	6.2	10.3
Sweetgum	1.7	3.3	5.0	4.2	7.9	12.0	5.2	9.9	15.0
Sycamore, American	1.7	2.5	4.7	4.1	6.1	11.4	5.1	7.6	14.2
Tangile	1.4	3.0	4.4	3.4	7.3	10.6	4.3	9.1	13.3
Teak	.8	1.4	2.3	1.8	3.4	5.4	2.3	4.2	6.8
Tupelo:									
Black	1.5	2.6	4.6	3.5	6.2	11.1	4.4	7.7	13.9
Water	1.4	2.5	4.2	3.4	6.1	10.0	4.2	7.6	12.5
Walnut, black	1.8	2.6	4.3	4.4	6.2	10.2	5.5	7.8	12.8
Willow:									
Black	.9	2.7	4.8	2.1	6.5	11.5	2.6	8.1	14.4
Pacific	1.0	3.0	4.6	2.3	7.2	11.0	2.9	9.0	13.8
Yellow-poplar	1.3	2.4	4.1	3.2	5.7	9.8	4.0	7.1	12.3

[1] These shrinkage values have been taken as ⅓ the shrinkage to the oven-dry condition as given in the last 3 columns of this table.
[2] These shrinkage values have been taken as ⅘ of the shrinkage to the oven-dry condition as given in the last 3 columns of this table.
[3] Average of butternut hickory, nutmeg hickory, water hickory, and pecan.

Values for longitudinal shrinkage are not given in table 39. The total longitudinal shrinkage of normal wood usually ranges from 0.1 to 0.3 percent of the green dimension. Exceptionally lightweight wood of any species tends to shrink abnormally in length.

Wood containing compression wood and tension wood also shrinks more along the grain than does normal wood. Compression wood occurs in softwoods and tension wood in hardwoods. Longitudinal shrinkage varies widely with the form of compression wood. In borderline forms that differ only slightly from normal wood, lengthwise shrinkage is but a little more than that of normal wood. Pronounced forms, on the other hand, shrink 5 to 10 times as much along the grain as normal wood of the conifers. In the same way, pieces of hardwoods with only a few tension wood fibers have nearly the same longitudinal shrinkage as normal wood, but if many of these fibers are present, longitudinal shrinkage is considerably greater than that of normal wood.

Compression wood or tension wood may occur in the same board with normal wood, so that internal stresses are set up that cause lengthwise distortion. If the boards contain moderate to pronounced forms of compression wood or moderate to large numbers of tension wood fibers, these stresses are large and serious warping usually results. Even borderline forms of compression wood or tension wood may interfere with the usefulness of pieces in products that permit only small tolerance with respect to warping. In addition, cross breaks may occur in bands of compression wood or tension wood.

Although theoretically the normal moisture content-shrinkage relation may be considered a direct one from zero shrinkage at fiber saturation point to maximum shrinkage at zero moisture content, actually the relationship is more like that shown in figure 71. For some shrinkage calculations, however, a straight-line relation may be assumed without too great an error. For example, assume that a piece

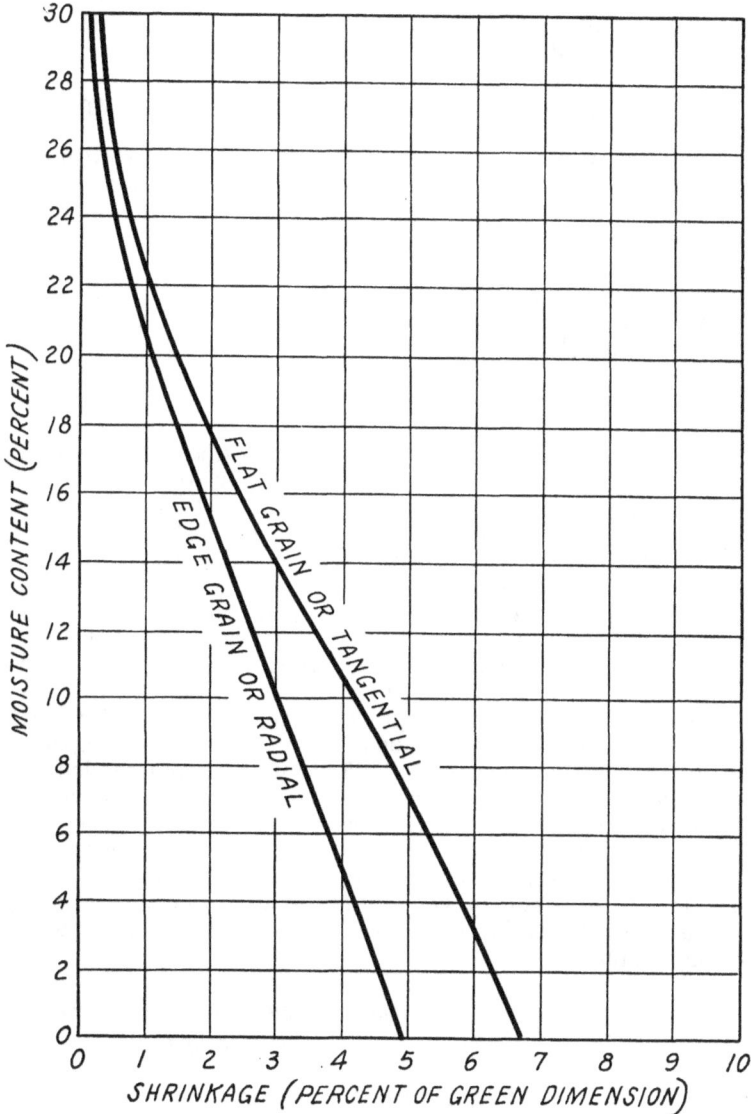

ZM 22048 F

FIGURE 71.—Typical moisture content-shrinkage curves. These curves are for Douglas-fir and southern yellow pine and may be used for estimating the amount of change in dimension that will take place with change in the moisture content of the wood.

of flat-sawed southern yellow pine sheathing at 12-percent moisture content is dried to 7 percent. According to the curve in figure 71 marked tangential, the shrinkage from the green condition to a moisture content of 7 percent would be 5 percent and that from green to 12 percent would be 3½ percent, for a difference of 1½ percent. If a straight-line relation is used, the shrinkage would equal five-thirtieths or one-sixth of the total tangential shrinkage of 6.67 percent, or 1.11 percent of its width. Since the shrinkage values and curves represent averages, the actual shrinkage of a board may vary somewhat from them.

DETERMINATION OF MOISTURE CONTENT

The amount of moisture in wood is ordinarily expressed as a percentage of the weight of the wood when oven-dry. Two methods of determining moisture content are described in the following paragraphs. The oven-dry method is probably the most nearly exact but is slow and necessitates cutting the wood; the electrical method is the more rapid and does not necessitate cutting the wood. A distillation method is used on woods containing volatile oils, such as pitchy southern yellow pine (5).

Oven-Drying Method

In the oven-drying method (5), cross sections, about 1 inch long in the direction of the grain, are cut from representative boards of a lot of lumber. These sections should be cut at least 1 foot from the ends of the boards to avoid the effect of end drying and should be free from knots and other irregularities, such as bark and pitch pockets.

Each section is immediately weighed, before any drying or adsorption of moisture has taken place, and is then placed in an oven heated to 214° to 221° F. and kept there until it reaches constant weight. If the section cannot be weighed immediately after it is cut, it should be wrapped in metal foil until it can be weighed. In the oven, a section will reach a constant weight in 12 to 48 hours. For weighing ordinary moisture content sections, balances having a capacity of about 200 grams and sensitive to 0.05 gram are recommended.

Both steam and electric ovens are in common use for drying moisture-determination sections. The sections, with either type of oven, should be open piled in order to permit good circulation of air, especially around the end-grain surfaces, and thus hasten drying.

The constant or oven-dry weight and the weight of the section when cut are used to determine the percentage moisture content with the following formula:

$$\text{Percent moisture content} = \frac{\text{weight when cut} - \text{oven-dry weight}}{\text{oven-dry weight}} \times 100$$

Electrical Methods

Electrical methods for determining the moisture content of wood make use of such electrical properties of wood as its electrical resistance, dielectric constant, and power-loss factor (1). Electrical moisture meters appeared on the market about 1930. Instruments are made that determine the moisture content through its effect upon the direct-current electrical resistance of wood and its effect on capacity and losses of a condenser in a high-frequency circuit in which the wood serves as the dielectric material of the condenser.

Although some meters are calibrated to cover a range of 4 to 120 percent, the operating range of most meters is 7 to 25 percent. Above and below these values the readings are inaccurate. Within the range of 7 to 25 percent, electrical moisture meters should read within ±1 percent of the moisture content, as determined by the oven-drying method. To obtain accurate readings, the instruments should be in good adjustment and used according to the manufacturer's instructions, which include corrections for various species and cover lumber up to 1½ inches thick or thicker lumber that is known to be of uniform moisture content.

Some judgment must be exercised in the use of electrical meters. They should not be used on dry lumber that has been wet by rain or exposed to damp conditions that have caused the surfaces to become wet. Preferably, meters should not be used on very cold or hot lumber. Most meters are calibrated at 70° F. An approximate temperature correction for resistance-type meters is the adding of 1 percent to the meter reading for each 20° below 70° and the subtracting of 1 percent for each 20° above 70° F.

The principal advantage of the electrical method over the oven-drying method is its speed and convenience. The time required to determine the moisture content of any piece of wood is only a few seconds. The electrical method is therefore adaptable to sorting lumber on the basis of moisture content and can be used to measure the moisture content of wood installed in a building. With this method, the piece of wood is not cut or mutilated except for the driving of a few small needles into the wood to serve as electrodes for the resistance-type meters.

Seasoning of Lumber

Moisture in Wood

The moisture in wood, commonly called "sap," may for most practical purposes in the drying of wood (3, 9) be considered as water alone. Table 40 gives some moisture content values for green heartwood and sapwood of various species. The values shown may be considered average, and considerable variation from these values may be expected in individual trees and single boards, particularly in sapwood.

Sawmill Practice

Methods of seasoning softwood lumber vary with the use requirements. It is common practice at most sawmills to kiln-dry all upper grade lumber intended for finish and flooring. Ordinarily dimension and lower grades used for framing lumber are sent to a yard for air-drying or, in the case of some species, are shipped without being seasoned. Timbers are generally not held long enough to be seasoned, though

TABLE 40.—*Average moisture content of green wood, by species*

Species	Moisture content [1]			Species	Moisture content [1]		
	Heart-wood	Sap-wood	Mixed heart-wood and sap-wood		Heart-wood	Sap-wood	Mixed heart-wood and sap-wood
SOFTWOODS				**HARDWOODS—continued**			
	Percent	*Percent*	*Percent*		*Percent*	*Percent*	*Percent*
Baldcypress	121	171		Beech, American	55	72	
Cedar:				Birch:			
Alaska-	32	166		Paper	89	72	
Atlantic white-			35	Sweet	75	70	
Eastern redcedar	33			Yellow	74	72	
Incense-	40	213		Buckeye, yellow			141
Northern white-			55	Butternut			104
Port-Orford-	50	98		California-laurel			65
Western redcedar	58	249		Cherry, black	58		
Douglas-fir:				Chestnut, American	120		
Coast type	37	115		Chinkapin, golden			134
Intermediate type	34	154		Cottonwood, black	162	146	
Rocky Mountain type	30	112		Dogwood, flowering			62
Fir:				Elm:			
Alpine			47	American	95	92	
Balsam			117	Cedar	66	61	
California red			108	Rock	44	57	
Grand	91	136		Hackberry	61	65	
Noble	34	115		Hickory:			
Pacific silver	55	164		Bitternut	80	54	
White	98	160		Mockernut	70	52	
Hemlock:				Pignut	71	49	
Eastern	97	119		Red	69	52	
Western	85	170		Sand	68	50	
Larch, western	54	119		Water	97	62	
Pine:				Holly, American			82
Eastern white			68	Hophornbeam, eastern			52
Lodgepole	41	120		Locust, black			40
Ponderosa	40	148		Madrone, Pacific			81
Red	32	134		Magnolia	80	104	
Southern yellow:				Maple:			
Loblolly	33	110		Silver (soft)	58	97	
Longleaf	31	106		Sugar (hard)	65	72	
Shortleaf	32	122		Oak:			
Sugar	98	219		California black	76	75	
Western white	62	148		Live			50
Redwood, old-growth	86	210		Northern red	80	69	
Spruce:				Southern red	83	75	
Eastern	34	128		Tan			89
Engelmann	51	173		Water	81	81	
Sitka	41	142		White	64	78	
Tamarack	49			Willow	82	74	
HARDWOODS				Osage-orange			31
				Persimmon, common			58
Alder, red		97		Sweetgum	79	137	
Apple	81	74		Sycamore, American	114	130	
Ash:				Tupelo:			
Black	95			Black	87	115	
Green		58		Swamp	101	108	
White	46	44		Water	150	116	
Aspen	95	113		Walnut, black	90	73	
Basswood, American	81	133		Willow, black			139
				Yellow-poplar	83	106	

[1] Based on weight when oven-dried.

some drying may take place between the time they are sawed and the time of shipment.

Sawmills cutting hardwoods commonly classify for size and grade at the time of sawing and then send all stock to the air-drying yard. After yard seasoning, the stock may be kiln-dried at the sawmill or shipped to a remanufacturing plant where it is kiln-dried before being made up into finished products, such as furniture, cabinet work, interior finish, and flooring.

Air-Drying

The principal advantages of air-dried wood over green wood (*3, 9*) are: Reduction in weight, with a resulting decrease in shipping costs; reduction or elimination of shrinkage, checking, honeycombing, and warping occurring in service; increase in strength; increase in nail-holding power; decrease in the tendency for blue stain and other forms of fungi to attack the wood; reduction in likelihood of attack by some forms of insects; and improvement in the capacity to hold paint or to receive preservative treatment.

Kiln-Drying

Among the advantages of kiln-drying over air-drying are the following: Greater reduction in weight and consequently in shipping charges; reduction in moisture content to any desired value, which may be lower than that obtainable through air-drying; reduction in drying time below that required in air-drying; the killing of any stain or decay fungi or insects that may be in the wood; setting of the resins in resinous woods; and less degrade, in most cases, if proper kiln schedules are used.

Seasoning Degrade

Obtaining lumber practically free of seasoning degrade in the higher grades is insured by adherence to approved grading rules on the part of the manufacturer and knowledge of the lumber and its grades on the part of the user. Degrade that sometimes develops in seasoning may be classified (*3*) into two main groups: (1) That caused by unequal shrinkage, which includes checks, honeycomb, warp (fig. 72), loosening of knots, and collapse; and (2) that caused by the action of fungi, namely, molds, stains, and decay. Chemical brown stain, frequently known as yard or kiln brown stain, may also occur in some softwoods. It is a yellow to dark-brown discoloration and is apparently caused by the oxidation of water-soluble materials in the wood. It has no effect upon strength and general usefulness of the wood except in the case of upper grade stock with a natural finish. So-called sticker stain is common in the air-drying of both softwoods and hardwoods and presumably is also caused by the concentration and oxidation of water-soluble materials in the wood.

Seasoning degrade, with the exception of chemical stains, can be largely eliminated by proper practice in either air-drying or kiln-drying. Drying lumber too rapidly will cause checking and splitting,

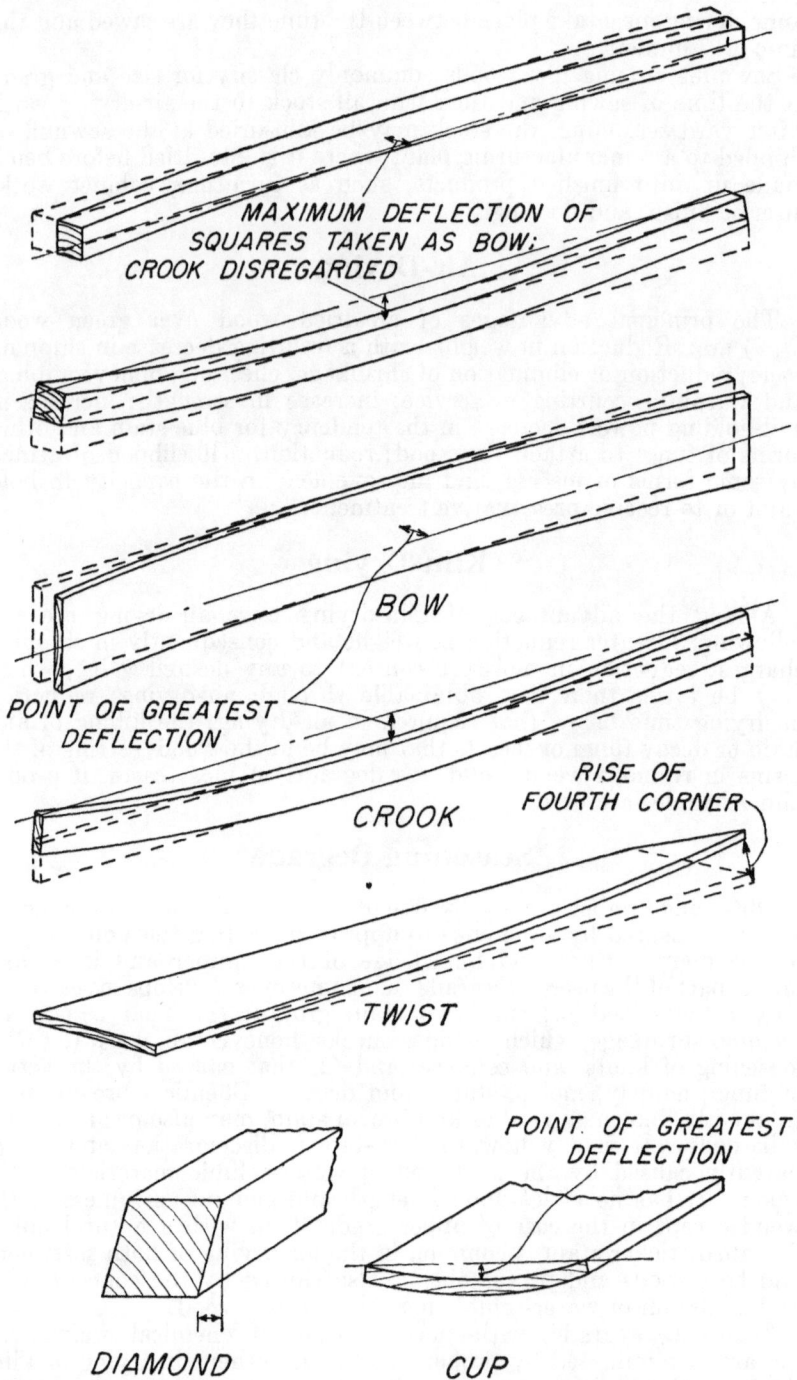

MAXIMUM DEFLECTION OF
SQUARES TAKEN AS BOW:
CROOK DISREGARDED

BOW

POINT OF GREATEST
DEFLECTION

RISE OF
FOURTH CORNER

CROOK

TWIST

POINT OF GREATEST
DEFLECTION

DIAMOND   CUP

M 89552 F

FIGURE 72.—Various kinds of warp.

whereas drying it too slowly under favorable temperatures will permit stain or decay to develop.   Honeycombing and collapse are more common in hardwoods than in softwoods and are more likely to occur during improper kiln-drying than during air-drying.

The grading rules of the various lumber associations specify the amount of degrade permitted for the various grades of lumber.

Moisture Content of Seasoned Lumber

The trade terms "shipping-dry," "air-dry," and "kiln-dried," although widely used, may not have identical meaning as to moisture content in the different producing regions.   The wide limitations of these terms as ordinarily used are covered in the following statements, which, however, are not to be construed as exact definitions:

Shipping-dry lumber.—Lumber that is partially dried to prevent stain and mold in transit.

Air-dry lumber.—Lumber that has been dried by exposure to the air either outdoors or in an unheated shed for any length of time. If exposed for a sufficient length of time, it may have a moisture content ranging from 6 percent, as in summer in the arid Southwest, to 24 percent, as in winter in the Pacific Northwest.   For the United States as a whole, the minimum moisture content range of thoroughly air-dry lumber is 12 to 15 percent, and the average is somewhat higher.

Kiln-dried lumber.—Lumber that has been kiln-dried, particularly the finish grades of softwoods and hardwoods, to a moisture content of 6 to 12 percent.   Kiln-dried softwood lumber of the common yard grades is likely to have a somewhat higher moisture content.

The importance of suitable moisture content values is recognized, and provisions covering them are now incorporated in some grading rules.   It should be noted, however, that the moisture content values in the general grading rules may or may not be suitable for a specific use, and, if not, a special moisture content provision should be made in the specifications.

Storage of Lumber at Yards

Lumber when received at a distributor's lumberyard may be practically green, partially seasoned, or throughly seasoned.   It can easily be protected from stain or decay if good storage practices are followed.

If green or partially seasoned, the lumber should be open piled on stickers and protected from sunshine and precipitation by a tight roof (3, 9, 15).   If the lumber is seasoned to a moisture content of less than 20 percent, it is good practice to pile it "solid," board on board, in a shed that will afford ample protection against sunshine and precipitation.   If it is desired to reduce the moisture content still further, however, the lumber should be open piled on stickers.

If dry lumber is stored in the open, it is best to pile it on stringers and to use stickers between the layers of lumber.   The pile should have a roof that will shed water.   Dry lumber can be piled solid in the open for relatively short periods but should be protected from rain with suitable pile covers.   Because it is difficult to keep rain out completely, long storage of solid-piled lumber is not recommended.

The foregoing relates primarily to such items as sheathing, shiplap, studs, and joists. Such kiln-dried items as exterior finish and siding can be stored in a closed but unheated shed. The equilibrium moisture content of such a shed would be higher than that of the atmosphere within an occupied house, however, and items kiln-dried to a moisture content suited for interior finish would absorb moisture if stored there. For this reason, interior trim, flooring, and cabinet work should be stored in a closed shed where some heat can be provided during cold or damp weather to maintain the lumber at the desired moisture content (4, 10).

The moisture content of lumber items in storage can be maintained by control of the temperature within the shed. If there is no source of moisture except that contained in the air, the proper shed temperature required to maintain a given moisture content can be determined by the use of figure 73. This chart shows equilibrium moisture content values of wood obtained on heating or cooling air at various temperature and relative humidity conditions. For example, if the outdoor temperature is 30° F., the relative humidity is 75 percent, and the desired moisture content of the lumber is 8 percent, proceed as follows: From the intersection of the (vertical) 30° temperature line and the (horizontal) 75 percent relative humidity line, extend a line midway between the adjacent (concave) vapor pressure lines until it intersects a line midway between the 7- and 9-percent moisture content lines indicated on the right-hand ordinate. The reading on the bottom scale below the point of the second intersection is about 45°. (See dotted lines in figure 73.) In other words, under the conditions stated, the moisture content of the flooring can be maintained at 8 percent merely by heating the air to 45°.

Heat is required in storage sheds when the outdoor relative humidity is high. If the temperature of the shed is kept almost 10° F. higher than the outdoor temperature, the lumber will usually be brought to or maintained at 10-percent moisture content. If the temperature is increased 20° the moisture content will be about 7 percent. During cold weather, if the storage shed contains any water lines the temperature should not be allowed to drop below 32°. A method of automatic control of equilibrium moisture content in a lumber shed is described in reference (4).

RECOMMENDED MOISTURE CONTENT

The percentages of moisture content recommended here for wood are selected primarily for the purpose of reducing changes in moisture content to a minimum, thereby minimizing dimensional changes after wood is put into service (8). The service conditions to which the wood will be exposed, whether outdoors, in unheated buildings, or in heated buildings, should be considered in determining seasoning requirements.

Timbers

Ordinarily, timber should be seasoned to as low a moisture content as it will ultimately come to in service, or as near this condition as practical. While this optimum is possible with lumber less than 3

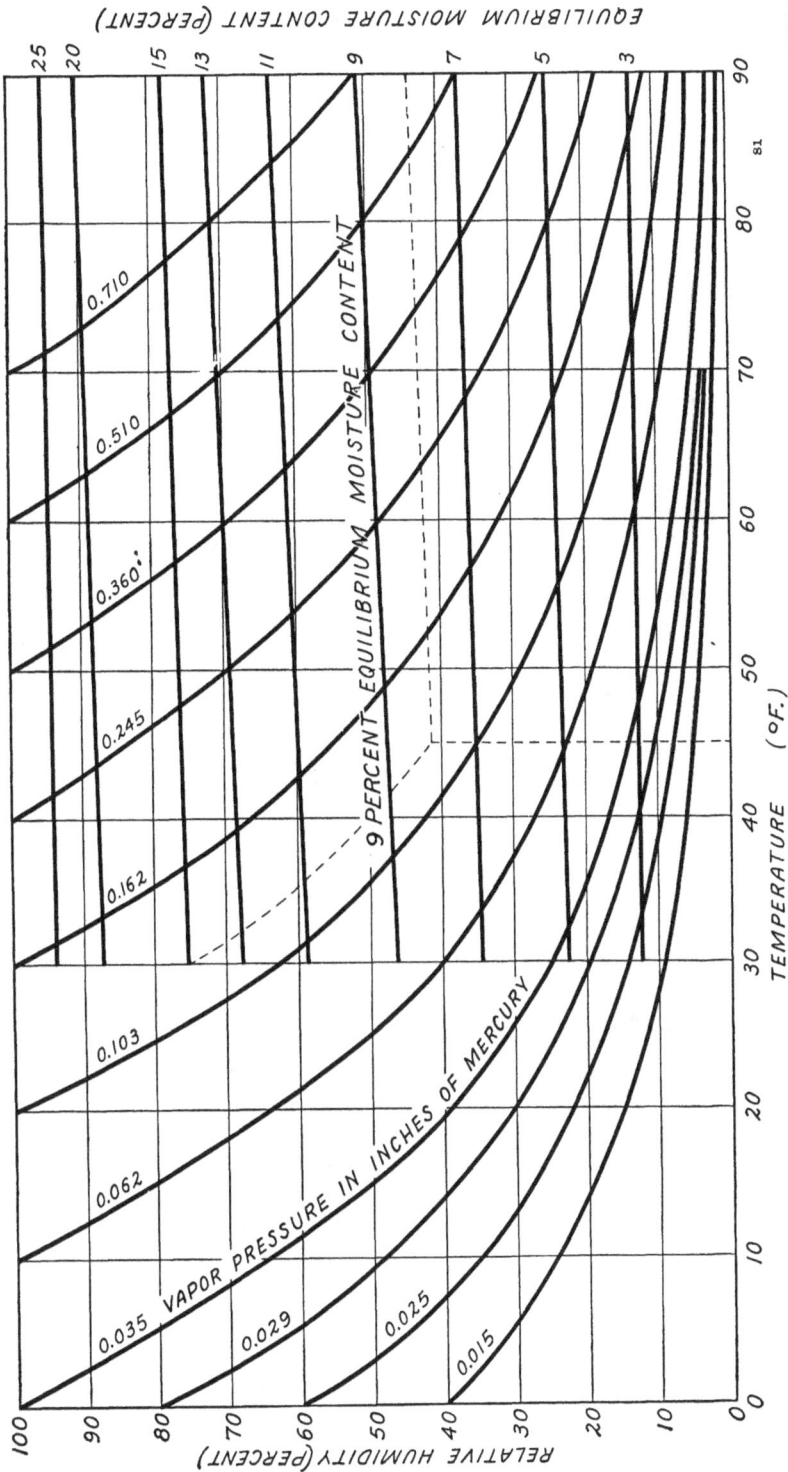

FIGURE 73.—Relationship of relative humidity, vapor pressure, equilibrium moisture content, and temperature.

inches thick, it is seldom possible to obtain fully seasoned timbers or thick joists and planks. Where it is necessary to use thick members, as in warehouses, bridges, trestles, and derricks, some shrinkage of the assembly should be expected, and the design of the structure should minimize shrinkage effects. In the case of built-up assemblies, such as roof trusses, it may be necessary from time to time to tighten the bolts or other fastenings as the members shrink.

Lumber for Exterior or Interior Service

The moisture content requirements for finish lumber or wood products used inside heated buildings are more exacting than those for lumber used outdoors or in unheated buildings. This is due to the higher character of the service required and also to the lower relative humidity conditions encountered within heated buildings than outdoors. Table 41 and figure 74 show the recommended moisture content values and tolerances for wood used in interior parts of heated buildings. The values for exterior trim and siding can be applied to lumber used outdoors and in unheated buildings. Table 41 also gives corresponding values for exterior parts of heated buildings.

General commercial practice is to kiln-dry some wood products, such as flooring (13) and furniture wood, to a slightly lower moisture content than service conditions demand, counting on a moderate increase in moisture content during the storage and manufacturing periods. This practice is intended to assure uniform distribution of moisture among the individual pieces. Common grades and dimension are sometimes but not ordinarily seasoned to the moisture content values indicated in table 41. When they are not, the design of the structure should take account of this condition in such a way as to minimize shrinkage effects.

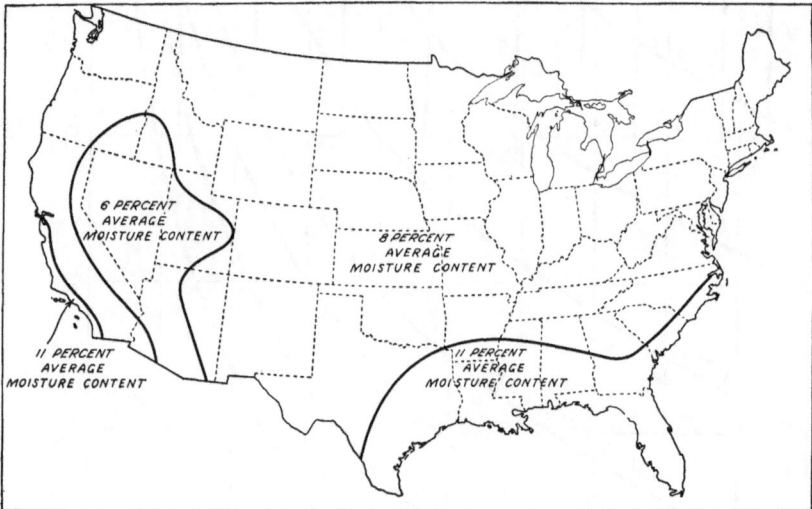

ZM 20143 F

FIGURE 74.—Recommended moisture content averages for interior-finishing woodwork for use in various parts of the United States.

TABLE 41.—*Recommended moisture content values for various wood items at time of installation*

Use of lumber	Moisture content (percentage of weight of oven-dry wood) for—					
	Dry southwestern States [1]		Damp southern coastal States [1]		Remainder of the United States [1]	
	Average [2]	Individual pieces	Average [2]	Individual pieces	Average [2]	Individual pieces
	Percent	*Percent*	*Percent*	*Percent*	*Percent*	*Percent*
Interior-finish woodwork and softwood flooring	5–7	4–9	10–12	8–13	7–9	5–10
Hardwood flooring	6–7	5–8	10–11	9–12	7–8	6–9
Siding, exterior trim, sheathing, and framing [3]	8–10	7–12	11–13	9–14	11–13	9–14

[1] For limiting range see fig. 74.

[2] To obtain a realistic average, test at least 10 percent of each item. If the amount of a given item is small, several tests should be made. For example, in an ordinary dwelling having about 60 floor joists at least 10 tests should be made on joists selected at random.

[3] Framing lumber of higher moisture content is commonly used in ordinary construction because lumber of the moisture content specified may not be available except on special order.

Veneer and Plywood

When veneers are glued together with cold-setting glues to make plywood, they absorb comparatively large quantities of moisture. To keep the final moisture content low and to minimize redrying of the plywood, the initial moisture content of the veneer should be as low as practical. Very dry veneer, however, is difficult to handle without cracking, so that the minimum practical moisture content is about 4 percent. After being glued, plywood intended for interior service should be conditioned to the moisture content values given in table 41 for interior lumber. If the plywood is glued by hot-press processes, the veneer may need conditioning to a moisture content dependent upon the water content of the glue. It is often desirable to add water to the plywood by conditioning (p. 241).

DESIGN FACTORS AFFECTING SHRINKAGE IN THE STRUCTURE

Framing Lumber in House Construction

The most effective method of minimizing settlement or dimensional change due to shrinkage in the house frame would be to use framing lumber seasoned to the moisture content it would reach in use (*6, 12, 14, 16*). There are practical aspects, however, that make it difficult, if not impossible, to meet this optimum condition, but, fortunately, some shrinkage of the frame may take place without visible evidence appearing after the house is completed. If the moisture content of the framing lumber at the time the wall and ceiling finish is applied is not more than about 5 percent above that which it will reach in service, there will be little or no evidence of defects caused by shrinkage of the frame. In heated houses in cold climates, joists over heated basements, studs, and perhaps ceiling joists may reach a moisture content as low as 6 to 7 percent. In mild climates the minimum moisture content will be higher.

The most common evidences of shrinkage are cracks in plastered walls, open joints and nail pops in dry wall construction, distortion of door openings, unlevel floors, loosening of joints and fastenings, or the occasional distortion in drying of some individual members that may push a wall out of alinement. Where incompletely seasoned framing lumber is used, the extent of vertical shrinkage after the house is completed is proportional to the amount of wood used as supports in a horizontal position, such as girders, floor joists, and plates. On this basis more evidence of shrinkage would appear in the second than in the first floor of a 2-story house; more in a 2-story than in a 1-story house; and more in a house with a heated basement than one without a basement.

If framing lumber has a higher moisture content when installed than that recommended for framing in table 41, some shrinkage may be expected. Framing lumber, even thoroughly air-dried stock, will generally have a moisture content higher than that recommended when it is delivered to the building site. If carelessly handled in storage at the site, it may take up more moisture. There are certain practices, however, that the builder can follow which will protect the stored lumber (see p. 334). He can also plan his work so that an appreciable amount of seasoning will take place during the early part of the construction period and thus minimize the effects of further drying and shrinkage after completion.

The builder's objective should be to follow a practice that provides time for the framing lumber to lose moisture before it is covered or enclosed. When the house has been framed, sheathed, and roofed, the framing is so exposed that in time it can dry to a lower moisture content than would ordinarily be expected in yard-dried lumber. If the application of the wall and ceiling finish is delayed for about 30 days during warm, dry spring or summer weather, air-dried yard stock should lose enough moisture so that any further drying in place will be relatively unimportant. During this period the builder can proceed with the application of outside wall covering, the installation of plumbing, heating, and wiring equipment, and any other work that does not enclose the framing. The basement floors should be installed during this period. In cool, damp weather, or if unseasoned lumber is used, the period of exposure should be extended. Closing the house and using the heating system will hasten the rate of drying.

Before wall finish is applied, the frame should be examined and any defects that may have developed during drying, such as warped or distorted studs, shrinkage of lintels over openings, or loosened joints, should be corrected.

Consideration should be given to the type of framing best suited to the building and particularly to construction methods that minimize the use of wood across the grain in vertical supports.

The platform type of construction (fig. 75) is intended to equalize but not to minimize shrinkage. This method is used generally for first floors and for the second story of houses having wood siding or shingle wall covering. Vertical shrinkage at the second-floor line presents a problem at chimneys and on outside walls of brick or stone veneer or stucco that are unbroken. Where the design calls for a break, such as an overhanging second floor or a change in materials, provision for shrinkage is readily made.

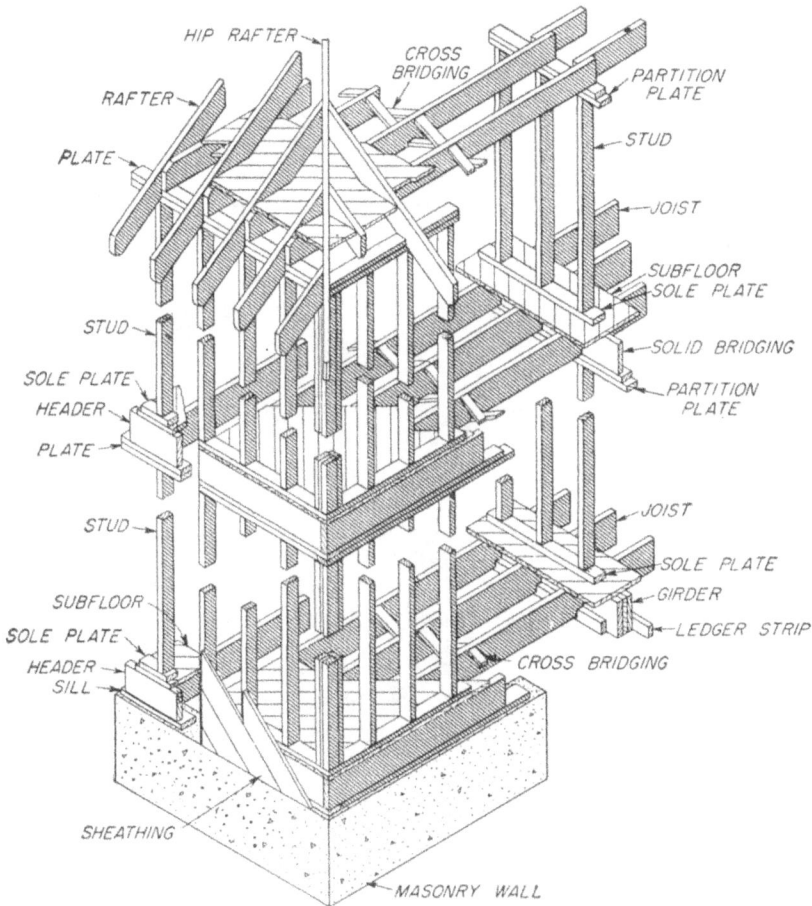

ZM 88257 F

FIGURE 75.—Platform type of frame construction.

Vertical shrinkage in the exterior walls is held to a minimum in the balloon (fig. 76) type of construction, and this system is preferable where exterior walls are of brick or stone veneer or stucco without a break at the intervening floor line.

Girders built up of seasoned 2-inch lumber will shrink less than solid girders, since it is not customary to season lumber thicker than nominal 2 inches. Potential shrinkage in joists and girders is less if the joists bear on ledger strips nailed to the side of the girder (fig. 77)

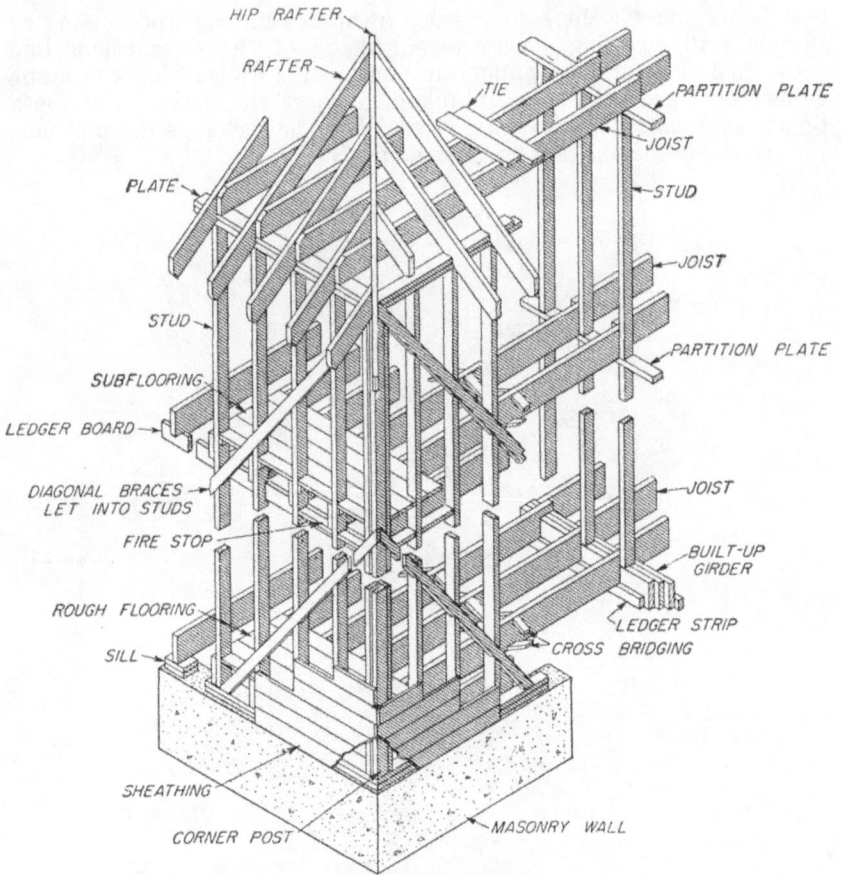

ZM 88258 F

FIGURE 76.—Balloon type of frame construction.

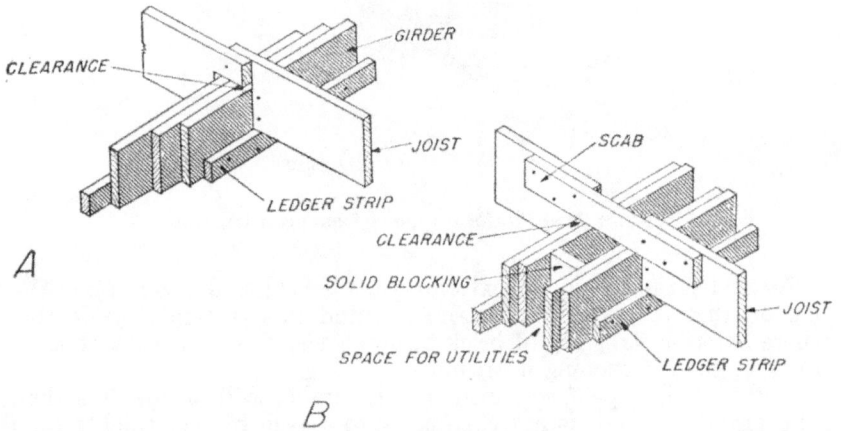

M 87429 F

FIGURE 77.—Joists bear on ledger strips to minimize shrinkage: *A*, Ledger strip nailed on girder to support joists; *B*, spaced girder.

than if the joists run over the top of the girder (fig. 78). The use of metal post caps instead of wood bolsters likewise reduces total shrinkage.

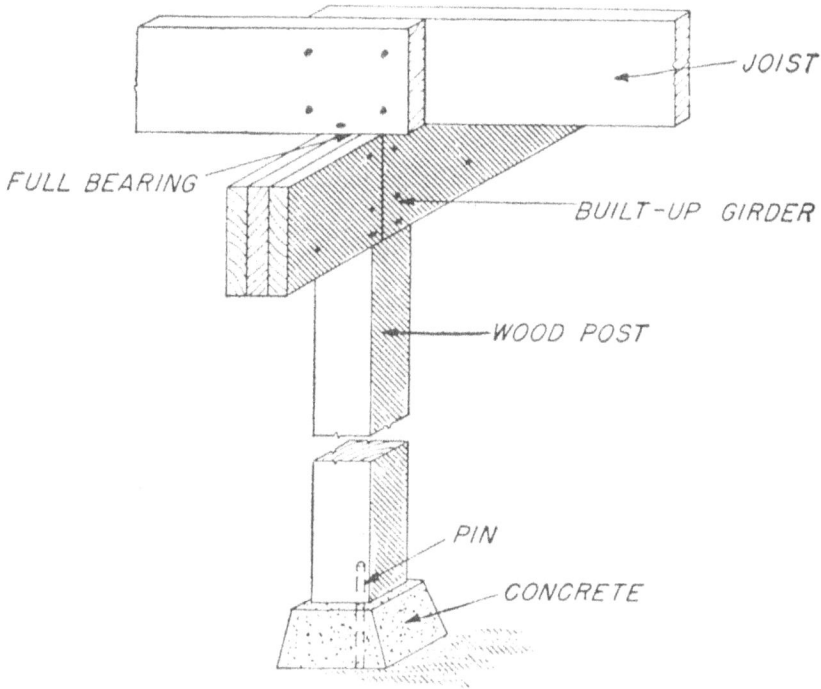

FIGURE 78.—Joists run over top of girder increase the vertical height of the wood used across the grain and increase subsequent shrinkage.

Heavy Timber Construction

In heavy timber construction, a certain amount of shrinkage is to be expected. If not provided for in the design, it may cause weakening of joints, affect floor levels, and be otherwise objectionable. One means of eliminating part of the shrinkage in mill buildings and similar structures is the use of metal post caps, whereby the upper column is separated from the lower column only by the metal in the post cap. This method eliminates the shrinkage that occurs if the wood girder is used as a bearing for the upper column. The same thing is accomplished by supporting the upper column on the lower column with wood corbels bolted to the side of the lower column to support the girders.

Where joist hangers are used, the top of the joist, when installed, should be slightly above the top of the girder; otherwise, when the joist shrinks in the stirrup, the floor over the girder will be higher than that bearing upon the joist. Laminated floor material can easily be properly seasoned and shrinkage minimized accordingly, because each piece is of relatively small cross section.

Interior Finish

The normal seasonal changes in the moisture content of interior finish are not enough to cause serious dimensional change if the stock is properly seasoned and the woodwork is carefully designed and assembled. Large members, such as ornamental beams, cornices, newel posts, stair stringers, and handrails, should be built up from comparatively small pieces. Wide, plain surfaces, such as table tops, counter tops, and panels, should be crossbanded. Door and window trim and base should be hollowbacked. Backband trim, if mitered at the corners, should be glued and splined before erection; otherwise butt joints should be used for the wide faces. Large, solid pieces, such as knotty pine panels, should be stained and finished as much as possible before erection and should be so designed and installed that the panels are free to move across the grain.

CARE OF LUMBER AND FINISH DURING CONSTRUCTION

Lumber and Dimension

Lumber received at the building site should be protected from wetting and other damage. Construction lumber in place in a structure before it is enclosed may be wet during a storm, but the wetting is mostly on the exposed surface, and the lumber can dry out quickly. Dry lumber may be solid piled at the site, but the piles should be at least 6 inches off the ground and covered with canvas or waterproof paper laid so as to shed water from the top, sides, and ends of the pile. If such protection is not provided, the lumber should be open piled on stickers to avoid the slow redrying that would result from solid piling.

Lumber that is received in the green or nearly green condition, or lumber that has been used for concrete forms, should be piled with stickers for more thorough drying before it is built into the structure.

Lumber, whether dry or green, should be protected from alternate wetting by rain and drying by direct sunshine in order to reduce checking and warping. Frequently in the construction of houses, the garage, if detached, can be built first and will serve as an excellent storage space for sheathing, siding, studs, and joists.

Finish Floor

Cracks develop in flooring if it absorbs moisture either before or after it is laid and then shrinks when the building is heated (13). Such cracks can be greatly reduced, if not entirely eliminated, by observing the following practices: (1) Specify flooring manufactured according to association rules and sold by dealers that protect it properly; (2) do not allow the flooring to be delivered on a damp or rainy day or before the masonry and plaster walls are dry; (3) have the heating plant installed before the flooring is delivered; (4) break open the flooring bundles and expose all sides of the flooring to the atmosphere; (5) close up the house at night and raise the temperature about 15° F. above the outdoor temperature for perhaps 3 days before laying the floor; (6) after laying the floor keep the house closed at night or during damp weather and supply some heat if necessary.

Better and smoother sanding and finishing can be done when the house is warm and the wood has been kept dry.

Interior Finish

In a building under construction, the relative humidity will average higher than it will in an occupied house because of the moisture that evaporates from wet concrete, brickwork, and plaster, and even from the structural wood members. The average temperature will also be lower, because workmen prefer a lower temperature than is agreeable in an occupied house. Under such conditions the finish tends to have a higher moisture content during construction than it will have later during occupancy.

Before any interior finish is delivered, the outside doors and windows should be hung in place so that they may be kept closed at night and in this way hold the conditions of the interior as close as possible to the higher temperature and lower humidity that ordinarily prevail during the day. Such protection may be sufficient during the dry summer weather, but during damp or cool weather it is highly desirable that some heat be maintained in the house, particularly at night (13). Whenever possible, the heating plant should be placed in the house before the interior trim goes in, so as to be available for supplying the necessary heat. Portable heaters also may be used. The temperatures during the night should be maintained at about 15° F. above outside temperatures and not be allowed to drop below about 70° during the summer or 62° to 65° when outside temperatures are below freezing.

After buildings have thoroughly dried, there is less need for heat, but unoccupied houses, new or old, should not be allowed to stand without some heat during the winter. A temperature of about 15° F. above outside temperatures and above freezing at all times will be sufficient to keep the woodwork, finish, and other parts of the house from being affected by dampness or frost.

Plastering

During a plastering operation in a moderate-sized 6-room house approximately 1,000 pounds of water are used, all of which must be evaporated before the house is ready for the interior finish. Failure to provide ventilation adequate to remove this evaporated moisture means trouble later because of the moisture absorbed by the framework. In houses plastered in cold weather the moisture may also cause paint on exterior finish and siding to blister. During warm, dry, summer weather with the windows wide open, this moisture is practically gone within a week after the final coat of plaster is applied. During damp, cold weather, drying is retarded accordingly. Adequate ventilation should be provided at all times of the year, as the evaporated moisture is airborne, and a large volume of air is required to carry away the amount of water involved.

When the heating system or portable heaters are used to prevent freezing of plaster and to hasten its drying, the windows should be properly adjusted to allow the escape of the evaporated moisture. Even in the coldest weather, the windows on the leeward side of the house should be opened 2 or 3 inches, preferably from the top.

LITERATURE CITED

(1) DUNLAP, M. E., and BELL, E. R.
   1949. ELECTRICAL MOISTURE METERS FOR WOOD.   U. S. Forest Prod. Lab. Rpt. R1660, 10 pp., illus. (Rev.) [Processed.]
(2) KOEHLER, A.
   1946. LONGITUDINAL SHRINKAGE OF WOOD.   U. S. Forest Prod. Lab. Rpt. R1093, 8 pp., illus. (Rev.) [Processed.]
(3) MATHEWSON, J. S.
   1930. AIR SEASONING OF WOOD.   U. S. Dept. Agr. Tech. Bul. 174, 55 pp., illus.
(4) ———
   1953. ELECTRICALLY OPERATED WOOD-ELEMENT HYGROSTATS FOR CONTROL OF MOISTURE FLUCTUATIONS IN LUMBER STORED IN CLOSED SHEDS.   U. S. Forest Prod. Lab. Rpt. R1140, 10 pp., illus. (Rev.) [Processed.]
(5) MCMILLEN, J. M.
   1950. METHODS OF DETERMINING THE MOISTURE CONTENT OF WOOD.   U. S. Forest Prod. Lab. Rpt. R1649, 8 pp., illus. (Rev.) [Processed.]
(6) NATIONAL LUMBER MANUFACTURERS ASSOCIATION.
   1929. HOUSE FRAMING DETAILS.   24 pp., illus. Washington, D. C.
(7) PECK, E. C.
   1947. SHRINKAGE OF WOOD.   U. S. Forest Prod. Lab. Rpt. R1650, 6 pp., illus. [Processed.]
(8) ———
   1950. MOISTURE CONTENT OF WOOD IN USE.   U. S. Forest Prod. Lab. Rpt. R1655, 8 pp., illus. (Rev.) [Processed.]
(9) ———
   1950. AIR DRYING OF LUMBER.   U. S. Forest Prod. Lab. Rpt. R1657, 18 pp., illus. (Rev.) [Processed.]
(10) ———
   1952. STORAGE AND HANDLING OF LUMBER (AT WHOLESALE AND RETAIL YARDS AND AT CONSUMER PLANTS).   U. S. Forest Prod. Lab. Rpt. R1919, 17 pp., illus. [Processed.]
(11) RASMUSSEN, E. F.
   1951. PROPERTIES OF WOOD RELATED TO DRYING.   U. S. Forest Prod. Lab. Rpt. R1900-1, 25 pp., illus. [Processed.]
(12) TEESDALE, L. V.
   1934. HOW PLASTERING AFFECTS THE MOISTURE CONTENT OF STRUCTURAL AND FINISH WOODWORK.   U. S. Forest Prod. Lab. Rpt. R1274, 4 pp., illus. (Rev.) [Processed.]
(13) ——— and MATHEWSON, J. S.
   1952. PREVENTING CRACKS IN NEW WOOD FLOORS.   U. S. Dept. Agr. Leaflet 56, 6 pp., illus.
(14) U. S. FOREST PRODUCTS LABORATORY.
   1947. TECHNIQUE OF HOUSE NAILING.   53 pp., illus.   (Issued by U. S. Housing and Home Finance Agency.)
(15) ———
   1952. SHRINKAGE AND SWELLING OF WOOD IN USE.   U. S. Forest Prod. Lab. Rpt. R736, 10 pp., illus. (Rev.) [Processed.]
(16) WEYERHAEUSER SALES CO.
   1950. HIGH COST OF CHEAP CONSTRUCTION.   71 pp., illus. St. Paul, Minn.

FIRE RESISTANCE OF WOOD CONSTRUCTION

Acceptable fire resistance in wood construction is obtained by taking advantage of the self-insulating qualities of wood and by employing good structural details. The performance of wood in structures under fire conditions may be further improved by impregnation treatments and surface coatings with fire-retardant materials.

IMPROVEMENT OF FIRE RESISTANCE OF WOOD

CONSTRUCTION THROUGH DESIGN

Wood construction for many years has been identified in standard building codes with 1 of 3 standard types (9).[22] These are the heavy-timber, ordinary (p. 339), and light-frame types. The first two have been widely used for industrial or commercial buildings and the third for dwellings and some smaller commercial buildings. With the recent development of the wood roof truss or arch and the glued laminated structural member, a new type of construction, the one-story industrial or commercial building, has appeared. General principles of design of these types of construction, particularly as they affect fire prevention and control, are presented here.

Heavy-Timber Construction

Heavy-timber construction, also known as mill-type construction, has been extensively built, and many structures of this type are still in use (fig. 79) (10). It is a multistory construction with exterior walls of masonry and interior columns, beams, and floors of wood in solid masses with a minimum of surface or projections exposed to fire. The present tendency is away from large, solid sawed timbers and toward glued laminated structural members with longer spans that permit larger unobstructed floor areas.

In heavy-timber construction, wood columns are not less than nominal 8 inches thick, with rounded or chamfered corners. Beams and girders are not less than nominal 6 inches thick and not less than nominal 10 inches deep. Built-up beams are permitted, when closely jointed by gluing or bolting. Beams made of two members with an intervening air space are not permitted in floor framing for heavy-timber construction. Floors are constructed of splined or matched planks nominal 3 inches thick with a nominal 1-inch finish floor or of planks nominal 4 inches wide set on edge with a nominal 1-inch finish floor. Roof decks are of similar construction but may have splined or matched planks nominal 2½ inches thick (nominal 2 inches thick if the building is sprinklered) or planks nominal 3 inches wide set on edge. Roof girders are not less than 6 inches (nominal) in least dimension.

---

[22] Italic numbers in parentheses refer to Literature Cited, p. 349.

M 93575 F

FIGURE 79.—Heavy-timber, or mill-type, construction.

Roof arches or trusses may be nominal 4 by 6 or larger members, except that spaced members may be nominal 3 inches thick if properly blocked or covered or single members nominal 3 inches thick if the building is sprinklered.

Heavy-timber construction is fire resistant because of the slow rate of burning of wood in massive form. The average rate of penetration of char under standard time-temperature fire conditions (1) is of the order of 1½ inches per hour.[23]

The resistance of heavy timber to fire is shown in the following results of fire tests of wood columns under design load:

Type of column [1] (timber, longleaf pine or Douglas-fir; crosssection, 11⅜ by 11⅜ inches) with—	Time to failure (minutes)
Steel plate cap	35–38
Cast iron cap and pintle	45–50
Reinforced concrete cap	78–112
Steel plate cap; cap and column protected with ⅜-inch gypsum wallboard	73
Cast iron cap; cap and column protected with 1-inch layer of portland cement plaster or metal lath	135

[1] Data for timber columns with reinforced concrete caps are from Handbook of Fire Protection (10); all other data are from Fire Exposure Tests of Loaded Timber Columns (12).

Ordinary Construction

The term "ordinary construction" was originally defined for buildings 2 to 4 stories high with exterior walls of masonry and interior wood-joist frames with members not less than 2 inches (nominal) thick. This type of construction has been widely employed in commercial or public buildings in villages and small cities. Ordinary construction differs from heavy-timber construction in that exterior walls are not so heavy and interior framing is less massive. These differences are reflected in smaller heights and areas allowed. It differs from light-frame construction in its larger allowable heights and areas, its self-supporting masonry walls, and in a number of interior requirements appropriate to public occupancy. There are detailed code requirements for firestops (nominal 2-inch-thick wood or the equivalent) in concealed spaces in walls or ceilings through which fire may spread. Large attic spaces are divided by "draft stops," partitions made of 2 thicknesses of 1-inch lumber or the equivalent.

One-Story Industrial Buildings

In recent years, the one-story industrial or commercial building has largely superseded other types of wood-frame industrial construction. Typically, this building has side walls 12 to 20 feet high, supporting wood roof trusses or arches with spans of 40 to 100 feet or more (fig. 80). Truss or arch members are often of glued laminated construction.

[23] Wood chars a little more rapidly when a fresh surface is exposed to fire than it does under a layer of charcoal, about 1/30 inch per minute in the first 5 minutes and 1/60 inch per minute after ½ hour of exposure (7).

M 80477 F

FIGURE 80.—One-story industrial building.

Larger buildings are simply multiples of such units placed side by side, separated by fire walls as required by building codes. In some instances, a ceiling is hung below the roof framing, but the space above the ceiling is usually unoccupied. This type of construction is popular because it provides large unobstructed working areas.

One-story industrial construction with masonry walls and roof framing members not less than nominal 2 inches thick often qualifies in building codes as ordinary construction. Smaller buildings without masonry walls may qualify as light-frame construction. The masonry walls and the interior wood construction generally are not massive enough to give fire resistance equivalent to that of heavy-timber construction.

Fire-fighting operations in such one-story buildings are facilitated by the limited height, the large unobstructed areas, and the general absence of inaccessible spaces. It is easy to provide any number of exits for personnel in case of fire. From the standpoint of fire safety, this construction ranks favorably with ordinary construction.

In construction of this type, provision for fire protection around basement heating units and maintenance of automatic fire separation between basement and main floor are commonly required by code. Unused attic spaces are accessible to firemen, often from the roof as well as from below. Metal connections at joints of roof trusses or at the crest of arches, embedded in the wood so that deep charring must take place before the connection is destroyed, provide greater fire resistance.

Light-Frame Construction

Most residential and some commercial or industrial buildings of wood are of light-frame or, as more commonly known, frame construction. Originally restricted to the conventional type of building with stud walls, joisted floors and ceilings, and raftered roofs, light-frame construction has more recently been diversified by the introduction of prefabricated, panelized, or stressed-skin structural elements.

Light-frame wood buildings do not have the fire resistance of the heavier wood frames. Therefore, in these buildings, attention to good construction details is important to retard the spread of fire and reduce hazards to occupants and property damage in case of fire. The fire resistance of light-frame construction depends in large part on construction details, of which firestops; separation of wood from masonry around chimneys and fireplaces; and walls, ceilings, floors, roofs, stairways, and doors are important (9, 13, 15).

Firestops

Firestops are obstructions provided in concealed air spaces and are designed to interfere with the passage of flames up or across a building. Fire in buildings spreads by the movement of high-temperature air and gases through open channels. In addition to halls, stairways, and other large spaces, heated gases also follow the concealed spaces between floor joists, between studs in partitions and walls of frame construction, and between the plaster and the wall where the plaster is carried on furring strips. Fire may quickly find its way through these hidden channels to every part of the structure if they are not obstructed at suitable points.

Wood of 2-inch nominal thickness or some noncombustible insulating material not less than 1 inch thick are effective firestops. Good practice includes: Firestops at exterior walls, each floor level, and the level where the roof connects with the wall; firestops at each floor level in partitions that are continuous through two or more stories; headers at the top and bottom of the space between stair carriages; mineral wool, asbestos, or an equivalent material packed tightly around pipes or ducts that pass through a floor or a firestop; and self-closing doors on vertical shafts, such as clothes chutes. Figure 81 shows applications of firestops in an exterior wall.

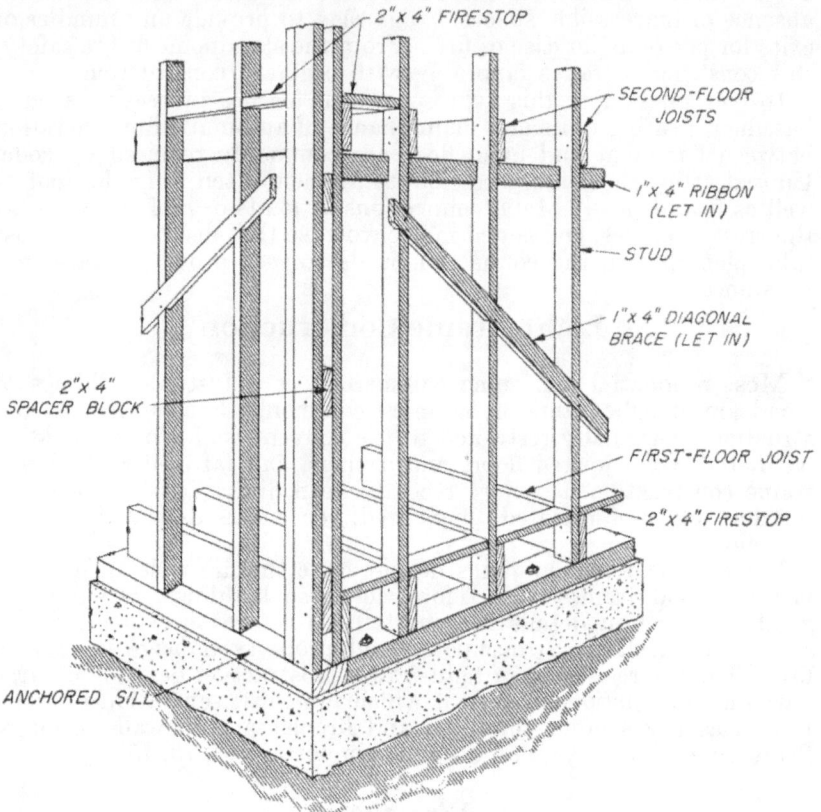

ZM 87434 F

FIGURE 81.—Firestops in frame construction.

Wood Construction Around Chimneys and Fireplaces

Good practice in the protection of wood from ignition by heat conducted through chimneys and fireplaces includes the following details: If smoke pipes from furnaces pass through walls, they are protected by thimbles at least 8 inches larger in diameter than the pipe. No smoke pipe passes through a floor or ceiling but joins the chimney on the same floor where it originates. Wood beams, joists, or rafters are separated from any chimney by a 2-inch space. Wood furring strips placed around chimneys to support base or other trim are insulated from the masonry by asbestos paper at least ⅛ inch thick, and metal wall plugs or approved noncombustible nail-holding devices attached to the wall surface are used for nailing. Wood construction is separated at least 4 inches from the back wall of any fireplace, and wood header beams supporting trimmer arches at fireplaces are placed at least 20 inches away from the face of the chimney breast. A wood mantel or other woodwork is placed not less than 8 inches from either side nor less than 12 inches from the top of any fireplace opening. Fireplace hearths are of noncombustible material, not less than 18 inches wide, measured from the face of the opening. All spaces between the masonry of chimneys and wood joists, beams, headers, or trimmers are filled with noncombustible material.

Partitions

A fire starting in one room of a building will be confined to that room for a variable period of time, depending on the fire resistance of the walls, partititions, ceilings, and doors. The fire resistance of wood frame walls and partititions depends to a considerable extent upon the materials used for faces, the method of joining wall and partition units, the quality of workmanship, and the type and quantity of any insulation that may be used.

The following tabulation gives the fire resistance of non-bearing built-up prefabricated wood partitions:

	Ultimate fire-re-sistance period [1] *(minutes)*
Solid partitions of 1¹¹⁄₁₆- by 3¼-inch, tongued-and-grooved, beaded wood boards placed vertically with staggered joints:	
1 board layer_____	10
Solid panels of ¾-inch wood boards, 2½ to 6 inches wide, grooved, and joined with wood splines, nailed together; boards placed vertically with staggered joints:	
2 board layers_____	15
2 board layers with asbestos paper weighing 30 pounds per 100 square feet between layers_____	25
3 board layers_____	40
3 board layers with center boards not vertical_____	35
Solid panels with ³⁄₁₆-inch plywood facings glued to 1¾-inch solid wood core of glued tongued-and-grooved construction for both sides and ends of core pieces, with tongued-and-grooved rails in the core about 2½ feet apart—	
2⅛-inch-thick panels_____	60
Hollow panels with plywood facings on both sides glued to wood frame; thickness of framing not less than 1¾ inches:	
¼-inch plywood faces_____	10
⅜-inch plywood faces_____	15
½-inch plywood faces_____	20
⅝-inch plywood faces_____	25
Hollow panels with plywood facings, ¼ inch thick on one side and ⅜ inch thick on the other side, glued to 2⅜- by ¾-inch wood studs set edgewise; panels filled with mineral wool batts weighing 2 pounds per square foot of filled space—3-inch-thick panel_____	45

[1] Data obtained from National Bureau of Standards BMS Report (*16*).

The following tabulation gives the fire resistance of well-constructed load-bearing wood-frame partitions:

	Ultimate fire-re-sistance period [1] (minutes)
Facings of boards without plaster:	
½-inch fiberboard weighing 0.7 pound per square foot	10
½-inch fiberboard weighing 1.1 pounds per square foot	15
½-inch flameproofed fiberboard weighing 1.6 pounds per square foot as treated	30
¾-inch tongued-and-grooved wood boards	20
¾-inch tongued-and-grooved wood boards with mineral-wool fill	35
¾-inch tongued-and-grooved wood boards with asbestos paper weighing 30 pounds per 100 square feet between boards and studs	45
⅜-inch gypsum wallboard	25
½-inch gypsum wallboard	40
½-inch gypsum wallboard with mineral wool fill	45
½-inch gypsum wallboard with mineral wool batts nailed to studs	60
Facings of plaster on wood lath:[2]—	
½-inch gypsum plaster (1:2, 1:3)	30
½-inch gypsum plaster (1:2, 1:3) with mineral wool fill	60
Facings of plaster on board plaster base:[2]	
½-inch gypsum plaster (1:2, 1:2) on ½-inch fiberboard weighing 0.7 pound per square foot	35
½-inch gypsum plaster (1:2, 1:2) on ⅞-inch flameproofed fiberboard weighing 2.8 pounds per square foot as treated	60
½-inch neat gypsum plaster on ⅜-inch plain gypsum lath	60
½-inch gypsum plaster (1:2, 1:2) on ⅜-inch perforated gypsum lath, 1 hole of ¾-inch or larger diameter per not more than 16 square inches of lath surface	60
½-inch gypsum plaster (1:2, 1:2) on ⅜-inch gypsum lath, plain, indented, or perforated other than as above	45

[1] Data are from National Bureau of Standards Report (16). Fire-resistance values are for nominal 2- by 4-inch wood studs (No. 1 common or better) set edgewise and having 2- by 4-inch wood plates at top and bottom and stayed transversely at midheight with wood blocks. Partitions were loaded on the basis of 360 pounds per square inch of the net area of the studs. "Mineral wool fill" can be taken as rock wool batts weighing not less than 1.0 pound per square foot of wall surface, glass wool batts at 0.6 pound per square foot, or rock wool blown in and weighing not less than 2.0 pounds per square foot of wall surface. "Wall surface" applies only to surface of filled space.
[2] Plaster proportions (in parentheses) are given as weights of dry plaster to dry sand, the first ratio being for the scratch coat and the second for the brown coat.

The fire tests upon which the fire-resistance values in the preceding tabulations are based were conducted in substantial accord with Standard Specifications for Fire Tests of Building Construction and Materials, No. A2–1934, American Standards Association.

Basement Ceilings

Since fires may start from heating plants located in basements, it is desirable that a fire-resistant separation of the basement from the remainder of the building be provided. Gypsum board, asbestos board, or plaster on metal or gypsum lath placed on the basement joists affords an effective means of increasing the fire resistance of the basement ceiling and of retarding the rapid spread of flames, but such material is usually omitted to reduce cost. Less expensive fire-retardant coatings applied to all exposed wood surfaces also provide protection from rapid spread of fire. Particular attention should be given to the wood floor members directly above and near the furnace.

If, as is common, a basement stairway is directly under the stairway leading from the first to the second floor, it is good practice to protect the underside of the upper stairway with fire-resistant coverings, as suggested for basement ceilings, and to place firestops between the wood carriages at the top and bottom.

Floors

The conventional floor construction of joists, subfloor, and finish floor offers considerable resistance to the penetration of fire and will retain some supporting capacity in fire exposures up to 15 minutes (16). Prefabricated floor panels, in which the supporting power depends upon stressed covers, and floor systems supported by box girders with thin plywood webs are more vulnerable to fire penetration.

Doors and Stairways

If a fire-resistant ceiling is placed on the basement joists, it is desirable to have a door leading to the basement with fire resistance equal to the combined resistance of the ceiling and floor over the basement.

Enclosed stairways retard rapid spread of fire from floor to floor. If the interior design calls for an open stairway below, it can often be closed at the top with a solid flush door. Hollow-core flush doors offer little resistance to the penetration of fire unless the space between covers is packed with an insulating material.

Roof Coverings

The better grades of wood shingles are edge grained and thick butted with 5 butts measuring at least 2 inches. Edge-grained shingles warp or curl less than flat-grained ones, thick-butted shingles less than thin ones, and narrow shingles less than wide ones. Fire sparks are less likely to lodge beneath and ignite tight, flat shingles than warped shingles. To follow accepted rules of good practice in laying shingles not only makes a long-lived and economical roof but markedly reduces fire hazards. Painting or staining shingles with the materials customarily used for that purpose has little effect on their fire resistance.

Surface Coverings and Finishes

Finishing or covering materials, such as paint, should be such as not to promote rapid spread of flame, give off poisonous gases, or otherwise cause an unusual or excessive fire hazard. Flame-spread tests suitable for evaluating surface coverings or treatments are available (2, 8, 19, 20).

Glued Laminated Members

The fire resistance of glued laminated structural members, such as arches, beams, and columns, is approximately equal to the fire resistance of solid members of similar sizes. Available information indicates that laminated members glued with phenol, resorcinol, or melamine glues are at least equal in fire resistance to a one-piece member of the same size and that laminated members glued with casein glues may have slightly less fire resistance.

In tests at the Forest Products Laboratory, when the edges of the laminations in sections of laminated members bonded with casein glue were exposed to a gas fire, slightly deeper charring resulted at the glue joints than between the glue joints. When the broad face of a lamination was exposed to the fire, the outer lamination loosened from the rest of the member when the zone of char penetrated to the depth of the glue line. The appearance of the joints and the results of shear tests indicated, however, that little if any weakening of the glue joints occurred beyond the charred surface as a result of the fire exposure. Also, the performance of casein-glued laminated members in actual fires is reported to demonstrate the integrity of casein-glued joints beyond the zone of char.

When the fire endurance required of a wood member is less than the time required for the zone of char to penetrate through the outer laminations, the presence of glue or the type of glue is unimportant. Thus, on the basis of a rate of penetration of char of 1½ inches per hour of exposure to fire, casein-glued laminated lumber, with outer laminations not less than 1½ inches thick (such as nominal 2-inch lumber laminations) would be equivalent in fire resistance to solid members of the same actual size with a fire resistance up to 1 hour.

For a casein-glued member whose thickness of laminations permits penetration of the zone of char to the outermost glue line in a period less than the required fire endurance of the member, the zone of char will penetrate a little faster in the laminated member, after it reaches the first glue line, than in a solid member. Available data indicate that for a beam made of ¾-inch casein-glued laminations the char may be as much as 10 percent deeper in the laminated beam than in a solid beam of the same size after exposure to fire for 1 hour.

Thus, for heavy-timber-construction classification, laminated members glued with phenol, resorcinol, or melamine glues or laminated members glued with casein glue and having nominal 2-inch outer laminations are considered equivalent to solid sawed members in comparable standard minimum sizes; for casein-glued members with face laminations less than 1½ inches thick, the dimension of the member in the direction at right angles to the glue lines should be increased at least 0.15 inch.

FIRE-RETARDANT TREATMENTS FOR WOOD

Two general kinds of treatments are available for improving the fire resistance of wood: (1) Impregnation treatments that deposit fire-retardant chemicals within the wood (17) and (2) coatings or layers of protective materials over the surface of the wood (19). Good fire-retardant effects can be obtained by both methods. For new wood construction and the repair of existing structures with new parts, pressure impregnation with effective fire-retarding chemicals offers the best possibility of obtaining lasting fire-retardant effects in high degree. For wood in existing structures, surface applications offer the principal means of increasing the fire-retardant properties. Lumber, timbers, and plywood impregnated with fire-retarding chemicals now find many uses in a great variety of buildings (5). During World War II, large quantities of impregnated lumber were used for the construction of dirigible hangars and for shoring aboard ships.

Impregnation Treatment

Wood is impregnated with fire-retardant chemicals by methods similar to those used to inject preservatives (p. 411). The wood is placed within a sealed treating cylinder and the treating solution forced in by means of pressure. Although additional information is needed on the effect of the treatments on the strength of wood, the amounts required for a specified performance, and the permanence of the treatments under adverse exposure, it is known that the fire-retarding effect of impregnation treatments is closely related to the quantity of chemical injected into the wood as well as to the chemical used (3, 11). Absorptions up to 1 pound of fire-retardant chemical per cubic foot of wood, quantities such as are used with toxic chemicals for decay prevention, have only a small effect on the combustion of wood. It is necessary to use much more chemical to obtain the best fire-retardant results. For a high degree of effectiveness, 5 to 6 pounds of the more effective chemicals per cubic foot of wood in thicknesses less than 2 inches are required, or approximately 400 to 500 pounds per thousand board-feet. Lumber in thicknesses greater than 2 inches requires proportionately less material (4, 18).

Effectively treated wood can be charred or disintegrated by continuous exposure to intense heat from an outside source, but when the heating is discontinued the burning ceases. The principal effects of fire-retardant impregnation treatments are to retard the normal increase in temperature under fire conditions, to decrease the rate of flame spread, to lessen the rate of flame penetration or destruction of wood in contact with fire, and to make fires more easily extinguished.

Impregnation treatments may be complete; that is, the treatment may extend completely through the piece; or they may be only partial, in which case only an outside zone of the piece is impregnated. Complete impregnation with an adequate quantity of fire-retardant chemicals makes wood sufficiently resistant to fire so that it will not of itself support combustion. Partial impregnation affords protection that is adequate for most purposes and under many conditions, although the central part is unimpregnated. It obviously is cheaper and generally more practical than complete impregnation. Only partial impregnation is possible with some species of wood and with most lumber and timbers of commercial size (6). When the lumber must be cut into smaller sized pieces or machined after treatment in such a way that the interior is exposed, however, the effectiveness of the treatment may be greatly reduced unless there has been practically complete penetration.

Many chemicals have a fire-retardant effect when wood is impregnated with them, but because of cost limitations or various objectionable characteristics, comparatively few are considered commercially practical. Among the most commonly used chemicals are monoammonium phosphate, diammonium phosphate, ammonium sulfate, sodium tetraborate (borax), boric acid, and zinc chloride. The ammonium phosphates are effective in checking both flaming and glowing. Borax, although effective in checking flaming, is not a good slow retardant. Boric acid, on the other hand, is exceptionally effective in stopping glow but is not so effective for retarding flaming. Because of the different characteristics of the compounds and because of cost considerations, fire-retardant treating formulations usually are mixtures of fire-retardant chemicals. Typical formulations are available (4, 5, 17). The fire-retarding effectiveness of numerous chemicals other than the six listed is reported in Experiments in Fire-proofing Wood (11).

Many chemicals that might be used as fire retardants may have undesirable effects on the wood or be otherwise objectionable for specific uses. Caution should be observed, therefore, in deciding upon any treatment that has not had thorough trial in practice. Among use characteristics of importance are permanence of the fire resistance imparted to wood, effect on strength of wood, tendency to corrode metals, effects on paint and glue, hygroscopicity, and toxicity to occupants of buildings.

There are a number of formulations of fire-retardant chemicals of high effectiveness that have been in commercial use over a sufficient period of years to demonstrate that they are practical (4, 5).

Coatings and Surface Treatments

Many coating materials protect wood against fire in varying degree (17). As with impregnation treatments, the amount of protection provided by a coating is related to the amount and thoroughness of application, and to the severity of the fire exposure.

Fire-retardant coatings are of varying composition and properties. Most available preparations are of value primarily for interior use and are not durable when exposed to the weather. They usually owe their effectiveness to one of the following water-soluble fire-retardant chemicals: Ammonium phosphate, borax, or sodium silicate. These fire retardants are combined with other constituents to provide other properties required or desirable in a paint, such as adherence to wood, appearance, and brushability. Some good proprietary fire-retarding coatings based on these chemicals are on the market.

The following nonproprietary formulations (17) are among those which have shown good results in laboratory tests, when applied in adequate amounts:

(1)—

	Percent by weight
Basic carbonate white lead	41
Borax	32
Raw linseed oil	22. 8
Turpentine	3. 6
Japan drier	. 6
	100. 0

3 or 4 thick coats, or approximately 1 gallon per 125 square feet of surface, are required for good protection.

(2)—

	Pounds
Sodium silicate solution (specific gravity 1.41 to 1.42, silica-soda ratio 3.2 to 3.4)_____	112
Kaolin_____	150
Water_____	100

3 to 4 coats are required to give good protection. 1 gallon will cover approximately 100 square feet (4 coats).

(3)—

	Parts by weight
Monoammonium phosphate_____	50
Sodium alginate gel 2 percent_____	50

Prepare the alginate gel by adding 2 parts by weight of dry sodium alginate to 98 parts by weight of hot water. Stir until a uniform gel is obtained. Grind in a pebble mill equal parts by weight of monoammonium phosphate and the alginate gel. A grinding period of 12 to 24 hours is sufficient. 2 to 3 coats are required for good protection. 1 gallon will cover about 70 square feet (3 coats).

So far as is known, no water-insoluble compound has been found that is equal in fire-retarding effectiveness to the three water-soluble chemicals listed previously. Compounds of low water solubility that have measurable fire-retardant properties are zinc borate, chlorinated paraffin, and chlorinated rubber, and exterior proprietary preparations are on the market containing these constituents. Although possessing moderate fire-retarding properties, these paints do not give a performance equal to that of the better interior paints.

LITERATURE CITED

(1) AMERICAN SOCIETY FOR TESTING MATERIALS.
  1950. METHODS OF FIRE TESTS OF BUILDING CONSTRUCTION AND MATERIALS. Amer. Soc. Testing Mater. Standard E–119–50. Pt. 4: 993–1001, illus.
(2) ————
  1950. TENTATIVE METHODS OF FIRE HAZARD CLASSIFICATION OF BUILDING MATERIALS. Amer. Soc. Testing Mater. Standard E–84–50T (Tentative). Pt. 4: 1020–1022.
(3) AMERICAN WOOD-PRESERVERS' ASSOCIATION.
  1944. REPORT OF COMMITTEE 9, FIREPROOFING. Amer. Wood-Preservers' Assoc. Proc. 40: 261–348, illus.
(4) ————
  1949. REPORT OF COMMITTEE P–5, FIRE-RETARDANT PRESERVATIVES. Amer. Wood-Preservers' Assoc. Proc. 45: 46–54.
(5) ANGELL, H. W.
  1951. PRODUCTION AND USE OF FIRE RETARDANT TREATED LUMBER. Forest Prod. Res. Soc. Proc. 5: 107–114, illus.
(6) HUNT, G. M., and GARRATT, G. A.
  1953. WOOD PRESERVATION. 402 pp., illus. New York.
(7) LAWSON, D. I., WEBSTER, C. T., and ASHTON, L. A.
  1952. FIRE ENDURANCE OF TIMBER BEAMS AND FLOORS. Struct. Engin. (England) 30 (2): 7 pp., illus.
(8) MCNAUGHTON, G. C., and VAN KLEECK, A.
  1953. FIRE-TEST METHODS USED IN RESEARCH AT THE FOREST PRODUCTS LABORATORY. U. S. Forest Prod. Lab. Rpt. R1443, 13 pp., illus. [Processed.]
(9) NATIONAL BOARD OF FIRE UNDERWRITERS.
  1949. NATIONAL BUILDING CODE. 258 pp., illus. New York.
(10) NATIONAL FIRE PROTECTION ASSOCIATION.
  1948. HANDBOOK OF FIRE PROTECTION. 10th ed., 1544 pp., illus. Boston
(11) TRUAX, T. R., HARRISON, C. A., and BAECHLER, R. H.
  1935. EXPERIMENTS IN FIREPROOFING WOOD—FIFTH PROGRESS REPORT. Amer. Wood-Preservers' Assoc. Proc. 31: 231–248, illus.

(12) UNDERWRITERS LABORATORIES, INC.
      1939. FIRE EXPOSURE TESTS OF LOADED TIMBER COLUMNS. Underwriters
            Labs. Bul. Res. 13, 23 pp., illus.
(13) U. S. FEDERAL BOARD OF VOCATIONAL EDUCATION.
      1931. LIGHT FRAME HOUSE CONSTRUCTION. U. S. Fed. Bd. Vocat. Ed.
            Bul. 145, Trade and Indus. Ser. 41, 216 pp., illus.
(14) U. S. NATIONAL BUREAU OF STANDARDS.
      1921. FIRE TESTS OF BUILDING COLUMNS. U. S. Natl. Bur. Standards
            Technol. Paper 184, 373 pp., illus.
(15) ——
      1932. RECOMMENDED MINIMUM REQUIREMENTS FOR SMALL DWELLING
            CONSTRUCTION. U. S. Dept. Com., Rpt. Bldg. Code Com., Bldg.
            and Housing 18, 107 pp., illus.
(16) ——
      1942. FIRE RESISTANCE CLASSIFICATION OF BUILDING CONSTRUCTIONS.
            Bldg. Mater. and Struct. Rpt. 92, 70 pp.
(17) U. S. NAVY.
      1944. CHEMICALS, FIRE-RETARDANT, FOR LUMBER AND TIMBER. Navy
            Dept. Specif. 51 C 38, 7 pp.
(18) ——
      1944. CHEMICALS, FIRE-RETARDANT, FOR LUMBER AND TIMBER (RECOM-
            MENDED TREATING PRACTICE). Navy Dept. Specif. 51 C 40,
            8 pp., illus.
(19) VAN KLEECK, A.
      1948. FIRE-RETARDING COATINGS. U. S. Forest Prod. Lab. Rpt. R1280,
            11 pp. [Processed.]
(20) ——
      1950. EVALUATION OF FLAME-SPREAD RESISTANCE OF FIBER INSULATION
            BOARDS. U. S. Forest Prod. Lab. Rpt. D1756, 12 pp., illus.
            [Processed.]

PAINTING AND FINISHING WOOD

The principal reason for coating woodwork is usually improvement and maintenance of appearance. Immediate considerations in coating wood for appearance are: (1) Paint and enamel are opaque and therefore conceal the grain and color of the wood, substituting color, sheen, and texture of their own. (2) Varnish, lacquer, oil, and wax are transparent and thus reveal the grain of a piece more fully by displacing the air at its surface with a medium of much higher refractive index. They intensify the natural color of the wood, the more so the deeper they penetrate into the wood. In conjunction with stain, they may alter the color of the wood. The sheen and texture of these coatings are substituted for those of the wood. (3) Stain without varnish or lacquer changes the color of wood without greatly altering its sheen or texture.

Ultimate considerations in coating wood for appearance are: (1) Unless protected by suitable coatings, wood exposed to the weather undergoes weathering, which greatly alters its appearance and may impair its usefulness (1, 27).[24] (2) Uncoated interior wood surfaces are porous enough to absorb liquids quickly, to stain and spot readily, and to hold dirt tenaciously. Nonporous coatings or coatings that tend to shed liquids will protect wood from discoloration and present a surface easily cleaned.

Other reasons for painting or finishing wood are to provide a brighter surface for better lighting in dark rooms, to retard changes in moisture content and dimensions of wood, and to make it more difficult for wood to take fire and for the flames to spread. Moisture-retardant coatings are discussed on pages 371–373, fire retardant coatings on pages 348–349.

Paints and finishes are not used primarily to prolong the life of wood (1). Ordinary paints and varnishes are ineffective as preservatives against decay fungi. In addition, wood that is damp enough for attack by fungi is likely to be difficult to keep painted (8). Decay of wood and its prevention are discussed on pages 381–386. The painting of wood that has been treated for the prevention of decay is discussed on page 359.

Exterior woodwork that stays reasonably dry, except for wetting of its exposed surfaces by rain or snow for short times, does not decay, but, if left uncoated, it is subject to weathering. In weathering, wood usually turns gray (at high altitudes, in the mountains, it may turn brown), smoothly planed surfaces become rough and cracked, and the wood wastes away at the exposed surfaces by about 0.25 inch a century. In addition, boards may cup, warp, and pull at their fastenings and, if too thin or too short, may split. Weathering can be prevented by keeping wood always well painted (fig. 82), but since the cost of repeated paintings soon exceeds the value of the wood, it is uneconomical to paint merely to keep wood from weathering. But if painting is to be done for the sake of appearance anyway, the protective power of paint makes it possible to use thinner pieces of wood than would otherwise be advisable.

---

[24] Italic numbers in parentheses refer to Literature Cited, p. 377.

M 92356 F

FIGURE 82.—Wood houses can be kept well painted for many decades if good paints and painting practices are used. This house, built in 1759, has been occupied and kept in good condition ever since. The exterior woodwork is undoubtedly the original, and it has never been necessary to remove the old paint before repainting.

Painting Characteristics of Woods

Wood may be finished by merely saturating its surface with drying-oil, sealer, or wax finishes, either with or without stain, and without forming a continuous coating of appreciable thickness over the surface. Such finishes may be called penetrating finishes or intra-surface finishes as distinguished from surface coatings, such as paint and enamel. A surface coating forms a continuous skin over the wood seldom less than 0.001 inch and preferably not more than 0.005 inch thick except in special cases, such as textured or sand finishes. It is applied over a suitable sealer or primer that makes the wood surface nonabsorptive of liquids, so that the finishing coats will remain on the surface. Hardwoods with relatively large pores (vessels) require a heavily pigmented wood filler, to plug the pores, before they can be painted smoothly.

Spreading Rates

The spreading rate of paint or other finish is the area of surface covered by a unit volume of paint. On smoothly planed surfaces of softwoods the spreading rate varies with species only for the first or priming coat of paint applied on the new surface; the spreading rate of subsequent coats does not vary with species (2).

Even for priming-coat paint the species of wood has less effect on spreading rate than the individuality of the painter or the kind and proportion of ingredients in the paint. Some painters brush paint out farther than others; the best painters apply paint in somewhat thicker coats than unskilled painters are inclined to do. Priming coats of house paint on new, smooth wood normally should be spread at the rate of approximately 600 square feet a gallon if they are to be followed by 2 coats of paint but not more than 450 square feet a gallon if only 1 coat is to follow. Less viscous materials, such as varnish, may be spread at 700 or 800 square feet a gallon. Some native soft-woods tend to take 10 to 12 percent more or less priming-coat paint than the average. By the time second- and third-coat paints have been applied, however, the difference in paint consumed amounts to only 3 or 4 percent above or below the average, a value that is prob-ably within the limits of uniformity with which a good workman will paint at different times. The woods that tend to take more than the average amount of paint are generally lightweight, and those that tend to take less paint are heavy and have a high proportion of sum-merwood. Rough surfaces, such as shingles, the sawed side of bevel siding, weathered wood, or surfaces on which repainting has been neglected for a long time, may consume twice the amount of priming-coat paint required for smoothly planed surfaces.

Drying of Paint

The time required for a coating of paint or other finish to dry—that is, to harden—usually depends almost entirely on the composition of the paint, the intensity of the sunlight, and the temperature and relative humidity of the atmosphere. Drying is retarded for a few hours, however, if the wood is wet when painted; its moisture content should therefore not exceed 20 percent. On wet redwood or cypress, priming-coat paint may remain liquid for days, especially when the paint contains little or no white lead or zinc sulfide pigment (16). On air-dry redwood, paint hardens as rapidly as on any other wood. On cypress boards that contain much more than the average amount of the oily extractive characteristic of cypress, the hardening of paints without white lead or zinc sulfide may be retarded even when the boards are reasonably dry, unless the drying takes place in strong sunlight. Eastern redcedar, Spanish cedar, and, to a less extent, Port-Orford-cedar also contain oily extractives that retard hardening of coatings or that may even, in confined places with little light and ventilation, resoften coatings after they have hardened (32).

The native hardwoods contain no constituent that delays hardening of paints if the wood is reasonably dry. On wet oak or chestnut, there may be some retardation of hardening when there is little sunlight or when the weather is cold.

Filling Porous Hardwoods Before Painting

For painting purposes, the hardwoods may be classified as follows (15):

Hardwoods with large pores	Hardwoods with small pores
Ash	Alder, red
Butternut	Aspen
Chestnut	Basswood
Elm	Beech
Hackberry	Cherry
Hickory	Cottonwood
Khaya (African mahogany)	Gum
Mahogany	Magnolia
Oak	Maple
Sugarberry	Poplar
Walnut	Sycamore

Birch has pores large enough to take wood filler effectively when desired but small enough, as a rule, to be painted satisfactorily without filling.

Hardwoods having small pores may be painted with ordinary house paints in exactly the same manner as softwoods. Hardwoods with large pores require wood filler before they can be covered smoothly with paint or enamel. Without filler the pores not only appear as depressions in the coating but become centers of early paint failure. For interior woodwork ordinary wood filler, which is a paste made of ground quartz, linseed oil or varnish, and paint drier, may be used. For exterior woodwork, the filler may be made by thinning a gallon of soft paste white lead with a gallon of turpentine or other paint thinner. The mixture is brushed over the wood, allowed to stand about 10 minutes, and wiped off with rags or cotton waste across the grain of the wood, in order to pack the filler into the large pores as thoroughly as possible and to remove excess filler from the surface. When the filler is dry, the wood is ready for painting.

Effect of Nature of Wood on Wearing of Paint

During the first year or two after an exterior wood surface has been painted properly the nature of the wood usually has very little effect on the behavior of the paint coating. Later on, when the coating has become sufficiently embrittled with age (7, 13), differences between woods become important and largely determine how rapidly the coating disintegrates (3, 20, 24). Under normal conditions of exposure—that is, when the abnormal conditions discussed on page 369 do not pertain—paint coatings on wood (6, 18, 25) deteriorate most rapidly where the greatest amount of sunshine falls (fig. 83), and deterioration proceeds in successive but overlapping stages as follows:

1. The soiling stage. The coating gradually becomes dirty.
2. The flatting stage. The coating loses its gloss.
3. The chalking stage. Dirt may be thrown off more or less completely, but colors appear to fade. Loose pigment may be wiped off the surface of the paint much like chalk.
4. The fissure stage. (Under the most favorable conditions, some paints may skip this stage and pass to the stage of disintegration.) Fissures are of two general types depending on the nature of the paint:
   a. Checking. The fissures are at first superficial but later may penetrate entirely through the coating. Fissures formed in checking are usually very inconspicuous.
   b. Cracking. The fissures pass entirely through the coating when first observed. The coating at the edges of the cracks sooner or later comes loose from the wood and curls outward. As a rule cracking comes at a more advanced stage of paint embrittlement than checking.
5. The stage of disintegration. Behavior in the fissure stage determines the form in which disintegration takes place (fig. 84):
   a. Erosion. When paints that do not suffer checking waste away by chalking or erosion fast enough to escape cracking, the coating finally becomes too thin to hide the wood and patches of the wood begin to be laid bare.
   b. Crumbling, which develops from checking. Tiny fragments of coating, cut off when the fissures become interwoven and penetrate entirely through the coating, fall away.
   c. Flaking, which develops from cracking and curling. The loosened edges of coating that curl outward finally fall off, after which the newly formed edges curl outward and the process continues. Flaking is usually a more rapid form of disintegration than crumbling.
6. The stage of advanced break-up or neglect. Disintegration has laid bare enough wood for the boards to begin to show serious weathering.

From the time that cracking or disintegration begins the nature of the wood painted dominates the further rate of wearing away of the coating. On softwoods cracking, crumbling, and flaking progress chiefly over the bands of summerwood (fig. 83). The wider the bands of summerwood intersected by the painted surface the more rapidly disintegration proceeds and the more conspicuous are the areas laid

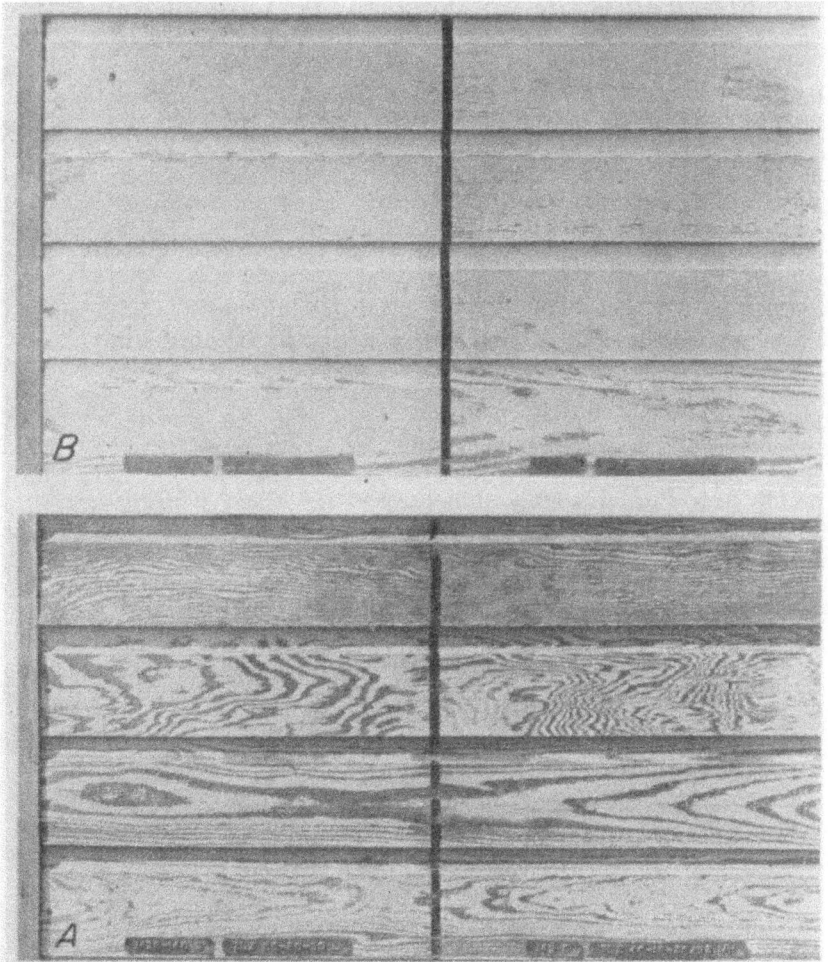

FIGURE 83.—In normal paint deterioration, the bands of summerwood are laid bare by crumbling or flaking of the coating long before the softer springwood is exposed. Paint disintegrates more rapidly on surfaces that face south and receive the most sunshine (*A*) than on surfaces that face north and receive less sunshine (*B*). On each of the 8-year-old test panels shown, the right-hand area was originally painted with a house paint of slightly lower quality than that used on the left-hand area.

bare. Repainted coatings behave in much the same way as coatings applied to new wood, provided the repaintings are spaced far enough apart to avoid unduly thick coatings.

Certain wood characteristics (*20, 24*) largely determine the rate of paint disintegration on softwoods: (1) On wood of light weight (low density) paint disintegrates relatively slowly because light wood contains little summerwood. (2) On narrow-ringed wood paint disintegrates more slowly than on wide-ringed wood because the bands

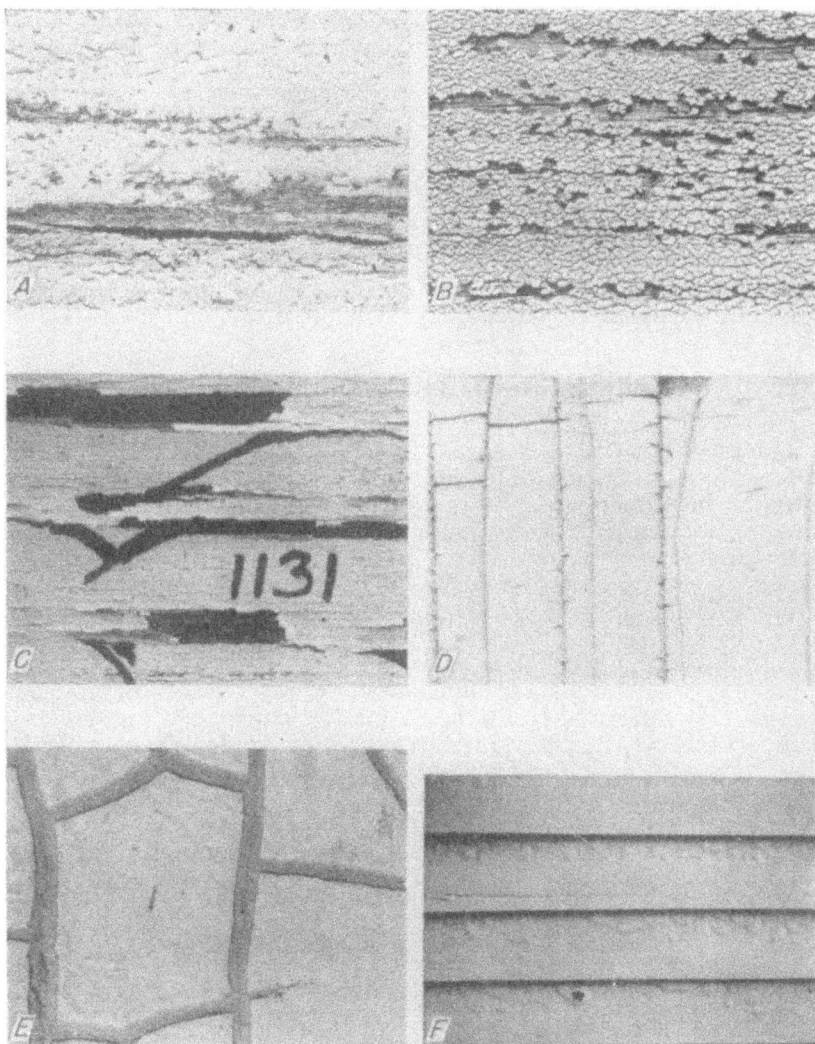

M 92357 F

FIGURE 84.—Three forms of normal paint disintegration: A, Coating thinned by erosion; B, checking penetrated to the wood surface, followed by crumbling; C, cracking, followed by curling and flaking away of loosened paint. Three forms of abnormal paint disintegration: D, Cross-grain cracking, which results from too thick coating; E, alligatoring, frequently caused by the use of dissimilar paints for successive paintings; F, blistering, in this case caused by chemical deterioration of one of the earlier layers of paint.

of summerwood are narrower. (3) On edge-grained surfaces, with narrower bands of summerwood, disintegration of paint is slower than on flat-grained surfaces. (4) Flat-grained boards may hold paint better on the bark side than on the pith side because on the pith side there is sometimes a tendency for the edges of the bands of summerwood to loosen and curl outward, thus dislodging the coating. (5)

High-grade lumber holds paint better than low-grade because knots and other grade-limiting features hold paint poorly and because the low grades usually come from the center of the tree where the wood is likely to be wider ringed.

Coatings begin to fail to protect wood against weathering when deep fissures develop or when disintegration sets in, though weathering seldom becomes serious until the coating is well into the stage of neglect. The consequences of inadequate protection depend upon the characteristics of the wood. Edge-grained boards are less seriously affected than flat-grained boards of the same species. Woods that hold their shape well and do not develop conspicuous checks or cracks on weathering are less seriously affected than other woods.

Classification of Softwoods for Painting

Almost all native softwoods of commercial importance can be kept well painted for long periods of time if proper care is given to selection of primers and paints and to choice of programs of maintenance in accordance with the painting characteristics of the wood. Some woods are less exacting than others in their requirements. Table 42 classifies native softwood species with respect to their painting characteristics for exterior use. The classification is based upon the average behavior of commercial shipments of select grades of the different woods; that is, upon lumber of a quality that is commonly used for house siding or millwork. Boards of any one species may vary widely in the ease with which they can be kept painted, so that, in general, boards selected for physical properties that are least exacting will serve more satisfactorily than the average run of the species rated in table 42. After several repaintings the differences among species diminish and may disappear entirely if such paint abnormalities as excessive coating thickness or incompatibility (p. 369) set in.

TABLE 42.—*Grouping of softwoods for exterior painting*

Group 1 [1]	Group 2 [2]	Group 3 [3]	Group 4 [4]
Alaska-cedar.	Eastern white pine.	Commercial white fir.	Douglas-fir.
Atlantic whitecedar.	Sugar pine.	Eastern hemlock.	Red pine.
Baldcypress.	Western white pine.	Eastern spruce.	Southern yellow pine.
Incense-cedar.		Engelmann spruce.	Tamarack.
Northern whitecedar.		Lodgepole pine.	Western larch.
Port-Orford-cedar.		Ponderosa pine.	
Redwood.		Sitka spruce.	
Western redcedar.		Western hemlock.	

[1] Woods on which paints of the widest range in kind and quality give good service.

[2] Woods on which care is needed to select a suitable priming paint, such as pure white lead or a zincless house-paint primer.

[3] Woods more exacting than those of group 2 but less exacting than those of group 4 in requirements of suitable priming paint, finish paint of high quality, and careful choice of program of maintenance.

[4] Woods that need most careful selection of suitable paints, and for which aluminum house paint is particularly desirable for priming when the wood is first painted, with a zincless house-paint primer as second choice if for any reason aluminum paint cannot be used.

Effect of Extractives in Wood on Paint Behavior

In general, the extractives in wood have much less effect on paint behavior than has been commonly supposed. They are far less important than the width of the bands of summerwood.

Distinction must be made between piny resins, which are characteristic of pines but occur also in other softwoods, and other extractives.

Piny resins consist of a solid rosin and a volatile liquid, such as turpentine. They affect paint unfavorably by exuding through coatings, leaving unsightly encrustations. Exudation is minimized by thorough seasoning of the lumber to drive out the turpentine and thus render the resin less mobile, but more careful seasoning is necessary to remove turpentine than to remove moisture (28, 32). The piny resins may also be responsible for the fact that paints containing zinc oxide fail somewhat more rapidly on the pines than they do on other softwoods having bands of summerwood of similar width.

Other extractives in wood may have a favorable effect on the durability of paint (16); there is evidence that such is the case with extractives in redwood and cypress. Colored extractives in redwood and redcedar that are soluble in water may exude through coatings if the wood beneath the coating becomes very wet; under normally dry conditions, however, these discolorations do not occur.

Effect of Impregnated Preservatives on Painting

Wood treated with the water-soluble preservatives in common use can be painted satisfactorily after it is redried. The life of the coating may in some instances be slightly less than it would have been on untreated wood, but the loss in durability is not enough to offer any practical objection to the use of treated wood for purposes where preservation against decay is necessary and the appearance of painted wood and protection against weathering are desired (12). Such treated wood when used indoors in textile mills, pulp mills, or other places where the relative humidity may be above 90 percent for long periods may give rise to discoloration of paint or to exudation of preservative solution.

Coal-tar creosote or other dark oily preservatives tend to stain through paint unless the treated wood has been exposed to the weather for many months before it is painted (33). Aluminum paint often serves well on wood freshly creosoted by an "empty-cell" process or cleaned after treatment by steaming or vapor drying. Fairly good results can also be obtained on creosoted wood with rough surfaces (sawed or weatherbeaten surfaces) by applying exterior water-thinned paints, such as casein paints or resin-emulsion paints.

Wood treated with oilborne chlorinated phenols can be painted without difficulty when the solvent oils have evaporated thoroughly from the surfaces. If light oils that evaporate rapidly are used for the treating solution, painting can be done soon after the treated wood has dried (21).

Plywood

The painting characteristics of plywood are essentially those of solid lumber of the kind used for the face ply. The most abundant and least expensive plywood is usually faced with rotary-cut veneer of woods in group 3 or group 4 of table 42. It therefore has flat-grained faces that display wide bands of summerwood and requires careful choice of primers and paints for best results. Moreover, the face plies of plywood are often more susceptible to checking or cracking than is lumber of the same kind of wood.

Plywood has been modified in different ways to improve the durability of paint coatings and to reduce the tendency to face checking (29). The methods are applicable to lumber also but have not yet been developed widely for that purpose.

Plywood made with face veneers that have been acetylated to make them more stable dimensionally holds paint coatings longer and resists checking and weathering better than ordinary plywood. Similarly plywood faced with veneers treated with phenolic resin (impreg, p. 467) has superior painting characteristics and stability against checking and weathering. Other coverings for plywood for the same purposes are paper impregnated with phenolic resin and finely ground wood waste bonded together and to the plywood with synthetic resin. Striating plywood by cutting parallel grooves, about eight to the inch, in the direction of the grain of the face plies breaks up the wide bands of summerwood and achieves significant improvement in durability of coatings and in resistance to checking. Consumption of paint is increased by about 10 percent, however.

CHOICE OF PAINTS FOR EXTERIOR WOODWORK

House paints of the best quality are recommended for nearly all needs for exterior painting. Inferior paints made to sell at materially lower prices must be applied in greater amounts to accomplish the same result, may not last so long, and are less reliable generally. Even when low initial cost of painting is the chief consideration, it usually proves cheaper to use high-grade paint in the minimum number of coats because the labor for a paint job generally costs several times as much as the paint.

For both low initial cost and great durability, whenever the colors are acceptable, paints of good quality made with iron oxide pigments are more economical than most other good paints. Two coats of exterior aluminum paint make an exceedingly durable, highly protective coating when a metallic appearance is not considered undesirable. For very brief service or where very frequent repainting is desired for some reason and protection against weathering is not necessary, whitewash or exterior water or water-emulsion paints are safer to use than linseed oil paints of very inferior quality. The water paints give their best service on rough rather than on smoothly planed wood.

For residences or buildings of architectural pretension white or light-colored paints are most popular for the major painted surfaces, supplemented if desired by deeply colored paints on minor surfaces.

When deep colors are appropriately dominant it is often better and more economical to use rough wood surfaces, such as shingles or the sawed side of siding, and apply exterior stains rather than gloss paints.

Composition of Paint

The quality of a house paint depends upon the quality of the liquids, the quality of the pigments, and the proportions of pigments and liquids of which it is made (*17, 18, 19, 22*).

Quality of the Liquids

Linseed oil is the most widely accepted liquid ingredient for exterior paints for wood. It is available in many degrees of refinement for different purposes. Proper choice of the kind of linseed oil most suitable for the pigments used is essential for making high-grade paint, but as a rule the purchaser must rely upon the manufacturer's judgment in selecting the kind of oil. Substitution of other drying oils for part of the linseed oil is common practice and may not impair the quality of the paint. Thus, many paints contain substantial proportions of soybean oil. Moreover, some of the modern oils are chemically treated to improve their drying characteristics. For the painter's needs in thinning paints for application, raw or boiled linseed oil of good quality should be used.

Raw or refined linseed oil is oil that still has about the same viscosity (consistency) that it had when first extracted from flax seed. Boiled oil has nearly the same viscosity, but has very small quantities of driers incorporated in it to make it dry faster, say in 16 hours instead of 72 hours. The driers are soaps of lead, manganese, or cobalt and are sometimes used with soaps of zinc or calcium. Concentrated solutions of such driers in mineral spirits are available for painters to use when needed, under the name of liquid paint drier. Bodied linseed oil is refined linseed oil that has been heated at a high temperature, usually between 550° and 625° F., long enough to increase its viscosity greatly. It can also be made by treating refined oil with chemicals, such as sulfur chloride.

Before 1943 the liquid part of house paints of best quality contained at least 85 percent of linseed oil, by weight. Such paints are called oil-rich paints, but there are few of them left on the market. Oil-rich paints may contain a little bodied oil, up to 10 or 15 percent, to improve flow and leveling of brush marks. Since 1943 most house paints have been oil-restricted, that is, the liquid part consists typically of 33 percent of raw oil, 33 percent of bodied oil, and 34 percent of volatile thinner. A gallon of oil-restricted paint costs less but normally covers less surface than a gallon of oil-rich paint. The liquid part of some brands of house paint have a composition between the extremes of the typical oil-restricted and oil-rich paints.

Varnish is rarely a desirable ingredient in the liquid of good house paint (except in aluminum paint, p. 367), but it is often used in inferior paints because large proportions of inexpensive volatile thinners can then be incorporated. On the other hand, good exterior varnish or bodied linseed oil or tung oil is necessary to make exterior enamels.

trim paints of the deeper colors, or exterior floor paints, which must stand mechanical wear as well as weather. In such enamels or floor paints of good quality as much as 50 percent by weight of the liquid may be volatile thinner.

The volatile thinner in most paints is mineral spirits, a petroleum product. Paint users thin paint, when necessary, with turpentine, mineral spirits, or prepared thinners sold by paint manufacturers. There should be no more than 1 or 2 percent by weight of water in the liquid of any good paint.

Quality of the Pigments

Differences in the behavior of linseed-oil paints during the successive stages in deterioration with age (p. 355) are determined largely by the nature of the pigments (25). No kind of paint now known is superior in all stages of deterioration; improvement in one stage is usually effected at some sacrifice in other stages. In addition, some paints require more careful spacing of repaintings for successful maintenance than other paints. For these reasons the user's judgment about the relative importance of the different stages, and particularly about the stage at which he expects to repaint, determines what kind of paint he can use to best advantage. For intelligent paint maintenance, therefore, the user should know the pigment composition of the paint and its bearing upon behavior during deterioration (17).

The opaque white pigments are of particular importance in white paints and in paints of all but the deepest colors. Most of the colored paints are tinted paints, that is, white paints in which minor proportions of colored pigments are incorporated. White lead, zinc oxide, and titanium dioxide are the principal opaque white pigments, although zinc sulfide is used in a few paints, and lead titanate and antimony oxide have been used in the past. There are also inexpensive pigments of low opacity, called extending pigments, that are present in modern mixed-pigment paints, often in large proportions. The white paints, white-base paints for the user to tint, and tinted paints now generally available may be grouped for convenience into pure white lead paint; mixed-pigment paints containing titanium dioxide, white lead, and zinc oxide; and mixed-pigment paints containing titanium dioxide and zinc oxide. Formerly there were also lead and zinc mixed-pigment paints.

In pure white lead paint the pigment is entirely basic carbonate white lead except for necessary addition of colored pigments when the paint is tinted, During the soiling and flatting stages of its life white lead paint may become dirtier than mixed-pigment paints, especially in congested urban locations or when applied in the late fall, winter, or early spring, or in deeply shaded places. Chalking of white lead paint and apparent fading, if it is tinted, usually set in fairly early, after which the dirt is thrown off promptly and evenly, leaving a clean surface, which may, however, contrast strongly with deeply shaded parts of the building that are still dirty and not yet faded. Checking begins comparatively early, usually inconspicuously, and gradually works deeply into the coating, but probably for that reason cracking and flaking rarely occur.

The outstanding merits of pure white lead paint are its exceptionally great resistance to adverse conditions of service (p. 369) and its wide tolerance for differing programs of maintenance. White lead paint can be repainted more frequently than other paints without suffering impaired service from excessive thickness of coating, or repainting may be postponed until the coating is far along in the stage of neglect without having a difficult and expensive surface to prepare for satisfactory repainting (*30*, *31*). Initial painting with white lead paint provides a very satisfactory foundation for subsequent painting with mixed-pigment paints, but it is seldom advisable to apply pure white lead paint over mixed-pigment paints because checking may then assume the conspicuous form known as alligatoring.

Mixed-pigment paints of the titanium-lead-zinc group vary widely in composition. During the soiling and flatting stages of paint life they remain much cleaner than pure white lead paint, particularly if their content of white lead is not too high. On the other hand, during the chalking stage there is a tendency for the white chalk to wash off and be deposited on nearby surfaces of darker color, such as brickwork or trim paints of dark color, unless the content of white lead is relatively high. If the paint is tinted, unduly free chalking also causes severe fading of the color. The titanium dioxide pigments, however, are made in different grades from freely chalking to chalk-resistant types. For white paints of greatest cleanliness in the soiling stage, the freely chalking type is used, but such paints should never be tinted and are usually so marked on the label. For tinted paints of moderate or low content of white lead, the chalk-resistant type or a blend of the freely chalking and chalk-resistant types of titanium dioxide is used.

In the fissure stage, titanium-lead-zinc paints high in white lead may develop checking; later on they crack and finally disintegrate by flaking. If the content of white lead is lowered, checking may be eliminated and cracking may be further delayed, though the cracking and subsequent flaking tend to be coarser and more conspicuous the longer they are postponed. Titanium-lead-zinc paints of low content of white lead and lead-free titanium-zinc paints under very favorable conditions may skip the fissure stage entirely and disintegrate by erosion. Lead-free titanium-zinc paints are not discolored by hydrogen sulfide, but that immunity may be lost if they are tinted because most colored pigments in common use contain metals that form black sulfides.

Mixed-pigment paints that contain zinc oxide are more exacting than pure white lead paint in their requirements for good maintenance over a long period of years (*29*). If the mixed-pigment paints are repainted too often the coating soon becomes unduly thick, cracks conspicuously, and disintegrates in large scales. On the other hand, if repainting is too long delayed, the advanced condition of cracking, curling, and flaking leaves a rough surface that requires scraping and sandpapering to prepare it for smooth repainting. With each successive painting the condition is likely to become worse until eventually it becomes necessary to remove all old paint completely with a painter's torch or with house-paint remover before further painting can be done satisfactorily.

Dull red, brown, and yellow pigment paints are made most economically with iron oxide pigments, either natural earths or manufactured pigments. For good retention of color and maximum durability, the content of iron oxide in the pigments may well be 50 percent or more by weight and should certainly not be less than 30 percent. Bright reds require organic lake pigments that are more expensive and less durable than iron oxide reds. Bright yellows and oranges are made with chrome yellows (lead chromate and basic lead chromate), which are fully as expensive as white pigments but are more durable. Greens are made with chrome green, which is a mixture of chrome yellow and prussian or chinese blue (iron ferrocyanide). Durable blacks are made with carbon pigments, such as lampblack; asphalt paints make cheaper but less durable blacks for exposure to the weather. The colored-pigment paints tend to be slow in drying, to make unduly soft coatings, and to be subject to resoftening under some conditions of service unless there is a substantial proportion of good varnish in the paint. When color permits, colored-pigment paints often retain color and protective power better if they contain some zinc oxide or white lead. If they do not, it is advisable to apply them over priming coats of white lead paint, housepaint primer, or aluminum paint.

Proportions of Pigments and Liquids

Durability as well as color and opacity are imparted to paint by pigments. For that reason there is an optimum proportion of pigments and nonvolatile liquids in good paint. In the oil-rich paints, in which most or all of the linseed oil is raw, refined, or boiled oil, the total pigment is usually from 27 to 30 percent by volume of the total nonvolatile ingredients. The total nonvolatile ingredients, which remain in the coating after it drys, are 87 percent or more by volume of the paint.

In typical oil-restricted paints, the total pigment commonly is 32 to 36 percent by volume of the total nonvolatile ingredients, and the total nonvolatile ingredients are not more than 73 percent by volume of the paint. Thus, it takes more oil-restricted paint than oil-rich paint to cover a given area of surface with a coating of prescribed thickness. As a rule the consistency of oil-restricted paints is such as to lead the painter to apply just about the extra amount of paint required.

The ratio of total pigment to total nonvolatile ingredients by volume is called the pigment volume. Paints in which the pigment volume is too low may become excessively dirty in the soiling stage and may crack and scale badly in the fissure stage of paint life. Paints in which the pigment volume is too high may chalk and fade badly and wear out too soon.

Many paint manufacturers customarily print the composition or formula on the labels of paints sold to the public; these formulas are nearly always expressed in percentage of ingredients by weight rather than by volume. Apart from the printing of such formulas there are no generally recognized standards of paint quality.

Priming Paints

The first coat of paint applied on new wood or over well-worn old paint, called the priming coat, should differ somewhat in composition from the paint applied subsequently for finish coats. A good priming paint should contain white lead but no zinc oxide, it should have about as much total pigment as finish-coat paint does, and the liquids should contain enough volatile thinner to wet wood quickly and enough bodied linseed oil to prevent undue penetration of liquids into the wood with consequent loss of oil by the priming paint. Such primers cannot be mixed on the job by adding liquids to finish-coat paint. For that reason most paint manufacturers sell special house-paint primers as one of the items in their lines of house paints (*23*).

Exterior Stains

Exterior wood stains are essentially paints greatly diluted with linseed oil, a volatile thinner, and sometimes creosote. Stains are cheaper than paints, but they do not protect wood against weathering and are therefore most suitable for rough surfaces, such as shingles or siding placed with the sawed side out. Good stains contain only pure, finely divided, opaque pigments and are free from transparent pigments. The pure iron oxide, chrome, and carbon pigments are very satisfactory; white pigments less so. The liquid part of the stain should be at least one-third linseed oil by volume, with two-thirds recommended by some authorities. Particular merits are claimed for stains made with pigments ground in very highly bodied linseed oil, because the pigments are more highly dispersed in such liquids and become more deeply embedded in rough wood surfaces. Houses with shingles or siding with the sawed side out, stained with colored-pigment stains, and with smoothly planed and painted wood trim often have the trim painted twice for every time that the side walls are stained.

APPLICATION OF PAINT

Wood should not be painted when it is wet. If no free water is present, however, the moisture content is of minor importance; wood painted at 16 to 20 percent moisture content holds paint slightly longer than wood painted at 10-percent moisture content. Paint dries very slowly at low temperatures. Painting should therefore not be done when there is danger of dew or frost or of a drop of more than 20° F. in temperature at night; application of paint at such times should cease several hours before sunset. In clear, warm weather coatings of paint can be applied within 24 hours of each other if necessary, but it is better practice to allow at least 2 or 3 days. On the other hand, it is generally inadvisable to allow more than 1 or 2 weeks to elapse between successive coats. When new houses reach the stage for exterior painting in late fall or early winter, it is poor practice to apply a priming coat only and then to wait for spring before finishing the job. A better way is to treat the exterior woodwork with water-repellent preservative (*21*) and then wait until spring to do all of the exterior painting at one time.

On new wood surfaces or on painted surfaces on which the old coating is not yet deeply cracked, paint may be applied with a brush or spray gun with equally serviceable results. Over old coatings that have passed well into the fissure or the flaking stage of paint deterioration, however, paint applied with a brush may make a somewhat smoother coating with better appearance and perhaps have a little better durability than sprayed paint.

Best practice in painting new wood surfaces calls for the application of three coats of paint, a priming coat, a second coat, and a finish coat. Nevertheless, much if not most painting of new houses is done with two coats only. If paint of sufficiently high quality is available and is properly applied, 2-coat work could be fully satisfactory (23), but most 2-coat work in recent years has been scanty in the quantity of paint applied and therefore less durable than a more generous paint job.

A coating of house paint 4.5 to 5.0 mils (1 mil is 0.001 inch) thick initially usually gives the best and longest service. A thinner coating wears out too soon, and a thicker one is inclined to crack too soon and later to scale badly. If a house-paint primer containing 25 percent of total pigment and 73 percent of total nonvolatile ingredients by volume is used with an oil-rich finish paint containing 26 percent of total pigment and 87 percent of total nonvolatile ingredients, a coating approximately 4.5 mils thick can be obtained as follows: Spread the primer at the rate of 1 gallon for 600 square feet of surface and then spread 2 coats of finish paint, each at the rate of 1 gallon for 850 square feet of surface. If the finish paint is oil-restricted and contains only 73 percent of total nonvolatile ingredients, each coat of it should be spread so that 1 gallon covers only 725 square feet of surface.

To obtain the same coating thickness in 2-coat work, the primer and the oil-rich finish paint must be applied about as generously as is practicable even in good weather for painting; the primer may be applied at 1 gallon to 450 square feet and the finish paint at 1 gallon to 500 square feet. If the painting must be done in cold weather or if the finish paint is oil-restricted a coating thickness of 4.5 mils is too much to expect of 2-coat work.

Most paints of high quality are sold in the prepared or ready-mixed form. Ordinarily they are ready for application as soon as the contents of the container have been thoroughly stirred. White-base paints for tinting may be colored by adding carefully the necessary quantities of colors ground in oil or the special preparations made by some manufacturers for that purpose. But as a rule, there should be no need for adding thinner, oil, or so-called "fortifying" oils. Unnecessary additions are made so often that it is wise to be on guard to avoid them. In cold weather paint thickens in viscosity and may then be thinned with no more than 1 pint of turpentine or mineral spirits to a gallon to ease the brushing, but no oil should be added because it increases the chances of the paint wrinkling as it dries.

Although pure white lead paint is available in the prepared form, it is more often purchased in a more concentrated form called soft paste white lead, which requires substantial addition of liquids to make it ready for application. Paint of highest quality can be mixed from soft paste white lead, but when mixed-pigment paints are sold in the paste form, they are usually of cheaper quality than the prepared paints of the same manufacturer. Use of paste paint has the advantage that mixtures can be made with greater pigment volume that that of the finish-coat paint. Thus, paint for the second of 3 coats on new wood, or for the first of 2 coats over old paint in reasonably good condition, can be mixed with 35 to 40 percent pigment volume, when the pigment volume in the finish coat is to be 27 to 36 percent, merely by thinning the paste with more turpentine and less oil than are used for the last coat. Such adjustment of pigment volume in successive coats favors better paint performance.

Soft paste white lead is sold by the pound rather than the gallon. The 100-pound pail contains about 3 gallons of paste, which may be mixed as follows: For the first or priming coat on new wood, add 4 gallons of raw linseed oil, 2 gallons of turpentine or mineral spirits, and 1 pint of liquid paint drier; for the second coat add 1.5 gallons of raw linseed oil, 1.5 gallons of turpentine or mineral spirits, and 1 pint of liquid paint drier; for the final coat add 2.75 gallons of raw linseed oil and 1 pint of liquid paint drier. Boiled linseed oil may be used in place of raw oil, but in that case add only ½ pint instead of 1 pint of drier. For 2-coat work on new wood mix as follows: For the first coat add 3 gallons of raw linseed oil, 3 quarts of turpentine or mineral spirits, and 1 pint of liquid paint drier; for the final coat add no more than 3 gallons of raw linseed oil and 1 pint of liquid paint drier.

To make the mixtures, first add only a small part, perhaps one-tenth, of the total oil to the soft paste white lead and stir until the paste is again smooth and uniform. Then twice pour in perhaps one-fifth of the extra oil and stir until smooth. Finally add the rest of the oil and the other liquids and stir well. If all the liquids are added to the paste at once, it is very difficult to make a smooth mixture.

When painting woods of groups 2, 3, and 4 (table 42) with mixed-pigment paints, it is particularly necessary to use a special priming paint for the first coat (4). Pure white lead paint or a good zincless titanium-lead house-paint primer may be used (23). For two-coat work, one or the other of these is about the only choice because, with a single finish coat, the primer must have about the same color. If three coats of paint can be applied, the experience of the Forest Products Laboratory indicates that the most effective special priming paint on woods of groups 3 and 4 is exterior aluminum paint for wood (9, 11, 14, 34). Aluminum priming paint for wood is made usually with 1.75 pounds of commercial paste aluminum in 1 gallon of very long-oil spar varnish made specifically for that purpose. The right kind of varnish must be used. Neither ordinary spar varnish, which is both unsuitable and too expensive, nor cheap bronzing liquid for interior use is acceptable for exterior aluminum paint. Suitable aluminum priming paints are sold for the one purpose of painting wood and are not recommended for other uses. When properly applied the aluminum primer hides the surface of the wood completely.

REPAINTING

The major problem in maintenance of paint coatings on exterior surfaces is to keep reasonably good appearance at all times without allowing the coating to become so thick that it cracks and scales badly and eventually has to be removed completely (*30, 31*). To do so over a long period of years, repainting must be spaced far enough apart for weathering to take off as much coating thickness as is restored at each repainting. Since no house paint fully meets these requirements, choice must be made between two alternatives. One, which is the older and better established practice, is to allow each paint job to go far enough into the stage of disintegration or even of neglect to entail a few years of somewhat shabby appearance before repainting is done. The other alternative is to repaint as soon as the coating passes far enough into the stage of disintegration to begin to look shabby, taking care to apply as little new paint as is necessary for a satisfactory job, and then to plan on having all old paint removed down to bare wood every 15 or 20 years. Old paint may be removed with a painter's blow torch, with the newer electric torches, or with house-paint remover, but in any case it is a costly job.

Coatings of pure white lead paint lose about ¾ mil of thickness annually on side walls fully exposed to sunshine at Madison, Wis., and their thickness can be built up to 15 mils before the normal performance is impaired. By reason of its relatively rapid wear, high critical thickness, and favorable behavior during the stages of disintegration and neglect, pure white lead paint makes the safest choice for those who prefer the first alternative of keeping the coating from ever getting thick enough to require complete removal. Suitable maintenance programs for pure white lead paint range from repainting with 1 coat every 3 years to repainting with 2 coats every 8 to 10 years, on the average.

Coatings of mixed-pigment paints that contain zinc oxide lose only about ½ mil of thickness annually at Madison, Wis., and they become too thick to perform normally long before the coating thickness reaches 10 mils. Since it is difficult to repaint satisfactorily without applying at least 2 mils of new paint, coatings of mixed-pigment paints cannot be repainted safely more often than once in 4 years and then with no more than 1 coat at a time. If 2 coats are applied at a time, repaintings should be 6 to 7 years apart, on the average. On the other hand, coatings of mixed-pigment paints can seldom be allowed to go too far into the stage of disintegration without serious danger of so much flaking or scaling that the surface becomes difficult to prepare for repainting and to repaint smoothly.

Since coatings wear away even less rapidly on the shaded parts of a house than they do on the sunny parts, in repainting correspondingly less new paint should be applied on the protected areas. Unless that is done, the paint may eventually crack, curl, and scale on the shaded parts before it begins to disintegrate on the sunny parts.

In the maintenance of paint over a long period it is advantageous to repaint always with paint of the same composition as that used for finish coat in the first painting. Although some combinations of dissimilar paints used successively are satisfactory or even beneficial, such as special house-paint primers with finish-coat paints, other combinations are harmful, such as white paint over most deeply colored paints. Often the same combination of paints in reverse order may serve well; thus, deeply colored paints may usually be applied over previous coatings of white paint. But a combination of dissimilar paints should be avoided whenever possible unless it is known to be satisfactory.

Sometimes a relatively small number of boards on a house lose their paint much sooner—within 1 or 2 years after painting—than the rest of the boards. In such cases the average life of subsequent coatings can often be increased by replacing these boards with new ones chosen for good painting characteristics (p. 358). On the other hand, if the less satisfactory boards hold their paint for 3 or more years, they may be brought nearly to equality with the others merely by removing all paint from them and priming them well with suitable special priming paint.

Abnormal Paint Wear

Under certain conditions of service, the normal course of paint deterioration (p. 355) gives way to one of several kinds of abnormal paint wear.

As the thickness of a coating is built up beyond 5 mils, any cracking during the stage of fissures and curling and flaking during the stage of disintegration become coarser in pattern, larger in size, and more conspicuous. Up to a critical thickness characteristic of the kind of paint, most of the larger cracks run roughly parallel to the direction in which the paint was last stroked with the brush during application. As a rule, that is also the direction of the grain of the wood underneath, but, when it is not, the cracking follows the direction of brushing rather than the grain of the wood. When the critical thickness of the coating is passed, however, a new and much coarser form of cracking sets in with its dominant direction at right angles to that of the normal cracking. Even paints that normally check and crumble without any cracking are subject to cross-grain cracking when the coating becomes too thick. Cross-grain cracking sets in earlier, becomes still more conspicuous, and leads to coarser scaling the thicker the coating becomes from further repaintings (30, 31).

Heterogeneous coatings formed by using dissimilar paints for successive paint jobs often behave abnormally. Difference in composition may result in variance in such properties as hardness, shrinkage with age, thermal expansion, and swelling on absorption of moisture. For example, paint coatings share with wood the property of absorbing water and swelling, and the extent to which they do so varies with the composition of the paint. Thus, a coating built up of dissimilar layers may be subject to severe internal stresses that impair its performance. Among the kinds of abnormal behavior that may result are alligatoring (outer layer of coating develops fissures and then shrinks until a lower layer is exposed), intercoat peeling or scaling, some forms of blistering, and cross-grain cracking, curling, and scaling (fig. 84).

Abnormal discolorations of paint may arise either in the soiling stage or in other stages of paint life. In damp places, mildew, which is a growth of fungi on the surface, may cause a blackening of the coating that is often mistaken for dirt. Less often there may be pink, red, purple, yellow, or brown discolorations according to the nature of the infecting organism. Plenty of white lead and zinc oxide in paint makes it resist mildew reasonably well. Resistance may also be imparted by adding preservatives, such as phenyl mercury compounds or chlorinated phenols, to the paint. The fungi of blue stain growing in damp sapwood may push their fruiting bodies through paint coatings, causing a bluish-black discoloration on the paint as well as in the wood.

Water-soluble colored ingredients in a few woods, such as redwood and redcedar, may discolor paint coatings if the wood becomes thoroughly wet (8); slight discoloration of this kind may occur temporarily from wetting with rain or with a garden hose after the coating reaches the stages of fissures and disintegration. A black discoloration with a metallic luster appears on paints that contain lead, iron, chromium, or manganese pigments when the air is contaminated with hydrogen sulfide. Such atmospheric pollution is rare, but it may occur near swamps, rivers overloaded with sewage, a few chemical processing factories and oil refineries, smoldering fires in underground coal mines, or natural sulfur springs. Sulfur dioxide, which is commonly present in industrial fumes, does not blacken paint, though too much of it shortens the life of most paints.

Paints may fail by blistering, often before the last paint job passes out of the soiling stage, for a number of different reasons, some of which are not yet well understood. Perhaps the most frequent cause of blistering is excessive moisture, but there are other common causes not connected with unusual conditions of dampness. A blast of heat suddenly directed at a coating on wood causes blistering by both softening the paint and expanding the air in the cavities of the wood underneath. Use of a painter's torch to remove paint depends on this principle. But modest warming of painted surfaces by sunshine can cause temperature blistering if there is a soft layer anywhere within the coating or if dampness in the wood blocks free movement of air through its capillaries until the pressure of the warm trapped air becomes high. New paint applied in a thick coating may dry with a hard skin on top and leave the bottom of the new layer soft enough for temperature blistering for some time, especially if the new paint is dark in color and therefore highly absorptive of the sun's heat. Old layers of paint blanketed from sunshine and oxygen by more recently applied paint may soften and reliquefy from a splitting of the fatty acids in linseed oil into azelaic and other simpler acids. If that process goes far enough, the backs of the blisters or the surfaces under them are covered with a glossy material that looks like varnish but is in fact largely azelaic acid or its metal salts. Such glossy-backed blistering is frequently seen on older houses that have been kept well painted.

Blistering of paint, often followed by peeling or scaling, is often brought about by prolonged conditions of excessive moisture (8). Moisture that gets into the side walls of houses is likely to be especially harmful. One way in which moisture enters side walls is by condensation during cold weather because there is too much humidity within the house or no vapor barrier on the warm side of the side walls (37). Another way is by excessive seepage of storm water through faulty or unguarded joints. During cold weather when the house is heated, the temperature gradient through the walls tends to move the free water through the wood toward the paint coating. Such movement of water and trapped air develops a pressure under the coating, and the coating is softened and weakened by absorption of water until it yields to the pressure in the form of blisters.

Pure white lead paint and other zincless white paints are markedly resistant to moisture blistering because they absorb less water, swell less, and thus are less seriously weakened than zinc-containing paints. Some manufacturers now sell blister-resisting house paints, sometimes called "breather type" house paints, for use where there is danger that ordinary mixed-pigment paints may become blistered. But to obtain the full benefit of the blister resistance of the special paints, it is necessary to have any previous paint of the ordinary kind removed completely.

Wood side walls have always been subject to moderate penetration by storm water, which gets in through both vertical and horizontal joints between boards and especially at joints around windows and doors and at corners. Early paint failures, whether or not they begin by blistering, must not be atrributed to faults in the construction or operation of a building unless it can be clearly established that more water enters the side walls than is customary in well-built houses.

BACK PRIMING

Concealed surfaces are occasionally painted to retard changes in the moisture content of the wood (*34*). More than one coat is rarely applied for such purposes, and using the same priming-coat paint that is applied to the exposed surfaces is often convenient. Siding, shingles, and millwork are the principal items of building lumber back primed at times in this way.

Although back priming has long been recognized as good practice, it is usually dispensed with because of the expense and inconvenience of doing part of the painting before the lumber has been erected. If lumber must be cut and fitted in the course of erection, back priming is best done immediately afterward, so that the freshly cut ends and surfaces can be coated. Prefabricated units are sometimes primed at the mill, but slow-drying paints delay production schedules and fast-drying paints are likely to prove incompatible with the paints and finishes used later at the site. Mill priming often must include priming of the surfaces to be left exposed if unbalanced protection and possible warping and twisting during shipment and storage are to be avoided, even though the kind of finish that may later be chosen for the exposed surfaces may not be known.

No coating that completely prevents changes in the moisture content of wood has been found; the best that can be done is to retard the rate of change (*10*). Good coatings furnish adequate protection against rapidly changing atmospheric conditions or against alternate rain and sunshine, but they are relatively ineffective against prolonged exposure to extreme conditions and do not prevent seasonal fluctuations in the moisture content of interior woodwork.

Table 43 gives the comparative effectiveness of various common coatings in protecting wood at 11-percent moisture content against absorption of moisture during 2 weeks' exposure to nearly saturated air. Table 43 gives also the comparative effectiveness of 1, 2, and 3 coats when new. With age the effectiveness at first increases somewhat and then decreases as the coating deteriorates, so that the coatings must be kept in good condition to maintain their effectiveness. A similar comparison of coatings for interior use only is given on page 377.

The degree of protection attainable with a priming coat alone is usually limited; at least 2 coats and often 3 are necessary to build up an adequate barrier against moisture. Mill priming, therefore, affords only a very moderate degree of protection against change in moisture content before installation, and after woodwork has been installed the benefit derived from back priming is probably limited.

Treatment of wood with water repellents or water-repellent preservatives in lieu of mill priming or back priming has been well established in the millwork industries for many years, and is finding use for siding and other items of lumber (*21*). Water repellents are usually applied by dipping. Although in moisture-excluding effectiveness they

TABLE 43.—*Moisture-excluding effectiveness of coatings, suitable for both exterior and interior use, during 2 weeks' exposure of wood (initially at 11-percent moisture content) to nearly saturated air*

[For ordinary exterior house paints in their customary uses moisture-excluding effectiveness greater than about 60 percent is not necessary. The data in this table should not be construed as an index of the serviceableness of such paints in general]

Coating No.	Description	Effectiveness [1] found for—		
		1 coat	2 coats	3 coats
		Percent	*Percent*	*Percent*
1	Aluminum powder in asphalt or pitch paint vehicle			98
2	Aluminum powder in No. 16 vehicle	39	88	95
3	Extra fine aluminum powder in No. 16 vehicle	78	92	94
4	Aluminum powder in "alkyd" type synthetic vehicle	15	81	93
5	White lead in a vehicle similar to No. 16	62	86	91
6	1 coat of No. 2 plus 2 coats of No. 17	39	86	91
7	White lead in No. 19 vehicle	24	85	91
8	Aluminum powder in No. 23 vehicle	9	61	90
9	Asphalt or pitch paint			90
10	Aluminum powder in bodied linseed oil vehicle	26	84	89
11	1 coat of No. 8 plus 2 coats of No. 17	9	62	86
12	White lead in No. 23 vehicle	7	62	83
13	Aluminum powder in linseed oil	14	57	77
14	Aluminum powder and red lead in linseed oil	7	65	75
15	Linseed oil house paint containing zinc oxide and other white pigments with or without tinting colors	30	69	73
16	Phenol aldehyde synthetic resin, 50 gallon varnish [2]	5	49	73
17	Linseed oil house paint containing no zinc oxide, such as common lead-and-oil paint	20	57	70
18	Red lead in linseed oil	15	56	67
19	Ester gum resin, 33 gallon spar varnish [2]	6	37	65
20	Graphite in linseed oil	4	58	64
21	Red linseed oil barn paint, pigment 98 percent pure iron oxide	25	53	56
22	Red linseed oil barn paint, pigment venetian red containing 40 percent of iron oxide	1	25	45
23	Ester gum resin, 75 gallon long-oil spar varnish	3	14	35
24	Linseed oil containing paint drier	3	5	21

[1] Perfect protection would be represented by 100 percent effectiveness; complete lack of protection, as with uncoated wood, by 0.

[2] A "50 gallon" varnish is made in the proportion of 50 gallons of drying oil to 100 pounds of resin; a "33 gallon" varnish is made with 33 gallons of drying oil to 100 pounds of resin.

stand no higher than three coats of linseed oil, coating No. 24 in table 43, in practice they often prove more effective than coatings of good paints because the water repellents have exceptional ability to keep water from penetrating through joints and spreading over wood surfaces.

NATURAL FINISH FOR EXTERIOR WOODWORK

Natural finishes that are transparent enough to allow the grain of the wood to show have been used sparingly in the past for exterior woodwork exposed to the weather because such finishes are much less durable and require more frequent renewal than house paints (26). Where surfaces are fully exposed to sun and rain, natural finishes may need renewal at least once a year. Nevertheless, there is an increasing demand for natural finishes on sidewalls of houses of some modern styles.

Available natural finishes may be classified into two types: Surface-coating finishes and penetrating finishes. The surface-coating finishes include varnishes and synthetics. The penetrating finishes include oils, sealers, and water repellents.

The surface-coating finishes form a continuous coating that is resinous in nature and usually glossy, possibly 1 mil or more thick, over the wood.

After several renewals such finishes are inclined to crack badly or to become opaque in spots, and then to curl and scale. It is then necessary to strip them from the wood with varnish remover before a satisfactory finish can be restored. Such finishes are often preferred for relatively small areas, such as front doors, but are chosen less often for large areas, such as wood siding.

The penetrating finishes soak into the wood and do not leave a film of measurable thickness on the surface. Neither do they result in a highly glossy finish. One generous coat may suffice, though commonly 2 coats are put on at the outset and renewed at intervals with 1 coat at a time. For best results care should be taken during application to wipe off any excess sealer that is not absorbed by the surface before it dries. If that is not done, the finish is no longer a sealer finish but becomes instead a varnish finish, that is, a coating.

A finish with even less gloss than the sealer finish may be obtained by applying a drying oil, such as boiled linseed oil or raw linseed oil to which a little liquid paint drier has been added. Oil finish should be applied and maintained in the same way as a wood-sealer finish, with care to see that unabsorbed oil is always wiped off before it dries.

Water-repellent finishes, which usually contain wax and other nonvolatile ingredients, give the wood an appearance similar to that given by the drying-oil finishes. They penetrate wood so well that it is seldom necessary to wipe off excess material.

Some of the penetrating finishes are made with a small quantity of pigment usually red or brown, which adds somewhat to the durability, evens up the variations in natural color of the wood, and obscures slight grayness from weathering of the wood if renewal of finish is delayed a little too long. In some commercial products the pigment is gray to give the appearance of weathered wood.

All of the natural finishes are subject to blackening from mildew in places where there may be lingering dampness. For that reason the natural finishes usually should contain fungicides, such as chlorinated phenols or phenyl mercury compounds, If they do not, or if additional protection seems advisable, the wood may be treated with a water-repellent preservative before the natural finish is applied.

UNCOATED EXTERIOR WOODWORK

Architectural design sometimes calls for the appearance of weather-beaten wood (*27*). The best way to attain such appearance is to let wood weather naturally, especially since uncoated exterior woodwork is not necessarily lacking in durability. For weathered construction it is advisable to choose woods that weather with a minimum of cupping, twisting, and conspicuous checking, such as those of group 1 in table 42. Edge-grained boards weather better than flat-grained boards, and thick boards better than thin ones. It is also best to exclude boards containing pith and to place flat-grained boards with the bark side (the side nearer the bark of the log) out. Weathered construction should be fastened especially firmly with noncorrosive metals, and joints should be designed to prevent retention of moisture as far as that is possible.

INTERIOR FINISHING

Interior finishing differs from exterior chiefly in that interior woodwork usually requires much less adequate protection against moisture and that more exacting standards of appearance and a greater variety of effects are expected. Good interior finishes used indoors should last much longer than paint coatings on exterior surfaces, but interior finishes are not made to withstand the rigors of outdoor exposure.

Opaque Finishes

Interior surfaces may, if desired, be painted with the materials and by following the procedures recommended for exterior surfaces. As a rule, however, smoother surfaces, better color, and a more lasting sheen are demanded for interior woodwork, and therefore enamels rather than paints are used. These finishes differ from paints in that linseed oil is replaced partly or entirely by bodied oils or by varnishes in order to make a coating that does not show brush marks and presents a harder surface with a desired degree of gloss. They may also be made of nitro-cellulose lacquers or synthetic resins and drying oils that dry much more rapidly than oleo-resinous enamels. Unless made with expensive pigments of extraordinary opacity, such as titanium dioxide or zinc sulfide, enamels are less opaque than paints of the same color because they cannot be made with so large a proportion of pigment.

Before enameling, the wood surface should be made extremely smooth. Imperfections, such as planer marks, hammer marks, and raised grain, are accentuated by enamel finish. Raised grain is especially troublesome on flat-grained surfaces of the heavier softwoods because the hard bands of summerwood are sometimes crushed into the soft springwood in planing and later are pushed up again when the wood changes in moisture content (*35, 36*). It is helpful to sponge softwoods with water, allow them to dry thoroughly, and then sandpaper them lightly with sharp sandpaper before enameling. In new buildings woodwork should be allowed adequate time to come to its equilibrium moisture content before finishing.

Hardwoods having large pores must be filled with wood filler before the priming coat. For all woods the priming coat may be white-lead paint mixed according to directions for exterior priming-coat paint, or special priming paints may be used. Knots in the white pines, ponderosa pine, or southern yellow pine should be shellacked or sealed with a special knot sealer after the priming coat is dry. A coat of knot sealer is sometimes necessary also over the white pines and ponderosa pine to prevent discoloration of light-colored enamels by colored matter apparently present in the resin of the heartwood of these species. One or two coats of enamel undercoat are next applied; this should completely hide the wood and should also present a surface that can easily be sandpapered smooth. For best results the surface should be sandpapered before applying the finishing enamel, but this operation is sometimes omitted. After the finishing enamel has been applied, it may be left with its natural gloss or rubbed to a dull finish.

Transparent Finishes

There are many good ways of applying transparent finishes to either hardwoods or softwoods. Most finishing consists in some combination of the following fundamental operations: Staining, filling, sealing, surface coating, rubbing, and polishing. Before finishing, planer marks and other blemishes of the wood surface that would be accentuated by the finish must be removed.

Both softwoods and hardwoods are often finished without staining, especially, if the wood is one with a pleasing and characteristic color. When used, however, stain often provides much more than color alone because it is absorbed unequally by different parts of the wood and therefore accentuates the natural variations in grain. With hardwoods such emphasis of the grain is usually desirable; the best stains for the purpose are dyes dissolved either in water or in oil. The water stains give the most pleasing results but raise the grain of the wood and require an extra sanding operation after the stain is dry. "Non-grain-raising" stains are now available that often approach the water stains in clearness and uniformity of color. With softwoods, stains color the springwood more strongly than the summerwood, reversing the natural gradation in color in a manner that is often garish. Pigment-oil stains, which are essentially thin paints, are less subject to this objection than are other stains and are therefore more suitable for softwoods. Alternatively the softwood may first be coated with clear sealer after which the pigment-oil stain may be applied to give more nearly uniform coloring.

Hardwoods having large pores must be filled before varnish or lacquer is applied if a smooth coating is desired. The filler may be transparent and without effect on the color of the finish, or it may be colored to contrast with the surrounding wood. Usually colored filler is darker than the rest of the wood.

Sealer is used to prevent absorption of subsequent surface coatings and to prevent the bleeding of some stains and fillers into surface coatings, especially lacquer coatings. Shellac is the oldest type of sealer. Varnish can be used as a sealer but is not so effective as shellac or other suitable spirit varnish.

Transparent surface coatings may be of wax, shellac, varnish, or nitrocellulose lacquer. Wax provides a characteristic sheen without forming a coating of sensible thickness and without greatly enhancing the natural luster of the wood. Coatings of a more resinous nature, especially shellac and varnish, accentuate the natural luster of some hardwoods and seem to permit the observer to look down into the wood to a certain extent. Shellac applied by the laborious process of French polishing probably achieves this impression of depth most fully, but the coating is easily marred by water and is expensive. Rubbing varnishes made with resins of high refractive index for light are nearly as effective as shellac. Lacquers have the advantage of drying rapidly and forming a hard surface but require more applications than varnish to build a lustrous coating.

Varnish and lacquer usually dry with a highly glossy surface. To reduce the gloss the surfaces may be rubbed with pumice stone and water or polishing oil. Waterproof sandpaper and water may be used instead of pumice stone. The final sheen varies with the fineness of the powdered pumice stone, coarse powders making a dull surface and fine powders a bright sheen. For very smooth surfaces with high polish the final rubbing is done with rotten-stone and oil. Varnish and lacquer can be made to dry dull in the first place, but the result is not quite the same as produced by rubbing.

Finishes for Floors

Finishes for wood floors are subject to severe mechanical wear and consequently deteriorate more rapidly than finishes on most other classes of woodwork. Floor finishes must, therefore, be renewed frequently. Discussions of floor finishes usually place undue emphasis upon the kind or quality of material used and pay far too little attention to the maintenance. Any of the common types of floor finishes will keep wood floors in excellent condition if properly maintained, but no finish will do so if it is neglected. The essential problem is to adapt a systematic program of maintenance to the requirements of the finishing material chosen and of the degree of wear to which the floor is subjected.

The opaque and transparent interior finishes already described are applicable to floor finishing, provided that the floor paints, floor varnishes, or lacquers are made with the careful balance between toughness and hardness necessary to make them resistant to mechanical wear. The essential point in maintaining any floor finish that forms a surface coating is to renew it before the old coating has actually worn down to the wood in conspicuous spots. Good floor paints or varnishes may require less frequent renewal than wood-sealer or sealer and wax finishes under similar conditions of wear, but each renewal with paint or varnish is more expensive in material and labor and keeps the floor from being used for a longer time than finishes that do not include substantial surface coating.

Floor-sealer finishes, which are absorbed by the wood and do not form substantial coatings, are widely used both for residences and for public and commercial buildings where traffic is heavy. They may furnish slightly less protection against moisture and dirt than coatings, but they can be applied with less labor and inconvenience. In particular, floor-sealer finishes can be renewed more rapidly, so that floors are kept out of service for a shorter time than coatings require. Like other wood sealers, floor sealers look and serve best when they are rubbed into the wood surface and any unabsorbed material is wiped off. Buffing the floor with fine steel wool shortly after the sealer has been applied gives excellent results, and machines are available for this purpose.

Floor wax is often applied over floor sealers, less often over varnish or paint finishes, for additional luster and for periodically cleaning and repolishing floors. Paste or liquid floor waxes require vigorous rubbing, preferably with a power-driven polishing machine. Water-emulsion waxes dry with reasonably good sheen without rubbing, though the finish can be further improved by rubbing.

EFFECTIVENESS OF MOISTURE-EXCLUDING COATINGS SUITABLE FOR INTERIOR USE

The comparative effectiveness of various common coatings for interior use in protecting wood at 11-percent moisture content against absorption of moisture during 2 weeks' exposure to nearly saturated air is given in the following tabulation:

	Effectiveness[1] (percent)
3 coats of aluminum powder in gloss oil (quick drying) or in varnish	92
3 coats of aluminum powder in shellac	92
Heavy coating of paraffin	91
3 coats of rubbing varnish	89
3 coats of shellac	87
3 coats of enamel (cellulose-lacquer vehicle)	76
3 coats of cellulose lacquer	73
3 coats of gloss oil bronzing liquid	12
3 coats of furniture wax	8

[1] Perfect protection would be represented by 100-percent effectiveness; complete lack of protection, as with uncoated wood, by zero.

LITERATURE CITED

(1) BROWNE, F. L.
  1925. ROLE OF PAINT AND VARNISH IN WOOD CONSERVATION. Drugs, Oils and Paints 41(3): 17 pp., illus.
(2) ———
  1926. SPREADING RATE OF OUTSIDE WHITE HOUSE PAINT ON DIFFERENT WOODS. Drugs, Oils and Paints 42(7): 22 pp., illus.
(3) ———
  1930. PROPERTIES OF WOOD THAT DETERMINE PAINT SERVICE OF EXTERIOR COATINGS. Paint, Oil and Chem. Rev. 89(12): 24 pp., illus.
(4) ———
  1930. EFFECT OF PRIMING-COAT REDUCTION AND SPECIAL PRIMERS UPON PAINT SERVICE ON DIFFERENT WOODS. Indus. Engin. Chem. 22(8): 29 pp., illus.

(5) BROWNE, F. L.
   1930. DRYING OF EXTERIOR PAINTS UNDER VARIOUS WEATHER CONDITIONS
         AND OVER DIFFERENT WOODS. Indus. Engin. Chem. 22(4): 400–
         401.
(6) ———
   1930. PROCEDURE USED BY THE FOREST PRODUCTS LABORATORY FOR
         EVALUATING PAINT SERVICE ON WOOD. Amer. Soc. Testing Mater.
         Proc. 30(II): 852–870.
(7) ———
   1931. ADHESION IN THE PAINTING AND IN THE GLUING OF WOOD. Indus.
         Engin. Chem. 23(3): 290–294.
(8) ———
   1933. SOME CAUSES OF BLISTERING AND PEELING OF PAINT ON HOUSE
         SIDING. U. S. Forest Prod. Lab. Rpt. R6, 9 pp., illus. (Rev.)
         [Processed.]
(9) ———
   1933. DURABILITY OF PAINT ON LONGLEAF AND SHORTLEAF PINE. South.
         Lumberman 146(1844): 20–22.
(10) ———
   1933 and 1936. EFFECTIVENESS OF PAINT PRIMERS AND PAINTS IN RE-
         TARDING ABSORPTION OF MOISTURE BY WOOD. Indus.
         and Engin. Chem. 25(8): 835–842, illus; 28(7): 798–809,
         illus.
(11) ———
   1934. ALUMINUM PRIMING PAINT, EFFECT ON THE DURABILITY OF HOUSE
         PAINTS ON WOOD. Indus. and Engin. Chem. 26(4): 369–376.
(12) ———
   1934. DURABILITY OF PAINT ON WOOD TREATED WITH ZINC CHLORIDE.
         Amer. Wood-Preservers' Assoc. Proc. 30: 410–430.
(13) ———
   1935. EFFECT OF CHANGE FROM LINOXYN GEL TO XEROGEL ON THE BE-
         HAVIOR OF PAINT. Colloid Symposium Monog. 11: 211–222.
(14) ———
   1935. SPECIAL PRIMING PAINTS FOR WOOD. Indus. and Engin. Chem.
         27(3): 292–298.
(15) ———
   1935. PAINTING CHARACTERISTICS OF HARDWOODS. Indus. and Engin.
         Chem. 27(1): 42–47.
(16) ———
   1936. EFFECT OF EXTRACTIVE SUBSTANCES IN CERTAIN WOODS ON THE
         DURABILITY OF PAINT COATINGS. Indus. and Engin. Chem.
         28(4); 7 pp., illus.
(17) ———
   1937. PROPOSED SYSTEM OF CLASSIFICATION OF HOUSE PAINTS. Indus.
         and Engin. Chem. 29(9): 1018–1026.
(18) ———
   1938. WHEN AND HOW TO PAINT HOMES AND FARM BUILDINGS. U. S.
         Forest Prod. Lab. Rpt. R962, 14 pp. (Rev.) [Processed.]
(19) ———
   1946. POINTERS FROM YOUR PAINT FARM. Successful Farming 44(9):
         28–29, 52, 54–55.
(20) ———
   1947. WOOD PROPERTIES AND PAINT DURABILITY. U. S. Dept. Agr. Misc.
         Pub. 629, 10 pp., illus.
(21) ———
   1949. WATER-REPELLENT PRESERVATIVES FOR WOOD. Architect. Rec.
         105(3): 131–132, 174, illus.
(22) ———
   1949. PAINTING THE FARM AND CITY HOME. U. S. Dept. Agr. Yearbook
         1949: 625–630.
(23) ———
   1950. TWO-COAT SYSTEM OF HOUSE PAINTING. U. S. Forest Prod. Lab.
         Rpt. R1259, 21 pp., illus. (Rev.) [Processed.]
(24) ———
   1951. WOOD PROPERTIES THAT AFFECT PAINT PERFORMANCE. U. S. Forest
         Prod. Lab. Rpt. R1053, 23 pp., illus. (Rev.) [Processed.]

(25) BROWNE, F. L.
1952. TESTING HOUSE PAINTS FOR DURABILITY. U. S. Forest Prod. Lab. Rpt. R1011, 15 pp., illus. (Reviewed and Reaffirmed.) [Processed.]

(26) ——
1952. NATURAL WOOD FINISHES FOR EXTERIORS OF HOUSES. Architect. Rec. 111(2): (Also issued as U. S. Forest Prod. Lab. Rpt. 1908, 9 pp.)

(27) ——
1952. WOOD SIDING LEFT TO WEATHER NATURALLY. Architect. Rec. 112(5): 197–199, illus.

(28) —— and HRUBESKY, C. E.
1931. EFFECT OF RESIN IN LONGLEAF PINE ON THE DURABILITY OF HOUSE PAINTS. Indus. and Engin. Chem. 23(8): 874–877.

(29) —— and LAUGHNAN, D. F.
1952. MODIFICATION OF WOOD AND PLYWOOD TO IMPROVE PAINTABILITY. Jour. Forest Prod. Res. Soc. 2(3): 3–24, illus.

(30) —— and LAUGHNAN, D. F.
1952. HOW OFTEN SHOULD A HOUSE BE PAINTED—AN EXPERIMENTAL STUDY OF PROGRAMS OF PAINT MAINTENANCE. Jour. Forest Prod. Res. Soc. 2 (5): 173–193.

(31) —— and LAUGHNAN, D. F.
1953. EFFECT OF COATING THICKNESS ON THE PERFORMANCE OF HOUSE PAINTS UNDER DIFFERENT PROGRAMS OF MAINTENANCE. Off. Digest Fed. Paint and Varnish Prod. Clubs 338: 137–159.

(32) —— and RIETZ, R. C.
1949. EXUDATION OF PITCH AND OILS IN WOOD. U. S. Forest Prod. Lab. Rept. R1735, 11 pp., illus. [Processed.]

(33) DUNLAP, M. E.
1926. PAINTING OF TREATED WOOD. Indus. and Engin. Chem. 18 (10): 1091.

(34) EDWARDS, J. D., and WRAY, R. I.
1927. PROTECTING WOOD WITH ALUMINUM PAINT. Indus. and Engin. Chem. 19 (9): 975–977.

(35) KOEHLER, A.
1930. CAUSES AND PREVENTION OF RAISED GRAIN. Timberman 31 (4): 162–166, illus.

(36) ——
1932. SOME OBSERVATIONS ON RAISED GRAIN. Amer. Soc. Mech. Engin. Trans. Wood Indus. 54 (4): 27–30.

(37) TEESDALE, L. V.
1947. REMEDIAL MEASURES FOR BUILDING CONDENSATION DIFFICULTIES. U. S. Forest Prod. Lab. Rpt. R1710, 14 pp., illus. [Processed.]

PROTECTION FROM WOOD-DESTROYING ORGANISMS

Under proper conditions, wood has proved itself good for centuries of service. Generally speaking, molds, stains, or decay appearing in lumber may be traced to lack of proper precautions in yarding logs and in piling, storing, and handling sawed material. The problems of fungus attack in milling and merchandising stages have been extensively studied by Government and other investigators, and preventive measures are well known and widely practiced in the industry. Decay should not be tolerated in lumber except in the lower Common grades, and stain should be limited to lumber used where appearance is secondary. It remains for the user to build with his material in such a way as to insure the structure against the inception and spread of decay.

Molds, stains, and decay in wood are caused by fungi, which are microscopic plants that must have organic material on which to live, and for some of them wood offers the required food supply. Their growth, however, is dependent upon suitably mild temperature (50° to 90° F.) and dampness. Most decay occurs in wood having a moisture content above the fiber saturation point. Wood that is continuously water soaked or continuously dry will not decay. Usually, wood maintained at 20-percent moisture content or less, typical of air-dried wood, is safe from fungus damage.

MOLDS AND STAINS

Molds and stains are confined largely to sapwood and are of various colors. Little direct staining of the wood is caused by molds, since the discoloration caused by them is largely superficial and is due for the most part to cottony or powdery surface growths, which vary from white or light colors to black. Such blemishes often are easily brushed or surfaced off.

Stains penetrate into the sapwood (not heartwood) and cannot be removed by surfacing. The discoloration of the wood occurs as specks, spots, streaks, or patches of varying intensities of color. The so-called "blue" stains, which vary from bluish to bluish black and brown, are the most common, although various shades of yellow, orange, purple, and red are sometimes encountered. The exact color of the stain depends on the infecting organisms and the species and moisture condition of the wood. The brown stain mentioned should not be confused with chemical brown stain (p. 323).

Under favorable moisture and temperature conditions staining and molding fungi may become established and develop rapidly in the sapwood of wood products shortly after they are cut. In addition, lumber and such products as veneer, furniture stock, and millwork may become infected at any stage of manufacture or subsequent use if they become sufficiently moist. Freshly cut or unseasoned stock that is piled during warm, humid weather may be noticeably discolored within 5 or 6 days. Recommended control measures are given by Scheffer and Lindgren (22)[25] and on pages 325–326.

[25] Italic numbers in parentheses refer to Literature Cited, p. 397.

Stains and molds should not be considered stages of decay, since the causal fungi do not attack the wood substance appreciably. Ordinarily, they affect the strength of the wood only slightly; their greatest effect is usually confined to those strength properties which determine shock resistance, or toughness (p. 99).

Stained and molded stock is practically unimpaired for many uses in which appearance is not a limiting factor. It is not entirely satisfactory for siding, trim, and other exterior millwork, however, because the infected wood has greater water absorptiveness. Also, incipient decay often is present in the discolored areas. Both of these factors increase the hazard of decay in wood that is rained on unless the wood has been treated with a suitable preservative.

DECAY

Decay-producing or wood-destroying fungi may, under conditions that favor their growth, attack either heartwood or sapwood, causing a condition that is variously designated as decay, rot, or dote (3). The fresh surface growths of decay fungi are usually fluffy or cottony —seldom powdery like the surface growths of molds. They may appear as fan-shaped patches, strands, or root-like structures, usually white or brown. Sometimes fruiting bodies are produced that take the form of toadstools, brackets, or crusts. The microscopic strands or hyphae that permeate the wood use parts of the wood itself as food. Some wood-destroying fungi live largely on the cellulose; others use the lignin as well as the cellulose.

The early or incipient stages of decay are often accompanied by a discoloration of the wood, which is more evident on freshly exposed surfaces of unseasoned wood than on dry wood. Many fungi produce incipient decay that differs only slightly from the normal color of the wood or gives a somewhat water-soaked appearance to the wood.

Typical or late stages of decay are easily recognized, because the wood has then undergone definite changes in color and properties, depending upon the organism and the substances it removes. The rotted wood may be white or brown in color—white if the decay fungus consumes lignin as well as cellulose, and brown if the cellulose primarily is removed. There are also intermediate types between the white and brown rots.

Brown, crumbly rot, in the dry condition, is sometimes called "dry rot," but the term is incorrect because wood must be damp in order to decay, although it may become dry subsequently. A few fungi, however, have water-conducting strands and are therefore capable of carrying water, usually from the soil into buildings or lumber piles, where they moisten and rot wood that would otherwise be dry. They are sometimes referred to technically as the "dry rot fungi," but they nevertheless must have access to water in order to function.

Decay Resistance of Wood

For a discussion of the natural resistance of wood to wood-destroying fungi and a grouping of species according to decay resistance, see page 45.

Certain wood-destroying fungi attack the heartwood, and rarely the sapwood, of living trees, whereas others confine their activities to logs or manufactured products, such as sawed lumber, structural timbers, poles, and ties. A few of the first group may cease their activities after the trees have been cut, as in pecky cypress or pecky incense-cedar.

Others may continue their destruction after the trees have been cut and worked into products, provided conditions remain favorable for their growth. In living trees the sapwood is less subject to attack than the heartwood, presumably because of the nature of the attacking fungi and the very high moisture content of living sapwood. The sapwood of logs or wood products, on the other hand, usually decays more readily than the heartwood, because the moisture content of the sapwood becomes more favorable for the growth of decay fungi and the heartwood often contains natural chemicals that retard the growth of fungi.

Effect of Decay on Strength of Wood

Incipient decay induced by some fungi is reflected immediately in pronounced weakening of the wood, whereas other fungi reduce strength much less (p. 99). For example, the decay produced by *Fomes pini* causes little or no reduction in strength in its incipient stage. This is the most common decay in living softwood trees. On the other hand, *Polyporus schweinitzii*, another common rot in standing softwoods, greatly reduces the strength of wood at a very early stage. In the later stages of decay any wood-destroying fungus will seriously reduce the strength of wood (*23*), as well as its fire resistance.

CONTROL OF MOLDS, STAINS, AND DECAY

Logs, Poles, Piling, or Ties

The species, section of the country, and time of the year will determine what precautions must be taken to avoid serious damage from fungi in poles, piling, ties, and similar thick products during seasoning or storage. In dry sections, rapid surface seasoning by first peeling off the bark and then decking the peeled product on high skids or piling it on high, well-drained ground in the open will retard molds, stains, and decay; but checking must be guarded against with other measures. In humid regions, such as the Gulf States, air-drying of these products often is not sufficiently rapid to avoid losses from fungi. Therefore, preseasoning treatments with antiseptic solutions sometimes are used, but main reliance is placed on early preservative treatment or use of the product (*22*).

For logs, rapid conversion into lumber or storage in water is the surest way to avoid fungus damage. Antiseptic sprays promptly applied to ends and places from which the bark is removed will protect most timber species during storage for 2 to 3 months, provided wood-infesting insects are not prevalent. For longer storage, the addition of an end coating is needed to prevent seasoning checks, through which infection can enter the log (*33*).

Lumber

Growth of fungi can be, prevented in lumber and other wood products that can be quickly dried to a moisture content of 20 percent or less and kept dry. Kiln-drying is the most satisfactory and effective method of rapidly reducing moisture content. Standard air-drying practices are usually sufficient; end racking is used for some species.

Dip or spray treatment of freshly cut lumber with suitable antiseptic solutions, such as organic mercury compounds, chlorinated phenols, and borax, alone or in mixture, will prevent fungus infection during air-drying. Successful control by this method depends not only upon immediate and adequate treatment but also upon the proper subsequent handling of the lumber (*33*).

Air-drying yards and sheds should be kept as sanitary as possible. Recommended practice includes locating yards and sheds on well-drained ground; removing debris, which serves as a source of infection, and weeds, which reduce air circulation; and employing proper piling methods that permit rapid drying of the lumber. Proper piling requires high, sloping foundations, the avoidance of close or solid piling of unseasoned stock, watertight roofs raised several inches above the top courses, and sound, dry stickers of heartwood or of preservative-treated wood (*14, 20*).

The user's best assurance of receiving lumber free from decay or stain is to buy stock marked by a lumber association in a grade that requires freedom from such quality-reducing features. If lumber is to be used under conditions conducive to decay, it should be the heartwood of a naturally durable species or should be adequately treated with a good wood preservative (p. 399).

Buildings

The lasting qualities of properly constructed wood buildings are apparent in all parts of the country. Serious decay problems are almost always a sign of faulty design or construction or of lack of reasonable care in the handling of the wood. The conditions that favor decay development are described on page 381.

Construction principles that assure long service and avoid decay hazards in buildings include: (1) Build with dry lumber; (2) use designs that will keep the wood dry and accelerate rain runoff; or (3) for parts exposed to decay hazards, use preservative-treated wood or heartwood of a decay-resistant species (*30*).

A building site that is dry or for which drainage is provided will reduce decay hazards. Stumps, wood debris, stakes, or wood concrete forms frequently lead to decay if left under or near a building and should be removed.

Unseasoned or infected wood should not be enclosed until thoroughly dried. Unseasoned wood may be infected because of improper handling at the sawmill, at retail yard, or after delivery on the job.

Untreated wood parts of substructures often are infected with decay if in contact with the soil. Sills in contact with dirt fill under porches, joists touching the ground under the house, and particularly siding or skirting in contact with earth must be guarded against. If contact with soil is necessary, as with foundation posts, heartwood of a decay-resistant species should be used or the wood should be impregnated with a reliable preservative (p. 399).

Sill plates and other wood in contact with concrete near the ground should be protected by a moistureproof membrane, such as heavy asphalt paper. In some cases, preservative treatment of the wood in actual contact with the concrete is advisable. Girder and joist openings in masonry walls should be big enough to assure an air space around the ends of these wood members; if the members are below the outside soil level, moistureproofing of the outer face of the wall is essential.

In basementless buildings, wetting of the wood by condensation may result in serious decay damage. A crawl space with at least an 18-inch clearance should be left under wood joists and girders. Condensation can be prevented either by providing openings on opposite sides of the foundation walls for cross ventilation or by laying roll roofing on the soil (30).

Porches, exterior steps, and platforms present a decay hazard that cannot be fully avoided by construction practices. Therefore, the use of preservative-treated wood usually is advisable.

The prevention of decay in walls and roofs rests largely in designs that prevent the entrance and retention of rain water. A fairly wide roof overhang, with gutters and downspouts that are never permitted to clog, is very desirable. Sheathing papers in the walls should be of a "breathing" or vapor-permeable type. Vapor barriers, if installed, should be near the warm face of insulated walls and ceilings. Roofs must be kept tight, and cross ventilation in attics is desirable (28). The use of sound, dry lumber is important in the exterior as well as substructure parts of buildings (31).

Where service conditions in a building are such that the wood cannot be kept dry, as in textile mills, pulp and paper mills, and cold-storage plants, properly treated lumber should be used unless heartwood of naturally decay-resistant species is available (13).

In making repairs necessitated by decay, every effort should be made to correct the conditions leading to the damage. If the conditions cannot be remedied, all infected parts should be replaced with treated or naturally decay-resistant wood, and it is advisable to remove surrounding material at least 2 feet beyond any evidence of decay. If the sources of moisture that caused the decay are entirely eliminated, it is necessary to replace only the weakened wood with dry lumber.

Other Wood Products and Structures

In general, the principles underlying the prevention of mold, stain or decay damage to veneer, plywood, containers, boats, and other wood products and structures are similar to those described for buildings—dry the wood rapidly and keep it dry, or treat the wood with accepted protective and preservative solutions. Interior grades of plywood should not be used where the plywood will be exposed to moisture, since the glues, as well as the wood, may be disintegrated by fungi and bacteria (6, 9). With either plywood or fiberboard of the exterior type, joint construction should be carefully designed to prevent the entrance of rain water.

Wood boats present both a serious decay hazard and certain problems that are not encountered in the ordinary use of wood. The parts especially subject to decay are the stem, knighthead, transom, and frameheads that are reached by rainwater from above or condensation moisture from below. Faying surfaces are more liable to decay than exposed surfaces, and in salt water service, hull members just below the weather deck are more vulnerable than those below the water line. In fresh water, members below the water line are most subject to decay. Recommendations for avoiding decay include: (1) Use only heartwood of durable species, preferably below 20 percent in moisture content and not already infected; (2) provide and maintain ventilation in the hull and avoid paint coatings that lock in moisture; (3) keep water out as much as is practicable, especially fresh water; and (4) where it is necessary to use sapwood or nondurable heartwood, impregnate the wood with a preservative (11).

INSECTS

The more common types of degrade caused by wood-attacking insects are shown in table 44 (26). Methods of controlling and preventing insect attack of wood are described in the following paragraphs.

Beetles

Bark beetles (2, 24) may damage log structures and other rustic construction on which the bark is left. They are reddish brown to black and vary in length from about ⅟₁₆ to ¼ inch. They bore through the outer bark to the soft inner part, where they make tunnels in which they lay their eggs. In making the tunnels, they push out fine brownish-white sawdustlike particles. If many beetles are present, their extensive tunneling will loosen the bark and permit it to fall off in large patches, making the structure unsightly.

To avoid bark-beetle damage, the logs should be cut during the dormant season (October or November, for instance) and piled at once off the ground in the open or under cover to promote the rapid drying of the inner bark before the beetles begin to fly in the spring. In almost every case, this will prevent damage by insects that prefer freshly cut wood. When the logs cannot be cut or handled as recommended, it is advisable to immerse them in or thoroughly spray them with a solution containing 0.075 percent (0.2 percent under severe conditions) of gamma benzene hexachloride (BHC) in No. 2 fuel oil (15, 17).

Ambrosia beetles, roundheaded and flatheaded borers, and some powderpost beetles (2, 7, 24, 26) that get into freshly cut timber can cause considerable damage to wood in rustic structures and some manufactured products. Utilization of proper cutting practices and spraying the material with a toxic chemical, as mentioned for bark beetles, will control these insects. Damage by ambrosia beetles can be prevented in freshly sawed lumber by dipping the product in a

TABLE 44.—*Classification of the more common types of degrade caused by wood-attacking insects*

Type of degrade	Description	How and where made	Condition of degraded timber
Pinholes	Holes with dark streak in surrounding wood, $\frac{1}{100}$ to $\frac{1}{4}$ inch in diameter, usually circular and open (not grouped in given space): Hardwoods:		
	Stained area 1 inch or more long.	By ambrosia beetles in living trees.	Wormholes, no living worms.
	Stained area less than 1 inch long.	By ambrosia beetles in recently felled trees and green logs.	Do.
	Softwoods: Stained area less than 1 inch long.	By ambrosia beetles in sapwood of peeled trees, green logs, and green lumber.	**Do.**
	Holes usually without streak in surrounding wood of both softwoods and hardwoods (usually grouped in given space):		
	Holes darkly stained, less than $\frac{1}{8}$ inch in diameter.	By ambrosia beetles in felled trees, green logs, and green lumber.	Do.
	Holes unstained, open, and variable in diameter:		
	Lined with a substance the color of wood; from $\frac{1}{100}$ to $\frac{1}{4}$ inch in diameter (hardwoods rarely with streaks in surrounding wood).	By timberworms in living and felled trees and green logs.	Do.
	Unlined, less than $\frac{1}{8}$ inch in diameter (not grouped in given space).	By ambrosia beetles in green logs and green lumber.	Do.
Grub holes	Holes $\frac{3}{8}$ to 1 inch in diameter, variable in shape, open or with boring dust:		
	Hole stained, usually open	By wood-boring grubs in living trees.	Do.
	Holes unstained, usually with boring dust present:		
	Borings fine, granular, or fibrous.	do	Do.
	Borings variable in character, sometimes absent.	By adults and larvae of beetles and other wood-boring insects in recently felled softwoods and hardwoods.	Do.
Powderpost [1]	Holes mostly $\frac{1}{16}$ to $\frac{1}{4}$ inch in diameter, circular to broadly oval, filled with granular or powdery boring dust, and unstained:		
	Hardwoods	By roundheaded borers and powderpost beetles in green or seasoned wood.	Wormholes, powderposted.
	Softwoods	By flatheaded borers in living, recently felled, and dead trees.	Do.
		By roundheaded borers, weevils, and Anobium powderpost beetles in seasoned wood.	Do.

See footnotes at end of table.

TABLE 44.—*Classification of the more common types of degrade caused by wood-attacking insects*—Continued

Type of degrade	Description	How and where made	Condition of degraded timber
Pitch pocket..	------------------------------------	By various insects in living trees.	Wormholes, no living worms.
Black check..	------------------------------------	By the grubs of various insects in living trees.	Do.
Bluing.........	Stained area over 1 inch long...........	By fungus following insect wounds in living trees and recently felled sawlogs.	Do.
Pith fleck.....	------------------------------------	By the maggots of flies or adult weevils in living trees.	Do.
Gum spot.....	------------------------------------	By the grubs of various insects in living trees.	Do.
Ring distortions.	------------------------------------	By defoliating larvae or flat-headed cambium miners in living trees.	Do.

¹ Powderpost or the so called "live worm" is of continuous injury; hence timber having this type of degrade cannot be used with safety before treatment with chemicals, kiln-drying, or steaming in a kiln. Powderpost can be prevented by proper handling of the stock.

solution containing 0.2 percent of gamma benzene hexachloride in the form of a water emulsion (*15, 17*). The addition of one of the sap-stain preventives mentioned for controlling molds, stains, and decay will keep the lumber bright.

Powderpost beetles (*24, 25, 26*) attack both hardwoods and soft-woods, both freshly cut and seasoned lumber and timber. The powderpost beetles that cause most damage to dry hardwood lumber belong to the *Lyctus* species. They attack the sapwood of ash, hickory, and oak, in particular, as it begins to season. Eggs are laid in pores of the wood, and the larvae burrow through the wood making tunnels from $\frac{1}{16}$ to $\frac{1}{12}$ inch in diameter, which they leave packed with a fine powder. Powderpost damage is indicated by holes left in the surface of the wood by the winged adults as they emerge and by the fine powder that may fall from the wood.

When selecting hardwood lumber for building or manufacturing purposes, any evidence of powderpost infestation should not be over-looked, for the beetles may continue to be active long after the wood is put to use. Sterilization of green wood with steam at 130° F. or sterilization of wood with a lower moisture content at 180° F. under controlled conditions of relative humidity for about 2 hours is effective for inch lumber. Thicker material requires a longer time. A 3-minute soaking in a petroleum oil solution containing 5 percent of pentachlorophenol, 5 percent of DDT, or 1 percent of lindane is also effective for checking infestation or preventing attack of lumber up to 1 inch thick. Small dimension stock also can be protected for about a year by brushing or spraying with these chemicals. In the case of infested furniture or finished woodwork in a building, these same insecticides may be used, but they should be dissolved in a refined petroleum oil, like mineral spirits. Since the beetles lay their eggs in the open pores of wood, infestation can be prevented by covering the entire surface of each piece of wood with a suitable finish.

Susceptible hardwood lumber used for manufacturing purposes should be protected from powderpost attack as soon as it arrives at the plant. Good plant sanitation helps check infestation. Damage to manufactured items frequently is traceable to infestations that occur before the products are placed on the market, particularly if a finish is not applied to the surface of the items until they are sold. Once wood is infested, the larvae will continue to work, even though the surface is subsequently painted, oiled, waxed, or varnished.

A roundheaded powderpost beetle, commonly known as the old house borer, causes most insect damage to seasoned pine floor joists. The larvae reduce the sapwood to a powdery or granular consistency and make a ticking sound while at work. When mature, the beetles make an oval hole about one-fourth inch in diameter in the surface of the wood and emerge. Infested wood should be drenched with a solution of 5 percent of DDT, 2 percent of chlordane, or 1 percent of lindane in a highly penetrating solvent; trichlorobenzene is preferable, though a petroleum oil is satisfactory. Borers working in wood behind plastered or paneled walls can be eliminated by having a licensed operator fumigate the building with a deadly poisonous gas like hydrocyanic acid.

Termites

Termites (*16, 18, 27, 29, 30*) superficially resemble ants in size, general appearance, and habit of living in colonies. About 56 species of them are known in the United States. From the standpoint of their methods of attack on wood, they can be grouped into two main classes: (1) The ground-inhabiting or subterranean termites; and (2) the wood-inhabiting or nonsubterranean termites.

Subterranean Termites

Subterranean termites are responsible for most of the termite damage done to wood structures in the United States. Their occurrence and damage are much greater in the Southern States than in the Northern States where low temperatures do not favor their development (fig. 85). The hazard of infestation is greatest (1) beneath basementless buildings that are poorly drained and ventilated and (2) in houses with basements in the parts of the foundation that are adjacent to enclosed porches, sun parlors, and terraces where filled earth is close to the timbers.

The subterranean termites develop their colonies and maintain their headquarters in the ground, from which they build their tunnels through earth and around obstructions to get at the wood they need for food. Each colony shuts itself off and lives in the dark, but the termites must have a constant source of moisture or they will die. It is the worker members of the colony that cause the destruction of wood. At certain seasons of the year male and female winged forms swarm from the colony, fly a short time, lose their wings, mate, and if successful in locating a suitable place they start new colonies. The appearance of "flying ants" or their shed wings, is an indication that a termite colony may be near and causing serious damage. Not all "flying ants" are termites; therefore, suspicious insects should be identified before money is spent for their eradication (fig. 86).

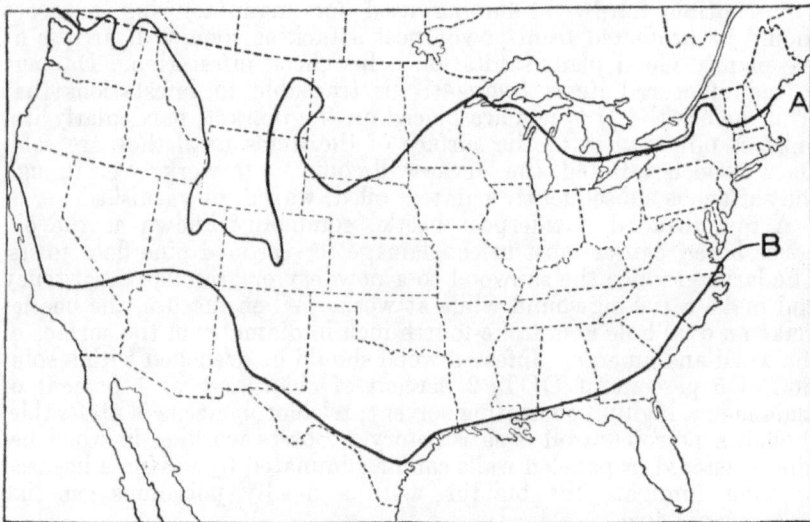

FIGURE 85.—*A*, The northern limit of recorded damage done by subterranean termites in the United States; *B*, the northern limit of damage done by drywood or nonsubterranean termites.

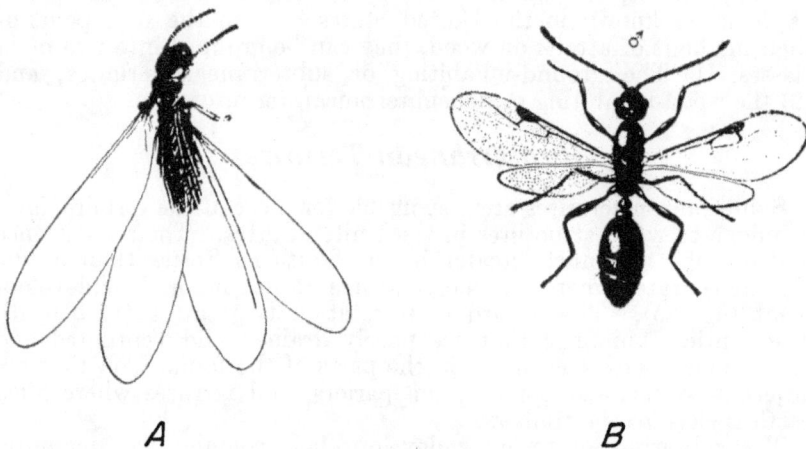

A   *B*

FIGURE 86.—*A*, Winged termite; *B*, winged ant (both greatly enlarged). The wasp waist of the ant and the long wings of the termite are distinguishing characteristics.

Subterranean termites do not establish themselves in buildings by being carried there in lumber but by entering from ground nests after the building has been constructed. Telltale signs of their presence are the earthen tubes or runways built by these insects over the surfaces of foundation walls to reach the wood above. Another sign is the swarming of winged adults early in the spring or fall. In the wood itself, the termites make galleries that follow the grain, leaving a shell of sound wood to conceal their activities. Since the galleries seldom show on the wood surfaces, probing with an ice pick or knife is advisable if the presence of termites is suspected.

The best protection where subterranean termites are prevalent is to build so as to prevent their gaining access to the building (*16*, *30*). The foundations should be of concrete or other solid material through which the termites cannot penetrate. With brick, stone, or concrete blocks, cement mortar should be used, for termites can work through some other kinds of mortar. Wood that is not impregnated with an effective preservative must be kept well away from the ground. If there is a basement, it should preferably be floored with concrete. Posts supporting the first-floor beams must be thoroughly treated if they bear directly on the ground or on wood blocking. Untreated posts should rest on concrete piers extending a few inches above the basement floor if that floor is of concrete; if the basement floor is of earth, the concrete piers should extend at least 18 inches above it. If the earth is not excavated beneath the building, and the floor is of wood construction, the floor joists and other woodwork, unless adequately treated with preservative, should be kept at least 18 inches from the earth and good ventilation should be provided beneath the floor.

Moisture condensation on the floor joists and subflooring, which may cause conditions favorable to decay and thus make the wood more attractive to termites, can be avoided by covering the soil below with roll roofing. In the case of ranch-style buildings or buildings with radiant heating, where concrete-slab floors are laid directly over a gravel fill, the soil directly beneath the expansion joints and around utility openings in the floor should be poisoned before the concrete is poured. Furthermore, insulation containing cellulose intended for use as a filler in expansion joints should be impregnated with a chemical toxic to termites. Sealing the top one-half inch of the expansion joint with coal-tar pitch also provides effective protection from ground-nesting termites.

All concrete forms, stakes, stumps, and waste wood should be removed from the building site, since they are possible sources of infestation. In the main, the precautions that are effective against subterranean termites are also helpful against decay.

Metal termite shields are used only to supplement good construction. They are employed mainly in the Deep South and in the tropical lands. Unless well designed and correctly installed, they offer a false sense of security and are a waste of money. Shields made of sisal kraft, roll roofing, and other building papers are not effective substitutes for metal because they are penetrable. Termiteproof shields are made of 26-gage galvanized iron or 16-ounce (per square foot) copper. They are used between the foundation and woodwork and fitted tightly around pipes and other equipment connecting wood with the ground. All metal joints must be soldered or tightly interlocked to be effective. The metal shields extend out from the foundation at an angle of 45° for a horizontal distance of at least 2 inches (*29*, *30*).

Where it is not feasible to install metal shields, and other protection is needed in addition to that obtained by structural methods, a thorough poisoning of the soil adjacent to the foundation walls and piers beneath the building with a recognized soil poison will prove to be beneficial (29, 30). A thorough treatment should last at least 5 years.

To control termites in a building, one of the main things to remember is to break any contact between the termite colony in the soil and the woodwork. This can be done by blocking mechanically the runways from soil to wood, by poisoning the soil, or by using both of these methods. When more difficult treatments are needed, such as drilling of masonry foundations, the detailed instructions in U. S. Department of Agriculture Farmers' Bulletin 1911 (29) should be followed. Possible reinfestations should be guarded against by frequent inspections for the telltale signs that were listed previously.

Nonsubterranean Termites

Nonsubterranean termites have been found only in a narrow strip of territory extending from central California around the southern edge of the United States to Virginia (fig. 85). Their principal damage is confined to an area in southern California and to parts of southern Florida, notably Key West.

The nonsubterranean or dry-wood termites are fewer in number, do not multiply so rapidly, and have colony life and habits somewhat different from those of the subterranean termites. The total amount of destruction they cause in the United States is very much less than that caused by the subterranean termites. Their ability to live in damp or dry wood without outside moisture or contact with the ground, however, makes them a definite menace in the regions where they occur. Their depredations are not rapid, but they can thoroughly riddle timbers with their tunnelings if allowed to work unmolested for a few years.

In constructing a building in localities where the nonsubterranean termites are prevalent, it is well to inspect the lumber carefully to see that it was not infested before arrival at the building site. If the building is constructed during the swarming season, the lumber should be watched during the course of construction, because infestation by colonizing pairs can easily take place at this season (18, 27). Since paint is a good protection against the entrance of dry-wood termites, all exposed wood should be kept adequately painted. Fine screen should be placed over any openings through which access might be gained to the interior unpainted parts of the building. As in the case of ground-nesting termites, old stumps, posts, or wood debris of any kind that could serve as sources of infestation should be removed from the premises.

If a building is found to be infested with dry-wood termites, the infested wood should be replaced if badly damaged. If the wood is only slightly damaged or difficult to replace, further termite activity can be arrested by blowing a teaspoonful of a poisonous dust, such as finely divided Paris green, arsenical dust, or 50 percent DDT, into each nest. Also effective are liquid insecticides, such as trichlorobenzene or the following dissolved in fuel oil at the strength indicated: DDT, 6 percent; chlordane, 2 percent; pentachlorophenol, 5 percent; and gamma benzene hexachloride, 0.4 percent. Detached houses heavily infested with nonsubterranean termites have been fumigated with success. This method is quicker and often cheaper than the use of poisonous liquids and dusts, but it does not prevent the termites from returning, since no poisonous residue is left in the tunnels. Moreover, fumigation is very dangerous and should be conducted only by licensed fumigators.

In localities where dry-wood termites do serious damage to posts and poles, the best protection for these and similar forms of outdoor timbers is full-length treatment with wood preservatives of recognized value.

Naturally Termite-Resistant Woods

Only a limited number of woods grown in the United States offer any marked degree of natural resistance to termite attack. The close-grained heartwood of California redwood and baldcypress has some resistance, especially when used above ground. Termites are known to attack these woods at times, however. Very resinous heartwood of southern yellow pine is practically immune, but wood of this character is not available in large quantities or suitable for many uses.

Carpenter Ants

Carpenter ants are large or small black or brown ants that are found usually in stumps, trees, or logs but sometimes are found doing damage in poles, structural timbers, or buildings. They use the wood for shelter rather than for food, usually preferring wood that is naturally soft or has been made soft by decay. They may enter a building directly, by crawling, or may be carried there in fuel wood. If left undisturbed they can in a few years enlarge their tunnels to the point where replacement or extensive repairs are necessary. The parts of dwellings mostly frequented by them are porch columns, porch roofs, window sills, and sometimes the wood plates in foundation walls. The logs of rustic cabins are also attacked.

Precautions that prevent attack by decay and termites are usually effective against carpenter ants. Decaying or infested wood, such as logs or stumps, should be removed from the premises, and crevices present in the foundation or woodwork of the building should be sealed. Particularly, leaks in porch roofs should be repaired, because the decay that may result makes the wood more attractive to the ants.

When carpenter ants are found in a structure, badly damaged timbers should be replaced. In wood not sufficiently damaged to require replacement, the ants can be killed by dusting with 5 percent chlordane; 4 percent rotenone (derris powder), or 10 percent DDT (10). Where the colony is located in decaying wood in porches, soak the wood with a solution containing 2 percent of chlordane or 5 percent of pentachlorophenol (a wood preservative as well as an insecticide) in petroleum oil.

MARINE BORERS

Damage by marine boring organisms to fixed or floating structures of wood in salt or brackish waters is practically worldwide. Slight attack is sometimes found in rivers even above the region of brackishness. The rapidity of attack depends upon local conditions and the kinds of borers present. Along the Pacific, Gulf, and South Atlantic coasts of the United States attack is rapid, and untreated piling may be completely destroyed in a year or less. Along the coast of the New England States the rate of attack is less rapid but still sufficiently rapid generally to require protection of wood where long life is desired.

The principal marine borers from the standpoint of wood destruction in the United States (1, 4, 12) are described in the following paragraphs.

Shipworms

Shipworms are mollusks of various species that superfically are wormlike in form. The group includes several species of *Teredo* and several species of *Bankia* that are readily distinguishable on close observation but are all very similar in several respects. In the early stages of their life they are minute, free-swimming organisms. Upon finding suitable lodgement on wood they quickly develop into a new form and bury themselves in the wood. A pair of boring shells on the head grows rapidly in size as the boring progresses, while the tail part or siphon remains at the original entrance. Thus, the animal grows in length and diameter within the wood but remains a prisoner in its burrow, which it lines with a shell-like deposit. It lives upon the wood borings and upon the organic matter extracted from the sea water that is continuously being pumped through its system. The entrance holes never grow large, and the interior of a pile may be completely honeycombed and ruined while the surface shows only slight perforations. When present in great numbers the borers grow only a few inches before the wood is so completely occupied that growth is stopped, but when not crowded they can grow to lengths of 1 to 4 feet according to species.

Limnoria

Limnoria are small crustaceans about ⅛ to ⅙ inch long that bore small burrows in the surface of piling. They can move freely from place to place if occasion requires but usually continue to bore in one place. When great numbers are present their burrows are separated by very thin walls of wood that are easily eroded by the motion of the water and objects floating upon it. This erosion causes the animals to burrow continually deeper; otherwise the burrows would probably not become more than 2 inches long or more than ½ inch deep. Since erosion is greatest between tide levels, piling heavily attacked by *Limnoria* characteristically wears within such levels to an hourglass shape. Untreated piling can be destroyed by *Limnoria* within a year in heavily infested harbors.

Martesia

Martesia are wood-boring mollusks that resemble clams in general appearance. Like the shipworms, they enter the wood when very small, leaving a small entrance hole, but grow larger as they burrow into the wood. They generally do not exceed 2½ inches in length and

1 inch in diameter but are capable of doing considerable damage. Their activities in the United States appear to be confined to the Gulf of Mexico.

Sphaeroma

Sphaeroma are somewhat similar to *Limnoria* but larger, sometimes reaching a length of ½ inch and a width of ¼ inch. They resemble in general appearance and size the common sow bug or pill bug that inhabit damp places. They are widely distributed but not so plentiful as *Limnoria* and do much less damage. Nevertheless piling in some structures has been ruined by them. Occasionally they have been found working in fresh water.

Resistance of Woods to Marine Borers

No wood is immune to marine-borer attack, and no commercially important wood of the United States has sufficient marine-borer resistance to justify its use untreated in any important structure in areas where borers are active. The heartwood of several foreign species, such as turpentine, greenheart, jarrah, azobe, totara, kasikasi, manbarklak, and several others, has shown resistance to marine-borer attack. Service records on these woods, however, do not always show uniform results and are affected by local conditions.

Protection of Permanent Structures from Marine Borers

The best practical protection for piling in sea water is heavy treatment with coal-tar creosote or creosote-coal tar solution. The treatment must be thorough, the penetration as deep as possible, and the retention high in order to give satisfactory results in heavily infested waters. It is best to treat such piling by the full-cell process "to refusal"; that is, to force in all the oil the piling can be made to hold without using treatments that cause serious damage to the wood. The retentions recommended on page 417 are minimum values. When maximum protection against marine borers is desired, as much more oil as is practicable should be injected. For highest retentions it is necessary to air-dry the piling before treatment.

Limnoria, Martesia, and *Sphaeroma* are not always stopped even by thorough creosote treatment. The average life of well-creosoted structures is many times the average life that could be obtained from untreated structures; nevertheless, well-creosoted structures are sometimes damaged seriously.

Shallow or erratic penetration affords but slight protection. The spots with poor protection are attacked, and from them the borers spread inward and destroy the untreated interior of the pile. Low retention fails to provide a reservoir of surplus preservative to compensate for depletion by evaporation and leaching.

When wood is to be used in salt water, avoidance of cutting or injury by sharp tools after treatment is even more important than when wood is to be used on land. No cutting or injury of any kind for any purpose should be permitted in the underwater part of the pile. Where piling is cut to grade above the waterline, the exposed surfaces should, of course, be protected from decay.

The life of treated piling is influenced by the thoroughness of the treatment, the care and intelligence used in avoiding damage to the creosoted shell during handling and installation, and the severity of borer attack. Exposure conditions, such as water temperature, salinity, dissolved oxygen, water depth, and currents, tend to cause wide variations in the severity of borer attack even within limited areas (5). The San Francisco Bay Marine Piling Committee estimated a range of 15 to 30 years for the life of creosoted piling on the Pacific coast (12). More recent service records (8) show average-life figures of from 22 to 48 years on well-treated Douglas-fir piling in San Francisco Bay waters. In South Atlantic and Gulf of Mexico waters, creosoted piles are estimated to have an average life of 10 to 12 years (1) and frequently last much longer (8). On the North Atlantic coast longer life is to be expected.

Metal armor and concrete jacketing have been used with varying degrees of success for the protection of marine piling (5). The metal armor may be in the form of sheets, wire, or nails. Scupper-nailing with iron or steel furnishes some protection, particularly against *Limnoria*. Sheathing of piling with copper or Muntz metal has been only partially successful, owing to difficulty in maintaining a continuous armor. Theft, damage in driving, damage by storm or driftwood, and corrosion have sooner or later let in the borers, and in only a few cases reported has long life been obtained. Attempts during World War II to electroplate wood piling with copper were not successful. Concrete casings are now in greater use than metal armor and appear to provide better protection when high-quality materials are used and are carefully applied. Applying concrete casings to creosoted piling appears to be a practice worthy of consideration where the importance of a structure and the severity of borer attack justify the expense of the two-way protection system.

Protection of Wood Boats from Marine-Borer Attack

Wood barges and lighters have been constructed with planking or sheathing pressure treated with creosote to provide hull protection from marine borers, and the results have been favorable. While coal-tar creosote is the most effective preservative yet developed for protecting wood against marine borers, it has disadvantages that make its use objectionable in many types of boats. Creosote adds considerably to the weight of the boat hull, and its odor is objectionable to boat crews. In addition, antifouling paints are difficult to apply over creosoted wood.

Some copper bottom paints (31) protect boat hulls against marine-borer attack, but the protection continues only while the coating remains unbroken. Since it is difficult to maintain an unbroken coating of antifouling paint, the U. S. Navy finds it desirable to impregnate the hull planking of some wood boats with certain copper-containing preservatives. Such preservatives, when applied with high retentions (1.5 to 2.0 pounds per cubic foot), have limited effectiveness against marine borers and should help to protect the hull of a boat during intervals between renewals of the antifouling coating. These copper preservatives do not provide protection equivalent to that furnished by coal-tar creosote, and their effectiveness in protecting boats is therefore best assured if the boats are dry docked at regular and frequent intervals and the maintenance of the antifouling coating is not neglected.

Marine-borer tests on plywood show that improved service can be obtained through preservative treatment of the plywood, acetylation of the veneer, and the use of special adhesives, particularly those containing siliceous extenders (19, 21). These developments have not yet been fully explored in the building of plywood boats, however.

LITERATURE CITED

(1) ATWOOD, W. G., and JOHNSON, A. A.
    1924. MARINE STRUCTURES, THEIR DETERIORATION, AND PRESERVATION. Natl. Res. Council Rpt. Com. on Marine Piling Invest., 534 pp., illus.
(2) BEAL, J. A., and MASSEY, G. L.
    1945. BARK BEETLES AND AMBROSIA BEETLES. Duke Univ. Forestry Bul. 10, 178 pp., illus.
(3) BOYCE, J. S.
    1923. DECAYS AND DISCOLORATIONS IN AIRPLANE WOODS. U. S. Dept. Agr. Bul. 1128, 52 pp., illus.
(4) CALMAN. W. T.
    1919. MARINE BORING ANIMALS INJURIOUS TO SUBMERGED STRUCTURES. Brit. Mus. Nat. Hist. Econ. Ser. 10, 35 pp., illus.
(5) CHELLIS, R. D.
    1948. FINDING AND FIGHTING MARINE BORERS. Engin. News Rec. 140 (12): 422–424, illus.; (14): 493–496, illus.
(6) CHRISTENSEN, C. M., and MOSES, C. S.
    1945. MOLDS AND BACTERIA THAT DELAMINATE PLYWOOD BONDED WITH CASEIN AND SOYBEAN GLUES. U. S. Dept. of Agr. Forest Path. Spec. Release 25, 23 pp., illus.
(7) CRAIGHEAD, F. C.
    1950. INSECT ENEMIES OF EASTERN FORESTS. U. S. Dept. Agr. Misc. Pub. 657, 679 pp., illus.
(8) DANIELS, A. S.
    1949. REPORT OF COMMITTEE U–6, MARINE AND FOUNDATION PILE SERVICE RECORDS. Amer. Wood Preservers' Assoc. Proc. 45: 294–305.
(9) ENGLERTH, G. H.
    1950. DECAY RESISTANCE OF PLYWOOD BONDED WITH VARIOUS GLUES. Forest Prod. Res. Soc. Proc. 4: 14 pp., illus.
(10) FURNISS, R. L.
    1944. CARPENTER ANT CONTROL IN OREGON. Oreg. Agr. Expt. Sta. Cir. 158, 12 pp., illus.
(11) HARTLEY, C., and MAY, C.
    1943. DECAY OF WOOD IN BOATS. U. S. Dept. Agr. Forest Path. Spec. Release 8, 12 pp., illus.
(12) HILL, C. L., and KOFOID, C. A.
    1927. MARINE BORERS AND THEIR RELATION TO MARINE CONSTRUCTION ON THE PACIFIC COAST. San Francisco Bay Marine Piling Com. Final Rpt., 357 pp., illus.
(13) HOXIE, F. J.
    1930. DECAY OF WOOD IN INDUSTRIAL BUILDINGS. Assoc. Factory Mutual Fire Insurance Co., Inspection Dept. 119 pp., illus. Boston.
(14) HUMPHREY, C. J.
    1917. TIMBER STORAGE CONDITIONS IN THE EASTERN AND SOUTHERN STATES WITH REFERENCE TO DECAY PROBLEMS. U. S. Dept. Agr. Bul. 510, 43 pp., illus.
(15) JOHNSTON, H. R.
    1g52. CONTROL OF INSECTS ATTACKING GREEN LOGS AND LUMBER. South, Lumberman 184 (2307): 37–39, illus.

(16) KOFOID, C. A., LIGHT, S. F., HORNER, A. C., RANDALL, M., HERMS, W. B., and ROWE, E. E.
 1934. TERMITES AND TERMITE CONTROL: A REPORT TO THE TERMITE INVESTIGATIONS COMMITTEE. 734 pp., illus. Berkeley, Calif.

(17) KOWAL, R. J.
 1949. CONTROL OF WOOD BORING INSECTS IN GREEN LOGS AND LUMBER. Forest Prod. Res. Soc. Proc. 3: 469–479.

(18) LIGHT, S. F.
 1929. TERMITES AND TERMITE DAMAGE. Calif. Agr. Expt. Sta. Cir. 314, 28 pp., illus.

(19) MACLEAN, J. D.
 1950. RESULTS OF EXPERIMENTS ON THE EFFECTIVENESS OF VARIOUS PRESERVATIVES IN PROTECTING WOOD AGAINST MARINE-BORER ATTACK. U. S. Forest Prod. Lab. Rpt. D1773, 19 pp., illus. [Processed.]

(20) PECK, E. C.
 1952. STORAGE AND HANDLING OF LUMBER. U. S. Forest Prod. Lab. Rpt. D1919, 17 pp., illus. [Processed.]

(21) RICHARDS, A. P., and CLAPP, W. F.
 1946. CONTROL OF MARINE BORERS IN PLYWOOD. Amer. Soc. Mech. Engin. Wood Indus. Div. Paper 46–A–43, 6 pp., illus.

(22) SCHEFFER, T. C., and LINDGREN, R. M.
 1940. STAINS OF SAPWOOD AND SAPWOOD PRODUCTS AND THEIR CONTROL. U. S. Dept. Agr. Tech. Bul. 714, 123 pp., illus.

(23) ——— WILSON, T. R. C., LUXFORD, R. F., and HARTLEY, C.
 1941. EFFECT OF CERTAIN HEART ROT FUNGI ON THE SPECIFIC GRAVITY AND STRENGTH OF SITKA SPRUCE AND DOUGLAS-FIR. U. S. Dept. Agr. Tech. Bul. 779, 24 pp., illus.

(24) ST. GEORGE, R. A.
 1943. PROTECTION OF LOG CABINS, RUSTIC WORK, AND UNSEASONED WOOD FROM INJURIOUS INSECTS. U. S. Dept. Agr. Farmers' Bul. 1582, 19 pp., illus. (Rev.)

(25) SNYDER, T. E.
 1926. PREVENTING DAMAGE BY LYCTUS POWDER POST BEETLES. U. S. Dept. Agr. Farmers' Bul. 1477, 12 pp., illus.

(26) ———
 1927. DEFECTS IN TIMBER CAUSED BY INSECTS. U. S. Dept. Agr. Bul. 1490, 46 pp., illus.

(27) ———
 1950. CONTROL OF NONSUBTERRANEAN TERMITES. U. S. Dept. Agr. Farmers' Bul. 2018, 16 pp., illus.

(28) TEESDALE, L. V.
 1949. THERMAL INSULATION MADE OF WOOD-BASE MATERIALS, ITS APPLICATION AND USE IN HOUSES. U. S. Forest Prod. Lab. Rpt. R1740, 40 pp., illus.

(29) U. S. DEPARTMENT OF AGRICULTURE.
 1949. DAMAGE TO BUILDINGS BY SUBTERRANEAN TERMITES AND THEIR CONTROL. U. S. Dept. Agr. Farmers' Bul. 1911, 38 pp., illus. (Rev.)

(30) ———
 1951. DECAY AND TERMITE DAMAGE IN HOUSES. U. S. Dept. Agr. Farmers' Bul. 1993, 26 pp., illus. (Rev.)

(31) U. S. FOREST PRODUCTS LABORATORY.
 1945. WOOD: A MANUAL FOR ITS USE IN WOODEN VESSELS. 235 pp., illus. (Issued by Navy Bureau of Ships.)

(32) VERRALL, A. F.
 1952. CONTROL OF WOOD DECAY IN BUILDINGS. Agr. Engin. 33 (4): 217–219, illus.

(33) ——— and SCHEFFER, T. C.
 1949. CONTROL OF STAIN, MOLD AND DECAY IN GREEN LUMBER AND OTHER WOOD PRODUCTS. For. Prod. Res. Soc. Proc. 3: 9 pp., illus.

WOOD PRESERVATION

Wood can be protected from attack by decay fungi, harmful insects, or marine borers by the application of selected chemicals, or wood preservatives. The degree of protection obtained depends on the kind of preservative used and the thoroughness of the method of application. Some preservatives are more effective than others, and some are more adaptable to certain use requirements. Furthermore, the wood is well protected only when the preservative substantially penetrates it, and some methods of treatment assure better penetration than others. There is also a difference in the treatability of various species of wood, particularly of their heartwood, which generally resists preservative treatment more than sapwood (*19*).[26]

Good wood preservatives, thoroughly applied with standard retentions (see pp. 417–418) and with the wood satisfactorily penetrated, substantially increase the life of wood structures, often as much as five times (*24*). On this basis the annual cost of treated wood in service is greatly reduced below that of similar wood without treatment (*16, 17*).

WOOD PRESERVATIVES

Wood preservatives fall into two general classes: Oils, such as creosote and petroleum solutions of pentachlorophenol; and waterborne salts that are applied as water solutions.

Preservative Oils

Preservative oils generally have high resistance to leaching and are therefore suitable for outdoor exposures. They do not cause the wood to swell, but some shrinkage may result if a loss of moisture occurs in the wood during the treating process. Creosote and solutions with the heavier, less volatile petroleum oils often help protect the wood from weathering but may adversely influence its cleanness, odor, color, paintability, and combustibility. Preservative oils sometimes travel from treated studs or subflooring along nails and discolor adjacent plaster or finish flooring.

Coal-Tar Creosote

Coal-tar creosote is a black or brownish oil made by distilling coal tar. The character of the tar used, the method of distillation, and the temperature range in which the creosote fraction is collected all influence the character of the creosote oil. The character of the various coal-tar creosotes available, therefore, may vary to a considerable extent. Small differences in character, however, do not prevent creosotes from giving good service, and satisfactory results in preventing decay may be expected, as a general rule, from any

---

[26] Italic numbers in parentheses refer to Literature Cited, p. 427.

coal-tar creosote that complies with or closely approaches the requirements of standard specifications.

Coal-tar creosote is the most important and most generally useful wood preservative (*25*). Its advantages are (1) its high toxicity to wood-destroying organisms, (2) its relative insolubility in water and its low volatility, which impart to it a great degree of permanence under the most varied use conditions, (3) its ease of application, (4) the ease with which its depth of penetration can be determined, (5) its general availability and relatively low cost (when purchased in wholesale quantities), and (6) its long record of satisfactory use.

Although for general outdoor service in structural timbers and for marine use there is, as yet, no better preservative than coal-tar creosote, for some special purposes it has certain properties that are disadvantages. Freshly creosoted timber can be ignited easily and will burn readily, producing a dense smoke. After the timber has seasoned some months, however, the more volatile parts of the oil disappear from near the surface and the creosoted wood usually is but little, if any, easier to ignite than untreated wood. Until this volatile oil has evaporated, ordinary precautions should be taken to prevent fires. On the other hand, after untreated wood has started to decay, often within a few years, it is easier to ignite than timber that has been kept sound by creosote treatment. The only recommendation that can be made with respect to fire hazard is that some preservative other than creosote should be used where this hazard is considered of utmost importance, unless the treated wood is also protected from fire.

The odor of creosoted wood is unpleasant to some persons. Foodstuffs that are sensitive to odors should not be stored where creosote odors are present. Creosote vapors are harmful to growing plants. Workmen sometimes object to the use of creosoted wood because it soils their clothes and because it burns the skin of the face and hands of some individuals, causing an injury similar to sunburn. There need be no fear, however, that creosoted timber has a serious effect on the health of workmen handling or working near it or on the health of the occupants of buildings in which it is used. Creosoted wood can very often be used in sills and foundation timbers and in floor sleepers embedded in or resting on concrete with little danger of the odor becoming objectionable.

The color of creosote and the fact that wood treated with it usually cannot be painted satisfactorily make it unsuitable for finish lumber or other lumber used where appearance and paint receptivity are of major importance.

A number of specifications prepared by different organizations are available for creosote oils of different kinds. Although the oil obtained under most of these specifications will probably be sufficiently effective in preventing decay, the requirements of some organizations are more exacting than others. Federal Specification TT–W–00556b for coal-tar creosote, adopted for use by the U. S. Government, will generally prove satisfactory and, under normal conditions, can be met without difficulty by most creosote producers. The requirements of this specification are similar to those of the American Wood-Preservers' Association Standard P1 for creosote (*1*), which is equally acceptable.

Crystal-Free Coal-Tar Creosote

Crystal-free coal-tar creosote is creosote from which some of the crystal-forming materials have been removed, so that the oil will flow freely at ordinary temperatures and will not deposit crystals in its container. Oils of this type are more convenient to handle than ordinary creosotes, but their general properties and effectiveness are similar to those of the ordinary coal-tar creosote. They are intended primarily for brush and spray applications. American Wood-Preservers' Association Standard P7 (1) and Federal Specification TT–W–560 cover crystal-free creosotes.

Anthracene Oils

Anthracene oils, sometimes referred to as carbolineums, are coal-tar distillates of higher specific gravity and higher boiling range than ordinary coal-tar creosotes, but their general properties and effectiveness as preservatives are similar to those of coal-tar creosotes. High-boiling distillates are advantageous to use in open-tank treatments that involve heating, since losses through evaporation are likely to be less than with low-boiling oils. Anthracene oils are usually sold under proprietary or trade names and are covered by American Wood-Preservers' Association Standard P7 (1) and Federal Specification TT–W–531.

Other Creosotes

Creosotes distilled from tars other than coal tar are used to some extent for wood preservation, although they are not included in current Federal or American Wood-Preservers' Association specifications. These include wood-tar creosote (table 45), oil-tar creosote, and water-gas-tar creosote. These creosotes protect wood from decay and insect attack but are generally less effective than coal-tar creosote (16).

Tars

Coal tars are seldom used alone for preserving wood because good penetration is usually difficult to obtain with them and because they are less poisonous to wood-destroying fungi than the coal-tar creosotes. Service tests have demonstrated that surface coatings of tar are of little value. Coal tar has been used in the pressure treatment of crossties, but satisfactory penetration of the wood is difficult to obtain with the highly viscose tar. When good absorptions and deep penetrations are obtained, however, it is reasonable to expect a satisfactory degree of effectiveness from treatment with coal tar. The tar is particularly effective in reducing checking in crossties in service.

Water-gas tar is used less extensively than coal tar, but, in certain cases where the wood was thoroughly impregnated with it, the results were good (table 45).

TABLE 45.—*Condition of round southern yellow pine experimental fence posts on the Harrison Experimental Forest, Saucier, Miss., after about 11½ to 16 years of service*

Preservative	Posts in test [1]	Form of preservative [2]	Retention of preservative [2]				Method of treatment	Condition of posts December 1952						
			Minimum	Maximum	Average	Standard deviation		Serviceable	Removed on account of—			Total removed		Average life [3]
									Decay	Decay termites	Termites			
	No.		*Lb. per cu.ft.*	*Lb. per cu.ft.*	*Lb. per cu.ft.*	*Lb. per cu.ft.*		*Pct.*	*Pct.*	*Pct.*	*Pct.*	*No.*	*Pct.*	*Yr.*
Posts set late in 1936 to February 1937	98	Solution	3.90	10.00	6.20	1.30	Pressure	19.4	15.3	47.0	18.3	79	80.6	14
Beta-napthol, 5 percent (by weight) in oil mixture.														
Borax-boric acid (50–50 mixture)	97	Salt	.64	1.32	.92	.11	do	7.2	15.5	41.2	36.1	90	92.8	12
Celcure (acid cupric chromate)	94	do	.75	1.05	.92	.08	do	90.4	7.5	2.1		9	9.6	
Chromated zinc chloride	97	do	.37	1.33	.87	.19	do	74.2	14.4	11.4		25	25.8	20
Coal tar	96	Oil	1.60	19.20	6.50	2.90	do	90.6	4.2	5.2		9	9.4	
Coal-tar creosote, grade 1	98	do	1.90	8.60	6.00	1.50	do	98.0	2.0			2	2.0	
Coal-tar creosote, 50 percent; used crankcase oil, 50 percent (by volume).	98	Solution [4]	1.60	14.80	5.40	2.20	do	99.0	1.0			1	1.0	
Coal-tar creosote, 10 percent; used crankcase oil, 90 percent (by volume).	100	do [5]	.10	23.20	7.10	3.80	do	30.0	63.0	6.0	1.0	70	7.0	15
Crankcase oil (used)	99	Oil [6]	2.50	16.80	7.60	2.90	do	19.2	77.8	3.0		80	80.8	14
Lignite coal-tar creosote	98	do	1.10	11.60	6.30	3.20	do	73.5	10.2	14.3	2.0	20	26.5	20
Mercuric chloride	100	Salt	.05	.15	.09	.03	Steeping	75.0	3.0	14.0	8.0	25	25.0	21
No-D-K (hardwood-tar creosote)	100	Oil	1.30	13.60	6.60	3.50	Pressure	52.0	11.0	33.0	4.0	48	48.0	17
Osmosar	96	Salt			7.30		Osmose	74.0	11.4	12.5	2.1	25	26.0	20
P. D. A. (phenyldichlorarsine) 0.84 percent (by weight) in gas oil.	94	Solution	4.10	11.00	5.90	.90	Pressure	61.7	4.2	24.5	9.6	36	38.3	18
Pentachlorophenol, 4.82 percent (by weight) in used crankcase oil.	98	do	2.90	9.50	6.70	1.60	do	100.0						
Pentachlorophenol, 3.02 percent (by weight) in used crankcase oil.	95	do	3.10	11.40	6.40	1.80	do	97.9	2.1			2	2.1	
Sodium bichromate	98	Salt	.59	1.14	.88	.11	do	49.0	5.1	18.3	27.6	50	51.0	17
Sodium chromate	90	do	.70	1.19	.93	.10	do	34.4	3.3	16.7	45.6	59	65.6	15

Tanalith (Wolman Salts)	97	do	.20	.47	.35	.07	do	80.4	12.4	3.1	4.1	19	19.6	22
Tetrachlorophenol, 2.9 percent (by weight) in used crankcase oil.	95	Solution	1.20	.10	7.10	3.10	do	94.7	5.3			5	5.3	
Tetrachlorophenol, 4.83 percent (by weight) in used crankcase oil.	96	do	3.50	9.40	5.80	1.50	Pressure	100.0						
Water-gas tar	96	Oil	1.20	19.00	6.30	3.00	do	98.0	1.0	1.0		2	2.0	
Zinc chloride	98	Salt	.67	1.11	.94	.10	do	84.7	4.1	11.2		15	15.3	23
Zinc meta arsenite	96	do	.25	.54	.42	.06	do	100.0						
Untreated posts (set Feb. 1937)	65						None		3.1	93.8	3.1	65	100.0	3.1
Posts set in 1938 and 1941														
Untreated posts (set Nov., Dec. 1938)	33				.35		None		33.3	63.6	3.1	33	100.0	3.7
Copper sulfate and sodium arsenate (set May 1941).	99	Salt			.16		Double diffusion.	99.0	1.0			1	1.0	
Osmoplastic ground-line treatment (set Feb. 1941).	99	Mixture			8.34		(8)	31.3	20.2	38.4	10.1	68	68.7	11

[1] Installation included 100 posts for each treatment. This number has since been reduced in some cases by fire and pillerage.

[2] Based on the 100 posts treated in each group, unless otherwise indicated.

[3] Average life of all untreated posts is 3.3 years; other values are estimates taken from a mortality curve. Where percentage of posts removed is 10 percent or less, no estimate on average life is given.

[4] Retention values based on 97 posts.

[5] Retention values based on 89 posts.

[6] Retention values based on 99 posts.

[7] Average application.

[8] Average application per post to a 15-inch band (3 inches above and 12 inches below ground line) and to top surface of post.

Creosote Solutions

For many years, either coal tar or petroleum oil has been mixed with coal-tar creosote, in various proportions, as a means of lowering preservative costs. These creosote solutions have a satisfactory record of performance, mostly when used in the treatment of crossties.

Federal Specification TT–W–566a, "Creosote-coal-tar Solution," places no limit on the quantities of coal-tar creosote and coal tar present in the solution, but the solution must meet definite requirements as to physical and chemical composition. American Wood-Preservers' Association Standard P2 (1) includes 4 creosote-coal-tar solutions that must contain, respectively, not less than 80, 70, 60, or 50 percent by volume of coal-tar distillate and must also meet requirements as to physical and chemical properties. Federal Specification TT–W–568 and American Wood-Preservers' Association Standard P3 (1) stipulate that creosote-petroleum oil solutions shall contain not less than 50 percent by volume of coal-tar creosote.

Creosote solutions not only cost less than straight creosote but also tend to reduce weathering and checking of the treated wood. The solutions may have a greater tendency to accumulate on the surface of the treated wood (bleed) and may penetrate the wood with greater difficulty, particularly since they generally are more viscous than straight creosote. Higher temperatures and pressures during treatment, when they can safely be used, will often improve penetration of solutions of high viscosity.

Since petroleum oil and coal tar are less toxic to wood-destroying organisms than straight creosote, their mixtures with creosote are also less toxic. The toxicity of a creosote-petroleum oil solution, compared to that of straight creosote, is generally less, proportionately, than the percentage of creosote in the solution. Creosotes of high toxicity are therefore considered desirable for use in creosote-petroleum solutions. Some preservative users, particularly in pole treatment, fortify the petroleum oils in creosote solutions with such toxic materials as pentachlorophenol and copper naphthenate to increase the preservative effectiveness of the solution.

Creosote-petroleum solutions with favorable performance records generally contain petroleum oils of the comparatively heavy, high-boiling, high-viscosity residuum types, such as covered by American Wood-Preservers' Association Standard P4 (1). The protective qualities of solutions containing lighter, distillate fuel oils have not been adequately demonstrated.

Pentachlorophenol Solutions and Copper Naphthenate Solutions

The use of petroleum oils fortified with chlorinated phenols, principally pentachlorophenol, or with copper naphthenate (13, 22) as wood preservatives became large scale following World War II, because of the shortage of coal-tar creosote. These preservatives were first used primarily for surface applications. The chlorinated phenols in volatile light-colored solvents, such as mineral spirits, were first used for window sash and millwork that required a clean, nonswelling, and paintable treatment. Commercial treatment of poles and lumber with pentachlorophenol solutions began in 1941.

The petroleum oil used in these solutions varies from the diesel-oil type to the heavier types ordinarily used in creosote-petroleum solutions. American Wood-Preservers' Association Standard P9 (1) covers petroleum oils for use in pentachlorophenol or copper naphthenate solutions.

Federal Specification TT–W–571c covers the treatment of wood with solutions of pentachlorophenol; other large-scale users, such as telephone companies and the Rural Electrification Administration, also have specifications covering the formulation and use of these solutions. Federal Specification TT–W–570 covers pentachlorophenol, and the American Wood-Preservers' Association Standard P8 (1) covers both pentachlorophenol and copper naphthenate.

Pentachlorophenol solutions for wood preservation generally contain 5 percent of this chemical; copper naphthenate solutions have a copper metal concentration varying from 0.5 to 3.0 percent for different use requirements. The performance of these preservatives is influenced by the character of the petroleum oil in the treating solution, and experience to date indicates that best results are obtained with fairly heavy oils of low volatility. Petroleum oils that remain in the wood in substantial quantities after treatment, however, are likely to interfere with painting.

Service and field tests on wood treated with petroleum oils containing 5 percent of pentachlorophenol or copper naphthenate equivalent to 0.5 percent or more of copper metal show that these preservatives provide a high degree of protection against decay fungi and termites when the wood is properly treated. It is too early to predict their utlimate protective properties in comparison with coal-tar creosote. They are known, however, to provide much less protection than creosote against marine borers.

Pentachlorophenol does not appreciably alter the natural color of the wood, but copper naphthenate gives it a green color. The properties of the wood treated with petroleum oils containing these preservatives, with respect to cleanness, paintability, color, odor, and combustibility, depend largely upon the properties of the petroleum oil used. Pentachlorophenol in the dry form and in solutions irritates the skin of workers, but with careful handling and the use of suitable protective clothing it is possible to avoid harmful effects. The use of "bloom" preventives, such as ester gun, is required in pentachlorophenol solutions with volatile solvents to prevent the formation of crystals of the preservative on the surface of the wood after treatment.

Water-Repellent Preservatives

Preservative oils containing water-repellent components are sold under various trade names, principally for the treatment of window sash and other millwork. Federal Specification TT–W–572 stipulates that such preservatives consist of volatile solvents, such as mineral spirits, that do not cause appreciable swelling of the wood and not less than 10 percent or more than 25 percent of nonvolatile matter, including the preservative and water-repellent components. The preservative chemicals may be not less than 5 percent of pentachlorophenol or not less than 2 percent of copper in the form of copper naphthenate.

Effective water-repellent preservatives retard moisture changes in the wood but do not prevent them. They therefore help reduce dimensional changes in the wood due to moisture changes when the wood is exposed to rainwater or dampness for short periods but not when it is in contact with the ground or other sources of prolonged dampness. As with any wood preservative, their effectiveness in protecting wood against decay and insects depends upon the thoroughness of the method of application.

Waterborne Preservatives

Standard wood preservatives used in water solution include zinc chloride, chromated zinc chloride, copperized chromated zinc chloride, Tanalith (Wolman Salts), acid copper chromate (Celcure), zinc meta arsenite, ammoniacal copper arsenite (Chemonite), chromated zinc arsenate (Boliden salt), and chromated copper arsenate (Greensalt or Erdalith). These preservatives are employed principally in the treatment of wood for uses where it will not be in contact with the ground or water and where the treated wood requires painting. As a general rule, they are less resistant to leaching and do not perform so satisfactorily as the preservative oils under conditions favorable to leaching. The leaching resistance of some of these preservatives has been developed to the extent that good performance can be expected in ground contact or in other wet installations, but they are still not considered equal in effectiveness to creosote when used under such conditions. On the other hand, waterborne preservatives are generally preferable to creosote for indoor use and can give indefinitely long life where not subject to leaching.

Waterborne preservatives leave the wood surface comparatively clean, paintable, and free from objectionable odor. With several exceptions, they must be used at low treating temperatures (100° to 160° F.) because of their instability at the higher temperatures common with preservative oils. This may involve some difficulty when higher temperatures are needed to obtain good treating results in such woods as Douglas-fir. Since water is added during treatment, the wood must be dried after treatment to the moisture content required for use.

Zinc chloride and chromated zinc chloride are frequently used as fire retardants for wood but at retentions higher than those used only for wood-preserving purposes.

Zinc Chloride and Its Improved Forms

Zinc chloride and chromated zinc chloride are covered in Federal Specifications TT–W–576a and TT–W–551, respectively, and in American Wood-Preservers' Association Standard P5 (1). Copperized chromated zinc chloride and chromated zinc chloride (FR) [27] are included in American Wood-Preservers' Association Standards P5

---

[27] (FR) for fire retardant.

and P10 (*1*), respectively. Copperized chromated zinc chloride is covered by Interim Federal Specification No. TT–W–00562.

Until 1938, zinc chloride was the most extensively used waterborne preservative in the United States. Its principal advantages are its low cost, cleanness, and lack of fire hazard. Zinc chloride solutions were at one time mixed with creosote for the treatment of crossties and construction timbers, but this use has been discontinued because the zinc chloride does not give so satisfactory performance as some of the other extenders for creosote already mentioned. Since 1938, zinc chloride has been gradually displaced by other waterborne preservatives.

Chromated zinc chloride was developed and promoted about 1934. The specifications require that it contain not less than 77.5 percent of zinc chloride and not less than 17.5 percent of sodium dichromate dihydrate. The claim that chromated zinc chloride has greater resistance to leaching than plain zinc chloride is supported by the results of some field tests but not by others.

More recently developed forms of chromated zinc chloride are copperized chromated zinc chloride (*14*) and chromated zinc chloride (FR) (*3*). Copperized chromated zinc chloride contains approximately 73 percent of zinc chloride, 20 percent of sodium dichromate, and 7 percent of cupric chloride. This preservative has been in commercial use for a limited time, but in laboratory and accelerated field tests it has shown better permanence than chromated zinc chloride. These tests indicate that the two preservatives compare favorably in other properties. Chromated zinc chloride (FR) contains 80 percent of chromated zinc chloride, 10 percent of boric acid, and 10 percent of ammonium sulfate. Retentions of from 1½ to 3 pounds per cubic foot of wood provide combined protection from fire, decay, and insect attack.

Tanalith (Wolman Salts)

Tanalith (Wolman Salts) is covered by U. S. Patent No. 1,957,873 (1934). Its approximate composition, according to Federal Specification TT–W–573 and American Wood-Preservers' Association Standard P5 (*1*), is 25 percent of sodium fluoride, 25 percent of disodium hydrogen arsenate, 37.5 percent of sodium chromate, and 12.5 percent of dinitrophenol. Service records on various types of above-ground wood structures treated with this preservative show good performance.

Acid Copper Chromate (Celcure)

Acid copper chromate (Celcure) is covered by U. S. Patents Nos. 1,684,222 (1926) and 2,041,655 (1936). It contains, according to Federal Specification TT–W–546, approximately equal quantities of copper sulfate and sodium dichromate and a sufficient quantity of free acetic acid or chromic acid to maintain the preservatives in solution under operating conditions. Tests on stakes and posts exposed to decay and termite attack indicate that wood well impregnated with Celcure gives good service. Tests by the Forest Products Laboratory and the U. S. Navy showed that wood thoroughly impregnated with at least 1 pound of Celcure per cubic foot has some resistance to marine borer attack. The protection against marine borers furnished by this preservative, however, is much less than that provided by a standard treatment with creosote.

Zinc Meta Arsenite (ZMA)

Zinc meta arsenite is covered by U. S. Patents Nos. 1,659,135 (1928) and 1,984,254 (1934). Federal Specification TT–W–581 and American Wood-Preservers' Association Standard P5 (1) require that the zinc meta arsenite in treating solutions be composed approximately of 60 parts of arsenious acid and 40 parts of zinc oxide, with sufficient acetic acid to keep the preservative in solution under operating conditions. Zinc meta arsenite has been in limited use for nearly 25 years. Service records on treated crossties, poles, and posts show that it provides good protection.

Ammoniacal Copper Arsenite (Chemonite)

Ammoniacal copper arsenite (Chemonite), which was developed about 1925, is covered by U. S. Patent No. 2,149,284 (1939) and has been in commercial use for about 14 years. According to Interim Federal Specification No. TT–W–00549b, Chemonite should contain approximately 57.7 percent of copper hydroxide, 40.7 percent of arsenic trioxide, and 1.6 percent of acetic acid. These ingredients should be dissolved in a solution of ammonia and water with the weight of the ammonia not less than 1.5 times nor more than 2.0 times the weight of the copper hydroxide in the final solution. The solution strength is adjusted as necessary according to the retention of dry chemicals desired in the treated wood. The net retention of preservative is calculated as pounds of copper hydroxide plus arsenic trioxide per cubic foot of wood treated within the proportions as limited in the specification.

Service records on Chemonite-treated structures show that this preservative provides good protection against decay and termites (8). Chemonite adds somewhat to the life of wood exposed to attack by marine borers, but it cannot be recommended as a substitute for creosote when long service is required or when the treated structure is to be used in heavily infested waters.

Chromated Copper Arsenate (Greensalt or Erdalith)

Chromated copper arsenate (Greensalt or Erdalith) was developed in India under the name of Ascu and is covered by U. S. Patent No. 2,106,978 (1938) (8). It has been used in the United States since 1938 to a limited extent for the treatment of poles, posts, and lumber. A standard of the American Wood-Preservers' Association stipulates that Greensalt contain 56 percent of potassium dichromate, 33 percent of copper sulfate, and 11 percent of arsenic pentoxide. Service data, on poles, posts, and stakes treated with this preservative and installed in the United States, show good protection against decay and termites over a period of 12 years or less. Favorable records for a somewhat longer period are available outside the United States.

Chromated Zinc Arsenate (Boliden Salt)

Chromated zinc arsenate (Boliden salt) was developed in Sweden and has been used there since 1936 for the commercial treatment of various wood products (*8*). It has been used commercially in the United States since 1951 and is covered by U. S. Patent No. 2,139,747 (1938) Interim Federal Specification No. TT–W–00538 stipulates that chromated zinc arsenate contain approximately 20 percent of arsenic acid, 21 percent of sodium arsenate, 16 percent of sodium dichromate, and 43 percent of zinc sulfate (*30*). Test data on stakes installed in Sweden in 1936, in South Africa in 1937, and in the United States and Panama Canal Zone in 1940 show that chromated zinc arsenate, in retentions of 0.33 pound (anhydrous basis) per cubic foot or higher, provides good protection against decay and against termites (*4, 8*).

PRESERVATIVE EFFECTIVENESS

Results of service tests on various treated products showing the effectiveness of different wood preservatives are published in the annual Proceedings of the American Wood-Preservers' Association. There are few service tests, however, that include a variety of preservatives under comparative conditions of exposure. Some Forest Products Laboratory stake tests (*4*) show such comparative data under similar exposure conditions, and table 45 shows comparative data on test posts treated with a number of preservatives in general use, as well as several in less extensive use.

PREPARING TIMBER FOR TREATMENT

For satisfactory treatment and good performance thereafter, the timber must be sound and suitably prepared for treatment. Except in a few specialized treating methods involving unpeeled or green material, the timber should be well peeled and either seasoned or otherwise conditioned before treatment. It is also highly desirable that all machining of timber be completed before treatment. Machining may include incising to improve the preservative penetration in woods that are resistant to treatment, cutting, framing, or boring of holes.

Peeling

Peeling round or slabbed timber is necessary to enable it to season quickly enough to avoid decay and insect damage and to permit the preservative to penetrate satisfactorily.[28] Strips of even the thin inner bark may prevent penetration. Patches of bark left on during treatment usually fall off in time and expose untreated wood, thus permitting decay to reach the interior of the timber. Careful peeling is especially important for wood that is to be treated by a superficial method. In the more thorough processes some penetration will take

[28] Processes in which a preservative is forced or diffuses through green wood lengthwise do not require peeling of the timber.

place both lengthwise and tangentially in the wood, and consequently small strips of bark are not quite so harmful. Machines of various types have been developed for peeling round timbers, such as poles, piling, and posts (10).

Seasoning

For treatment with waterborne preservatives by certain diffusion methods, the wood should be as green and as full of water as possible. For treatment by other methods, however, and particularly for treatment with preservative oils, seasoning before treatment is desirable.

Plants treating timber by pressure processes can condition green timber by other means than air-drying to make it more absorptive and thus avoid a long delay and possible deterioration of the timber before treatment. Nevertheless, air-drying, despite the greater time, labor, and storage space required, is a widely used method of conditioning and is generally the cheapest and most effective, even for pressure treatment. Kiln-drying is often used for the seasoning of lumber before treatment and has been used experimentally as a means of seasoning poles (16, 21).

How long the lumber must be air-dried depends on the climate, location, and condition of the seasoning yard, methods of piling, season of the year, and size, species, and character of the timbers (16). It is not uncommon in commercial practice to air-dry railway ties and larger timbers for 8 to 12 months if conditions permit the timber to be kept that long without deterioration, but the desirability of so long a drying period is not fully established. A great deal can be accomplished even with large pieces in 3 to 6 months of good seasoning weather. Small pieces season more quickly than large pieces and sapwood more quickly than heartwood. The most satisfactory seasoning practice for any specific case will depend on the individual drying conditions and the preservative treatment to be used and must be determined by experience. Treating specifications therefore are not generally specific as to moisture content requirements.

It is not necessary that the timber be seasoned to a uniform moisture content throughout, but the part that is to be penetrated by the preservative must have enough water removed to make room for the preservative to enter. It is also important that the timbers be seasoned sufficiently, so that any checking that may occur subsequent to treatment will not expose unpenetrated wood.

Hardwood ties and timbers tend to check and split seriously during air-drying. These defects can be reduced by careful piling to avoid rapid drying and by applying antichecking irons or dowels before seasoning. Split ties are also salvaged by pressing the split ends together in a hydraulic jack and holding them with bolts or spiral dowels (16).

To prevent decay and other forms of fungus infection during air-drying, the timbers should be cut and seasoned when conditions are less favorable for fungus development (see p. 381). If this is impossible, chances for infection can be minimized by prompt conditioning of the green timbers (see p. 414), careful piling and roofing during air-drying, and pretreatment of green timber with preservatives that will protect the timber during air-drying (20).

Incising

Wood that is resistant to penetration by preservatives is often incised (fig. 87) before treatment to permit deeper and more uniform penetration. To accomplish this, sawed or hewed timber is passed through a machine having horizontal and vertical rollers equipped with teeth that sink into the wood to a predetermined depth, usually ½ to ¾ inch. The teeth are spaced so as to give the desired distribution of preservative with the minimum number of incisions. A machine of different design is required for incising the butts of poles. The effectiveness of incising depends on the fact that preservatives usually penetrate into wood much farther in a longitudinal direction than in a direction perpendicular to the faces of the timber. The incisions expose end-grain surfaces and thus permit longitudinal penetration. Incising is practiced chiefly on Douglas-fir, western hemlock, and western larch ties and timbers for pressure treatment and on poles of cedar and other species. It is especially effective in improving penetration in the heartwood areas of sawed or hewed surfaces.

Cutting and Framing

All cutting, framing, and boring of holes should be done before treatment (1). Cutting into the wood in any way after treatment will frequently expose the untreated interior of the timber and permit ready access to decay fungi or insects. It is much more practical than is commonly supposed to design wood structures so that all cutting and framing may be done before treatment. Railroads are are now following the practice extensively and find it not only practical but economical (26). Many wood-preserving plants are equipped to carry on such operations as the adzing and boring of crossties; gaining, roofing, and boring of poles; and the framing of material for bridges and for many specialized structures, such as water tanks and barges.

Treatment of the wood with preservative oils involves little or no dimensional change. In the case of treatment with waterborne preservatives, however, some change in the size and shape may occur even though wood is reseasoned to its original moisture content. If precision fitting is necessary, the wood is cut and framed before treatment to its approximate final dimensions in order to allow for slight surfacing, trimming, and reaming of bolt holes. Grooves and bolt holes for timber connectors are cut before treatment and can be reamed out if necessary after treatment, although this is not always necessary.

APPLYING PRESERVATIVES

Wood-preserving methods are of two general types: (1) Pressure processes, in which the wood is impregnated in closed vessels under pressures considerably above atmospheric pressure, and (2) nonpressure processes, which vary widely as to procedures and equipment used. Pressure processes generally provide a closer control over preservative retentions and penetrations, and to this extent they usually provide greater protection than nonpressure processes. Some of the nonpressure methods, however, are better than others and are occasionally as effective as pressure processes in providing good preservative retentions and penetrations.

M 93345 F

FIGURE 87.—Pneumatically controlled timber incisor with large-size timber leaving machine.

PRESSURE PROCESS

In commercial practice, wood is most often treated by surrounding it with preservative in high-pressure apparatus and applying pressure to the preservative (25). A number of pressure processes (1, 16, 19) differ from one another in a few details, but the general principle is the same in all. The wood, placed on steel cars, is run into a long steel cylinder (fig. 88), which is then closed and filled with preservative. Pressure then forces preservative into the wood until it has absorbed the desired amount. A considerable amount of preservative is absorbed resulting in a relatively deep penetration. Two principal types of pressure treatment, the full-cell or Bethell process and the empty-cell process, are in common use.

Full-Cell Process

The full-cell process is used when the retention of a maximum quantity of preservative is desired. For timbers to be treated with creosote and exposed to marine borers it is standard. Water-borne preservatives are generally applied by the full-cell process, and control over preservative retention is obtained by regulating the

M 92948 F

FIGURE 88.—Exterior view of treating cylinders at wood-preserving plant under construction.

concentration of the treating solution. The steps in this process are essentially as follows:

1. The charge of timber is sealed in the treating cylinder, and a preliminary vacuum is applied for ½ hour or more to remove the air from the cylinder and as much as possible from the wood.

2. The preservative, previously heated to somewhat above the desired treating temperature, is admitted to the cylinder without the admission of air.

3. After the cylinder is filled, pressure is applied until the required retention of oil is obtained.

4. When the pressure period is completed, the preservative is withdrawn from the cylinder.

5. A short final vacuum may be applied immediately afterward to free the charge from dripping preservative.

When the timber is steamed before treatment, the preservative is admitted at the end of the vacuum period that follows steaming. When the timber has received preliminary conditioning by the Boulton or boiling-under-vacuum process, the cylinder can be filled and the pressure applied as soon as the conditioning period is completed.

Empty-Cell Process

The object of empty-cell treatment is to obtain deep penetration with a relatively low net retention of preservative. For treatment with oil preservatives, the empty-cell process should always be used if it will result in the desired retention. Two empty-cell treatments, the Rueping and the Lowry processes, are commonly employed, both of which use the expansive force of compressed air to drive out part of the preservative absorbed during the pressure period.

The Rueping empty-cell process has been widely used for many years, in both Europe and the United States. The following general procedure is employed:

1. Air under pressure is forced into the treating cylinder, which contains the charge of timber. The air penetrates some species easily, requiring but a few minutes' application of pressure. In the treatment of the more resistant species, common practice is to maintain air pressure from ½ to 1 hour before admitting the preservative, but the necessity for long air-pressure periods does not seem fully established. The air pressures employed generally range between 25 and 100 pounds per square inch, depending on the net retention of preservative desired and the resistance of the wood.

2. Following the application of preliminary air pressure, the preservative is forced into the cylinder. During the filling process the air in the treating cylinder interchanges with preservative at the same pressure in an equalizing or Rueping tank, or it is gradually allowed to escape from the treating cylinder as the preservative is pumped in, at a rate that keeps the pressure within the cylinder constant. When the treating cylinder is filled with preservative, the treating pressure is raised above that of the initial air and is maintained until the point of refusal is reached, or until the gross absorption is sufficient to leave the required net retention of preservative in the wood after the completion of the treatment.

3. At the end of the pressure period the preservative is drained from the cylinder, and a final vacuum removes the surplus preservative from the wood. The amount recovered may be from 20 to 60 percent of the gross amount injected.

The Lowry process is often called the empty-cell process without initial air. The chief difference between it and the Rueping process, which employs initial air pressures above atmospheric, is that in the Lowry process the preservative is admitted to the cylinder without either an initial air pressure or a vacuum, and the air originally in the wood at atmosphere pressure is imprisoned during the filling period. After the cylinder is filled with the preservative, pressure is applied, and the remainder of the treatment is the same as described for the Rueping treatment.

The Lowry process has the advantage that the equipment for the full-cell process can be used without other accessories, while the Rueping process usually requires additional equipment, such as an air compressor and an extra cylinder or Rueping tank for the preservative or a suitable pump to force the preservative into the cylinder against the air pressure. Both processes, however, have advantages, and both are widely and successfully used.

Conditioning Green Timber for Pressure Treatment

When green timber is to be treated under pressure (19), either of two commonly used methods for conditioning may be selected. One is the steaming-and-vacuum process, which is employed mainly for southern yellow pine, and the other is the Boulton or boiling-under-vacuum process, which is used for Douglas-fir and to some extent for hardwoods. In the steaming process the green timber is steamed in the treating cylinder for several hours, usually at 20 pounds gage pressure (259° F.), and when the steaming is completed a vacuum is immediately applied. During the steaming period the outer part of the wood is heated to a temperature approaching that of the steam, and the subsequent vacuum lowers the boiling point so much that part of the water is evaporated or is forced out of the wood by the steam produced when the vacuum is applied. The steaming and vacuum periods employed depend upon the size, species, and moisture content of the wood. The steaming method usually reduces the moisture content of green wood somewhat, and the heating assists greatly in getting the preservative to penetrate, but it does not dry the wood to the "seasoned" condition. If the steaming period is sufficient, it will also sterilize the wood.

In the Boulton or boiling-under-vacuum method of partial seasoning, the timber is heated in the preservative oil under vacuum, usually at temperatures of about 180° to 220° F. This temperature range, lower than that of the steaming process, is a considerable advantage in treating woods that are especially susceptible to injury from high temperatures. The Boulton method, however, usually removes only a very limited amount of moisture from the heartwood.

A third method of conditioning known as "vapor drying" (15) has been patented and has recently come into commercial use for the seasoning of railroad ties, poles, and other products. The green wood is seasoned by subjecting it in the treating cylinder to the vapors produced by boiling an organic chemical, such as xylene, and removing the mixed vapors of water and the chemical from the drying chamber. A small quantity of the chemical remains in the wood, but the balance is recovered and reused. The wood is treated by standard pressure methods after the seasoning is completed.

TREATING CONDITIONS AND SPECIFICATIONS

Specifications on the treatment of various wood products by pressure processes and on the hot-and-cold-bath treatment of cedar poles have been developed by the American Wood-Preservers' Association (1). These specifications limit pressures, temperatures, and time during conditioning and treatment in order to avoid conditions that will cause serious injury to the wood. They also contain minimum requirements as to preservative retentions and penetrations and recommendations for handling wood after treatment, as means of providing a quality product. The specifications are rather broad in some respects, allowing the purchaser some latitude in specifying the details of his individual requirements. The purchaser should exercise great care, however, to avoid limiting the operator of the treating plant so closely that he cannot do a good treating job, and to avoid requiring treating conditions so severe that they will damage the wood. Federal Specification TT–W–571 (27) recommends treatment practices for use on U. S. Government orders for pressure-treated wood products; other purchasers have specifications similar to those of the American Wood-Preservers' Association.

Treating Pressures and Preservative Temperatures

The pressures used in the full-cell and Lowry (empty-cell) treatments vary from about 100 to 200 pounds per square inch, depending on the species and the ease with which the wood takes the treatment, but are most commonly about 150 to 175 pounds per square inch. Pressures applied in the Rueping (empty-cell) treatment are usually between 150 and 200 pounds per square inch, depending more or less on the preliminary air pressure employed. Many woods are sensitive to high treating pressures, especially when hot. American Wood-Preservers' Association Standards (1), for example, permit a maximum pressure of 150 pounds per square inch in the treatment of Douglas-fir and 125 pounds per square inch for western redcedar poles. In commercial practice even lower pressures are frequently used on such woods.

Specifications of the American Wood-Preservers' Association commonly require that the temperature of preservative oils during the pressure period shall not be more than 210° F. and shall average at least 180° F. Since high temperatures are much more effective than low temperatures for treating resistant wood, better results are obtained by using average temperatures between 190° and 200° F. with both preservative oils and water solutions that are not injured by high temperatures. With a number of waterborne preservatives, however, especially those containing chromium salts, maximum temperatures are limited to 160° F. or less to avoid partial precipitation of the preservative.

Preservative Retention and Penetration

Preservative retentions are generally specified in terms of the weight of preservative per cubic foot of wood treated, based on total weight of preservative retained and the total volume of wood treated in a charge. Some specifications, however, stipulate a minimum retention of preservative as determined from chemical analysis of borings from the treated wood (9).

The preservatives and minimum retentions recommended for various products in Federal Specification TT–W–571 (28) are shown in tables 46, 47, and 48. Since the figures given in tables 46, 47, and 48 are minimum retentions, it may often be desirable to use retentions somewhat higher. For lumber of small sizes, retentions should be higher than those for larger timbers in order to compensate for the increased surface area per unit volume (18). It is customary to increase retentions of waterborne preservatives about 50 percent when the treated wood is used under moist conditions that may result in leaching of the preservatives.

Somewhat higher retentions of preservative oils are justified in products to be installed under severe climatic or exposure conditions. Heavy-duty poles and other items with a high replacement cost are also often treated with higher than the minimum retentions. It may be necessary to increase retentions in order to assure satisfactory penetration, particularly when the sapwood is either unusually thick or is somewhat resistant to treatment. To reduce bleeding of the preservative, however, it may be desirable to use preservative-oil

TABLE 46.—*Recommended minimum net retentions of creosote and solutions containing creosote for various wood products* [1]

Product and service condition	Coal-tar creosote	Creosote-coal-tar solutions	Creosote-petroleum oil solution
	Lb. per cu. ft.	Lb. per cu. ft.	Lb. per cu. ft.
Ties (crossties, switch ties, and bridge ties)	8	8	9
Lumber and structural timbers:			
For use in coastal waters:			
Douglas-fir (coast type)	14	14	
Southern yellow pine	2 20	2 20	
For use in fresh water, in contact with ground, or for important structural members not in contact with ground or water	10	10	12
For other use not in contact with ground or water	6	6	7
Piles:			
For use in coastal waters:			
Douglas-fir (coast type)	14	14	
Southern yellow pine	2 20	2 20	
For land or fresh-water use	12	12	14
Poles	8		
Posts	6	6	7

[1] As stipulated in Federal Specification TT–W–571. Requirements of American Wood-Preservers' Association standards as to penetrations and treating conditions are to be followed with minor exceptions noted.

[2] The greater retention specified for southern yellow pine is due to the fact that this species has thicker sapwood and it is thus easier to get a heavier retention than with Douglas-fir.

TABLE 47.—*Recommended minimum net retentions of oilborne preservatives for various wood products* [1]

Product and service condition	Pentachlorophenol, 5 percent in petroleum oil	Copper naphthenate (0.75 percent copper metal) in petroleum oil
Lumber and structural timber:	Lb. per cu. ft.	Lb. per cu. ft.
For use in contact with ground	10	10
For use not in contact with ground or water	6	6
Poles	8	8
Posts	6	6

[1] As stipulated in Federal Specification TT–W–571. Requirements of American Wood-Preservers' Association standards as to penetrations and treating conditions are to be followed with minor exceptions noted.

retentions lower than the stipulated minimum. Treatment to refusal is usually required for woods that are resistant to treatment and will not absorb sufficient preservative to meet the minimum retention requirements.

Penetrations vary widely, even in pressure-treated material. In most species, heartwood is more difficult to penetrate than sapwood. In addition, species differ greatly in the degree to which their heartwood may be penetrated; the red heartwood of beech and the heartwood of white oak and Rocky-Mountain-type Douglas-fir, for example, are exceptionally resistant to penetration by commercial treating processes, even when incised. Penetrations in unincised

TABLE 48.—*Recommended minimum net retentions of waterborne preservatives for various wood products* [1]

Product and service condition	Celcure (acid cupric chromate)	Chemonite (ammoniacal copper arsenite)	Chromated zinc chloride	Tanalith	Boliden salt (chromated zinc arsenate)	Copperized chromated zinc chloride
Lumber and structural timber:	*Lb. per cu. ft.*	*Lb. per cu. ft.*	*Lb. per cu. ft.*	*Lb. per cu. ft.*	*Lb. per cu. ft.*	*Lb. per cu. ft.*
For use under moderate leaching conditions_____	0.75	0.50	1.15	0.55	1.00	1.15
For use not in contact with ground or water_____	.50	.30	.75	.35	.50	.75
Posts_____	.75	.50	1.15	.55	1.00	1.15

[1] As stipulated in Federal Specification TT–W–571. Requirements of American Wood-Preservers' Association standards as to penetrations and treating conditions are to be followed.

heart faces of these species may occasionally be as deep as ¼ inch, but usually are less and often are not more than ¹⁄₁₆ inch. Long experience has shown that even these slight penetrations have value, although deeper penetrations are highly desirable. The heartwood of coast-type Douglas-fir, southern yellow pine, and various hardwoods, while highly resistant, can be made to take transverse penetrations of ¼ to ½ inch on the average and sometimes considerably more. Incising is beneficial for many resistant species. The white heartwood of beech and the heartwood of the red oaks, black tupelo, and ponderosa pine can usually be penetrated without difficulty.

Complete penetration of the sapwood should be the ideal in all pressure treatments. It can often be accomplished in small-size timbers of various commercial woods, and with skillful treatment it may often be obtained in piling, ties, and structural timbers. Practically, however, the operator cannot always insure complete penetration of sapwood in every piece when treating large pieces of round material having thick sapwood, for example, poles and piling. Specifications therefore permit some tolerance, as in American Wood-Preservers' Association Standard C4 (*1*) on southern yellow pine poles, which, for a retention of creosote of 8 pounds per cubic foot, requires that 2.5 inches or 85 percent of the sapwood thickness be penetrated in not less than 18 out of 20 poles sampled in a charge. The requirements vary somewhat depending on the species and specified retentions.

HANDLING AND SEASONING OF TIMBER AFTER TREATMENT

Treated timber should be handled with sufficient care to avoid breaking through the treated part. The use of pikes, cant hooks, picks, tongs, or other pointed tools that dig deeply into the wood should be prohibited. Handling heavy loads of lumber or sawed timber in rope or cable slings may crush the corners or edges of the outside pieces. Breakage or deep abrasions may also result from throwing the lumber or dropping it any considerable distance. When damage results through carelessness or otherwise, the exposed places should be re-treated as thoroughly as conditions permit. Treated wood that must be stored before use should be protected in order to avoid checking or mechanical injury.

Although cutting wood after treatment is highly undesirable, it cannot always be avoided. When cutting is necessary, the damage may be partly overcome in timber for land or fresh-water use by thoroughly brushing the cut surfaces with coal-tar creosote. Two coats of hot creosote should be applied if practicable, but cold application will be better than none. Brush-coating cut surfaces, however, gives little protection against marine borers. A special device is available for applying pressure treatment to bolt holes bored after treatment. For wood treated with water-soluble preservatives, where the use of creosote on the cut surfaces is not practicable, a strong solution of the preservative in use may be substituted.

For treating the end surfaces of piles where they are cut off after driving, at least two generous coats of creosote should be applied with great care and thoroughness. A coat of asphalt or similar material may well be applied over the creosote, followed by some protective sheet material, such as metal, roofing felt, or saturated fabric, fitted over the pile head and brought down the sides far enough to protect against damage to the top treatment and the entrance of storm water. American Wood-Preservers' Association Standard M4–52 (1) contains instructions for the care of pressure-treated wood after treatment.

Timber treated with creosote and other oils may be used immediately after treatment, if desired, but a period of seasoning may help dry the oily surfaces and make the timber less unpleasant to handle. The removal of petroleum-oil solvents is particularly important in the case of wood that requires painting after treatment with oilborne preservatives, such as pentachlorophenol.

With waterborne preservatives, seasoning after treatment is important for the timbers that are to be used in buildings or in any other place where shrinkage after placement in the structure would be undesirable. Injecting waterborne preservatives puts large amounts of water into the wood, and considerable shrinkage is to be expected as subsequent seasoning takes place. For best results, the wood should be dried to approximately the moisture content it will ultimately reach in service.

With some waterborne preservatives, seasoning after treatment is recommended for all treated timber in order to complete the chemical reactions that are intended to take place within the wood and, thus, to increase the resistance of the preservative to leaching by water.

INSPECTION

Inspection of timber for quality and grade before treatment is desirable. Grade-marked lumber and timber can be specified and obtained in many instances. Such material is graded at the mills and is usually dependable without additional inspection. When inspection previous to treatment is impractical, the purchaser can usually inspect for quality and grade after treatment; if this is to be done, however, it should be made clear in the purchase order.

Currently, there are no generally accepted procedures for accurately determining retention of various preservatives, other than measurement of quantity of preservative used in treatment. As previously mentioned, some specifications (9) provide for a determination of minimum retentions by chemical analysis of borings from the treated wood, but such sampling and analysis have not been used to determine the overall retention of a charge, as it is commonly specified. Furthermore, the nature and quantity of the preservative in the wood change as the period of service increases, so that samples of treated wood taken sometime after treatment may not be representative of those taken when the treatment was completed. The purchaser, therefore, must either accept the statements or affidavit of the treating-plant operator or have an inspector at the treating plant to observe the treatments and insure compliance with the specifications. Railroad companies and other corporations that purchase large quantities of treated timber usually maintain their own inspection services. Commercial inspection and consulting service is available for purchasers willing to pay an inspection fee but not using enough treated timber to justify employing inspectors of their own. Care should be taken to choose experienced, competent, and reliable inspectors, since premature failures of treated structures can often be attributed to faulty inspection.

Penetration measurements should be made at the treating plant if inspection service is provided but can be made by the purchaser at any time after the timber has been treated. They give about the best single measure of the thoroughness of the treatment. Rejection of treated timber for insufficient penetration is hardly enforceable, however, unless there was a previous agreement with the plant operator on minimum penetration requirements.

The depth of penetration of creosote and other dark-colored preservatives can be determined directly by observing a boring removed by an increment borer. The boring should usually be taken at about midlength of the piece, or at least several feet from the end of the piece, in order to avoid the unrepresentative end portion that is sometimes completely treated by end penetration. Since preservative oils tend to creep over cut surfaces, the observation should be made promptly after the boring is taken. Holes made for penetration measurements should be tightly filled with a thoroughly treated wood plug.

The penetration of preservatives that are practically colorless must be determined by chemical dips or sprays that show the penetration by color reactions.

EFFECT OF TREATMENT ON STRENGTH

Coal-tar creosote, creosote-coal tar mixtures, and creosote-petroleum oil mixtures are practically inert to wood and have no chemical influence upon it that would affect its strength. Likewise, solutions containing standard waterborne preservatives, in the concentrations commonly used in preservative treatment, and solutions of toxic chemicals in petroleum apparently have no important effect on the strength of wood.

Although wood preservatives are not regarded as harmful in themselves, the treatment necessary to inject them into the wood, if unusually severe or if not properly carried out, may result in considerable loss in strength of the wood. Among the factors that influence the effect of the treating process on strength are (1) the species of wood, (2) the size and moisture content of the timbers treated, (3) the heating medium used and its temperature, (4) the length of the heating period used in conditioning the wood for treatment and the length of time the wood is in the hot preservative, and (5) the amount of pressure used. Of these factors, the most important are the severity and duration of the heating and pressure conditions used. Information on the effect of temperature on the strength of wood is covered on pages 89–93. Working stresses for treated timber are discussed on pages 161–162.

NONPRESSURE PROCESSES

The numerous nonpressure processes differ widely in the retentions and penetrations of preservative obtained and consequently in the degree of protection that they provide to the treated wood. When similar retentions and penetrations are achieved, wood treated by a nonpressure method should have a service life comparable to that of wood treated by pressure. It should be recognized, nevertheless, that the results of nonpressure treatments, particularly those involving superficial applications, are not so generally satisfactory as the results of pressure treatment. The less effective superficial nonpressure processes, however, serve a useful purpose when more thorough treatments are impractical.

Nonpressure methods, in general, consist of: (1) Superficial applications of preservative oils by spraying, brushing, or brief dipping; (2) soaking in preservative oils or steeping in water solutions; (3) diffusion processes with waterborne preservatives; (4) various adaptations of the hot-and-cold-bath process; (5) vacuum treatment; and (6) a variety of miscellaneous processes (16).

Superficial Applications

The simplest treatment is to apply the preservative—creosote or other oils—to the wood with a brush or a spray nozzle. Oils that are thoroughly liquid when cold should be selected, unless it is possible to heat the preservative. The oil should be flooded over the wood, rather than merely painted upon it, and every check and depression in the wood should be thoroughly filled with the preservative, since any untreated wood left exposed provides ready access for fungi. At least two coats should be applied, the second one after the first has dried. Rough lumber may require as much as 10 gallons of oil to 1,000 square feet of surface, but surfaced lumber requires considerably less. The transverse penetrations obtained will usually be less than one-sixteenth of an inch.

Brush and spray treatments should be used only when more effective treatments cannot be employed. The additional life obtained by such treatments over that of untreated wood will be affected greatly by the conditions of service; for wood in contact with the ground, it may be from 1 to 3 years. Brushing with solutions of water-soluble preservatives is seldom worth while for wood that is exposed to the weather, to soil, or to water and is likely to be less effective than brushing with preservative oils in protected situations.

Dipping for a few seconds to several minutes in a preservative oil gives greater assurance than brushing or spraying that all surfaces and checks are thoroughly coated with the oil, and it usually results in slightly greater penetrations. It is a common practice to treat window sash, frames, and other millwork, either before or after assembly, by dipping for approximately 3 minutes in a water-repellent preservative (*23*). The amount of preservative used may vary from about 6 to 17 gallons per thousand board-feet (0.5 to 1.5 pounds per cubic foot) of millwork treated. The penetration of preservative into end surfaces of ponderosa pine sapwood is, in some cases, as much as 1 to 3 inches. End penetration in such woods as southern yellow pine and Douglas-fir, however, is much less, particularly in the heartwood. Transverse penetration of the preservative applied by brief dipping is very shallow, usually only a few hundredths of an inch. Since the exposed end surfaces at joints are the most vulnerable to decay in millwork products, good end penetration is especially advantageous. Dip applications provide very limited protection to wood used in contact with the ground or under very moist conditions (*4*). They do have value, however, for exterior woodwork and millwork that is painted, that is not in contact with the ground, and that is exposed to moisture only for brief periods at a time (*29*).

Cold Soaking and Steeping

Cold soaking well-seasoned wood for several hours or days in low-viscosity preservative oils or steeping green or seasoned wood for several days in waterborne preservatives, such as zinc chloride, are limited-purpose processes that have been tried with varying degrees of success on fence posts, lumber, and timbers (*7, 28*).

Pine posts treated by cold soaking in a solution containing 5 percent of pentachlorophenol in No. 2 fuel oil for 24 to 48 hours or longer, in most cases, were in good condition after approximately 10 years of service test. These posts were well penetrated in the sapwood and had preservative-solution retentions of from 2 to 6 pounds per cubic foot. Most species are not treated so satisfactorily as the pines by cold soaking, and test posts of such woods as birch, aspen, and sweetgum treated by this method have failed prematurely (*7*). Preservative retentions and penetrations obtained by cold soaking lumber for several hours are considerably better than those obtained by brief dipping of similar species. Preservative retentions, however, seldom equal those obtained in pressure treatment unless the wood is infected with fungus at the time of treatment (*5*).

Steeping with waterborne preservatives has very limited use in the United States but has been employed for many years in Europe. In treating seasoned wood both the water and the preservative salt in the solution soak into the wood. With green wood, the preservative enters the water-saturated wood by diffusion. Preservative retentions and penetrations vary over a wide range, and the process is not generally recommended when more reliable treatments are practical. Tests on posts also show variable results depending upon the preservative, the species of wood, and the severity of the exposure conditions. When mercuric chloride [29] is used as a preservative, as is fairly common in Europe, the steeping process is known as Kyanizing. Zinc chloride, copper sulfate, sodium fluoride, and other waterborne preservatives also may be used. Preservative-solution concentrations vary with different preservatives; 1 percent solution is common with mercuric chloride and 5 percent for zinc chloride. The wood is soaked in the solution from 1 to 2 weeks.

Diffusion Processes

In addition to the steeping process, diffusion processes are used with green or wet wood. These processes employ waterborne preservatives that will diffuse out of the water of the treating solution into the water of the wood. The end-diffusion process is a simple, low-cost process of limited effectiveness that can be applied to unpeeled, unseasoned fence posts. A newer diffusion process of considerable promise, called double diffusion, is under development by the Forest Products Laboratory (2). It consists of steeping green wood first in one chemical and then in another. The two chemicals diffuse into the wood and then react to deposit toxic compounds with high resistance to leaching. Service tests on posts treated by this method and installed as early as 1941 are showing very satisfactory results (6).

Other diffusion processes consist of applying preservatives to the butts or around the ground line of posts or poles. In standing-pole treatments the preservative may be injected with a special tool into the pole, applied as a paste or bandage, poured into holes bored in the pole at the ground-line zone, or poured on the surface of the pole and into an excavation several inches deep around the ground line of the pole. These treatments have recognized value for application to untreated standing poles.

Adaptations of the Hot-and-Cold-Bath Process

The hot-and-cold-bath treatment with coal-tar creosote and other oil preservatives is the most effective of the nonpressure processes, and the thoroughness of the treatment obtainable most nearly approaches that of the pressure processes. The treatment consists of heating the wood in the preservative in an open tank for several hours, and then quickly submerging it in cold preservative and allowing it to remain several hours in this cold bath. Two methods may be used: (1) Transferring the wood at the proper time from a hot-bath tank to a cold-bath tank; or (2) draining the hot preservative from a single tank and quickly refilling the tank with cold preservative. The same result can also be accomplished, although more slowly, by shutting off the heat at the proper time and allowing the wood and the hot preservative to cool together.

The principle involved is the same in each procedure. During the hot bath, the heating causes the air in the wood to expand and

[29] This preservative is very poisonous and therefore dangerous to handle.

some to be forced out. Heating the wood also helps improve the penetration of the preservative. When the cooling takes place, whether it is sudden or slow, the air in the wood contracts, and a partial vacuum is thus created that results in the forcing of liquid into the wood by the pressure of the atmosphere. Some preservative is absorbed during the hot bath, but most of the preservative is taken up by the wood during the cooling bath.

The chief use of the hot-and-cold-bath process is for treating poles of some of the thin-sapwood species, such as western redcedar, for telegraph, telephone, and power lines (fig. 89). The process is also useful for fence posts and for lumber or timbers for other purposes when circumstances do not permit the more effective pressure treatments. Coal-tar creosote and pentachlorophenol solutions are the preservatives ordinarily chosen for posts and poles, but water solutions of preservatives, such as zinc chloride solution, may also be employed if the solution is kept at uniform strength and the water-soluble preservative is not adversely affected by heating. For the preservatives that cannot safely be heated, the process must be modified.

With coal-tar creosote, hot-bath temperatures up to 235° F. may be employed, but usually a temperature of 210° to 220° is sufficient. In the commercial treatment of cedar poles, temperatures of from 212° to 235° are specified with creosote and pentachlorophenol solution (1). If the temperature is too high, a considerable percentage of the oil may be lost by evaporation, especially if a creosote with a relatively low boiling range is used. In the cold bath the desired temperature is usually approximately 100°. This temperature usually keeps the oil fluid but is much lower than that of the hot bath. If the oil is not sufficiently fluid at 100°, however, the temperature of the cold bath should be increased. Specifications for treating poles set a cold-bath temperature limit between 90° and 150° (1).

The length of both baths must be governed by the ease with which the timber takes treatment. With well-seasoned timber that is moderately easy to treat, a hot bath of 2 or 3 hours and a cold bath of like duration will probably be sufficient, but much longer periods are required with resistant woods. With preservative oils, the object is to obtain as deep penetration as possible, but with a minimum amount of oil. If the penetration is not sufficient, either the hot or the cold bath should be lengthened. If the penetration is satisfactory, but too much oil is absorbed, the cold bath should be shortened.

In the hot-and-cold-bath treatment of posts of some woods, such as southern yellow pine, particularly those posts containing molds, blue stain, and incipient decay, preservative retentions are often excessively high. One method of reducing preservative retentions is to employ a final heating or "expansion" bath with the creosote at 200° to 220° F. for an hour or two and to remove the wood while the oil is hot. This second heating expands the oil and air in the wood, and some of the oil is thus recovered. The expansion bath also leaves the wood cleaner than when it is removed directly from cold oil. Another means for reducing retentions of creosote is to use water, zinc chloride solution, or steam as a heating medium before the cold bath in creosote. By such heating methods, some water is absorbed that tends to reduce retention of preservative, and some sacrifice in penetration may also result.

M 92433 F

FIGURE 89.—Tanks for butt treatment of poles by hot-and-cold-bath process.

The best combination of treating conditions in any instance will vary with the character and the condition of the timber and must be learned by trial. The penetration cannot always be controlled at will by manipulating the treating conditions, however, for some woods are highly resistant to treatment and at best may permit a depth of only one-eighth inch or less, especially in heartwood faces. The sapwood of most species is less resistant to penetration than the heartwood and transverse penetrations of ¼ to 1 inch are often obtained in sapwood.

Certain preservative chemicals in water solution cannot safely be heated to high temperatures, because, at such temperatures, reactions may take place that precipitate some of the chemicals out of solution or otherwise reduce the value of the solution. Such solutions are not practical for use in the hot bath. This difficulty has been solved in some open-tank plants by heating the wood with steam instead of in the preservative. The tank is covered after the wood is in place, and steam is applied at atmospheric pressure for a few hours. At the conclusion of the steaming period, cold preservative solution is pumped into the tank and absorption takes place as the wood cools [see U. S. Patent No. 2,235,822 (1941)].

Preservatives consisting of toxic chemicals in volatile organic solvents cannot safely be heated in open tanks because of the loss of solvent by evaporation and the considerable danger of fire or explosion. Steaming is unsuitable because it wets the surface of the wood and thus interferes with the penetration of preservative oils. The obstacle is sometimes overcome by heating the wood in a dry kiln or hot box and then immersing it quickly in cold preservative. A kiln-heating process for this purpose, in which the relative humidity in the kiln is controlled during the heating process, is covered by U. S. Patent 1,991,811 (1935).

Vacuum Processes

Two vacuum processes, in which the wood is either subjected to a vacuum alone or to steaming and a vacuum before it is submerged in a cold bath of preservative, have had some commercial use in the treatment of lumber, timbers, and millwork. In the treatment of millwork, a vacuum is applied, the wood is immersed in an unheated water-repellent preservative, and, after the wood is removed from the preservative, a final vacuum is used to remove surplus preservative from the surface (11). In the treatment of mine timbers and lumber, the wood is subjected to an initial vacuum, steamed, subjected to a second vacuum, and immersed in an unheated waterborne preservative (12). Good penetrations and retentions of the preservative can be obtained by these processes in woods that are not resistant to treatment. Even with easily penetrated woods the results will vary within wide limits according to the magnitude and time of initial vacuum, the time the wood is immersed in the preservative, and, to some extent, the magnitude and time of the final vacuum. In treating millwork by the vacuum method, the treating time used and the retentions and penetrations of preservative obtained are similar to those with the brief dipping treatment of millwork. The vacuum treatment, however, leaves the wood in drier condition, so that there is no delay at the plant in the further processing of the millwork.

Miscellaneous Nonpressure Processes

A number of other nonpressure methods of various types have been used to a limited extent. Several of these involve the application of waterborne preservatives to living trees (*31*). The Boucherie process for the treatment of green, unpeeled poles (*16*) has been used for many years in Europe. The process consists of attaching liquid-tight caps to the butt ends of the poles and through a pipe line or hose leading to the cap, a waterborne preservative, such as copper sulfate, is forced into the pole under hydrostatic pressure.

The tire-tube process (*32*) is a simple adaptation of the Boucherie process used for treating green, unpeeled fence posts. In this treatment a section of used inner tube is fastened tightly around the butt end of the post to make a bag that is used to hold a solution of waterborne preservative, such as zinc chloride.

LITERATURE CITED

(1) AMERICAN WOOD-Preservers' ASSOCIATION.
  1952. MANUAL OF RECOMMENDED PRACTICE. STANDARD SPECIFICATIONS. (A looseleaf handbook, revised annually.)
(2) BAECHLER, R. H.
  1953. HOW TO TREAT FENCE POSTS BY DOUBLE DIFFUSION. U. S. Forest Prod. Lab. Rpt. R1955, 4 pp., illus. [Processed.]
(3) BESCHER, R. H., HENRY, W. T., and DREHER, W. A.
  1948. A STUDY OF PERMANENCE OF COMMERCIAL FIRE RETARDANTS. Amer. Wood-Preservers' Assoc. Proc. 44: 369–379, illus.
(4) BLEW, J. O., JR.
  1952. COMPARISON OF WOOD PRESERVATIVES IN STAKE TESTS (1952 progress report). U. S. Forest Prod. Lab. Rpt. R1761, 7 pp., illus. [Processed.]
(5) ———
  1952. TREATABILITY OF SOUTHERN YELLOW PINE AS INFLUENCED BY FUNGUS INFECTION. Jour. Forest Prod. Res. Soc. 2 (3): 85–86, illus.
(6) ———
  1952. COMPARISON OF WOOD PRESERVATIVES IN MISSISSIPPI POST STUDY (1952 progress report). U. S. Forest Prod. Lab. Rpt. R1757, 11 pp. [Processed.]
(7) ———
  1950. TREATING WOOD IN PENTACHLOROPHENOL SOLUTIONS BY THE COLD-SOAKING METHOD. U. S. Forest Prod. Lab. Rept. R1445, 9 pp., illus. (Rev.) [Processed.]
(8) BUCKMAN, S. J., and COMMITTEE.
  1949. REPORT OF COMMITTEE P–4, NON-STANDARD PRESERVATIVES. Amer. Wood-Preservers' Assoc. Proc. 45: 28–45.
(9) EDISON ELECTRIC INSTITUTE.
  1952. SPECIFICATIONS FOR THE FULL-LENGTH, NONPRESSURE, PRESERVATIVE TREATMENT OF POLES. 12 pp.
(10) FOBES, E. W.
  1949. BARK PEELING MACHINES AND METHODS. U. S. Forest Prod. Lab. Rept. D1730, 22 pp., illus. [Processed.]
(11) GARLICK, G. G.
  1948. NEW VACUUM WOOD TREATING PROCESS. Timberman 49 (8): 58–62.
(12) GRIFFITH, E. M.
  1950. INITIATION AND OPERATION OF A TIMBER TREATMENT SYSTEM FOR A GROUP OF PENNSYLVANIA ANTHRACITE MINES. 5 pp., illus. Glen Alden Coal Co., Wilkes-Barre, Pa. [Processed.]

(13) HATFIELD, I.
     1944. INFORMATION ON PENTACHLOROPHENOL AS A WOOD-PRESERVING
           CHEMICAL.   Amer. Wood-Preservers' Assoc. Proc. 40: 47–67.
(14) HENRY, W. T., and KEPFER, R. J.
     1949. COPPERIZED CHROMATED ZINC CHLORIDE.   Amer. Wood-Preservers'
           Assoc. Proc. 45: 66–75, illus.
(15) HUDSON, M. S.
     1947. VAPOR DRYING. THE ARTIFICIAL SEASONING OF WOOD IN VAPOR
           OF ORGANIC CHEMICALS.   Forest Prod. Res. Soc. Proc. 1: 32 pp.,
           illus.
(16) HUNT, G. M., and GARRATT, G. A.
     1953. WOOD PRESERVATION.   402 pp., illus., New York.
(17) MACLEAN, J. D.
     1925. THE RELATIVE COST OF TREATED AND UNTREATED TIMBER.   Amer.
           Wood-Preservers' Assoc. Proc. 21: 111–122.
(18) ───────
     1929. ABSORPTION OF WOOD PRESERVATIVES SHOULD BE BASED ON THE
           DIMENSIONS OF THE TIMBER.   Amer. Wood-Preservers' Assoc.
           Proc. 25: 129–142.
(19) ───────
     1952. PRESERVATIVE TREATMENT OF WOOD BY PRESSURE METHODS.
           U. S. Dept. Agr. Handb. 40, 160 pp., illus.
(20) MAYFIELD, P. B., and COMMITTEE.
     1948. REPORT OF COMMITTEE 6–5 SEASONING PRACTICE.   Amer. Wood-
           Preservers' Assoc. Proc. 44: 313–320.
(21) ─────── and COMMITTEE.
     1949. REPORT OF GENERAL COMMITTEE ON SEASONING PRACTICE.   Amer.
           Wood-Preservers' Assoc. Proc. 45: 236–241.
(22) MINICH, A., and GOLL, M.
     1949. TECHNICAL ASPECTS OF COPPER NAPHTHENATE AS A WOOD-PRE-
           SERVING CHEMICAL.   Amer. Wood-Preservers' Assoc. Proc. 44:
           72–81.
(23) NATIONAL WOODWORK MANUFACTURERS ASSOCIATION.
     1951. MINIMUM STANDARDS FOR WATER REPELLENT PRESERVATIVES FOR
           MILLWORK.   2 pp., illus.   Chicago.
(24) SHIPLEY, G. B.
     1951. WOOD PRESERVERS' CONTRIBUTION TO CONSERVATION.   Amer.
           Wood-Preservers' Assoc. Proc. 47: 339–345, illus.
(25) STEER, H. B.
     1952. WOOD PRESERVATION STATISTICS 1951.   Amer. Wood-Preservers'
           Assoc. Proc. 48: 363–404, illus.   (Published annually.)
(26) STIMSON, E.
     1928. ECONOMY OF FRAMING STRUCTURAL TIMBERS BEFORE TREATMENT.
           Amer. Wood-Preservers' Assoc. Proc. 24: 106–113.
(27) U. S. FEDERAL SUPPLY SERVICE.
     1950. WOOD PRESERVATIVE, RECOMMENDED TREATING PRACTICES.   Fed.
           Specif. TT–W–571, 11 pp.   (See yearly index for specifications
           on individual preservatives.)
(28) U. S. FOREST PRODUCTS LABORATORY.
     1945. PRESERVATION OF TIMBER BY THE STEEPING PROCESS.   U. S
           Forest Prod. Lab. Rpt. R621, 10 pp., illus.   (Rev.) [Processed.]
(29) VERRALL, A. F.
     1949. DECAY PROTECTION FOR EXTERIOR WOODWORK.   South. Lumber-
           man 178 (2237): 74, 76–78, 80, illus.
(30) VINSON, J. C., and COMMITTEE.
     1950. REPORT OF COMMITTEE P–4, NON-STANDARD PRESERVATIVES.   Amer.
           Wood-Preservers' Assoc. Proc. 46: 33–49.
(31) WILFORD, B. H.
     1944. CHEMICAL IMPREGNATION OF TREES AND POLES FOR WOOD PRESER-
           VATION.   U. S. Dept. Agr. Cir. No. 717, 30 pp., illus.
(32) WIRKA, R. M.
     1945. TIRE-TUBE METHOD OF FENCE POST TREATMENT.   U. S. Forest
           Prod. Lab. Rpt. R1158, 16 pp., illus.   (Rev.) [Processed.]

POLES, PILING, AND TIES

PRINCIPAL SPECIES USED FOR POLES

The more important factors to be considered in judging the suitability of a species for poles are:

1. Abundance. An adequate timber supply must be available if a species is to be seriously considered as a potential source for poles.

2. Treatability. Since poles generally require preservative treatment, it is desirable to use woods that can be satisfactorily penetrated during commercial wood-preserving processes.

3. Location of timber. Transportation costs from forest to treating plant or from treating plant to destination may be more favorable to one species than to another.

4. Strength. Poles must be sufficiently strong to support the weight of the wire, ice coatings on the wire, and wind pressure.

5. Weight. Light weight is desirable from the standpoint of shipping costs and ease of handling and erection. Since the lighter species of wood are not so strong as the heavier species, the larger sizes of poles required in some instances may offset this advantage.

6. Form. The poles should be reasonably straight, with a gradual taper, and with strength-reducing features, such as knots, limited so as to insure sufficient strength. The good appearance resulting from smoothness, straightness, and uniform taper is particularly desirable in urban districts.

The principal species used for poles in telephone, telegraph, and powerlines are southern yellow pine, the cedars, lodgepole pine, Douglas-fir, western larch, and ponderosa pine (*9, 15*).[30]

Southern Yellow Pine

The southern yellow pines now furnish the major part of the timber used for poles throughout most of the United States. The disappearance of chestnut, the diminishing supply of northern whitecedar, and the cost of western redcedar in the East and South are factors in the increased use of treated southern yellow pine poles. Southern yellow pine accounted for about 75 percent of the total number of poles treated in 1951. The supply of southern yellow pine poles in lengths over 40 feet is diminishing and is being supplemented by Douglas-fir, larch, and other western species.

Southern yellow pine poles are pressure treated full length with preservatives.

Cedar

Western redcedar and northern whitecedar are extensively used for poles; western redcedar furnishes the bulk of the cedar used for this purpose. The Lake States furnish most of the northern whitecedar poles, and Idaho, Washington, and British Columbia most of the western redcedar poles. The cedars have comparatively thin sapwood, and the heartwood is naturally decay resistant. The butts of

---

[30] Italic numbers in parentheses refer to Literature Cited, p. 443.

western redcedar and northern whitecedar poles are usually treated with creosote or other oil preservatives to increase their decay resistance. A considerable number of poles of these species are also treated full length, particularly for use under climatic conditions that favor decay or "shellrot" in the tops of butt-treated poles.

Lodgepole Pine

Lodgepole pine is widely used in the Rocky Mountain area because of its desirable shape, treatability with preservatives, and availability. Lodgepole pine poles are given either full-length or butt treatment, depending upon the climatic conditions. In the dry sections of the Rocky Mountain area, decay in the tops of poles is not rapid, and butt treatment only may be sufficient.

Douglas-Fir

Until World War II, the use of Douglas-fir poles was confined principally to the Pacific coast. The limited supply of southern yellow pine poles in long lengths, however, has created a market for Douglas-fir poles throughout the United States. Douglas-fir poles compare favorably with southern yellow pine poles in strength. They are generally treated full length.

Other Species

Western larch, ponderosa pine, Atlantic whitecedar, jack pine, red pine, eastern redcedar, redwood, spruce, and hemlock are occasionally used for poles.

Western larch poles produced in Montana and Idaho came into use following World War II because of their favorable size, shape, and strength properties. Western larch requires preservative treatment full length for use in most areas but is butt treated for use in dry areas of the Rocky Mountain States.

Ponderosa pine has been used to some extent because of its availability, favorable shape, and good qualities for being treated.

Redwood, Atlantic whitecedar, jack pine, red pine, eastern redcedar, spruce, and hemlock poles are used to a slight extent, for the most part locally in the areas where they are produced. All of these species generally require preservative treatment.

Other species having limited use, mostly local, include tamarack, baldcypress, black locust, ash, elm, and cottonwood. With the exception of black locust, none of these last long without preservative treatment.

WEIGHT OF SPECIES USED FOR POLES

The weight of poles is a factor in freight charges and, except where machines are used, in the ease with which they can be handled and erected. Table 49 gives the weights per cubic foot for the more important pole woods grown in the United States. The weight values for green wood were determined by the method given in the section on weight of round timbers (p. 58).

TABLE 49.— *Weight per cubic foot and specific gravity of the more important species used for poles*

Species	Green	Air-dry (15-percent moisture content)	Specific gravity [1]	Species	Green	Air-dry (15-percent moisture content)	Specific gravity [1]
	Lb. per cu. ft.	Lb. per cu. ft.			Lb. per cu. ft.	Lb. per cu. ft.	
Baldcypress	50	32	0.42	Pine—Continued			
Douglas-fir:				Red	49	31	0.41
Coast type	39	34	.45	Shortleaf [2]	52	35	.46
Rocky Mountain type	34	30	.40	Redcedar:			
Hemlock, western	41	29	.38	Eastern	37	33	.44
Larch, western	50	38	.51	Western	27	23	.31
Pine:				Redwood	50	25	.38
Jack	40	30	.40	Spruce (red, Sitka, and			
Loblolly [2]	53	36	.47	white)	34	28	.37
Lodgepole	39	29	.38	Whitecedar:			
Longleaf [2]	55	41	.54	Atlantic	25	23	.31
Ponderosa	45	28	.38	Northern	27	22	.29

[1] Based on volume when green and weight when oven-dry.
[2] These 3 species belong to the southern yellow pine group.

VOLUME OF POLES AND PILES

Volumes of poles or piles may be computed by measuring the lengths and diameters and applying a formula or a volume table.

The values in table 50 were computed by the cone-frustum formula

$$V = 0.001818L(D^2 + d^2 + Dd)$$

where V is the volume in cubic feet, L the length in feet, D the top diameter in inches, and d the butt diameter in inches.

The values are accurate for pieces of uniform taper and circular cross section. Errors from irregularities of shape and taper may be minimized by taking diameter measurements at short intervals and computing the volume of each short section separately.

American Wood-Preservers' Association Standard Specification F3–51, Standard Volumes of Round Forest Products, provides two methods for the calculation of pole and pile volume, one of which uses the cone-frustum formula but with the application of correction factors for some species (10).

ENGINEERING DESIGN OF POLES

Poles are selected on the basis of the loads that must be supported and are specified by class, length, and species. The principal load is in bending from horizontal wind forces on ice-covered wires. The horizontal pull from dead-end or unbalanced wires is important in some cases. Compressive stresses from the vertical loads on the pole are of little consequence unless transformers or other heavy objects are supported. If this is the case, consideration must be given to bending stresses resulting from the vertical loads when the pole is deflected from the vertical.

TABLE 50.—*Cubical contents of poles and piling*

[To determine the volume of a pole or pile find the average diameter at each end of the piece to the nearest half inch. Multiply the number in the table corresponding to these 2 diameters by the length of the piece in feet. The result is in cubic feet]

Diameter of large end (inches)	Cubic feet per lineal foot when diameter of small end in inches is—											
	4	4½	5	5½	6	6½	7	7½	8	8½	9	9½
4	0.087											
4½	.099	0.110										
5	.111	.123	0.136									
5½	.124	.137	.150	0.165								
6	.138	.151	.165	.180	0.196							
6½	.153	.167	.181	.197	.213	0.230						
7	.169	.183	.198	.214	.231	.249	0.267					
7½	.186	.200	.216	.232	.250	.268	.287	0.307				
8	.204	.219	.235	.251	.269	.288	.307	.328	0.349			
8½	.222	.238	.254	.271	.290	.309	.329	.350	.371	0.394		
9	.242	.258	.275	.292	.311	.330	.351	.372	.395	.418	0.442	
9½	.262	.279	.296	.314	.333	.353	.374	.396	.419	.442	.467	0.492
10	.284	.300	.318	.337	.356	.377	.398	.420	.444	.468	.493	.519
10½	.306	.323	.341	.360	.380	.401	.423	.446	.470	.494	.520	.546
11	.329	.347	.365	.385	.405	.427	.449	.472	.496	.521	.547	.574
11½	.353	.371	.390	.410	.431	.453	.476	.500	.524	.550	.576	.603
12	.378	.397	.416	.437	.458	.480	.504	.528	.553	.579	.605	.633
12½	.404	.423	.443	.464	.486	.509	.532	.559	.582	.609	.636	.664
13	.431	.450	.471	.492	.515	.538	.562	.587	.613	.640	.667	.696
13½	.459	.479	.500	.521	.544	.568	.592	.618	.644	.671	.700	.729
14	.487	.508	.529	.551	.575	.599	.624	.650	.676	.704	.733	.762
14½	.517	.538	.560	.582	.606	.630	.656	.682	.710	.738	.767	.797
15	.547	.569	.591	.614	.638	.663	.689	.716	.744	.772	.802	.832
15½	.579	.600	.623	.647	.671	.697	.723	.750	.779	.808	.838	.869
16	.611	.633	.656	.680	.705	.731	.758	.786	.815	.844	.875	.906
16½	.644	.667	.690	.715	.740	.767	.794	.822	.851	.881	.912	.944
17	.678	.701	.725	.750	.776	.803	.831	.860	.889	.920	.951	.983
17½	.713	.737	.761	.787	.813	.840	.869	.898	.928	.959	.990	1.023
18	.749	.773	.798	.824	.851	.879	.907	.937	.967	.999	1.031	1.064
18½	.786	.810	.836	.862	.890	.918	.947	.977	1.008	1.039	1.072	1.106

19	1.149	1.114	1.081	1.049	1.018	.987	.958	.929	.901	.875	.849	.824
19½	1.192	1.158	1.124	1.091	1.059	1.029	.999	.969	.941	.914	.888	.862
20	1.237	1.202	1.168	1.134	1.102	1.071	1.040	1.011	.982	.955	.928	.902
20½	1.282	1.247	1.212	1.179	1.146	1.114	1.083	1.053	1.024	.996	.969	.942
21	1.329	1.293	1.258	1.224	1.190	1.158	1.127	1.096	1.067	1.038	1.010	.984
21½	1.376	1.339	1.304	1.269	1.236	1.203	1.171	1.140	1.110	1.081	1.053	1.026
22	1.424	1.387	1.351	1.316	1.282	1.249	1.217	1.185	1.155	1.125	1.097	1.069
22½	1.473	1.436	1.399	1.364	1.329	1.296	1.263	1.231	1.200	1.170	1.141	1.113
23	1.523	1.485	1.449	1.413	1.378	1.344	1.310	1.278	1.247	1.216	1.187	1.158
23½	1.574	1.536	1.499	1.462	1.427	1.392	1.359	1.326	1.294	1.263	1.233	1.204
24	1.626	1.587	1.549	1.513	1.477	1.442	1.408	1.374	1.342	1.311	1.280	1.251
24½	1.679	1.639	1.601	1.564	1.528	1.492	1.458	1.424	1.391	1.359	1.329	1.299
25	1.732	1.693	1.654	1.616	1.579	1.544	1.509	1.474	1.441	1.409	1.378	1.347

Cubic feet per lineal foot when diameter of small end in inches is—

	10	10½	11	11½	12	12½	13	13½	14	14½	15
10	0.545										
10½	.573	0.601									
11	.602	.630	0.660								
11½	.631	.660	.690	0.721							
12	.662	.691	.722	.753	0.785						
12½	.693	.723	.754	.786	.819	0.852					
13	.725	.756	.787	.819	.853	.887	0.922				
13½	.759	.790	.821	.854	.888	.922	.958	0.994			
14	.793	.824	.856	.890	.924	.959	.994	1.031	1.069		
14½	.828	.860	.892	.926	.960	.996	1.032	1.069	1.108	1.147	
15	.864	.896	.929	.963	.998	1.034	1.071	1.109	1.147	1.187	1.227
15½	.900	.933	.967	1.001	1.037	1.073	1.110	1.149	1.188	1.228	1.269
16	.938	.971	1.005	1.040	1.076	1.113	1.151	1.189	1.229	1.269	1.311
16½	.977	1.010	1.045	1.080	1.117	1.154	1.192	1.231	1.271	1.312	1.354
17	1.016	1.050	1.085	1.121	1.158	1.196	1.234	1.274	1.314	1.356	1.398
17½	1.057	1.091	1.127	1.163	1.200	1.239	1.278	1.318	1.359	1.400	1.443
18	1.098	1.133	1.169	1.206	1.244	1.282	1.322	1.362	1.404	1.446	1.489
18½	1.140	1.176	1.212	1.249	1.288	1.327	1.367	1.408	1.449	1.492	1.536
19	1.184	1.219	1.256	1.294	1.333	1.372	1.413	1.454	1.496	1.539	1.584
19½	1.228	1.264	1.301	1.339	1.379	1.419	1.459	1.501	1.544	1.588	1.632

TABLE 50.—*Cubical contents of poles and piling*—Continued

Diameter of large end (inches)	Cubic feet per lineal foot when diameter of small end in inches is—										
	10	10½	11	11½	12	12½	13	13½	14	14½	15
20	1.273	1.309	1.347	1.386	1.425	1.466	1.507	1.549	1.593	1.637	1.682
20½	1.319	1.356	1.394	1.433	1.473	1.514	1.556	1.599	1.642	1.687	1.732
21	1.365	1.403	1.442	1.481	1.522	1.563	1.605	1.649	1.693	1.738	1.784
21½	1.413	1.451	1.490	1.530	1.571	1.613	1.656	1.699	1.744	1.789	1.836
22	1.462	1.500	1.540	1.580	1.622	1.664	1.707	1.751	1.796	1.842	1.889
22½	1.511	1.550	1.590	1.631	1.673	1.716	1.759	1.804	1.849	1.896	1.943
23	1.562	1.601	1.642	1.683	1.725	1.769	1.813	1.858	1.903	1.950	1.998
23½	1.613	1.653	1.694	1.736	1.779	1.822	1.867	1.912	1.958	2.006	2.054
24	1.665	1.706	1.747	1.789	1.833	1.877	1.922	1.968	2.014	2.062	2.111
24½	1.719	1.759	1.801	1.844	1.888	1.932	1.978	2.024	2.071	2.119	2.168
25	1.773	1.814	1.856	1.899	1.943	1.988	2.034	2.081	2.129	2.178	2.227

Required factors of safety are sometimes applied as multipliers of the expected load. The selected multiple of the design load is then used with the rated ultimate strength of the pole.

Pole classes set up by the American Standards Association (4) are used in the National Electrical Safety Code (21, 22) and in numerous State and other codes. Sizes of poles in those classes are so arranged that all poles of the same class, regardless of length or species, will resist the same horizontal force applied 2 feet from the top end of the pole. The pole species are placed in six groups, each group having its own rating for ultimate stress in the outer fiber in bending. Stress ratings apply to treated or untreated poles. Required sizes 6 feet from the butt are calculated on the basis that the ultimate stress will be reached at the ground line with the following horizontal forces:

Class:	Pounds	Class:	Pounds
1	4, 500	5	1, 900
2	3, 700	6	1, 500
3	3, 000	7	1, 200
4	2, 400		

Three classes smaller than the preceding are also established on the basis of required top sizes only (for attachment of crossarms or other services).

Fiber stresses for poles are discussed more fully in a Forest Products Laboratory publication (26).

These loads and stresses are related to poles with knots and other strength-reducing features limited as provided in the specifications of the American Standards Association (4). Since permitted strength-reducing features are essentially the same in all species, the stress ratings show about the same relations among species as are shown in tests of small clear specimens. The design of the smaller pole lines is usually more or less standardized on the basis of a certain class of pole for a certain type of line, without specific reference to loads or stresses. Such designs are often more conservative than the American Standards Association specifications, to provide for future load increases or for other reasons.

PRESERVATIVE TREATMENT OF POLES

Federal Specification TT–W–571c (23) and American Wood-Preservers' Association Standard C4 (12) cover the general treatment of poles.

The great majority of the poles of species other than the cedars are given full-length, empty-cell, pressure treatment. In 1951, 89 percent of the poles were treated with coal-tar creosote and 7 percent were treated with pentachlorophenol solution. Waterborne preservatives have been employed to some extent. Poles so treated are cleaner than creosoted poles, but it has not been demonstrated that they will have equivalent serviceability. A minimum retention of 8 pounds of coal-tar creosote or of a 5-percent solution of pentachlorophenol per cubic foot of wood by an empty-cell process is commonly specified. Retentions of 10 or 12 pounds per cubic foot are frequently specified, especially in the South. The higher retentions cost more but give greater assurance of satisfactory penetration and long life.

When butt treatment only is required, poles are treated with preservative oils by the hot-and cold-bath, open-tank process. In cedar poles, the best penetration is obtained by incising them before treatment. Special machines are used for incising, so as to give the distribution of preservative with the minimum number of incisions (see p. 411). Specifications covering butt treatment of the incised cedar poles by the hot-and-cold bath have been adopted by the American Wood-Preservers' Association (11). Lodgepole, pine, Douglas-fir, and western larch poles are commonly incised before treatment by the hot-and-cold-bath process.

Some treated poles exude preservative sufficiently to make the surface oily in spots. This bleeding is not likely to reduce the life of the poles, but it may prove objectionable to men who work on the poles or to anyone who may come in contact with a bleeding pole. Methods of completely preventing bleeding have not been established. The use of lower retentions, low-residue creosotes or selected oil solutions of pentachlorophenol as preservatives, and a final heating or expansion bath following treatment, however, will help to produce clean poles.

TREATMENT TO STOP DECAY IN STANDING POLES

Poles in service that have started to decay are sometimes treated to arrest the decay and extend their life. Several ground-line treatments for standing poles are employed in the United States (6). These methods require excavating around the pole to a depth of 12 to 18 inches and then applying a preservative to the exposed surface.

One patented method involves the application of the preservative in paste form. A waterproof paper bandage is used to hold the preservative against the pole. In another patented method, the preservative paste is injected into the wood with a heavy, hollow needle. Both of these methods use waterborne preservatives, which will diffuse in the wet wood of the pole.

Another commonly used method involves first spreading sodium fluoride in the trench around the pole and partially refilling the trench with soil. Coal-tar creosote is then poured against the surface of the pole about 18 inches above the ground line and allowed to run into the checks and around the ground line. Petroleum oil containing 5 percent of pentachlorophenol has been used instead of the sodium fluoride and creosote.

Butt-treated poles that have started to decay in the untreated tops are often sprayed with a No. 2 fuel oil containing 5 percent of pentachlorophenol.

LIFE OF POLES

The life of poles is so greatly influenced by their species; size and amount of sapwood; the soil, climate, and use conditions; the type of preservative used and the method of treatment; and the loads carried that an average life figure is not reliable and wide variations from the average are to be expected in individual installations. Service records are not available in sufficient quantity to determine a mathematical average.

Roughly, it may be said that average-sized poles of northern white-cedar, western redcedar, Port-Orford-cedar, redwood, and baldcypress, when the sapwood is less than 1 inch thick, will last about 12 to 17 years without treatment. Good butt treatment with creosote will give a life of 20 to 25 years or more in territory where failure from top decay is infrequent and dry-wood termites are not active. Poles of baldcypress, Port-Orford-cedar, and redwood with sapwood much thicker than 1 inch, if untreated, will have a much shorter life.

Douglas-fir and western larch poles, when used untreated, are likely to last only about 9 years or less. Southern yellow pine poles used without treatment under the warm, damp conditions prevailing throughout most of the Southern States would be likely to fail in a year or two. On the other hand, with a retention of 8 to 10 pounds or more of creosote thoroughly injected by an empty-cell process, southern yellow pine poles of good quality should last 35 years or more. Equally good results can be expected from other pole species when similarly treated.

In the dry Rocky Mountain region, lodgepole pine thoroughly butt-treated with coal-tar creosote should have an average life of 20 years or more. With full-length treatment and good sapwood penetration, lodgepole pine poles should compare favorably with similarly treated southern yellow pine and Douglas-fir poles.

CHOICE OF SPECIES FOR PILING

The properties desirable in piles include sufficient strength and straightness to withstand driving and to carry the weight of structures built on them, and in some instances to resist bending stresses. Decay resistance or ease of penetration by preservatives is also important except in piles for temporary use or piles that will be in fresh water and entirely below the permanent water level.

Southern yellow pine, Douglas-fir, and oak are among the principal species used for piling (9), but western redcedar and numerous other species are also used.

Specifications for timber piles covering kinds of wood, general quality, resistance to decay, dimensions, tolerance, manufacture, inspection, delivery, and shipment have been published in the American Railway Engineering Association Manual (2). Specifications for timber piles have also been prepared by the American Association of State Highway Officials (1), the American Society for Testing Materials (3), and the Federal Supply Service (24).

BEARING LOADS FOR PILES

Bearing loads on piles are sustained by earth friction along the sides of the pile, by bearing of the tip on a solid stratum, or by a combination of the two. Wood piles, because of their tapered form, are particularly efficient in supporting loads by side friction. Bearing values that depend upon side friction are related to the stability of the soil and generally do not approach the ultimate strength of the pile (5, 18, 20). Engineering handbooks give formulas (18, pp. 730–741) for estimating bearing values from the penetration under blows of known energy from the driving hammer. Some engineers prefer to estimate bearing capacity from experience or observation of the behavior of pile foundations under similar conditions or from the results of static-loading tests.

Where wood piles sustain foundation loads by bearing of the tip on a solid stratum, loads may be limited by the compressive strength of the wood parallel to the grain. Working stresses in compression parallel to grain should not exceed 75 percent of the basic stress values of table 25. If a large proportion of the length of a pile extends above ground, its bearing value may be limited by its strength as a long column. Side loads may also be applied to piles extending above ground. In such instances, however, bracing is often used to reduce the unsupported column length or to resist the side loads.

ECCENTRIC LOADING AND CROOKED COLUMNS

The reduction in strength of a wood column resulting from crooks, eccentric loading, or any other condition that will result in combined bending and compression is not so great as might be expected. Tests have shown that a timber, when subjected to combined bending and compression, develops a higher stress at both the elastic limit and maximum load than when subjected to compression only (19). This does not imply that crooks and eccentricity should be without restriction, but it should relieve anxiety as to the influence of crooks, such as those common in piling.

DRIVING OF PILES AS AFFECTED BY SEASONING

Under usual conditions of service wood piles will be wet, but they may be driven in either the green or the seasoned condition. Because of the increased strength resulting from drying, seasoned piles either treated or untreated are likely to stand driving better than are green or unseasoned ones. This is particularly true of treated piles, inasmuch as tests have demonstrated that, while the strength of green wood may be considerably reduced by treatment, thoroughly seasoned treated wood may be but little weaker than seasoned untreated wood. Under the same drying conditions, however, untreated wood loses moisture more rapidly than does treated wood.

DECAY RESISTANCE AND PRESERVATIVE TREATMENT OF PILING

Piles in general have rather wide sapwood and consequently low decay resistance. High natural decay resistance will be found only when the piles have thin sapwood and are of species that have decay-resistant heartwood.

Since wood that remains completely submerged in water does not decay, decay resistance is not necessary in piling so used, but it is necessary in any part of the piling that may extend above the permanent water level. When piling that supports the foundations of bridges or buildings is to be cut off above the permanent water level, it should, as a safety precaution, be pressure creosoted, usually by the empty-cell process, with not less than 12 pounds of creosote per cubic foot (*7*, *13*, *23*) (see p. 417). The untreated surfaces exposed at the cut-offs should also be given protection by thoroughly brushing the cut surfaces with coal-tar creosote. A coat of pitch, asphalt, or similar material may then be applied over the creosote and a protective sheet material, such as metal, roofing felt, or saturated fabric, fitted over the pile head.

Piling driven into earth that is not constantly wet is subject to about the same service conditions as apply to poles, but is generally expected to last longer and therefore requires higher preservative retentions than poles (see p. 416).

Piling used in salt water is, of course, subject to destruction by marine borers even though it does not decay below the waterline. The best practical protection against marine borers is heavy treatment with coal-tar creosote or creosote-coal-tar solution. The piling should be treated by the full-cell process "to refusal"; that is, to force in all the oil the piling can be made to hold without using treatments that cause serious damage to the wood (see p. 412) (*8*).

STRENGTH AND OTHER REQUIREMENTS FOR TIES

Many species of wood are used for ties. Their relative suitability depends largely upon their strength, wearing qualities, treatability with wood preservatives, and to some extent their natural resistance to decay and tendency to check, although availability and cost must also be considered.

The chief strength properties considered in a wood for crossties are (1) bending strength, (2) end hardness and strength in compression parallel to grain, which are indicative of resistance to spike pulling and the lateral thrust of spikes, and (3) side hardness and compression perpendicular to the grain, which indicate resistance to wear under the rail or the tieplate.

The composite tie-strength figures given in table 51 consist of a combination of the strength properties listed above as necessary for satisfactory service (*17*). Average values of specific gravity are also given. Table 51 shows that a species of high specific gravity usually has a high composite tie-strength value, and hence in the absence of other strength data, the specific gravity is a good criterion.

The relative importance of the several mechanical properties involved for ties changes with conditions of track installation and maintenance and with service conditions. For example, the requirement for ties under heavy traffic will be greatly different than for light traffic. Consequently, the composite tie values should be regarded as indicative, rather than an exact measure of the relative mechanical suitability of species for crossties.

Sizes of ties range from 6 by 7 to 7 by 9 inches; lengths are usually 8 or 8½ feet. With heavier traffic and higher speeds of trains, the present tendency is toward increasing use of the larger sizes.

Specifications for crossties covering general quality, resistance to wear, resistance to decay, design, manufacture, inspection, delivery, and shipment have been published in the American Railway Engineering Association Manual (*2*) and in Federal Specification MM–T–371a (*25*).

TABLE 51.—*Composite strength figures and specific gravity values of crosstie woods*

Species	Composite-strength figure	Specific gravity [1]	Species	Composite-strength figure	Specific gravity [1]
HARDWOODS			HARDWOODS—continued		
			Tupelo:		
Ash, commercial white	108	0.54	Black	79	0.46
Aspen	46	.35	Water	81	.46
Aspen, Bigtooth	53	.35	Walnut, black	100	.51
Beech, American	97	.56			
Birch:			SOFTWOODS		
Sweet	108	.60			
Yellow	94	.55	Baldcypress	68	.42
Catalpa, northern	52	.38	Cedar:		
Cherry:			Eastern redcedar	78	.44
Black	84	.47	Northern white-	40	.29
Pin	52	.36	Port-Orford-	67	.40
Chestnut, American	59	.40	Western redcedar	52	.31
Cottonwood:			Douglas-fir:		
Black	45	.32	Coast type	78	.45
Eastern	49	.37	Intermediate type	71	.41
Elm:			Rocky Mountain type	65	.40
American	73	.46	Fir, commercial white	58	.36
Rock	103	.57	Hemlock:		
Slippery	81	.48	Eastern	63	.38
Hackberry	74	.49	Western	64	.38
Hickory:			Larch, western	82	.51
Pecan	127	.59	Pine:		
True	141	.65	Jack	59	.40
Honeylocust	133	.60	Loblolly [2]	73	.47
Locust, black	161	.66	Lodgepole	56	.38
Maple:			Longleaf [2]	90	.54
Silver	67	.44	Ponderosa	54	.38
Sugar	112	.56	Red	58	.41
Oak:			Shortleaf [2]	72	.46
Commercial red	100	.57	Pinyon	60	.48
Commercial white	103	.59	Redwood	72	.38
Poplar, balsam	36	.30	Spruce:		
Sassafras	66	.42	Englemann	47	.32
Sweetgum	71	.44	Red, Sitka, and white	57	.37
Sycamore, American	69	.46	Tamarack	71	.49

[1] Based on volume when green and weight when oven-dry.
[2] These species belong to the southern yellow pine group.

DECAY RESISTANCE AND PRESERVATIVE TREATMENT OF TIES

Although the majority of ties used are given preservative treatment before installation, a few are used untreated, and for these natural decay resistance is important. In ties given preservative treatment, variations in natural decay resistance are of less importance than susceptibility to treatment.

The majority of ties treated are pressure treated with coal-tar creosote, creosote-coal-tar solutions, or creosote-petroleum mixtures. Minimum retentions of 8 to 9 pounds per cubic foot, injected by an empty-cell process, are most commonly employed.

Federal Specification TT–W–571c (*23*) includes recommendations for the treatment of crossties, switch ties, and bridge ties. Specifications covering the preservative treatment of ties have been adopted and published by the American Wood-Preservers' Association (*14*) and the American Railway Engineering Association (*2*).

LIFE OF TIES

The service conditions under which ties are exposed are severe. The life of ties in service therefore depends on their ability to resist decay and the extent to which they are protected from mechanical destruction by breakage, loosening of spikes, and rail or plate wear. Under sufficiently light traffic, heartwood ties of naturally durable wood, even if of low strength, may give 10 or 15 years' service without preservative treatment, but under heavy traffic without adequate mechancial protection, the same ties might fail through mechanical wear in 2 or 3 years. The life of treated ties is affected also by the preservative used and the thoroughness of treatment. As a result, the life of individual groups of ties may vary widely from the general average depending on the local circumstances.

With these preceding limitations, the following rough estimates are given: (1) Ties well penetrated with 8 pounds or more per cubic foot of coal-tar creosote, creosote-petroleum mixture, or creosote-coal-tar solution should last 25 years or more when protected against mechanical destruction. (2) Ties treated with zinc chloride should last 10 to 15 years except in very wet or very dry climates, where their life would be shorter. (3) Untreated heartwood ties of cedar, cypress, or redwood may last 12 to 15 years and sometimes longer, under light traffic conditions. (4) Heartwood ties of Douglas-fir, white oak, tamarack, and western larch should last 6 to 10 years without treatment. (5) Regardless of species, ties that contain a large proportion of sapwood are not likely to last longer than 2 to 4 years without preservative treatment.

Records on the life of treated and untreated ties are published from time to time in the annual proceedings of the American Railway Engineering Association and the American Wood-Preservers' Association.

RELATION BETWEEN PERCENTAGE TIE RENEWALS AND AVERAGE LIFE

If, after the installation of a group of ties, accurate records are kept of the percentage of the group that are removed each year, figure 90 can be used to estimate the average life that will be obtained.

There is a fairly definite relation (*16*) between the rate of renewals and the average life of any particular group of ties of similar quality and subjected to similar conditions of service. The accuracy with which the average life can be predicted by means of figure 90 is greater with large groups of ties than with small groups and increases as the percentage of renewals increases. Since very early renewals are often caused by unusual conditions, the average life indicated by renewal percentages below 10 or 15 is not so reliable as that indicated by somewhat higher percentages. The average life of a group of ties is usually reached when about 60 percent have been removed.

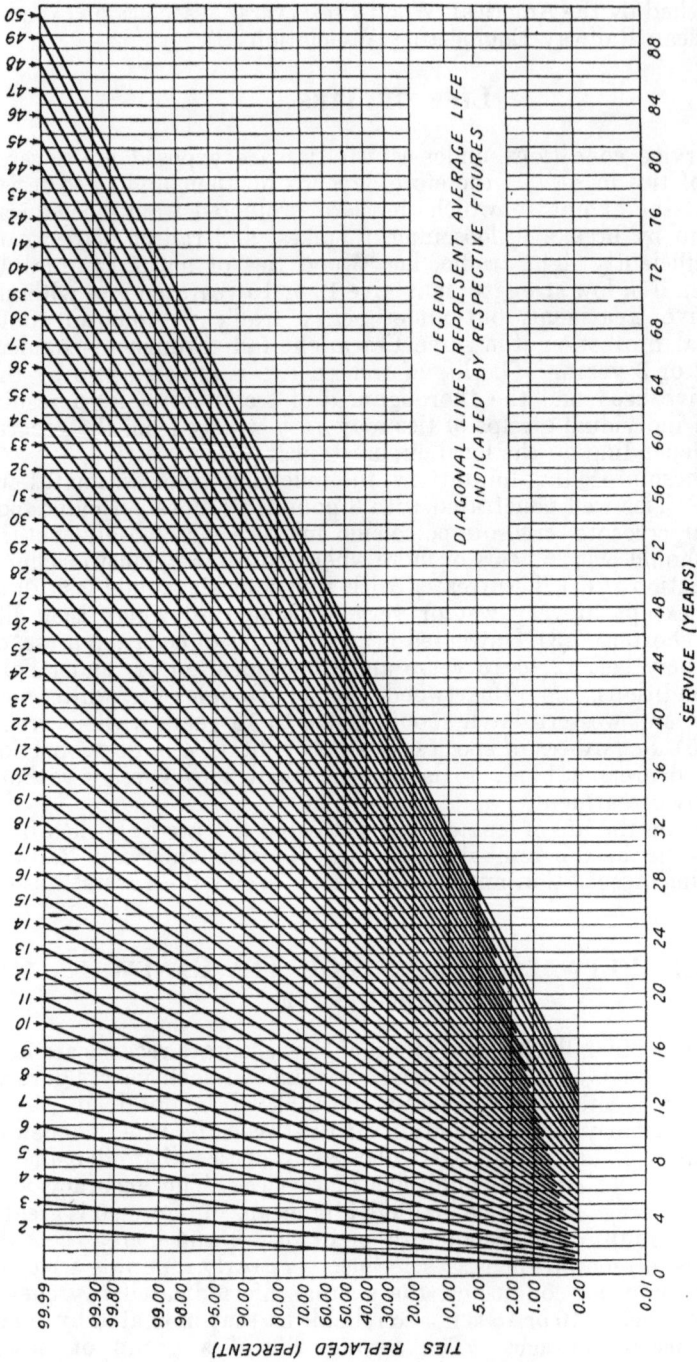

FIGURE 90.—Chart for determining the probable average life of ties from the percentage replaced in a given time. To use the chart find the percentage-renewal figure at the left; follow this line horizontally until it intersects the vertical line corresponding to the number of years the group of ties has been in service; then follow the nearest diagonal line to the upper edge of the chart where the probable average life of the group of ties in years will be found.

LITERATURE CITED

(1) AMERICAN ASSOCIATION OF STATE HIGHWAY OFFICIALS.
   1953. STANDARD SPECIFICATIONS FOR HIGHWAY BRIDGES. 328 pp. Washington, D. C.
(2) AMERICAN RAILWAY ENGINEERING ASSOCIATION.
   1950. MANUAL OF THE AMERICAN RAILWAY ENGINEERING ASSOCIATION. Illus., Chicago. (Looseleaf manual, revised annually.)
(3) AMERICAN SOCIETY FOR TESTING MATERIALS.
   1952. TENTATIVE SPECIFICATIONS FOR ROUND TIMBER PILES. Amer. Soc. Testing Mater. Standard D25–52T. Pt. 4: 681–685, illus.
(4) AMERICAN STANDARDS ASSOCIATION.
   1948. STANDARD SPECIFICATIONS AND DIMENSIONS FOR WOOD POLES. Amer. Standards Assoc. Specif. 05.1–1948, 14 pp., illus.
(5) AMERICAN WOOD-PRESERVERS' ASSOCIATION.
   1932. USE OF CREOSOTED WOOD PILING FOR BUILDING FOUNDATIONS. Amer. Wood-Preservers' Assoc. Proc. 28: 10 pp.
(6) ———
   1942. REPORT OF COMMITTEE 5- 5–2-POLES, NON-PRESSURE METHODS. Amer. Wood-Preservers' Assoc. Proc. 38: 400–409.
(7) ———
   1951. STANDARD FOR CREOSOTED-WOOD FOUNDATION PILES. Amer. Wood-Preservers' Assoc. Specif. C12–51, 1 p.
(8) ———
   1951. STANDARD FOR PRESSURE-TREATED PILES AND TIMBERS IN MARINE CONSTRUCTION. Amer. Wood Preservers' Assoc. Specif. C18–51, 1 p.
(9) ———
   1952. WOOD PRESERVATION STATISTICS 1951. Amer. Wood-Preservers' Assoc. Proc. 48: 363–404, illus. (Published annually.)
(10) ———
   1952. STANDARD VOLUMES OF ROUND FOREST PRODUCTS. Amer. Wood-Preservers' Assoc. Specif. F3–51, 8 pp.
(11) ———
   1952. PRESERVATIVE TREATMENT OF INCISED POLE BUTTS BY THE NON-PRESSURE PROCESS, WESTERN REDCEDAR AND NORTHERN WHITE-CEDAR POLES. Amer. Wood-Preservers' Assoc. Specif. C7–53, 3 pp.
(12) ———
   1952. PRESERVATIVE TREATMENT OF POLES BY PRESSURE PROCESSES. Amer. Wood-Preservers' Assoc. Specif. C4–53, 7 pp.
(13) ———
   1952. PRESERVATIVE TREATMENT OF PILES BY PRESSURE PROCESSES. Amer. Wood-Preservers' Assoc. Specif. C3–53, 5 pp.
(14) ———
   1952. PRESERVATIVE TREATMENT OF CROSSTIES AND SWITCH TIES BY PRESSURE PROCESSES. Amer. Wood-Preservers' Assoc. Specif. C6–53, 7 pp.
(15) ANDERSON, I. V.
   1947. TRENDS IN THE UTILIZATION OF POLE SPECIES AND THEIR EFFECT ON FOREST MANAGEMENT. Soc. Amer. Foresters Proc. 1947: 128–140, illus.
(16) MacLEAN, J. D.
   1926. PERCENTAGE RENEWALS AND AVERAGE LIFE OF RAILWAY TIES. Engin. News-Rec. 97 (9): 336–339.
(17) MARKWARDT, L. J.
   1930. COMPARATIVE STRENGTH PROPERTIES OF WOODS GROWN IN THE UNITED STATES. U. S. Dept. Agr. Tech. Bul. 158, 39 pp., illus.
(18) MERRIMAN, M., MERRIMAN, T., and WIGGIN, T. H.
   1930. AMERICAN CIVIL ENGINEERS' HANDBOOK. Ed. 5., 2,263 pp., illus. New York. (Reprinted 1942.)
(19) NEWLIN, J. A., and TRAYER, G. W.
   1924. STRESSES IN WOOD MEMBERS SUBJECTED TO COMBINED COLUMN AND BEAM ACTION. Natl. Advisory Com. Aeronaut. Rpt. 188, 13 pp., illus.

(20) TERZAGHI, C.
    1929. SCIENCE OF FOUNDATIONS—ITS PRESENT AND FUTURE. Amer.
        Soc. Civ. Engin. Trans. 93 (1704): 270–301, illus.
(21) U. S. DEPARTMENT OF COMMERCE.
    1948. NATIONAL ELECTRIC SAFETY CODE. Natl. Bur. Standards Handbook
        H30, 408 pp., illus. Washington, D. C.
(22) ————
    1944. DISCUSSION OF NATIONAL ELECTRICAL SAFETY CODE. Natl. Bur.
        Standards Handbook 39, 162 pp., illus. Washington, D. C.
(23) U. S. FEDERAL SUPPLY SERVICE.
    1950. WOOD PRESERVATIVE, RECOMMENDED TREATING PRACTICES. Fed.
        Specif. TT–W–571, 11 pp.
(24) ————
    1947. PILES: WOOD. Fed. Specif. MM–P–371, 10 pp., illus., and
        Amend. 1, 1 p., Oct. 20, 1949.
(25) ————
    1950. TIES, RAILROAD, WOOD (CROSS AND SWITCH). Fed. Specif. MM–T–
        371a, 4 pp.
(26) WILSON, T. R. C., and DROW, J. T.
    1953. FIBER STRESSES FOR WOOD POLES. U. S. Forest Prod. Lab. Rpt.
        D1619, 10 pp. illus. (Rev.) [Processed.]

THERMAL INSULATION

The inflow of heat through outside walls and roofs in hot weather or its outflow during cold weather has important effects upon the comfort of the occupants of a building. During cold weather, such heat flow also governs fuel consumption to a great extent. Wood itself is a good insulator but commercial insulating materials are often incorporated into exterior walls, ceilings, and floors to increase resistance to heat passage (10).[31]

Commercial insulating materials are manufactured in a number of forms and types (6). Each has advantages for specific uses, some are more resistive to heat flow than others, and no one type is best for all applications (3).

INSULATING MATERIALS

For purposes of description, materials commonly used for insulation may be grouped in the following general classes: (1) Rigid insulation, (a) structural, (b) nonstructural; (2) flexible insulation, (a) blanket, (b) batt; (3) fill insulation; (4) reflective insulation; and (5) miscellaneous types.

Rigid Insulation

Structural Insulating Board

Structural insulating board is made by reducing wood, cane, or other vegetable fibers to a pulp and then assembling the fibrous pulp into large lightweight or low-density boards that combine strength with thermal and acoustical insulating properties. Such boards may also be used for decorative purposes. The finished board as generally made is 6 to 12 feet long, 4 feet wide, and ½ to 1 inch thick. Boards of greater thickness are produced by laminating sheets of standard thickness together. The most common fiberboard products are general-use board, sheathing, lath, tile board, planks, and roof insulation. A more detailed discussion of these products is given in the section on building fiberboards (p. 457).

General-use insulating board is available in sizes 4 feet wide, 6 to 12 feet long, and ⅜ to 1 inch thick. Sheathing is used on walls and is available in widths of 2 and 4 feet, lengths of usually 8 feet, and thicknesses of ½ and ²⁵⁄₃₂ inches. It may be obtained with or without surface coatings, which usually have an asphalt base. Lath made of fiberboard is 18 inches wide, 48 inches long, and ½ to 1 inch thick. Fiberboard has better insulating properties than most materials used for a plaster base and is commonly used for that purpose on exterior walls and under roofs or attics. Some fiberboard lath has an asphalt coating that serves as a vapor barrier. Tile board produced from

---

[31] Italic numbers in parentheses refer to Literature Cited, p. 454.

fiberboard is made in small squares or rectangular patterns, generally with interlocking edges, and is used for interior finish, particularly on ceilings. It is available for thermal insulation purposes in sizes from 12 by 12 inches to 16 by 32 inches and in thicknesses of ½, ¾, and 1 inch.

Plank is another type of interior finish and is frequently used in conjunction with tile board. It is 8 to 16 inches wide, 6 to 12 feet long, and ½ inch thick. Tile board and plank may be obtained as manufactured or with a factory-applied surfacing. These materials are often used where a decorative finish is desired and insulating properties are of secondary importance. They are sometimes combined with sound insulation. Roof insulation is used on flat roofs under composition roofing and under certain types of roofing on pitched roofs. It is also used under concrete floors. It is 23 to 24 inches wide, 47 to 48 inches long, and ½ to 2 inches thick.

Slab Insulation

Nonstructural rigid insulation is often called "slab insulation." The slabs or blocks are small, rigid units sometimes 1 inch thick but generally thicker and vary in size up to 24 by 48 inches. The types made from wood-base materials are cork blocks, wood fiber bonded with cement, magnesite, or other adhesive, and fiberboard slabs. Cork blocks are made by bonding small pieces of cork together in blocks or slabs ranging from 12 by 36 inches to 36 by 36 inches and 1 to 6 inches thick. They are used widely for cold-storage insulation and for insulating flat roofs of industrial and commercial buildings.

Wood-fiber blocks are made by bonding wood fibers similar to excelsior, but coarser, with some suitable bonding agent, such as magnesite cement. They are made in thicknesses of 1 to 3 inches and in various widths and lengths. The principal uses are for roof-deck insulation in industrial buildings, structural floor and ceiling slabs, and nonbearing partitions.

Fiberboard slabs are made by laminating fiberboard products to produce rigid blocks and are used for cold-storage insulation. Mineral-wool slabs or blocks are made of both rock wool and glass wool with suitable binders for low-temperature insulation and specialty uses. Other types of blocks and slabs include cellular glass, insulation board, cellular-rubber products, and vermiculite or expanded mica with asphalt binder.

Flexible Insulation

Flexible insulation is manufactured in two types, (1) blanket or quilt and (2) batt. Blanket insulation is furnished in rolls or strips of convenient length and in various widths suited to standard stud and joist spacing. The usual thicknesses are from ½ to 2 inches. The body of the blanket is made of loosely felted mats of mineral or vegetable fibers, such as rock, slag, or glass wool, wood fiber, cotton, eel grass, and cattle hair. Organic fiber mats are usually chemically treated to make them resistant to fire, decay, insects, and vermin. Most blanket insulation is provided with a covering sheet of paper on one or both sides and with tabs on the edges for fastening the blanket in place. The covering sheet on one side may be of a type intended to serve as a vapor barrier. In some cases the covering sheet is surfaced with aluminum foil or other reflective insulation.

Batt insulation also is made of loosely felted fibers, generally of mineral-wool products. It is made in small units, in widths suitable for fitting between standard framing spaces, and in thicknesses of 2, 3, and 3⅝ inches. Some batts have no covering; others are covered on one side with a paper similar to that used for blanket insulation.

Fill Insulation

Loose-fill insulation is usually composed of materials used in bulk form, supplied in bags or bales, and intended to be poured or blown into place or packed by hand. It is used to fill stud spaces or to build up any desired thicknesses on horizontal surfaces. Loose-fill insulation includes rock, glass, and slag wool, wood fibers, shredded redwood bark, granulated cork, ground or macerated wood pulp products, vermiculite, perlite, powdered gypsum, sawdust, and wood shavings.

Reflective Insulation

Most materials reflect radiant heat, and certain ones have this property to a high degree. Some radiate or emit less heat than others. For reflective insulation, high reflectivity and low emissivity are required, as provided by aluminum foil, sheet metal coated with an alloy of lead and tin, and paper products coated with a reflective oxide composition. Aluminum foil is available in sheets mounted on paper, in corrugated form supported on paper, or mounted on the back of gypsum lath or paper-backed wire lath. Reflective insulation is installed with the reflective surface facing or exposed to an air space preferably three-fourths inch or more in width. Reflective surfaces in contact with other surfaces lose their reflective properties.

Miscellaneous Insulation

Some insulation does not fit in the classifications described, such as (1) confettilike material mixed with adhesive and sprayed on the surface to be insulated, and (2) multiple layers of corrugated paper.

Lightweight aggregates, such as vermiculite and perlite, are sometimes used in plaster as a means of reducing heat transmission.

Lightweight aggregates made from blast-furnace slag, burned-clay products, and cinders are used in concrete and concrete blocks. The conductivity of concrete products made of such lightweight aggregates is substantially lower than that of gravel and stone aggregates.

METHODS OF HEAT TRANSFER

Heat seeks to attain a balance with surrounding conditions, just as water will flow from a higher to a lower level. When occupied buildings are heated to maintain inside temperature in the comfort range, there is a difference in temperature between inside and outside.

Heat will therefore be transferred through walls, floors, ceilings, windows, and doors at a rate that bears some relation to the temperature difference and to the resistance to heat flow of intervening materials. The transfer of heat takes place by 1 or more of 3 methods—conduction, convection, and radiation.

Conduction is defined as the transmission of heat through solid materials; for example, the conduction of heat along a metal rod one end of which is heated in a fire. Convection is the term applied to the transfer of heat by air currents from a warm zone to a colder zone; for example, air moving across a hot radiator carrying heat to other parts of the room or space. Heat also may be transmitted from a warm body to a cold body by wave motion through space, and this process is called radiation because it represents radiant energy. The waves do not heat the space through which they move, but when they come in contact with a colder surface or object a part of the radiant energy is absorbed and converted into sensible heat, and a part is reflected. Heat obtained from the sun is radiant heat.

Heat transfer through a structural unit composed of a variety of materials may include any one or more of the three methods described. In a frame house having an exterior wall of plaster, lath, 2- by 4-inch studs 16 inches on center, sheathing, sheathing paper, and bevel siding, heat is transferred from the room atmosphere to the plaster by radiation, conduction, and convection, and through the lath and plaster by conduction. Heat transfer across the stud space is by radiation from the back of the lath to the colder sheathing, and by convection, the air warmed by the lath moving upward on the warm side of the stud space, and that cooled by the sheathing moving downward on the cold side. Heat transfer through sheathing, sheathing paper, and siding is by conduction. Some small air spaces will be found back of the siding, and the heat transfer across these spaces is principally by radiation. Through the studs from lath to sheathing, heat is transferred by conduction and from the outer surface of the wall to the atmosphere, it is transferred by convection and radiation.

The thermal conductivity of a material is an inverse measure of the insulating value of that material. The customary measure of heat conductivity is the amount of heat in British thermal units that will flow in 1 hour through 1 square foot of a layer 1 inch thick of a homogeneous material, per 1° F. temperature difference between surfaces of the layer. This is usually expressed by the symbol k. Where a material is not homogeneous in structure, such as one containing air spaces like hollow tile, the term conductance is used instead of conductivity. The conductance, usually designated by the symbol C, is the amount of heat in British thermal units that will flow in 1 hour through 1 square foot of the material or combination of materials per 1° F. temperature difference between surfaces of the material.

Resistivity or resistance, a direct measure of the insulating value, is the reciprocal of transmission (conductivity or conductance) and is represented by the symbol R. The overall coefficient of heat transmission through a wall or similar unit, including surface resistances, is represented by the symbol U and defines the movement in British thermal units per hour, per square foot, per 1° F. The resistance of the unit would be $R=\dfrac{1}{U}$.

Heat Loss Through Different Types of Walls

The heat loss through walls and roofs made of different materials can be compared by comparing the overall coefficients of heat transmission, or U values, of the construction assemblies. To determine the U value by test would be impractical in most cases, but it is a simple matter to calculate this value for most combinations of materials commonly used in building construction whose thermal properties are known.

Table 52 gives conductivity and conductance values with corresponding resistivity and resistance values used in calculating the thermal properties of construction units (1).

To compute the U value: Add the resistance of each material, exposed surface, and air space in the given section, using values given in table 52. The sum of these resistances divided into 1 (reciprocal of the sum) gives the coefficient U. For reflective insulation, the value given in the table for an air space bounded by aluminum foil is used.

Example: The overall U value through the stud space of a conventional frame wall consisting of plaster, gypsum lath, air space, wood sheathing, sheathing paper, and siding where heat flow is horizontal is calculated as follows:

Interior surface resistance	0. 61
Gypsum lath and plaster	. 42
Air space	. 91
Wood sheathing, sheathing paper, and bevel siding	2. 00
Exterior surface resistance (wind movement 15 miles per hour)	. 17
Overall resistance	4. 11

The overall coefficient of thermal transmission through the stud space becomes:

$$U = \frac{1}{4.11} = 0.243$$

For the U value through the stud, substitute the resistivity of wood based on the depth of the stud for the resistance value of the air space. For example, a species of wood that has a k value of 0.9 at 10-percent moisture content has a resistivity of 1.11. The resistance of a nominal 2- by 4-inch stud is:

$$3\tfrac{5}{8} \times 1.11 = 4.02$$

Substituting the value of 4.02 for the air-space value of 0.91 gives an overall resistance value of 7.22 or a U value of 0.138. Assuming the area of the stud represents 15 percent of the wall area, the corrected transmission value becomes:

$$\frac{0.243 \times 85 + 0.138 \times 15}{100} = 0.227$$

If 1-inch blanket insulation, bounded by two ¾-inch air spaces, is used in the stud space, the resistance becomes 5.52 instead of the air-space value of 0.91; and the overall resistance of the wall through the stud space is 8.72, compared with 4.11 without insulation. The replacement of $^{25}\!/_{32}$-inch wood sheathing with fiberboard of the same thickness increases the overall resistance from 4.11 to 5.47.

The heat-transmission formula is applied to top-floor ceilings in the same manner as for side walls, but the coefficients used for surface and air-space resistances depend upon the direction of heat flow.

TABLE 52.— *Values* [1] *recommended for computing overall coefficients of thermal transmission* [2]

Material	Description	Conductivity k	Conductance C	Resistivity $\frac{1}{k}$	Resistance $\frac{1}{C}$
Air spaces:					
Bounded by ordinary materials.	¾ inch or more in width:				
	Vertical		1.10		0.91
	Horizontal, heat flow up		1.31		.76
	Horizontal, heat flow down		.945		1.06
Bounded by aluminum foil.	Vertical		.46		2.17
Do	Horizontal, heat flow up		.54		1.85
Do	Horizontal, heat flow down		.20		5.00
Do	Space 1½ inches or more divided by material reflective on both sides:				
	Vertical		.23		4.35
	Horizontal, heat flow up		.27		3.70
	Horizontal, heat flow down		.09		11.11
	30° slope, heat flow up		.25		4.00
	30° slope, heat flow down		.13		7.69
Exterior finish, frame walls:					
Brick veneer	4 inches thick (nominal)		2.27		.44
Stucco	1 inch thick	12.50		0.08	
Wood shingles			1.28		.78
Southern yellow pine lap siding.			1.28		.78
Wood sheathing, paper, and siding.	Fir sheathing—²⁵⁄₃₂ inch, yellow pine lap siding.		.50		2.00
Asbestos shingles			6.00		.17
Plywood siding	3-ply, ⅜ inch		[3] 2.70		[3] .37
Stone veneer	Sandstone or limestone	12.00		.08	
Insulating materials:					
Fill and blankets	Made from mineral or vegetable fiber, or animal hair; enclosed or open.	.27		3.70	
Corkboard	No added binder	.30		3.33	
Insulating board	Made from vegetable fiber	.33		3.03	
Sawdust and shavings	Various species	.41		2.44	
Macerated paper	Ground newsprint and other pulp products.	.27		3.70	
Shredded wood and cement.	Slab insulation made from shredded wood with cement binder.	.46		2.17	
Shredded redwood bark.	Fill insulation made from redwood bark at density of 5 pounds per cubic foot.	.26		3.84	
Paper and asbestos fiber.	Shredded paper and asbestos fiber with emulsified asphalt binder.	.28		3.57	
Mineral wool	Fiber made from rock, slag or glass.	.27		3.70	
Vermiculite	Expanded	.48		2.08	
Cotton	Batt or blanket	.24		4.17	

See footnotes at end of table.

TABLE 52.— *Values* [1] *recommended for computing overall coefficients of thermal transmission* [2]— Continued

Material	Description	Conductivity k	Conductance C	Resistivity $\frac{1}{k}$	Resistance $\frac{1}{C}$
Interior finish:					
Composition wallboard..	³⁄₁₆ to ³⁄₈ inch thick	0.50		2.00	
Gypsum plaster		3.30		.30	
Gypsum wallboard	³⁄₈ inch thick		3.70		0.27
Gypsum lath and plaster	Plaster assumed ½ inch		2.40		.42
Insulating fiberboard (½ inch).	Plain or decorated		.66		1.52
Insulating board lath (½ inch) and plaster.	Plaster assumed ½ inch		.60		1.67
Insulating board lath (1 inch) and plaster.	....do		.31		3.18
Metal lath and plaster...	Plaster assumed ¾ inch		4.40		.23
Plywood	3-ply, ¼ inch thick		[3] 4.00		[3] .25
Wood lath and plaster...			2.50		.40
Masonry materials:					
Brick	Common, assumed 4 inches thick.		1.25		.80
Do	Face, assumed 4 inches thick		2.27		.44
Cement mortar		12.00		.08	
Clay tile:					
4-inch	Hollow		1.00		1.00
6-inch	....do		.64		1.57
8-inch	....do		.60		1.67
10-inch	....do		.58		1.72
12-inch	....do		.40		2.50
Concrete	Lightweight aggregate [4]	2.50		.40	
Do	Cinder aggregate	4.09		.22	
Do	Sand and gravel aggregate	12.00		.08	
Concrete blocks:					
8-inch	Hollow, lightweight aggregate [4]		.50		2.00
12-inch	....do [4]		.47		2.13
4-inch	Hollow, cinder aggregate		1.00		1.00
8-inch	....do		.60		1.66
12-inch	....do		.53		1.88
8-inch	Hollow, sand and gravel aggregate.		1.00		1.00
12-inch	....do		.80		1.25
Gypsum tile:					
3-inch	Hollow		.61		1.64
4-inch	....do		.46		2.18
Tile and terrazzo	For flooring	12.00		.08	
Stone		12.00		.08	
Roofing materials:					
Asbestos shingles			6.00		.17
Asphalt shingles			6.50		.15
Built-up roofing	Assumed thickness ³⁄₈ inch		3.53		.28
Heavy roll roofing			6.50		.15
Slate	Assumed thickness ½ inch		20.00		.05
Wood shingles			1.28		.78
Sheathing:					
Gypsum, (½ inch)			2.82		.35
Insulating fiberboard (²⁵⁄₃₂ inch).			.42		2.37
Plywood	3-ply, ⁵⁄₁₆ inch thick		[3] 3.22		[3] .31

See footnotes at end of table.

TABLE 52.—*Values [1] recommended for computing overall coefficients of thermal transmission [2]—Continued*

Material	Description	Conductivity k	Conductance C	Reistivity 1/k	Resistance 1/C
Sheathing—Continued					
Wood (25/32 inch)_____	For known species and moisture content see values for wood (fig. 5).	_____	1.02	_____	0.98
Wood (25/32 inch) plus building paper.	_____	_____	.86	_____	1.16
Surfaces:					
Inside_____	Ordinary materials, still air:				
	Vertical_____	_____	1.65	_____	.61
	Horizontal, heat flow up____	_____	1.95	_____	.51
	Horizontal, heat flow down__	_____	1.21	_____	.83
Outside_____	Ordinary materials, 15 miles-per-hour wind.	_____	6.00	_____	.17
	Reflective materials, 15 miles-per-hour wind.	_____	5.25	_____	.19

[1] Values are expressed in British thermal units per hour per square foot per 1° F. temperature difference on opposite sides of the material. Conductivity values (k) are per inch of thickness; conductance values (C) are for thickness or construction stated.

[2] Table based chiefly on data from 1952 American Society of Heating and Ventilating Engineers Guide.

[3] Calculated values based on tests made at the Forest Products Laboratory.

[4] Expanded slag, burned clay, or pumice.

HEAT LOSS THROUGH DIFFERENT TYPES OF DOORS AND WINDOWS

In determining heat loss for houses, the loss through doors and windows should be included in the computations. Table 53 gives heat transmission values for doors and windows.

WHERE TO INSULATE

Insulation is used to retard the flow of heat through ceilings, walls, and floors if wide temperature differences occur on opposite sides of these structural elements (4, 6, 7, 10). In dwellings, for example, insulation should be used in the ceiling of those rooms just below an unheated attic. If the attic is heated, the insulation should be placed in the attic ceiling and in the dwarf walls extending from the roof to the floor. All exterior walls should be insulated. Floors over unheated basements, crawl spaces, porches, or garages should also be insulated.

Condensation

Two types of condensation create a problem in buildings during cold weather; that which collects on the inner surfaces of windows, ceilings, and walls, and that which collects within walls or roof spaces (8, 9). Surface condensation is quite common in industrial buildings where relative humidity is high. In a factory or warehouse, water dripping from a ceiling may seriously damage manufactured materials and machinery. "Sweating" walls and windows also are a

TABLE 53.— *Coefficients of transmission* (U) *commonly used for doors, windows and glass blocks* [1]

Item	Door [2]	Door plus storm door	Window
1¼₆-inch door	0. 59	0. 32	
1⅜-inch door	. 51	. 30	
1⅝-inch door	. 46	. 28	
2¼-inch door	. 38	. 25	
Single			1. 13
With storm sash, spaced 1 inch			. 53
Double glazing [3]			. 61
Glass blocks [4]			. 56

[1] Table based upon data from 1952 American Society of Heating and Ventilating Engineers' Guide.
[2] For doors containing thin wood panels or glass, use the same U values as shown for single windows.
[3] 2 thicknesses of glass in 1 sash spaced ¼ inch apart.
[4] Blocks are 7¾ by 7¾ by 3⅞ inches.

serious nuisance. Condensation may collect on the indoor surface of exterior walls of houses, particularly behind furniture or in outside closets, causing damage to finish, furniture, and flooring. It may also collect on windows, particularly those unprotected by storm sash. Water running off the windows may create conditions favorable to decay in wood sash, cause rust in steel sash, and damage window finish and walls and floors below the windows.

To prevent surface condensation, the relative humidity in the building must be reduced or the surface temperature must be raised above the dew point of the atmosphere. Adding insulation to a wall or roof reduces heat transfer through the unit, and the inside surface temperature is increased accordingly. The amount of insulation required for given conditions can be calculated. Storm sash or double glazing reduces condensation on windows.

Moisture sometimes condenses within the walls or roof spaces of buildings if relative humidity is comparatively high during cold weather. In walls, this type of condensation may result in decay of wood, rusting of steel, and damage to exterior paint coatings (2). In roofs and ceilings it may cause stained finish and loosened plaster and increase the chances of decay in structural members.

Under certain conditions when outdoor temperatures are low, water vapor will pass through permeable inner-surface materials and condense within a wall or roof space on some cold surface having a temperature below the dew point of the atmosphere on the warm side. When the condensing surface is considerably below the dew point, differences in vapor pressure between the cold and warm sides cause vapor to move from the high-vapor-pressure zone to the low-pressure zone. The rate of movement is more or less proportional to the difference in vapor pressure and inversely proportional to the resistance of interposed materials. The amount of condensation that collects on the condensing surface depends upon the resistance of intervening materials, differences in vapor pressure, and time. There will also be some difference in vapor pressure between the condensing surface and the outdoor atmosphere. Some part of the water vapor reaching the condensing surface will therefore escape outside through materials that are permeable. Materials commonly used for side wall coverings are usually permeable. Roofing materials are generally highly resistant to vapor transmission. Wood shingles applied over narrow roof boards are not resistant to vapor transmission.

Insulation can cause increased condensation under certain conditions. Heat flow is reduced by insulation, and consequently the temperature of those parts of a wall or roof on the cold side of the insulation is lower during cold weather than if no insulation were used. This in turn means a greater difference in vapor pressure between the warm side and the condensing surface and a greater amount of condensation. Insulation is important, however, as a means of conserving heat and creating comfortable living conditions and its influence on condensation can be largely mitigated.

The rate of vapor transmission through inner surfaces may be controlled by the use of materials having high resistance to vapor movement. Such vapor barriers should be located on or near the warm surface so that the temperature of the barrier will always be above the dew point of the heated space.

For new construction, the barrier may be any one of several materials, such as asphalt-impregnated and surface-coated paper applied over the face of the studs, gypsum lath with aluminum-foil backing, fiberboard lath with vapor-resistive coating, blanket insulation with vapor-resistive cover, and reflective insulation. For existing construction, certain types of paint coatings add materially to the resistance to vapor transfer. One coat of aluminum primer followed by two decorative coats of flat paint or lead and oil seems to offer satisfactory resistance (5, 8, 10).

Ventilation

Attics and roof spaces are generally provided with suitable openings for ventilation, partly as a means of summer cooling and partly as a means of preventing winter condensation (10). For gable roofs, louvered openings are provided in the gable ends, allowing at least 1 square foot of louver opening for each 300 square feet of projected ceiling area. For hip roofs, inlet openings are usually provided under the overhanging eaves with a globe ventilator at or near the peak for an outlet. The inlets should equal 1 square foot to each 900 square feet of projected ceiling area and the outlets 1 square foot to each 1,600 square feet. Ventilation for flat roofs should be developed to suit the method of construction.

LITERATURE CITED

(1) AMERICAN SOCIETY OF HEATING AND VENTILATING ENGINEERS.
    1952. HEATING, VENTILATING, AIR CONDITIONING GUIDE. Ed. 30, 1496 pp., illus. New York.
(2) BROWNE, F. L.
    1927. SOME CAUSES OF BLISTERING AND PEELING OF PAINT ON HOUSE SIDING. Amer. Paint and Varnish Mfrs. Assoc., Sci. Sec. Cir. 317: 480–486.
(3) CLOSE, P. D.
    1946. BUILDING INSULATION. Ed. 3, 372 pp., illus. Chicago.
(4) DILLER, J. D.
    1946. DECAY, A HAZARD IN BASEMENTLESS HOUSES ON WET SITES: PROTECTION OF THE SUBSTRUCTURE TIMBERS FROM EXCESSIVE MOISTURE BY THE USE OF AN ASPHALT ROLL ROOFING AS A SOIL COVER. Amer. Builder 68 (7): 92, 122, 124.

(5) DUNLAP, M. E.
   1949. CONDENSATION CONTROL IN DWELLING CONSTRUCTION. U. S.
      Housing and Home Finance Agency, 73 pp., illus.
(6) SHUMAN, L.
   1950. INSULATION, WHERE AND HOW MUCH. U. S. Housing and Home
      Finance Agency, Tech. Reprint Ser. 4, 9 pp., illus.
(7) ———
   1950. INSULATION OF CONCRETE FLOORS IN DWELLINGS. U. S. Housing
      and Home Finance Agency, Tech. Reprint Ser. 8, 8 pp., illus.
(8) TEESDALE, L. V.
   1947. REMEDIAL MEASURES FOR BUILDING CONDENSATION DIFFICULTIES.
      U. S. Forest Prod. Lab. Rpt. R1710, 27 pp., illus. [Processed.]
(9) ———
   1941. CONDENSATION PROBLEMS IN MODERN BUILDING. U. S. Forest
      Prod. Lab. Rpt. R1196, 9 pp., illus. [Processed.]
(10) ———
   1949. THERMAL INSULATION MADE OF WOOD-BASE MATERIALS. U. S.
      Forest Prod. Lab. Rpt. R1740, 40 pp., illus. [Processed.]

BUILDING FIBERBOARDS

A building fiberboard is a sheet material manufactured of refined or partially refined vegetable fiber, principally from wood, although fiber from bagasse, waste paper, straw, licorice root, and cornstalks is used to some extent. Binding agents and other materials may be added during manufacture to increase strength, resistance to moisture, fire, or decay or to improve some other property. Among the materials added are rosin, alum, asphalt, paraffin, various cements, preservative and fire-resistant chemicals, synthetic resins, and drying oils.

Building fiberboards are manufactured primarily for use as panels, insulation, and cover materials in buildings, for components of cabinets, cupboards, doors for millwork and furniture, and other constructions where flat sheets of moderate strength are desired. The use of fiberboards as a structural material is increasing.

These board products are manufactured in densities ranging from 2 to about 90 pounds a cubic foot. The lower density boards are used primarily as insulation and the higher density boards as structural materials or wearing surfaces. Since the classification of building boards used in this handbook is based on density, rather than on method of manufacture or form of raw material used in manufacture, it is essentially a classification based on end uses, rather than manufacturing processes.

Building boards may be essentially fibrous in character, as are most of the conventional insulation boards and hardboards; or they may be composed of distinct particles bonded together in a flat sheet with a synthetic resin. Some boards are not definitely either a fibrous or a particle type but are a combination of the two types. Most of the fibrous boards are manufactured by an adaptation of a papermaking process; that is, the fiber raw material is reduced to a pulp and formed into a mat from a water slurry on the screen of a paper machine. The wet mat is then dried in a continuous drier or simultaneously compressed and dried into a compact sheet. In a modification of the conventional wet-felting process, known as the semidry process, the fibers are carried in an air suspension rather than a water suspension and are air felted into the mat. The resin-bonded particle boards are made by blending wood particles or chips with various resins and consolidating the mixture under heat and pressure.

CLASSIFICATION OF BUILDING FIBERBOARDS

The following classification by density may be used to define the different classes of building boards:

	Density (gm. per cc.)
Semirigid insulation board	0.02 to 0.15
Rigid insulation board	.15 to .40
Intermediate-density board	.40 to .80
Hardboard, untreated or treated	.80 to 1.15
Special densified hardboards	1.35 to 1.45

Classification of building fiberboards by density is not ideal, particularly when there is a considerable overlap in both properties and uses.   When details of manufacture are considered, however, a classification by density can be used with better success than classifications based on strength, thickness, or some other mechanical property.

The strength and physical properties of the different types of building fiberboards are given in table 54.   These values can be considered as only approximate, since the properties of different proprietary products vary, depending on the kind of fiber and its degree of refinement and the binders, impregnants, and method of manufacture used. Published data from actual tests should be consulted for comparisons of different products or when the contemplated use requires a particular strength or physical property.   The test procedures given in reference (1)[32] may be used to evaluate the different properties of building fiberboards.

SEMIRIGID INSULATION BOARDS

Semirigid insulation board is the term applied to fiberboard products manufactured primarily for use as insulation.   These very low density fiberboards have about the same heat-flow characteristics as conventional blanket or batt insulation but have sufficient stiffness and strength to maintain their position and form without being attached to the structure proper.   They may be bent around curves or corners, and when cemented, mechanically fastened, or placed between framing members will hold their shape and position even though subjected to considerable vibration.

The semirigid insulation boards are manufactured in sheets from ½ to 1½ inches thick.   When greater thicknesses are required, two or more sheets are cemented together.   Sheet sizes vary from 1 by 2 feet to 4 by 4 feet.

Semirigid insulation boards are used for heat insulation in truck and bus bodies, automobiles, refrigerators, railway cars, on the outside of duct work, and wherever vibrations are so severe that loose-fill or batt insulation may pack or shift.   They are also used as sound insulation for the walls of telephone booths and around speakers in radios, public-address systems, and phonographs and as cushioning in utility-quality furniture and for packaged material.

RIGID INSULATION BOARDS

About two-thirds of the footage of building fiberboard produced may be classed as rigid insulation boards.   These boards are used in housing and many allied applications where their economy, structural strength, and insulation (p. 445) and sound-absorbing qualities satisfy a particular use requirement.   They are usually subdivided into the following use classifications: Class A, general-use boards; Class B, lath for plaster base; Class C, roof insulation board; Class D, interior boards, factory-finished; Class E, sheathing; and Class F, interior board, flame-resistant, finished surface.   Physical and strength properties of rigid insulation board are summarized in table 54.

[32] Italic numbers in parentheses refer to Literature Cited, p. 465.

TABLE 54.—*Physical properties and general classifications of building fiberboards* [1]

General classification of building fiberboards	Thicknesses generally available (In.)	Specific gravity	Density (Lb. per cu. ft.)	Thermal conductivity (k)[2]	Modulus of rupture (P.s.i.)	Modulus of elasticity in bending (1,000 P.s.i.)	Tensile strength parallel to surface (P.s.i.)	Tensile strength perpendicular to surface (P.s.i.)	Compression strength parallel to surface (P.s.i.)	Water absorption (24-hour immersion) By weight (Pct.)	By volume (Pct.)	Maximum linear expansion[3] (Pct.)	Hardness
Semirigid insulation board[4]	½–1½	0.02-0.15	1.5-9	0.24-0.27									
Rigid insulation board[4]	[5]⅜–1	0.15-0.40	9-25	0.27-0.40	250-700	25-90	200-500	10-25		5-10		0.5	
Intermediate-density fibrous board	3/16–½	0.40-0.80	25-50	0.40-0.80	400-4,000	90-700	800-2,000		700-3,400			.75	
Intermediate-density resin-bonded particle board	¼–¾	0.40-0.80	25-50	0.40-0.80	400-1,000		100-350			100-150	45-60	.4	
Fibrous hardboard (untreated)	1/16–5/16	0.88-1.04	50-65	0.80-1.40	3,000-6,000	400-800	1,000-3,000		2,000-4,000	9-18			
Fibroushardboard (treated)	⅛–5/16	0.95-1.15	60-70	1.50	7,500-10,500	800-1,000	4,000-5,500		4,200-5,300	5-13			
Resin-bonded particle hardboard	¼–1	0.80-1.05	50-70		3,000-3,500	400-600	300-600	300	3,500-4,000	15-40		.85	[6]2,100
Special densified hardboard	¼–2	1.35-1.45	85-89	1.85	10,000-12,500	1,250	7,700	500	26,500	0.3-1.2			[7]90

1 Values presented in this table must be considered as approximate, since properties between different products vary, depending on density, base materials, binders, impregnants, and method of manufacture; for accurate values for any product, actual tests are required.

2 British thermal units of heat passing through 1 square foot of 1-inch-thick material in 1 hour for each 1° F. difference in temperature between the surfaces of the material.

3 Expansion produced by the change in moisture content from equilibrium at 50 percent relative humidity to equilibrium at 97 percent relative humidity.

4 Includes general-use boards (class A); lath for plaster base (class B); roof insulation (class C); interior board, factory-finished (class D); sheathing (class E); and interior board, flame-resistant finished surface (class F).

5 Sheathing boards are generally available in thicknesses of ½ and 25/32 inch.

6 Load in pounds required to imbed a 0.444-inch-diameter steel ball to ½ its diameter.

7 Hardness as determined on the "M" scale of the Rockwell hardness tester.

Basically, the material in the different products from any one manufacturer may be the same. Because of end use, the board may be modified to some extent by adding impregnants, coating or painting its surface, or cutting it to some special size. The edges and ends may be tongued and grooved, shiplapped, V jointed, beveled, or shaped instead of square for tight or decorative construction. Acoustical board is a special type of insulation board that has multiple holes drilled or punched in the surface to break the sound-wave pattern. Additional information on rigid insulation boards is given in Commercial Standard 42–49 (4) and Federal Specification LLL–F–321b (8).

General-Use Boards

Some rigid insulation boards are manufactured for multiple uses or for remanufacture into boards for specific uses. These boards are classified as general-use boards. They are ⅜ to 1 inch thick and are usually available in sizes of 4 by 8 feet or larger. Some of these products may be used as exterior wall coverings for temporary buildings. Some have satisfactory durability when protected with paint, so that they may be used for panels, siding, or other exterior covering in low-cost housing. A few products, including some sheathing boards, are available that are impregnated with decay-inhibiting chemicals to increase their resistance to deterioration when used where the moisture content and temperature are favorable for decay. Some general-use boards have a prime coat of paint that reduces bleed-through of asphalt impregnants when ordinary paints are used. These prime-coated boards may be given successive coats of paint in the same way as other materials.

Fiberboard Lath

Rigid insulation board is often used as lath for plaster base. Such lath is usually manufactured in ½-inch-thick sheets, 18 by 48 inches in area, with shiplapped or otherwise matched edges. Some brands are provided with metal clips that aid in transferring forces from one sheet to another. Some are impregnated with an asphaltic or similar material, so that the board has high strength retention when wet. Fiberboard lath may have a heavy bituminous coating that acts as a vapor barrier.

Fiberboard lath is nailed to the studding of a house with the vertical joints staggered in the same manner as for gypsumboard lath. Large sheets 4 by 8 feet, are occasionally used as plaster base. When this is done, strips of expanded metal lath should span all joints to reduce plaster cracking at joints.

Fiberboard lath is superior to other plaster base in thermal and sound insulation properties, and, because of its thermal insulation qualities, dirt patterns are claimed to develop more slowly on the final painted plaster surface.

Roof Insulation

Roof insulation boards (Class C) also comprise a substantial part of the production of rigid insulation board. They are manufactured in sheets 2 by 4 feet in area and ½ to 2 inches thick. Roof insulation boards are used to insulate flat or slightly sloped, built-up roofs. They may be used in single and multiple thicknesses and are laid on the deck of the roof as a base for the layers of mopped roll roofing.

Sheathing Board

The biggest growth in the use of rigid insulation board has been as sheathing for houses. Sheathing boards are manufactured usually in a $2\frac{5}{32}$-inch thickness and are either 2 by 8 feet or 4 by 8 feet in area. The 2- by 8-foot sheathing boards are furnished with shiplapped, V-jointed, or otherwise matched edges and are applied horizontally to the outside framework of a house. The 4- by 8-foot board is applied with the long dimension vertical and, when adequately nailed, imparts considerable rigidity to the wall of a house. When properly attached 4- by 8-foot sheathing is approved by the Federal Housing Authority for use in a house without further bracing. With 2- by 8-foot sheathing, additional bracing of the wall is required.

Fiberboard sheathing generally is impregnated or coated with materials that increase its strength retention when wet. In general, it furnishes low-cost coverage, is easy to apply, offers good protection from air infiltration, and has a greater insulation value than heavier sheathing material. The principal disadvantage in using fiberboard sheathing is that, if beveled siding is used for exterior cover, joints in the siding can occur only at studs, because the fiberboard sheathing has little resistance to nail withdrawal. If some other siding material, such as wood shingles, is used, furring strips of wood are usually provided to which the covering material is nailed. Some manufacturers recommend the use of a "shingle backer construction" with special nails to attach the shingles.

The moisture content at which framing lumber, for use with either fiberboard or wood sheathing, should be installed is discussed on page 329.

Interior Board

Interior-finish boards are classified as Class F boards if they are finished with a paint or impregnated with chemicals that retard the spread of flame and as Class D boards if they have a natural finish or are painted with an ordinary paint. Both classes are used for interior wall and ceiling coverage and are manufactured in a wide range of sizes in tile and plank form. Thicknesses most generally available are ½, ¾, and 1 inch. Tiles are manufactured in 8-, 12,- and 16-inch squares and 12- by 24-inch and 16- by 32-inch rectangles and are generally applied to ceilings. All edges are tongued and grooved, or otherwise matched, so that adjacent tiles will produce an even surface. The edges also are usually beveled for decorative purposes. Planks are manufactured in widths of 8, 10, 12, and 16 inches and in lengths of 6, 8, 10, and 12 feet and are usually used for walls. They are usually applied vertically. Their edges are tongued and grooved, or otherwise matched, and may have decorative beveling and beading.

Interior-finish insulation board is used in new construction (drywall) and for repair and remodeling of existing structures. The boards can be cemented to existing plaster that is not loose or uneven, or attached to furring strips by concealed fasteners. Because interior-finish boards have good thermal insulation characteristics, dirt patterns are likely to develop more slowly than on conventionally plastered surfaces. Moreover, large expanses of plaster in ceilings are likely to crack in time, but, since ceilings of insulation board tile have numerous joints, they do not develop other cracks. Interior-finish insulation board usually should not be applied in large sheets with masked joints, as is done with gypsum board, because the linear expansion caused by increase in moisture content with seasonal changes in relative humidity may produce objectionable compression buckling of the panel.

The principal disadvantage of interior-finish insulation board for use as wall covering is its softness and poor resistance to scuffing or abrasion. For that reason, insulation board planking is often used for the upper parts of walls and denser and more abrasion-resistant materials, such as plywood, for the lower parts.

Acoustical board is a special form of interior board tile manufactured with numerous holes, about one-eighth inch in diameter, in the exposed surface of the tile. These tile are to improve the acoustical properties of rooms and are usually applied to ceilings.

INTERMEDIATE-DENSITY FIBROUS BOARDS

Building fiberboards of the intermediate-density classification include boards weighing between about 25 and 50 pounds a cubic foot. They are sometimes called wallboards, but because rigid insulation board and hardboard are also used as wall covering, the term "intermediate-density" is more restrictive. The large range in strength and other physical properties for intermediate-density fiberboard is shown in table 54. Included in the intermediate-density boards are the laminated-paper wallboards and the partially densified hardboards.

The primary use for intermediate-density fibrous boards is for interior wall covering. Because they are denser than the rigid insulation boards, they usually have more resistance to abrasion or other surface damage than lower density boards. These boards also have limited use in furniture manufacture. They can be used to advantage as drawer bottoms and dividers in chests, as backs in cabinets and similar items, and as core stock in utility-grade furniture.

Intermediate-density fibrous boards are manufactured in $\frac{3}{16}$- to $\frac{1}{2}$-inch-thick sheets up to 4 by 12 feet in size. When used as paneling, they may be cemented or nailed in place. Decorative molding often is used at joints. Standard woodworking tools can be used for fabrication. The boards take and hold paint well, though the surfaces sometimes must be sealed to prevent paint from soaking into them. The laminated-paper wallboards often have a large linear expansion with changes in moisture content. If large increases in moisture content may be expected in service, the panels should be attached in such a way that they are free to expand; otherwise, they may buckle between supporting members and even pull loose from the framing. Pertinent standards and specifications for intermediate-density fibrous boards are cited in Commercial Standard 112–43 (3) and Federal Specification UU–W–101a (6).

INTERMEDIATE-DENSITY RESIN-BONDED PARTICLE BOARDS

Production of intermediate-density resin-bonded particle boards is increasing, because more manufacturers are realizing that it is a profitable way to utilize residues from millwork, furniture, or similar operations. They are marketed for use as interior paneling in houses and for use in furniture and millwork. Because the resin-bonded particle boards of intermediate density stay flat with changes in moisture content, they are being used as alternates for solid lumber cores in veneered construction. The linear expansion with changes in moisture content may be large, but it is not greater than the values for wood across the grain.

FIBROUS HARDBOARDS

Most hardboards can be classified as fibrous in nature. They include boards made by the conventional wet-felting process, by forming a wet mat as in the conventional process but drying it before pressing, and by air-felting of dry or partially dry fibers. Hardboard manufactured by the conventional process is characterized by 1 smooth face and 1 face with a screen impression. The "screen back" results because a screen must be used to allow steam to escape from the wet fiber sheet when hot-pressed. With the modification of the conventional process in which the mats are dried before hot-pressing it is possible to produce a hardboard with two smooth faces. In the air-felting process, a board with two smooth faces is produced if the moisture content of the fiber is kept low; otherwise, it is necessary to produce a screen-back board.

Hardboards are classified as untreated or treated. The treated boards are often called tempered boards. The tempering process consists of impregnating pressed boards in drying oils and then baking them in an oven. This treatment increases the strength and water resistance above that of untreated hardboard. The comparison of the strength properties given in table 54 indicates the superiority of the treated hardboards over the untreated boards. The untreated hardboards are manufactured in densities ranging from 50 to 65 pounds a cubic foot; the treated hardboards usually are slightly heavier, from 60 to 70 pounds a cubic foot.

Hardboards are manufactured in $\frac{1}{8}$- to $\frac{5}{16}$-inch-thick sheets as large as 4 by 16 feet. They are dense, uniform, usually smooth surfaced can be worked with ordinary woodworking tools, take finishes well, can be bent to single curvature, and can be punched or die-cut. Some material is embossed or scored during the pressing operation to simulate a leather or tile surface.

Hardboards have many of the same uses thin plywood has. In housing, hardboards are used for interior paneling, floor surfacing, and exterior siding (in the form of either lapped siding or panels). It millwork, they are used for insert panels in conventional doors; surface panels for flush doors; backs, counter tops, and dividers for cabinets; and bottoms in drawers. Hardboards are used as facing material for reusable concrete forms, especially where a smooth finished surface is desired. Other uses of hardboard include window displays, store and office fixtures, paneling for railway passenger cars, and furniture paneling, drawer bottoms, dividers, and exterior surfaces. Hardboards are used for crossbands in veneered construction when decorative veneers are glued to solid lumber cores, and as facing for plywood. In manufacturing operations, templates and assembly jigs are made from hardboards.

Added information on fibrous hardboards is given in Hardboard Industry in the United States (*2*). The applicable specification is cited in Federal Specification LLL–F–311b (*7*).

An important part of the hardboard industry is the remanufacture of hardboards by scoring and painting to produce panel material with a highly decorative, factory-applied paint or enamel finish. These hardboard materials are called "prefinished wall panels" and are sold for wall covering in kitchens and bathrooms of houses where dry-wall construction is used. Metal moldings and a means of attachment are provided, so that the prefinished wall panels can be used successfully in these rooms where high humidity is common without objectionable buckling due to expansion of the hardboard. CS 176–51 (*5*) is the appropriate commercial standard for this material.

Resin-Bonded Particle Hardboards

Production of resin-bonded particle boards of densities between 50 and 70 pounds a cubic foot is limited. These boards are becoming available in greater quantity, however, and may be used for many of the same applications as the conventional fibrous hardboards. They are uniformly light in color and have a surface that is hard and resists damage. Hardboards of this type are made in panels as thick as 1 inch and as large as 4 by 12 feet. Their approximate strength and physical properties are given in table 54. The boards take paint and stain finishes well and can be glued and worked by methods and equipment used for fabricating with wood. Veneers or lithographed-paper overlays simulating wood grain are sometimes glued to the surface of the panels.

The resin-bonded particle hardboards have adequate strength (table 54) for interior applications in housing or for furniture. They are not usually recommended for use in exterior or other severe exposure, because they shrink and swell more than desirable. Because the wood particles are compressed to more than their usual density during pressing, these boards are also subject to springback (thickness swelling) when moisture content changes occur in service. This springback may be as much as 20 percent in some products. Special design considerations must be made when the sizes of the panels are large enough for the linear change to be appreciable.

Special Densified Hardboards

A new group of very dense hardboard products has been developed for specialized uses. These hardboards are manufactured from fibrous mats formed in the same manner as for the conventional hardboards and are available in densities from 85 to 89 pounds a cubic foot and in thicknesses up to 2 inches. A summary of their important strength properties is presented in table 54.

These high-density hardboards can be used to make dies, for either press forming or stretch forming of metals, that have adequate wearing qualities and are economical. Because these boards are lighter in weight than metals, manufacturing jigs made from them are handled and moved more easily then metal jigs. Because their electrical resistance is high, and because they are durable and can be cut and drilled easily, they can be used satisfactorily for bases and panels in electrical apparatus.

LITERATURE CITED

(1) AMERICAN SOCIETY FOR TESTING MATERIALS.
    1952. TENTATIVE METHODS OF TEST FOR EVALUATING THE PROPERTIES OF FIBER BUILDING BOARDS. Amer. Soc. Testing Mater. Standard D1037–52T: 831–846, illus.

(2) LEWIS, W. C.
    1952. HARDBOARD INDUSTRY IN THE UNITED STATES. Jour. Forest Prod. Res. Soc. 2 (4): 3–6, 68.

(3) U. S. DEPARTMENT OF COMMERCE.
    1943. HOMOGENEOUS FIBER WALLBOARD. Com. Standard CS 112–43, 8 pp

(4) ———
    1949. STRUCTURAL FIBER INSULATING BOARD. Com. Standard CS 42–49, 17 pp., illus.

(5) ———
    1951. PREFINISHED WALL PANELS. Com. Standard CS 176–51, 10 pp.

(6) U. S. FEDERAL SUPPLY SERVICE.
    1935. WALLBOARD, COMPOSITION. Fed. Specif. UU–W–101a, 4 pp.

(7) ———
    1939. FIBERBOARD, HARD PRESSED, STRUCTURAL. Fed. Specif. LLL–F–311, 4 pp.

(8) ———
    1942. FIBERBOARD, INSULATING. Fed. Specif. LLL–F–321b, 11 pp.

MODIFIED WOODS AND PAPER-BASE LAMINATES

Materials with properties quite different from those of the original wood are obtained by chemically treating wood, compressing it, or combining the two treatments. Likewise, sheets of paper treated with chemicals can be laminated into thick, hard structural materials. These modified woods and paper-base laminates are considerably more expensive than normal wood because of the cost of chemicals and the amount of processing required to produce them. They should therefore be used only for special purposes where their increased cost is offset by improvements in critical properties and better service.

MODIFIED WOODS

Chemical Treatments

Wood is treated with chemicals to increase its decay, fire, and moisture resistance. Application of water-resistant chemicals to the surface of wood, or impregnation of the wood with such chemicals dissolved in volatile solvents reduces the rate of subsequent swelling and shrinking of the wood but does not affect the final dimension changes caused by seasonal humidity changes. Paints, varnishes, lacquers, and wood-penetrating water repellents have this type of effect. These chemical treatments for wood are discussed elsewhere in this handbook.

Resin-Treated Wood (Impreg)

Permanent stabilization of the dimensions of wood is needed for certain specialty uses. This can be accomplished by depositing a bulking agent within the swollen structure of the wood fibers (26).[33] The most successful bulking agents that have been commercially applied are highly water-soluble, thermosetting, phenol-formaldehyde resin-forming systems, with initially low molecular weights (1, 20, 21, 23). No thermoplastic resins have been found that effectively stabilize the dimensions of wood.

Wood treated with a thermosetting fiber-penetrating resin that is cured without compression is known as impreg. The wood (preferably veneer) is soaked in the aqueous resin-forming solution or impregnated with the solution under pressure until the resin content is equal to 25 to 35 percent of the weight of the dry wood. It is allowed to stand under nondrying conditions for a day or two to permit uniform distribution of the solution throughout the wood. The resin-containing wood is dried at moderate temperatures to remove the water and then heated to higher temperatures to set the resin (21, 23).

Uniform distribution of the resin has been effectively accomplished with solid wood only in the case of the sapwood of readily penetrated species in lengths up to 6 feet (15). The drying of resin-treated solid wood also introduces checking and honeycombing (15). For these reasons, it is recommended that the treatments be confined to veneer and the treated and cured veneer be used to build up the desired objects.

---

[33] Italic numbers in parentheses refer to Literature Cited, p. 477.

Any species of veneer except the resinous pines can be used. The stronger the original wood, the stronger will be the product.

Impreg has a number of properties differing from those of normal wood and ordinary plywood. These are given in table 55, together with similar generalized findings for other modified woods. Table 56 gives data for the strength properties of birch impreg (2, 4).

Impreg found limited use during World War II in plywood panels used to make housings for special electrical control equipment where high electrical resistivity was needed. It shows promise for the facing of exterior plywood to eliminate face checking. It also could be used to advantage in master patterns, such as master shoe lasts, in which dimensional stability is of major importance. It has been given large-scale tests for use in patterns and die models for automobile parts.

TABLE 55.— *Properties of modified woods*

Property	Impreg	Compreg	Staypak
Specific gravity	15 to 20 percent greater than normal wood.	Usually 1.0 to 1.4 (22)	1.25-1.40 (16).
Equilibrium swelling and shrinking.	¼ to ⅓ that of normal wood (21, 23).	¼ to ⅓ that of normal wood at right angles to direction of compression, greater in direction of compression but very slow to attain (22, 24).	Same as normal wood at right angles to compression, greater in direction of compression but very slow to attain (16).
Springback	None	Very small when properly made.	Moderate when properly made (16).
Face checking	Practically eliminated (23).	Practically eliminated for specific gravities below 1.3 (22, 24).	About the same as in normal wood.
Grain raising	Greatly reduced	Greatly reduced for uniform-texture woods, considerable for contrasting-grain woods.	About the same as in normal wood.
Surface finish	Similar to normal wood	Varnished-like appearance for specific gravities above about 1.0. Cut surfaces can be given this surface by sanding and buffing (22).	Varnished-like appearance. Cut surfaces can be given this surface by sanding and buffing (16).
Permeability to water vapor.	About ⅒ that of normal wood (23).	No data but presumably much lower than impreg.	No data but presumably lower than impreg.
Decay and termite resistance.	Considerably better than normal wood (23).	Considerably better than normal wood (22).	Normal, but decay occurs somewhat slower.
Acid resistance	Considerably better than normal wood (23).	Better than impreg because of impermeability.	Better than normal wood because of impermeability but not as good as compreg.
Alkali resistance	Same as normal wood	Somewhat better than normal wood because of impermeability.	Somewhat better than normal wood because of impermeability.
Fire resistance	Same as normal wood (23).	Same as normal wood for long exposure, somewhat better for short exposure (20, 22).	Same as normal wood for long exposures, somewhat better for short exposures.
Heat resistance	Greatly increased	Greatly increased	No data.

TABLE 55.— *Properties of modified woods*— Continued

Property	Impreg	Compreg	Staypak
Electrical conductivity.	$\frac{1}{10}$ that of normal wood at 30 percent relative humidity; $\frac{1}{1000}$ that of normal wood at 90 percent relative humidity (*33*).	Slightly more than impreg at low relative humidity values due to entrapped water (*33*).	No data.
Heat conductivity.	Slightly increased (*23*).	Increased about in proportion to specific gravity increase.	No data, but should increase about in proportion to specific gravity increase.
Compressive strength.	Increased more than proportional to specific gravity increase (*2, 4*).	Increased considerably more than proportional to specific gravity increase (*2, 4*).	Increased about in proportion to specific gravity increase parallel to grain, increased more perpendicular to grain (*2*).
Tensile strength.	Decreased significantly (*2, 4*).	Increased less than proportional to specific gravity increase (*2, 4*).	Increased about in proportion to specific gravity increase (*2*).
Flexural strength.	Increased less than proportional to specific gravity increase (*2, 4*).	Increased less than proportional to specific gravity increase parallel to grain, increased more perpendicular to grain (*2, 4*).	Increased proportional to specific gravity increase parallel to grain, increased more perpendicular to grain (*2*).
Hardness.	Increased considerably more than proportional to specific gravity increase (*4, 32*).	10 to 20 times that of normal wood (*32*).	10 to 18 times that of normal wood (*26*).
Impact strength: Toughness.	About $\frac{1}{2}$ of value for normal wood (*2, 4*).	$\frac{1}{2}$ to $\frac{3}{4}$ of value for normal wood (*2, 4*).	Same to somewhat greater than normal wood (*2*).
Izod.	About $\frac{1}{8}$ of value for normal wood (*2, 4*).	$\frac{1}{2}$ to $\frac{3}{4}$ of value for normal wood (*2, 4*).	Same to somewhat greater than normal wood (*2*).
Abrasion resistance (tangential).	About $\frac{1}{2}$ of value for normal wood (*4, 35*).	Increased about in proportion to specific gravity increase (*4, 35*).	Increase about in proportion to specific gravity increase (*35*).
Machinability.	About the same as normal wood, but dulls tools more.	Requires metalworking tools and metalworking-tool speeds.	Requires metalworking tools and metalworking-tool speeds.
Moldability.	Cannot be molded, but can be formed to single curvatures at time of assembly.	Can be molded by compression and expansion molding methods (*20, 24*).	Cannot be molded.
Gluability.	Same as normal wood.	Same as normal wood after light sanding, or, in the case of thick stock, machining surfaces plane.	Same as normal wood after light sanding, or, in the case of thick stock, machining surfaces plane.

Resin-Treated Compressed Wood (Compreg)

Compreg is similar to impreg except that it is compressed before the curing of the resin within the structure (*22, 24*). The resin-forming chemicals act as plasticizers for the wood so that it can be appreciably compressed under plywood-assembly pressures and compressed to a specific gravity of 1.35 under pressures of about 1,000 pounds per square inch (*20, 22*). Some of its properties are similar to those of impreg, and others vary considerably (tables 55 and 56). Its advantages over impreg are its natural lustrous finish that can be developed on any cut surface by sanding with fine sandpaper and buffing, its greater strength properties, and the fact that it can be molded (tables 55 and 56) (*2, 4, 5*).

Compreg can be molded by gluing blocks of uncured impreg with a phenolic glue so that the glue lines and resin within the plies are only partially set, cutting to the desired length and width but 2 to 3 times the desired thickness and compressing in a split mold at about

TABLE 56.—*Strength properties of normal and modified laminates* [1] *of yellow birch and a laminated paper plastic*

Property	Normal [2] laminated wood	Impreg [3] (impregnated, uncompressed)	Compreg [3] (impregnated, highly compressed)	Staypak [2] (unimpregnated, highly compressed)	Papreg [4] (impregnated, highly compressed)
Thickness (*t*) of laminate_____in.	0.94	1.03	0.63	0.48	0.126, 0.512
Moisture content at time of test_____percent__	9.2	5.0	5.0	4.0	_____
Specific gravity (based on weight and volume at test)_____	.7	.8	1.3	1.4	1.4

PARALLEL LAMINATES

Flexure—grain parallel to span (flatwise): [5]					
Proportional limit stress_____p. s. i__	11,500	15,900	26,700	20,100	15,900
Modulus of rupture_____p. s. i__	20,400	18,800	36,300	39,400	36,600
Modulus of elasticity_____1,000 p. s. i__	2,320	2,380	3,690	4,450	3,010
Flexure—grain perpendicular to span (flatwise): [5]					
Proportional limit stress_____p. s. i__	1,000	1,300	4,200	3,200	10,500
Modulus of rupture_____p. s. i__	1,900	1,700	4,600	5,000	24,300
Modulus of elasticity_____1,000 p. s. i__	153	220	626	602	1,480
Compression parallel to grain (edgewise): [6]					
Proportional limit stress_____p. s. i__	6,400	10,200	16,400	9,700	7,200
Ultimate strength_____p. s. i__	9,500	15,400	26,100	19,100	20,900
Modulus of elasticity_____1,000 p. s. i__	2,300	2,470	3,790	4,670	3,120
Compression perpendicular to grain (edgewise):[6]					
Proportional limit stress_____p. s. i__	670	1,000	4,800	2,600	4,200
Ultimate strength_____p. s. i__	2,100	3,600	14,000	9,400	18,200
Modulus of elasticity_____1,000 p. s. i__	162	243	571	583	1,600
Compression perpendicular to grain (flatwise),[5] maximum crushing strength_____p. s. i__	_____	4,280	16,700	13,200	42,200
Tension parallel to grain (lengthwise):					
Ultimate strength_____p. s. i__	22,200	15,800	37,000	45,000	35,600
Modulus of elasticity_____1,000 p. s. i__	2,300	2,510	3,950	4,610	3,640
Tension perpendicular to grain (crosswise):					
Ultimate strength_____p. s. i__	1,400	1,400	3,200	3,300	20,000
Modulus of elasticity_____1,000 p. s. i__	166	227	622	575	1,710
Shear strength parallel to grain (edgewise):[6] Johnson, double shear across laminations p. s. i__	2,980	3,460	7,370	6,370	17,800
Cylindrical, double shear parallel to laminates p. s. i__	3,030	3,560	5,690	3,080	3,000
Shear modulus:					
Torsion method_____1,000 p. s. i__	182	255	454	_____	_____
Plate shear method (F. P. L. test)__1,000 p. s. i__	_____	_____	_____	385	909

See footnotes at end of table.

Table 56.—*Strength properties of normal and modified laminates [1] of yellow birch and a laminated paper plastic—*Continued

PARALLEL LAMINATES

Property	Normal [2] laminated wood	Impreg [3] (impregnated, uncompressed)	Compreg [3] (impregnated, highly compressed)	Staypak [2] (unimpregnated, highly compressed)	Papreg [4] (impregnated, highly compressed)
Toughness (F. P. L. test, edgewise [6])_____in-lb__	235	125	145	250	_____
Do_____in.-lb. per in. of width	250	120	230	515	_____
Impact strength (Izod)—grain lengthwise:					
Flatwise (notch in face)					
ft.-lb. per in. of notch____	14.0	2.3	4.3	12.7	4.7
Edgewise (notch in face)					
ft.-lb. per in. of notch____	11.3	1.9	[7] 3.2	_____	0.67
Hardness (Rockwell, flatwise [5]) M-numbers_____	_____	−22	84	_____	110
Load to imbed 0.444-inch steel ball to ½ its diameter_____lbs__	1,600	2,400	_____	_____	_____
Hardness modulus (H_M)[8]_____p. s. i__	5,400	9,200	41,300	43,800	35,600
Abrasion-Navy wear-test machine (flatwise),[5] wear per 1,000 revolutions_____in__	0.030	0.057	0.018	0.015	0.018
Water absorption (24-hour immersion), increase in weight_____percent__	43.6	13.7	2.7	4.3	2.2
Dimensional stability in thickness direction:					
Equilibrium swelling_____percent__	9.9	2.8	8.0	29	_____
Recovery from compression_____percent__	_____	0	0	4	_____

CROSS-BANDED LAMINATES

Flexure—face grain parallel to span (flatwise): [5]					
Proportional limit stress_____p. s. i__	6,900	8,100	14,400	11,400	12,600
Modulus of rupture_____p. s. i__	13,100	11,400	22,800	25,100	31,300
Modulus of elasticity_____1,000 p. s. i__	1,310	1,670	2,480	2,900	2,240
Compression—parallel to face grain (edgewise): [6]					
Proportional limit stress_____p. s. i__	3,300	5,200	8,700	5,200	5,000
Ultimate strength_____p. s. i__	5,800	11,400	23,900	14,000	18,900
Modulus of elasticity_____1,000 p. s. i__	1,360	1,500	2,300	2,700	2,370
Tension—parallel to face grain (lengthwise):					
Ultimate strength_____p. s. i__	12,300	7,900	16,500	24,500	27,200
Modulus of elasticity_____1,000 p. s. i__	1,290	1,460	2,190	2,570	2,700
Toughness (F. P. L. test edgewise [6])					
in.-lb. per in. of width__	105	40	115	320	_____

[1] Laminates made from 17 plies of ¹⁄₁₆-inch rotary-cut yellow birch veneer.

[2] Veneer conditioned at 80° F. and 65 percent relative humidity before assembly with phenol resin film glue.

[3] Impregnation, 25 to 30 percent of water-soluble phenol-formaldehyde resin based on the dry weight of untreated veneer.

[4] High-strength paper (0.003-inch thickness) made from commercial unbleached black spruce pulp (Mitscherlich sulfite), phenol resin content 36.3 percent, based on weight of treated paper. Izod-impact, abrasion, flatwise-compression, and shear specimens, all of ½-inch-thick papreg.

[5] Load applied to the surface of the original material (parallel to laminating-pressure direction).

[6] Load applied to the edge of the laminations (perpendicular to laminating-pressure direction).

[7] Values as high as 10.0 foot-pounds per inch of notch have been reported for compreg made with alcohol-soluble resins and 7.0 foot-pounds with water-soluble resins.

[8] Values based on the average slope of load-penetration plots, where H_M is an expression for load per unit of spherical area of penetration of the 0.444-inch steel ball expressed in pounds per square inch:

$$H_M = \frac{P}{2\,\pi r h} \text{ or } 0.717\,\frac{P}{h}.$$

300° F. (*20, 22*). Only a small flash squeeze-out at the parting line between the two halves of the mold need be machined off. This technique was used for molding motor-test "club" propellers and airplane antenna masts during World War II.

A more generally satisfactory molding technique, known as expansion molding, was recently developed (*27*). The method consists of rapidly precompressing dry but uncured single sheets of resin-treated veneer in a cold press after preheating them to 200° to 240° F. The heat-plasticized wood responds to compression before cooling. The heat is insufficient to cure the resin, but the subsequent cooling sets the resin temporarily. These compressed sheets are cut to the desired size, and the assembly of plies is placed in a split mold having the final desired dimensions. Because the wood was precompressed, the filled mold can be closed and locked. When the mold is heated the wood is again plasticized and tends to recover its uncompressed dimensions. This exerts an internal pressure in all directions against the mold equal to about half of the original compressing pressure. On continued heating, the resin is set, and after cooling, the object may be removed from the mold in finished form. Metal inserts or metal surfaces can be molded to compreg or compreg handles molded onto tools by this means. Compreg bands have been molded to the outside of turned wooden cylinders without compressing the core. Compreg tubes and small airplane propellers have been molded in this way.

Solid compreg was used chiefly during World War II for trainer-plane adjustable-pitch propellers, motor-test propellers, antenna masts, spar and connector plates, refrigerator blocks for ships, and tooling jigs. The flight propellers were carved from glued-up blocks of compreg made from compreg panels 1¼ inches thick. Compreg, because of its high compressive strength, makes excellent spar and connector plates. For this purpose, the crossbanded product is usually used. The combined load-bearing and thermal-insulating properties of compreg are advantages in use as supporting blocks for refrigerators. Compreg is extremely useful for aluminum drawing and forming dies, drilling jigs, and jigs for holding parts in place while welding because of its excellent strength properties, dimensional stability, low thermal conductivity, and ease of fabrication.

Solid compreg shows promise for use in silent gears, pulleys, water-lubricated bearings, fan blades, shuttles, bobbins and picker sticks for looms, instrument bases and cases, clarinets, electrical insulators, tool handles, knife handles, and various novelties. Compreg may replace fabric-reinforced plastics in a number of uses because of its better strength properties and lower cost. It should be significantly cheaper because veneer is cheaper than fabric on a weight basis and about 50 percent less resin is used per unit weight of compreg than for fabric laminates.

Partially compressed compreg appears suitable for use as facing materials for ordinary plywood. These facing materials could be used in house, trailer, and boxcar exterior panels, boat siding, interior panels, furniture, and flooring. A test floor was laid at the U. S. Forest Products Laboratory consisting of 10-inch-square tongued-and-grooved panels, composed of 3-ply yellow-poplar compreg faces compressed from $\frac{3}{16}$-inch to $\frac{1}{8}$-inch thickness, a single $\frac{1}{16}$-inch-thick back ply of impreg to balance the construction from a swelling standpoint, and a $\frac{1}{2}$-inch-thick 5-ply Douglas-fir core. All plies in the panel were glued with phenol glues. The floor has a very attractive greenish-tan natural finish and is in excellent shape after over 5 years' service. Such a floor is more expensive than the normal oak strip floor but should require less service to keep in good condition. The original finish can be restored by sanding with fine sandpaper and buffing.

Veneer of any nonresinous species can be used for making compreg. Most properties depend upon the specific gravity to which the wood is compressed rather than the species used (2). Up to the present time, however, compreg has been made almost exclusively from yellow birch or sugar maple.

Untreated Compressed Wood (Staypak)

Resin-treated wood in both the uncompressed (impreg) and compressed (compreg) forms is more brittle than the original wood. To meet the demand for a tougher compressed product than compreg, a compressed wood containing no resin was developed. It will not lose its compression under swelling conditions as will ordinary compressed untreated wood. This material, named staypak, is made by modifying the compressing conditions to cause the lignin cementing material between the cellulose fibers to flow sufficiently to eliminate internal stresses (16, 25).

Staypak is not as water resistant as compreg, but it is about twice as tough and has higher tensile and flexural strength properties, as shown in tables 55 and 56. The natural finish of staypak is almost equal to that of compreg. Under weathering conditions, however, it is definitely inferior to compreg. For outdoor use a good synthetic resin varnish or paint finish should be applied to staypak.

Staypak can be used in the same way as compreg where extremely high water resistance is not needed. It shows promise for use in propellers, tool handles, forming dies, connector plates, and picker sticks and shuttles for weaving, where high impact strength is needed. As staypak is not impregnated, it can be made from solid wood as well as from veneer. It should hence be considerably cheaper than compreg.

Staypak is not being manufactured at the present time. Several companies, however, are prepared to make it if the demand becomes appreciable.

PAPER-BASE LAMINATES

Paper-base laminates are plastic laminates made by compressing together sheets of resin-impregnated paper into coherent solid panels. Such panels have been made for years for electrical insulating purposes (34) but for structural use they are comparatively new. The chief advantages of these panels over ordinary wood-flour-filled, resin-molded materials are their greater toughness and strength, and the fact that they can be made with a higher filler-to-resin ratio and can be pressed at lower pressures without the use of confining molds.

High-Strength Laminates (Papreg)

Improving the paper used has helped develop paper-base laminates suitable for structural use. Cooking the pulp under milder conditions and operating the paper machines so as to give optimum orientation of the fibers in one direction, together with the desired absorbency, contribute markedly to improvements in strength (9, 10, 11, 12, 13, 19).

Phenol resins are the most suitable resins for impregnating the paper from the standpoint of high water resistance, low swelling and shrinking, and high strength properties (except for impact) (3, 6). They are also considerably cheaper than other resins that give comparable properties. Water-soluble resins of the type used for impreg impart the highest water resistance and compressive strength properties to the product, but they unfortunately make the product brittle (low impact strength). Advanced phenol resins fail to penetrate the fibers as well as water-soluble resins and thus impart less water resistance and dimensional stability to the product, but the product is considerably tougher. In practice, compromise alcohol-soluble phenol resins are generally used.

Table 56 gives the strength properties of high-strength paper-base laminates, which have been given the class name papreg (3, 14). The strength properties compare favorably with those for the wood laminates, compreg and staypak, and are considerably superior to those for fabric laminate, except for the edgewise Izod impact test. Fabric laminates have one other advantage, namely, that of being moldable to greater double curvatures. As paper is considerably cheaper than fabric and can be molded at considerably lower pressures, the paper-base laminates should have an appreciable price advantage over fabric laminates.

Papreg was used during World War II for molding nonstructural and semistructural airplane parts, such as gunner's seats and turrets, ammunition boxes, wing tabs, and the surfaces of cargo aircraft flooring and "catwalks." It was tried to a limited extent for the skin surface of airplane structural parts, such as wing tips. One major objection to its use for such parts is that it is more brittle than aluminum and requires special fittings. An experimental papreg tail fin with modified fittings was made for a trainer airplane that met all of the specified loading tests required of its aluminum counterpart (7).

Papreg has been used to some extent for heavy-duty truck floors, industrial processing trays for nonedible materials, and in combination with melamine-treated papers for decorative table tops. An experimental floor of papreg on a plywood base has been laid in the lobby of the Forest Products Laboratory. Papreg also appears suitable for pulleys, gears, bobbins, and many other objects for which fabric laminates are used. Because it can be molded at low pressures and is made from thin papers, it is advantageous for use where very large single sheets or accurate control of panel thickness are required.

Lignin-Filled Laminates

The relatively high cost and limited availability of phenolic resins have resulted in considerable effort to find cheaper, more available impregnating and bonding agents. One of the most successful tried is a lignin-filled laminate made with lignin recovered from the spent liquor of the soda pulping process. Lignin is precipitated from solution within the pulp or added in a preprecipitated form before the paper is made (*28, 29, 30, 31*). The lignin-filled sheets of paper can be laminated without the addition of other resins, but their water resistance is considerably enhanced when some phenolic resin is used. The water resistance can also be improved by merely impregnating the surface sheet with phenolic resin. It is also possible to introduce lignin, together with phenolic resin, into untreated paper sheets with an impregnating machine (*18*).

The lignin-filled laminates are always dark brown or black. Their strength properties except for toughness are, in general, somewhat lower than those of papreg. The Izod impact values are usually twice those for papreg. In spite of the fact that lignin is somewhat thermoplastic, the loss in strength on heating to 200° F. is proportionately no more than for papreg (*14, 31*).

Lignin-filled laminates are not as yet made on a very large scale. They have a number of potential uses, however, where a cheaper laminate with less critical properties than papreg can be used.

Paper-Face Overlays

Paper has recently found considerable use as an overlay material for veneer or plywood. Overlays can be classified into three different types according to their use, namely, (a) masking, (b) decorative, and (c) structural (*8, 17*). Masking overlays are used to cover minor defects in plywood, such as face checks and patches, minimize grain raising, and provide a more uniform paintable surface, thus making possible the use of lower grade veneer. Paper for this purpose need not be of high strength, as the overlays need not add strength to the product. For adequate masking a single surface sheet with a thickness of 0.015 to 0.030 of an inch is desirable. Paper impregnated with phenolic resins to the extent of about 25 percent of the weight of the paper gives the best all-around product. Higher resin contents make the product too costly and tend to make the overlay more transparent. Appreciably lower resin contents give a product with low scratch and abrasion resistance, especially when the panels are wet or exposed to high relative humidities.

The paper faces can be applied at the same time that the veneer is assembled into plywood in a hot press. Undue thermal stresses that might result in checking are not set up if the machine direction of the paper overlays is at right angles to the face plies of the plywood.

Important uses for masking overlays are on exterior plywood siding for houses, boxcars, and trucks where they minimize checking and improve paint-holding properties. Paper-faced plywood is admirably suited for reusable concrete forms, especially if faced with paper having somewhat higher than normal resin contents, because the paper face provides a smooth, nonsticking surface and prevents splintering. Overlays are applied to tops of tables on which parachutes are assembled, where absence of splintering is imperative. They also are suitable for tops of tables for messhalls and for covering various industrial trays because of their low absorbency of liquids and freedom from checking. Overlays are of value on paneling used where greater than normal moisture resistance is needed, such as interior walls of barns, and where smoothness of surface is important, such as bins and chutes.

In order to produce a highly glazed surface, paper-base overlays used for decorative purposes are made with higher resin contents than the masking overlays. Attractive colors and patterns are of major importance in the appearance of these overlays. Unfortunately if phenolic resins are used, the amber color of these makes it impossible to obtain many of the desired pastel shades. Melamine resins are, in general, the most suitable from the standpoint of obtaining light colors combined with high water resistance, stability, and abrasion resistance, and the top plies of most decorative laminates are made from paper treated with these resins. The core plies are usually treated with phenolic resins, because they are considerably cheaper and easier to impregnate into the paper. The back of the decorative overlay usually consists of a melamine backing sheet or a special paper that inproves adhesion.

A typical assembly of thin impregnated sheets, which is normally pressed separately and then glued to solid wood or plywood core, is as follows: A melamine-impregnated back balancing sheet, several plies of core sheet impregnated with phenol resin, a colored printed sheet carrying the design, and a thin, transparent surface sheet with a high content of melamine resin. The printed sheet may be impregnated with melamine resin or unimpregnated. If it is impregnated, the face consists of only two plies, a printed sheet with the design turned upward and a thin clear cover sheet highly impregnated with resin. The clear sheet also may be used as a barrier and the design sheet laid on top with the design facing down. It is important not to place the design on the outer surface, because it may be worn away more easily.

Covering decorative veneers of cabinet woods with a thin sheet of paper impregnated with melamine resin gives the veneer resistance to scuffing and to marring by spilled water or alcoholic drinks. Decorative overlays for table tops are frequently made with a thin aluminum-foil sheet just beneath the barrier sheet to give added resistance to marring or charring by cigarettes by conducting away excessive heat.

Service records of paper-base plastic overlays used extensively for decorative table, bar, and counter tops, and bathroom paneling have, in general, been excellent.

In structural overlays, laminated resin-treated faces contribute to the strength properties of the product and form a structural sandwich material. These overlays are made from high-strength papers and phenol resins of the type used for making papreg. The impregnated paper may be laminated and assembled with a wood core in a single operation, or bonded subsequent to lamination. If the resin content is 35 to 50 percent of the weight of the impregnated paper, the uncured sheets are self-bonding to wood and do not require the application of glue. If the papreg is laminated before gluing to the wood core, the back should be sanded before gluing to remove the surface glaze. A typical use for such sandwich material is for aircraft cargo and truck flooring.

LITERATURE CITED

(1) BURR, H. K., and STAMM, A. J.
1945. COMPARISON OF COMMERCIAL WATER-SOLUBLE PHENOL-FORMALDE-HYDE RESINOIDS FOR WOOD IMPREGNATION. U. S. Forest Prod. Lab. Rpt. 1384, 12 pp., illus. (Rev.) [Processed.]

(2) ERICKSON, E. C. O.
1947. MECHANICAL PROPERTIES OF LAMINATED MODIFIED WOOD. U. S. Forest Prod. Lab. Rpt. 1639, 16 pp., illus. [Processed.]

(3) ———— and BOLLER, K. H.
1945. STRENGTH AND RELATED PROPERTIES OF FOREST PRODUCTS LABORATORY LAMINATED PAPER PLASTIC (PAPREG) AT NORMAL TEMPERATURES. U. S. Forest Prod. Lab. Rpt. 1319, 13 pp. illus. (Rev.) [Processed.]

(4) ———— and FAULKES, W. F.
1949. BASIC PROPERTIES OF YELLOW BIRCH LAMINATES MODIFIED WITH PHENOL AND UREA RESINS. U. S. Forest Prod. Lab. Rpt. 1741, 14 pp. [Processed.]

(5) FINDLEY, W. N., WORLEY, W. J., and KACALIEFF, C. D.
1946. EFFECT OF MOLDING PRESSURE AND RESIN ON RESULTS OF SHORT-TIME TESTS AND FATIGUE TESTS OF COMPREG. Amer. Soc. Mech. Engin. Trans. 68 (4): 317–327, illus.

(6) INSTITUTE OF PAPER CHEMISTRY.
1943. A STUDY OF RESINS, OTHER THAN PHENOL-FORMALDEHYDE RESINS, FOR USE IN PAPER-BASE PLASTICS. Office Prod. Res. and Devlpmt., War Prod. Bd. 24 pp.

(7) JUNGMAN, E. C.
1945. DEVELOPMENT OF A COUNTERPART VERTICAL FIN OF PAPREG FOR THE AT–6 AIRPLANE. U. S. Forest Prod. Lab. Rpt. 1594, 12 pp., illus. [Processed.]

(8) LAND, J. C., GROSS, E. A., and MORAN, T. H.
1945. PLASTIC OVERLAYS FOR DOUGLAS-FIR VENEERS AND PLYWOOD. Douglas-fir Plywood Assoc. Rpt. 33, 22 pp. Tacoma, Wash. [Processed.]

(9) MACKIN, G. E., SEIDL, R. J., and BAIRD, P. K.
1943. EFFECT ON PAPREG OF SIX WATER-ALCOHOL RATIOS USED AS DILUENTS OF IMPREGNATING RESINS. U. S. Forest Prod. Lab. Rpt. 1387, 4 pp., Illus. [Processed.]

(10) ———— SEIDL, R. J., and BAIRD, P. K.
1943. CERTAIN PROPERTIES OF PAPREG AFFECTED BY CALENDERING PRESSURE ON MITSCHERLICH BASE PAPER. U. S. Forest Prod. Lab. Rpt. 1389, 4 pp., illus. [Processed.]

(11) ———— SEIDL, R. J., and BAIRD, P. K.
1943. CERTAIN PROPERTIES OF PAPREG AFFECTED BY WET-PRESS PRESSURE ON MITSCHERLICH BASE PAPER. U. S. Forest Prod. Lab. Rpt. 1575, 3 pp., illus. [Processed.]

(12) McGOVERN, J. N., and KELLER, E. L.
1943. SULFITE PULPS FOR HIGH-STRENGTH LAMINATED PAPER PLASTICS: PULPING VARIABLES AND PROPERTIES OF BLACK SPRUCE PULPS, PAPERS, AND PLASTICS. U. S. Forest Prod. Lab. Rpt. 1393, 6 pp., illus. [Processed.]

(13) ———— and KELLER, E. L.
1943. SULFITE PULPS FOR PAPREG: BASE PAPERS AND LAMINATES FROM BLACK SPRUCE, BALSAM FIR, WESTERN HEMLOCK, AND GRAND FIR. U. S. Forest Prod. Lab. Rpt. 1399, 7 pp. [Processed.]

(14) MEYER, H. R., and ERICKSON, E. C. O.
1945. FACTORS AFFECTING THE STRENGTH OF PAPREG. U. S. Forest Prod. Lab. Rpt. 1521, 7 pp., illus. [Processed.]

(15) MILLETT, M. A., and STAMM, A. J.
1947. WOOD TREATED WITH RESIN-FORMING SYSTEMS, A STUDY OF SIZE AND SPECIES LIMITATIONS. Mod. Plastics 25 (1): 125–127.

(16) SEBORG, R. M., MILLETT, M. A., and STAMM, A. J.
1945. HEAT-STABILIZED COMPRESSED WOOD (STAYPAK). Mech. Engin. 67 (1): 25–31.

(17) SEIDL, R. J.
    1947. PAPER AND PLASTIC OVERLAYS FOR VENEER AND PLYWOOD. Forest
        Prod. Res. Soc. Proc. 1: 8 pp., illus.
(18) ——— BURR, H. K., FERGUSON, C. N., and MACKIN, G. E.
    1944. PROPERTIES OF LAMINATED PLASTIC MADE FROM LIGNIN AND LIGNIN-
        PHENOLIC RESIN-IMPREGNATED PAPERS. U. S. Forest Prod. Lab.
        Rpt. 1595, 7 pp., illus. [Processed.]
(19) ——— MACKIN, G. E., and BAIRD, P. K.
    1943. CERTAIN PROPERTIES OF PAPREG AS AFFECTED BY LAMINATING
        PRESSURE, RESIN CONTENT, AND VOLATILE CONTENT. U. S. Forest
        Prod. Lab. Rpt. 1394, 9 pp., illus. [Processed.]
(20) STAMM, A. J.
    1948. MODIFIED WOODS. Mod. Plastics Encyclopedia. 40 pp., illus.
        New York.
(21) ——— and SEBORG, R. M.
    1950. FOREST PRODUCTS LABORATORY RESIN-TREATED WOOD (IMPREG).
        U. S. Forest Prod. Lab. Rpt. 1380, 8 pp. (Rev.). [Processed.]
(22) ——— and SEBORG, R. M.
    1951. FOREST PRODUCTS LABORATORY RESIN-TREATED, LAMINATED, COM-
        PRESSED WOOD (COMPREG). U. S. Forest Prod. Lab. Rpt. 1381,
        12 pp., illus. (Rev.). [Processed.]
(23) ——— and SEBORG, R. M.
    1939. RESIN-TREATED PLYWOOD. Indus. Engin. Chem. 31 (7): 897, 902.
(24) ——— and SEBORG, R. M.
    1941. RESIN-TREATED, LAMINATED COMPRESSED WOOD. Amer. Soc. Chem.
        Engin. Trans. 37: 17 pp., illus.
(25) ——— SEBORG, R. M., and MILLETT, M. A.
    1948. U. S. PATENT 2,453,679. U. S. Pat. Office, Off. Gaz. 616: 544.
(26) ——— and TARKOW, H.
    1949. WOOD, A LIMITED SWELLING GEL. Jour. Phys. Coll. Chem. 53:
        251–260.
(27) ——— and TURNER, H. D.
    1945. U. S. PATENT 2,391,489. U. S. Pat. Office, Off. Gaz. 581: 568.
(28) U. S. FOREST PRODUCTS LABORATORY.
    1943. LIGNIN-FILLED, LAMINATED PAPER PLASTICS. U. S. Forest Prod.
        Lab. Rpt. 1576, 8 pp., illus. [Processed.]
(29) ———
    1943. PREPARATION OF LIGNIN-FILLED PAPER FOR LAMINATED PLASTICS.
        U. S. Forest Prod. Lab. Rpt. 1577, 11 pp. [Processed.]
(30) ———
    1943. PHYSICAL AND MECHANICAL PROPERTIES OF LIGNIN-FILLED LAMINATED
        PAPER PLASTIC. U. S. Forest Prod. Lab. Rpt. 1579, 3 pp., illus.
        [Processed.]
(31) ——— and HOWARD SMITH PAPER MILLS.
    1946. LIGNIN-FILLED LAMINATED-PAPER PLASTICS. Paper Trade Jour. 122
        (14): 143–150, illus.
(32) WEATHERWAX, R. C., ERICKSON, E. C. O., and STAMM, A. J.
    1948. A MEANS OF DETERMINING THE HARDNESS OF WOOD AND MODIFIED
        WOODS OVER A BROAD SPECIFIC GRAVITY RANGE. Amer. Soc.
        Testing Mater. Bul. 153: 84–89, illus.
(33) ——— and STAMM, A. J.
    1945. ELECTRICAL RESISTIVITY OF RESIN-TREATED WOOD (IMPREG AND
        COMPREG), HYDROLIZED-WOOD SHEET (HYDROXYLIN) AND LAMI-
        NATED RESIN-TREATED PAPER (PAPREG). U. S. Forest Prod.
        Lab. Rpt. 1385, 9 pp., illus. [Processed.]
(34) WEST, C. J.
    1946. PAPER-BASE PLASTICS. Inst. Paper Chem. Bibliog. Ser. 167, 102 pp.
(35) YOUNGQUIST, W. G., and MUNTHE, B. P.
    1948. ABRASIVE RESISTANCE OF WOOD AS DETERMINED WITH THE U. S.
        NAVY WEAR-TEST MACHINE. U. S. Forest Prod. Lab. Rpt. 1732,
        8 pp., illus. [Processed.]

GLOSSARY

Adhesive.—A substance capable of holding materials together by surface attachment. It is a general term and includes cements, mucilage, and paste, as well as glue.

Air-dried.—(*See* Seasoning.)

American lumber standards.—American lumber standards embody provisions for softwood lumber dealing with recognized classifications, nomenclature, basic grades, sizes, description, measurements, tally, shipping provisions, grade marking, and inspection of lumber. The primary purpose of these standards is to serve as a guide or basic example in the preparation or revision of the grading rules of the various lumber manufacturers' associations. A purchaser must, however, make use of association rules, as the basic standards are not in themselves commercial rules.

Annual growth ring.—The growth layer put on in a single growth year, including springwood and summerwood.

Balanced construction.—A construction such that the forces induced by uniformly distributed changes in moisture content will not cause warping. Symmetrical construction of plywood (which see) in which the grain direction of each ply is perpendicular to that of adjacent plies is balanced construction.

Bark.—Outer layer of a tree, comprising the inner bark, or thin, inner living part (phloem), and the outer bark, or corky layer, composed of dry, dead tissue.

Bark pocket.—An opening between annual growth rings that contains bark. Bark pockets appear as dark streaks on radial surfaces and as rounded areas on tangential surfaces.

Basic stress.—(*See* Stress.)

Bastard sawn.—Hardwood lumber in which the annual rings make angles of 30° to 60° with the surface of the piece.

Beam.—A structural member supporting a load applied transversely to it.

Bending, steam.—The process of forming curved wood members by steaming or boiling the wood and bending it to a form.

Bent wood.—(*See* Bending, steam.)

Bird peck.—A small hole or patch of distorted grain resulting from birds pecking through the growing cells in the tree. In shape bird peck usually resembles a carpet tack with the point toward the bark, and it is usually accompanied by discoloration extending for a considerable distance along the grain and to a much lesser extent across the grain. The discoloration produced by bird peck causes what is commonly known as mineral streak (which see).

Birdseye.—Small localized areas in wood with the fibers indented and otherwise contorted to form few to many small circular or elliptical figures remotely resembling birds' eyes on the tangential surface. Common in sugar maple and used for decorative purposes, rare in other hardwood species.

Blue stain.—(*See* Stain.)

Boards.—(*See* Lumber.)

Bolt.—(1) A short section of a tree trunk; (2) in veneer production, a short log of a length suitable for peeling in a lathe.

Bow.—The distortion in a board that deviates from flatness lengthwise but not across its faces.

Boxed heart.—The term used when the pith falls entirely within the four faces of a piece of wood anywhere in its length. Also called boxed pith.

Brashness.—A condition that causes some pieces of wood to be relatively low in shock resistance for the species and, when broken in bending, to fail abruptly without splintering at comparatively small deflections.

Breaking radius.—The limiting radius of curvature to which wood or plywood can be bent without breaking.

Bright.—Unstained.

Broad-leaved trees.—(*See* Hardwoods.)

Brown stain.—(*See* Stain.)

Building fiberboard.—A broad generic term inclusive of sheet materials of widely varying densities manufactured of refined or partially refined wood (or other vegetable) fibers. Bonding agents and other materials may be added to increase strength, resistance to moisture, fire, or decay, or to improve some other property.

Built-up timbers.—An assembly made by joining layers of lumber together with mechanical fastenings so that the grain of all laminations is essentially parallel. (*See* Laminated wood.)

Bump.—A bark-covered protuberance whose bark pattern conforms with the bark pattern of the log. A bump has a height-to-length ratio of at least 1 inch in height to 12 inches in length.

Burl.—(1) A hard, woody outgrowth on a tree, more or less rounded in form, usually resulting from the entwined growth of a cluster of adventitious buds. Such burls are the source of the highly figured burl veneers used for purely ornamental purposes. (2) In wood or veneer, a localized severe distortion of the grain generally rounded in outline, usually resulting from overgrowth of dead branch stubs, varying from one-half inch to several inches in diameter; frequently includes one or more clusters of several small contiguous conical protuberances, each usually having a core or pith but no appreciable amount of end grain (in tangential view) surrounding it.

Butt joint.—(*See* Joint.)

Buttress root.—An adventitious root serving as an added prop or support for a tree. Adventitious roots arise from points on the trunk that are not the normal points of origin for roots.

Cambium.—The one-cell-thick layer of tissue between the bark and wood that repeatedly subdivides to form new wood and bark cells.

Casehardening.—A stressed condition in a board or timber characterized by compression in the outer layers accompanied by tension in the center or core, the result of too severe drying conditions.

Cell.—A general term for the minute units of wood structure, including wood fibers, vessel segments, and other elements of diverse structure and function.

Cellulose.—The carbohydrate that is the principal constituent of wood and forms the frame work of the wood cells.

Check.—A lengthwise separation of the wood that usually extends across the rings of annual growth and commonly results from stresses set up in wood during seasoning.

Chemical brown stain.—(*See* Stain.)

Close grain.—(*See* Grain.)

Coarse grain.—(*See* Grain.)

Collapse.—The flattening of single cells or rows of cells in heartwood during the drying or pressure treatment of wood, characterized by a caved-in or corrugated appearance.

Compartment kiln.—(*See* Kiln.)

Compound curvature.—Wood bent to a compound curvature has curved surfaces no element of which is a straight line.

Compreg.—Wood in which the cell walls have been impregnated with synthetic resin and compressed so as to give it reduced swelling and shrinking characteristics and increased density and strength properties.

Compression failure.—Deformation of the wood fibers resulting from excessive compression along the grain either in direct end compression or in bending. It may develop in standing trees due to bending by wind or snow or to internal longitudinal stresses developed in growth, or it may result from stresses imposed after the tree is cut. In surfaced lumber compression failures appear as fine wrinkles across the face of the piece.

Compression set.—Permanent compression of a group of wood fibers resulting from restraint from swelling while taking on moisture or compressive stresses on the wetter core of wood during drying imposed by adjoining fibers or an external mechanical agency.

Compression wood.—Abnormal wood formed on the lower side of branches and inclined trunks of softwood trees. Compression wood is identified by its relatively wide annual rings, usually eccentric, relatively large amount of summerwood, sometimes more than 50 percent of the width of the annual rings in which it occurs, and its lack of demarcation between springwood and summerwood in the same annual rings. Compression wood shrinks excessively lengthwise, as compared with normal wood.

Conifer.—(*See* Softwoods.)

Crook.—A distortion of a board in which there is a deviation edgewise from a straight line from end to end of the board.

Crossband.—To place the grain of layers of wood at right angles in order to minimize shrinking and swelling; also, in plywood of five or more plies, a layer of veneer whose grain direction is at right angles to that of the face plies.

Cross break.—A separation of the wood cells across the grain. Such breaks may be due to internal strains resulting from unequal longitudinal shrinkage or to external forces.

Cross grain.—(*See* Grain.)

Cup.—A distortion of a board in which there is a deviation flatwise from a straight line across the width of the board.

Curly grain.—(*See* Grain.)

Cut stock.—A term for softwood stock comparable to dimension stock in hardwoods. (*See* Dimension stock.)

Decay.—The decomposition of wood substance by fungi.

> *Advanced (or typical) decay.*—The older stage of decay in which the destruction is readily recognized because the wood has become punky, soft and spongy, stringy, ringshaked, pitted, or crumbly. Decided discoloration or bleaching of the rotted wood is often apparent.

> *Incipient decay.*—The early stage of decay that has not proceeded far enough to soften or otherwise perceptibly impair the hardness of the wood. It is usually accompanied by a slight discoloration or bleaching of the wood.

Decayed knot.—(*See* Knot.)

Delamination.—Separation of plies through failure of the adhesive; often used in reference to the durability of the glue line.

Density.—The weight of a body per unit volume. When expressed in the c. g. s. (centimeter-gram-second) system, it is numerically equal to the specific gravity of the same substance.

Density rules.—Rules for estimating the density of wood based on percentage of summerwood and rate of growth. The rules at present apply only to southern yellow pine and Douglas-fir and differ slightly.

Dew point.—Temperature at which a vapor begins to deposit as a liquid. Applies especially to water in the atmosphere.

Diagonal grain.—(*See* Grain.)

Diffuse-porous wood.—Certain hardwoods in which the pores tend to be uniform in size and distribution throughout each annual ring or to decrease in size slightly and gradually toward the outer border of the ring.

Dimension.—(*See* Lumber.)

Dimension stock.—A term largely superseded by the term hardwood dimension lumber. It is hardwood stock processed to a point where the maximum waste is left at a dimension mill, and the maximum utility is delivered to the user. It is stock of specified thickness, width, and length, in multiples thereof. According to specification it may be solid or glued up; rough or surfaced; semifabricated or completely fabricated.

Dimensional stabilization.—Reduction in swelling and shrinking of wood, caused by changes in its moisture content with changes in relative humidity, through special treatment.

Dote.—"Dote," "doze," and "rot" are synonymous with "decay" and are any form of decay that may be evident as either a discoloration or a softening of the wood.

Dressed lumber.—(*See* Lumber.)

Dry-bulb temperature.—Temperature of air as indicated by a standard thermometer. (*See* Psychrometer.)

Dry kiln.—(*See* Kiln.)

Dry rot.—A term loosely applied to any dry, crumbly rot but especially to that which, when in an advanced stage, permits the wood to be crushed easily to a dry powder. The term is actually a misnomer for any decay, since all fungi require considerable moisture for growth.

Durability.—A general term for permanence or resistance to deterioration. Frequently used to refer to the degree of resistance of a species of wood to attack by wood-destroying fungi under conditions that favor such attack. In this connection the term "decay resistance" is more specific.

Early wood.—(*See* Springwood.)

Edge grain.—(*See* Grain.)

Edge joint.—(*See* Joint.)

Empty-cell process.—Any process for impregnating wood with preservatives or chemicals in which air, imprisoned in the wood under pressure, expands when pressure is released to drive out part of the injected preservative or chemical. The distinguishing characteristic of the empty-cell process is that no vacuum is drawn before applying the preservative. The aim is to obtain good preservative distribution in the wood and leave the cell cavities only partially filled.

Encased knot.—(*See* Knot.)

End joint.—(*See* Joint.)

Equilibrium moisture content.—The moisture content at which wood neither gains nor loses moisture when surrounded by air at a given relative humidity and temperature.

Extractives.—Substances in wood, not an integral part of the cellular structure, that can be removed by solution in hot or cold water, ether, benzene, or other solvents that do not react chemically with wood components.

Factory and shop lumber.—(*See* Lumber.)

Fiber, wood.—A comparatively long (one twenty-fifth or less to one-third inch), narrow, tapering wood cell closed at both ends.

Fiberboard.—(*See* Building fiberboard.)

Fiber saturation point.—The stage in the drying or wetting of wood at which the cell walls are saturated and the cell cavities are free from water. It is usually taken as approximately 30 percent moisture content, based on weight when oven-dry.

Figure.—The pattern produced in a wood surface by annual growth rings, rays, knots, deviations from regular grain such as interlocked and wavy grain, and irregular coloration.

Fine grain.—(*See* Grain.)

Finish.—Wood products to be used in the joiner work, such as doors and stairs, and other fine work required to complete a building, especially the interior.

Fireproofing.—Making wood resistant to fire. Wood cannot be treated chemically so that it will not char or decompose at temperatures of about 280° F. and higher. What effective fireproofing does is to make wood difficult to ignite, keep it from supporting its own combustion, and delay the spread of flame over the wood surface.

Fire-retardant chemical.—A chemical or preparation of chemicals used to reduce flammability or to retard spread of fire.

Flakes.—(*See* Rays, wood.)

Flat grain.—(*See* Grain.)

Flitch.—A portion of a log sawed on two or more sides and intended for remanufacture into lumber or sliced or sawed veneer. The term is also applied to the resulting sheets of veneer laid together in sequence of cutting.

Framing.—Lumber used for the structural members of a building, such as studs and joists.

Full-cell process.—Any process for impregnating wood with preservatives or chemicals in which a vacuum is drawn to remove air from the wood before admitting the preservative. This favors heavy absorption and retention of preservative in the treated portions.

Gelatinous fibers.—Abnormal fibers in certain hardwoods. They are associated with "Tension wood," which see.

Girder.—A large or principal beam used to support concentrated loads at points along its length.

Grade.—The designation of the quality of a manufactured piece of wood or of logs.

Grain.—The direction, size, arrangement, appearance, or quality of the fibers in wood or lumber. To have a specific meaning the term must be qualified.

   Close-grained wood.—Wood with narrow, inconspicuous annual rings. The term is sometimes used to designate wood having small and closely spaced pores, but in this sense the term "fine textured" is more often used.

   Coarse-grained wood.—Wood with wide conspicuous annual rings in which there is considerable difference between springwood and summerwood. The term is sometimes used to designate wood with large pores, such as oak, ash, chestnut, and walnut, but in this sense the term "coarse textured" is more often used.

   Cross-grained wood.—Wood in which the fibers deviate from a line parallel to the sides of the piece. Cross grain may be either diagonal or spiral grain or a combination of the two.

Grain—Continued

Curly-grained wood.—Wood in which the fibers are distorted so that they have a curled appearance, as in "birdseye" wood. The areas showing curly grain may vary up to several inches in diameter.

Diagonal-grained wood.—Wood in which the annual rings are at an angle with the axis of a piece as a result of sawing at an angle with the bark of the tree or log. A form of cross grain.

Edged-grained lumber.—Lumber that has been sawed so that the wide surfaces extend approximately at right angles to the annual growth rings. Lumber is considered edge grained when the rings form an angle of 45° to 90° with the wide surface of the piece.

Fine-grained wood.—(*See* Close-grained wood.)

Flat-grained lumber.—Lumber that has been sawed in a plane approximately perpendicular to a radius of the log. Lumber is considered flat grained when the annual growth rings make an angle of less than 45° with the surface of the piece.

Interlocked-grained wood.—Wood in which the fibers are inclined in one direction in a number of rings of annual growth, then gradually reverse and are inclined in an opposite direction in succeeding growth rings, then reverse again.

Open-grained wood.—Common classification of painters for woods with large pores, such as oak, ash, chestnut, and walnut. Also known as "coarse textured."

Plainsawed lumber.—Another term for flat-grained lumber.

Quartersawed lumber.—Another term for edge-grained lumber.

Spiral-grained wood.—Wood in which the fibers take a spiral course about the trunk of a tree instead of the normal vertical course. The spiral may extend in a right-handed or left-handed direction around the tree trunk. Spiral grain is a form of cross grain.

Straight-grained wood.—Wood in which the fibers run parallel to the axis of a piece.

Vertical-grained lumber.—Another term for edge-grained lumber.

Wavy-grained wood.—Wood in which the fibers collectively take the form of waves or undulations.

Green.—Freshly sawed lumber, or lumber that has received no intentional drying; unseasoned. The term does not apply to lumber that may have become completely wet through waterlogging.

Growth ring.—(*See* Annual growth ring.)

Hardwoods.—Generally one of the botanical groups of trees that have broad leaves in contrast to the conifers or softwoods. The term has no reference to the actual hardness of the wood.

Heartwood.—The wood extending from the pith to the sapwood, the cells of which no longer participate in the life processes of the tree. Heartwood may be infiltrated with gums, resins, and other materials that usually make it darker and more decay resistant than sapwood.

Hemicellulose.—A celluloselike material in wood that is easily decomposable, as by dilute acid, yielding several different simple sugars.

Hollow-core construction.—A panel construction with faces of plywood, hardboard, or similar material bonded to a framed-core assembly of wood lattice, paperboard rings, or the like, which support the facing at spaced intervals.

Honeycomb.—A construction of thin sheet material, such as resin impregnated paper or fabric, which has been corrugated and bonded, each sheet in opposite phase to the phases of adjacent sheets, to form a core material whose cross section is a series of mutually continuous cells similar to natural honeycomb.

Honeycombing.—Checks, often not visible at the surface, that occur in the interior of a piece of wood, usually along the wood rays.

Horizontally laminated.—(*See* Laminated wood.)

Imperfect manufacture.—Includes all defects or blemishes that are produced in manufacturing, such as chipped grain, loosened grain, raised grain, torn grain, skips in dressing, hit and miss (series of surfaced areas with skips between them), variation in sawing, miscut lumber, machine burn, machine gouge, mismatching, and insufficient tongue or groove.

Impreg.—Wood in which the cell walls have been impregnated with synthetic resin so as to reduce materially its swelling and shrinking. Impreg is not compressed.

Insulation, thermal.—Materials that retard the transfer of heat when placed between two heat-conducting materials.

Intergrown knot.—(*See* Knot.)

Interlocked-grained wood.—(*See* Grain.)

Joint.—The junction of two pieces of wood or veneer.

>*Butt joint.*—An end joint formed by abutting the squared ends of two pieces. Because of the inadequacy and variability in strength of butt joints when glued, they are not generally glued.

>*Edge joint.*—The place where two pieces of wood are joined together edge to edge, commonly by gluing. The joints may be made by gluing two squared edges as in a plain edge joint or by using machined joints of various kinds, such as tongued-and-grooved joints.

>*End joint.*—The place where two pieces of wood are joined together end to end, commonly by scarfing and gluing.

>*Lap joint.*—A joint made by placing one adherend partly over another and bonding the overlapped portions.

>*Scarf joint.*—An end joint formed by joining with glue the ends of two pieces that have been tapered or beveled to form sloping plane surfaces, usually to a feather edge, and with the same slope of the plane with respect to the length in both pieces. In some cases, a step or hook may be machined into the scarf to facilitate alinement of the two ends, in which case the plane is discontinuous and the joint is known as a stepped or hooked scarf joint.

>*Starved joint.*—A glue joint that is poorly bonded because an insufficient quantity of glue remained in the joint. Starved joints are caused by the use of excessive pressure or insufficient viscosity of the glue, or a combination of these, which results in the glue being forced out from between the surfaces to be joined.

Joist.—One of a series of parallel beams used to support floor and ceiling loads and supported in turn by larger beams, girders, or bearing walls.

Kiln.—A heated chamber for drying lumber, veneer, and other wood products.

>*Compartment kiln.*—A dry kiln in which the total charge of lumber is dried as a single unit. It is designed so that, at any given time, the temperature and relative humidity are uniform throughout the kiln. The temperature is increased as drying progresses, and the relative humidity is adjusted to the needs of the lumber.

>*Progressive kiln.*—A dry kiln in which the total charge of lumber is not dried as a single unit but as several units, such as kiln-truckloads, that move progressively through the kiln. The kiln is designed so that the temperature is lower and the relative humidity higher at the entering end than at the discharge end.

Kiln brown stain.—(*See* Stain.)

Kiln-dried.—(*See* Seasoning.)

Knot.—That portion of a branch or limb which has been surrounded by subsequent growth of the wood of the trunk or other portion of the tree. As a knot appears on the sawed surface it is merely a section of the entire knot, its shape depending upon the direction of the cut.

>*Decayed knot.*—A knot that, due to advanced decay, is softer than the surrounding wood.

>*Encased knot.*—A knot whose rings of annual growth are not intergrown with those of the surrounding wood.

>*Intergrown knot.*—A knot whose rings of annual growth are completely intergrown with those of the surrounding wood.

>*Large knot.*—A knot more than 1½ inches in diameter.

>*Loose knot.*—A knot that is not held firmly in place by growth or position and that cannot be relied upon to remain in place.

>*Medium knot.*—A knot more than ¾ inch but not more than 1½ inches in diameter.

>*Pin knot.*—A knot that is not more than ½ inch in diameter.

>*Round knot.*—A knot whose sawed section is circular or oval.

>*Small knot.*—A knot more than ½ inch but not more than ¾ inch in diameter.

>*Sound knot.*—A knot that is solid across its face, at least as hard as the surrounding wood, and shows no indication of decay.

>*Spike knot.*—A knot cut approximately parallel to its long axis so that the exposed section is definitely elongated.

Knot cluster.—Three or more knots in a compact, roughly circular group, with the grain between them highly contorted, originating from adventitious buds. Two or more knots laterally arranged and without contortion of the fibers between them do not constitute a knot cluster.

Laminate.—A product made by bonding together two or more layers (laminations) of material or materials.

Laminate, paper-base.—A multilayered panel made by compressing sheets of resin-impregnated paper together into a coherent solid mass.

Laminated wood.—An assembly made by bonding layers of veneer or lumber with an adhesive so that the grain of all laminations is essentially parallel. (*See* Built-up timbers.)

Horizontally laminated wood.—Laminated wood in which the laminations are so arranged that the wider dimension of each lamination is approximately perpendicular to the direction of load.

Vertically laminated wood.—Laminated wood in which the laminations are so arranged that the wider dimension of each lamination is approximately parallel to the direction of load.

Lap joint.—(*See* Joint.)

Late wood.—(*See* Summerwood.)

Lignin.—The second most abundant constituent of wood, located principally in the middle lamella, which is the thin cementing layer between the wood cells. The chemical structure of lignin has not been definitely determined.

Log.—A section of the trunk of a tree in suitable length for sawing into commercial lumber.

Longitudinal.—Generally, the direction along the length of the grain of wood.

Loose knot.—(*See* Knot).

Lumber.—The product of the saw and planing mill not further manufactured than by sawing, resawing, passing lengthwise through a standard planing machine, crosscutting to length, and matching.

Boards.—Yard lumber less than 2 inches thick and 1 or more inches wide.

Dimension.—Lumber from 2 inches to but not including 5 inches thick and 2 or more inches wide.

Dressed size.—The dimensions of lumber after shrinking from the green dimension and being surfaced with a planing machine usually ⅜ or ½ inch less than the nominal or rough size; for example, a 2- by 4-inch stud actually measures 1⅝ by 3⅝ inches under American lumber standards for softwood lumber.

Factory and shop lumber.—Lumber intended to be cut up for use in further manufacture. It is graded on the basis of the percentage of the area that will produce a limited number of cuttings of a specified minimum size and quality.

Matched lumber.—Lumber that is edge dressed and shaped to make a close tongued-and-grooved joint at the edges or ends when laid edge to edge or end to end.

Nominal size.—As applied to timber or lumber, the rough-sawed commercial size by which it is known and sold in the market.

Patterned lumber.—Lumber that is shaped to a pattern or to a molded form in addition to being dressed, matched, or shiplapped, or any combination of these workings.

Rough lumber.—Lumber as it comes from the saw.

Shiplapped lumber.—Lumber that is edge dressed to make a lapped joint.

Shipping-dry lumber.—Lumber that is partially dried to prevent stain and mold in transit.

Structural lumber.—Lumber that is 2 or more inches thick and 4 or more inches wide, intended for use where working stresses are required. The grading of structural lumber is based on the strength of the piece and the use of the entire piece.

Surfaced lumber.—Lumber that is dressed by running it through a planer.

Timbers.—Lumber 5 or more inches in least dimension. Timbers may be classified as beams, stringers, posts, caps, sills, girders, purlins, etc.

Timbers, round.—Timbers used in the original round form, such as poles, piling, posts, and mine timbers.

Yard lumber.—Lumber of all sizes and patterns that is intended for general building purposes. The grading of yard lumber is based on the intended use of the particular grade and is applied to each piece with reference to its size and length when graded, without consideration to further manufacture.

Matched lumber.—(*See* Lumber.)

Medullary rays.—(*See* Rays, wood.)

Millwork.—Generally, all building materials made of finished wood and manufactured in millwork plants and planing mills. Includes such items as inside and outside doors, window and door frames, blinds, porch work, mantels, panel work, stairways, moldings, and interior trim. Does not include flooring, ceiling, or siding.

Mineral streak.—An olive to greenish-black or brown discoloration of undetermined cause in hardwoods, particularly hard maples; commonly associated with bird pecks and other injuries; occurs in streaks usually containing accumulations of mineral matter.

Modified wood.—Wood processed to impart properties quite different from those of the original wood by means of chemical treatment, compression, or treatment with or without heat.

Moisture content of wood.—The amount of water contained in the wood, usually expressed as a percentage of the weight of the oven-dry wood.

Moisture gradient.—A condition of graduated moisture content between successive thickness zones of wood that may be losing or absorbing moisture. During seasoning the gradations are between the relatively dry surface zones and the wet zones at the center of the piece.

Molded plywood.—(*See* Plywood.)

Naval stores.—A term applied to the oils, resins, tars, and pitches derived from oleoresin contained in, exuded by, or extracted from trees chiefly of the pine species (genus *Pinus*) or from the wood of such trees.

Niggerizing.—A process of treating wood with paraffin or a coke-oven coal tar to increase its resistance to acids.

Nominal-size lumber.—(*See* Lumber.)

Old growth.—Timber in or from a mature naturally established forest. When the trees have grown during most if not all of their individual lives in active competition with their companions for sunlight and moisture, this timber is usually straight and relatively free of knots.

Open grain.—(*See* Grain.)

Oven-dry wood.—Wood dried to constant weight in an oven or above the temperature of boiling water (usually 101° to 105° C. or 214° to 221° F.).

Overlaid veneer.—A single sheet of veneer overlaid and bonded on one or both sides with paper, resin-impregnated paper, or metal.

Overlays, paper-plastic.—One or more sheets of paper impregnated with resin and used as face material usually for plywood, but sometimes for lumber or other products. The paper-plastic material, when properly molded to the surface of the plywood, forms an integral part of the panel and cannot be peeled off. Overlays can be classified as masking, decorative, or structural, depending on their purpose.

Papreg.—Any of various paper products made by impregnating sheets of specially manufactured high-strength paper with synthetic resin and laminating the sheets to form a dense, moisture-resistant product.

Parenchyma.—Short cells having simple pits and functioning primarily in the metabolism and storage of plant food materials. They remain alive longer than the tracheids, fibers, and vessel segments, sometimes for many years. Two kinds of parenchyma cells are recognized, those in vertical strands, known more specifically as wood parenchyma, and those in horizontal series in the rays, known as ray parenchyma.

Peck.—Pockets or areas of disintegrated wood caused by advanced stages of localized decay in the living tree. It is usually associated with cypress and incense-cedar. There is no further development of peck once the lumber is seasoned.

Phloem.—(*See* Bark.)

Pin-knot.—(*See* Knot.)

Pit.—A relatively unthickened portion of a cell wall where a thin membrane may permit liquids to pass from one cell to another. A "bordered" pit has an overhanging rim that is not present in a "simple" pit.

Pitch pocket.—An opening extending parallel to the annual growth rings containing, or that has contained, pitch, either solid or liquid.

Pitch streak.—A well-defined accumulation of pitch in a more or less regular streak in the wood of certain conifers.

Pith.—The small, soft core occurring in the structural center of a tree trunk, branch, twig, or log.

Pith fleck.—A narrow streak, resembling pith on the surface of a piece; usually brownish, up to several inches in length; resulting from burrowing of larvae in the growing tissues of the tree.

Plainsawed.—(*See* Grain.)

Planing mill products.—Products worked to pattern, such as flooring, ceiling, and siding.

Plank.—A broad board, usually more than 1 inch thick, laid with its wide dimension horizontal and used as a bearing surface.

Plasticizing wood.—Softening wood by hot water, steam, or chemical treatment to increase its moldability.

Plywood.—A crossbanded assembly made of layers of veneer or of veneer in combination with a lumber core or plies joined with an adhesive. Two types of plywood are recognized, namely, (1) veneer plywood and (2) lumber-core plywood. The grain of adjoining plies is usually laid at right angles, and almost always an odd number of plies are used to obtain balanced construction.

   Molded plywood.—Plywood that is glued to the desired shape either between curved forms or more commonly by fluid pressure applied with flexible bags or blankets (bag molding) or other means.

   Postformed plywood.—The product formed by reshaping, by means of steaming or other plasticizing agent, flat plywood into a curved shape.

Pocket rot.—Advanced decay that appears in the form of a hole, pocket, or area of soft rot usually surrounded by apparently sound wood.

Pore.—(*See* Vessels.)

Porous woods.—Another name for hardwoods, which frequently have vessels or pores large enough to be seen readily without magnification.

Postformed plywood.—(*See* Plywood.)

Preservative.—Any substance that, for a reasonable length of time, is effective in preventing the development and action of wood-rotting fungi, borers of various kinds, and harmful insects that deteriorate wood.

Progressive kiln.—(*See* Kiln.)

Psychrometer.—An instrument for measuring the amount of water vapor in the atmosphere and having both wet-bulb and dry-bulb thermometers. Since the bulb of the wet-bulb thermometer is kept moistened, that thermometer will be cooled as a result of evaporation and consequently will show a temperature lower than that of the dry-bulb thermometer. The evaporation is less in moist air, hence the difference between the thermometer readings is greater when the air is dry; this difference constitutes a measure of the dryness of the surrounding air.

Quartersawed.—(*See* Grain.)

Radial.—Coincident with a radius from the axis of the tree or log to the circumference. A radial section is a lengthwise section in a plane that passes through the centerline of the tree trunk.

Raised grain.—A roughened condition of the surface of dressed lumber in which the hard summerwood is raised above the softer springwood but not torn loose from it.

Rate of growth.—The rate at which a tree has laid on wood, measured radially in the trunk or in lumber cut from the trunk. The unit of measure in use is number of annual growth rings per inch.

Rays, wood.—Strips of cells extending radially within a tree and varying in height from a few cells in some species to 4 or more inches in oak. The rays serve primarily to store food and transport it horizontally in the tree.

   Medullary rays.—Wood rays that originate at the pith.

Relative humidity.—Ratio of the amount of water vapor present in the air to that which the air would hold at saturation at the same temperature. It is usually considered on the basis of the weight of the vapor but, for accuracy, should be considered on the basis of vapor pressures.

Resin passage (or duct).—Intercellular passages that contain and transmit resinous materials. On a cut surface, they are usually inconspicuous. They may extend vertically parallel to the axis of the tree or at right angles to the axis and parallel to the rays.

Ring-porous woods.—A group of hardwoods in which the pores are comparatively large at the beginning of each annual ring and decrease in size more or less abruptly toward the outer portion of the ring, thus forming a distinct inner zone of pores, known as the springwood, and an outer zone with smaller pores, known as the summerwood.

Rot.—(*See* Decay.)

Rotary-cut veneer.—(*See* Veneer.)

Rough lumber.—(*See* Lumber.)

Round knot.—(*See* Knot.)

Round timbers.—(*See* Lumber.)

Sandwich construction.—(*See* Structural sandwich construction.)

Sap.—All the fluids in a tree, special secretions and excretions, such as oleoresin, excepted.

Sapwood.—The living wood of pale color near the outside of the log. Under most conditions the sapwood is more susceptible to decay than heartwood.

Sawed veneer.—(*See* Veneer.)

Saw kerf.—(1) Grooves or notches made in cutting with a saw; (2) that portion of a log, timber, or other piece of wood removed by the saw in parting the material into two pieces.

Scarf joint.—(*See* Joint.)

Seasoning.—Removing moisture from green wood in order to improve its serviceability.

> *Air-dried.* Dried by exposure to air, usually in a yard, without artificial heat.

> *Kiln-dried.*—Dried in a kiln with the use of artificial heat.

Second growth.—Timber that has grown after the removal, whether by cutting, fire, wind, or other agency, of all or a large part of the previous stand.

Set.—A permanent deformation, either in compression or tension, of wood fibers that are kept from shrinking or swelling, with loss or gain of moisture, by adjoining fibers that are at a different moisture content.

Shake.—A separation along the grain, the greater part of which occurs between the rings of annual growth.

Shear.—Slipping of one part of a piece of wood upon another along the grain.

Sheathing.—The structural covering, usually of boards or fiberboards, placed over exterior studding or rafters of a structure.

Shiplapped lumber.—(*See* Lumber.)

Shipping-dry lumber.—(*See* Lumber.)

Shop lumber.—(*See* Lumber.)

Side cut.—The term used when the pith is not present in a piece.

Sliced veneer.—(*See* Veneer.)

Softwoods.—Generally, one of the botanical groups of trees that in most cases have needlelike or scalelike leaves; the conifers; also the wood produced by such trees. The term has no reference to the actual hardness of the wood.

Sound knot.—(*See* Knot.)

Specific gravity.—The ratio of the weight of a body to the weight of an equal volume of water at 4° C. or other specified temperature.

Spike knot.—(*See* Knot.)

Spiral grain.—(*See* Grain.)

Split.—A lengthwise separation of the wood, due to the tearing apart of the wood cells.

Springwood.—The portion of the annual growth ring that is formed during the early part of the season's growth. It is usually less dense and weaker mechanically than summerwood.

Stain.—A discoloration in wood that may be caused by such diverse agencies as micro-organisms, metal, or chemicals. The term also applies to materials used to impart color to wood.

> *Blue stain.*—A bluish or grayish discoloration of the sapwood caused by the growth of certain dark-colored fungi on the surface and in the interior of the wood; made possible by the same conditions that favor the growth of other fungi.

> *Brown stain.*—A rich brown to deep chocolate-brown discoloration of the sapwood of some pines caused by a fungus that acts much like the blue-stain fungus.

> *Chemical brown stain.*—A chemical discoloration of wood, which sometimes occurs during the air-drying or kiln-drying of several species, apparently caused by the concentration and modification of extractives.

> *Sap stain.*—(*See* Blue stain.)

Starved joint.—(*See* Joint.)

Staypak.—Wood that is compressed in its natural state (that is, without resin or other chemical treatment) under controlled conditions of moisture, temperature, and pressure that practically eliminate springback or recovery from compression.

Stickers.—Strips or boards used to separate the layers of lumber in a pile and thus permit air to circulate between the layers.

Strength.—The term in its broader sense embraces collectively all the properties of wood that enable it to resist different forces or loads. In its more restricted sense, strength may apply to any one of the mechanical properties, in which event the name of the property under consideration should be stated, thus: strength in compression parallel to grain, strength in bending, hardness, and so on.

Strength ratio.—A ratio representing the strength of a piece of wood remaining after allowance is made for the maximum effect of the permitted knots, cross grain, shakes, or other strength-reducing features.

Stress.—Force per unit of area.

Basic stress.—The working stress for material free from strength-reducing features, such as knots, checks, and cross grain. It has in it all the factors appropriate to the nature of structural timber and the conditions under which it is used except those that are accounted for in the strength ratio.

Working stress.—Stress for use in the design of a wood member that is appropriate to the species and grade. It is obtained by multiplying the basic stress for the species and strength property by the strength ratio of the grade.

Structural lumber.—(*See* Lumber.)

Structural sandwich construction.—A layered construction comprising a combination of relatively high strength facing materials intimately bonded to and acting integrally with a low density core material.

Structural timbers.—Pieces of wood of relatively large size, the strength of which is the controlling element in their selection and use. Trestle timbers (stringers, caps, posts, sills, bracing, bridge ties, guard rails); car timbers (car framing, including upper framing, car sills); framing for buildings (posts, sills, girders, framing joints); ship timbers (ship timbers, ship decking); and crossarms for poles are examples of structural timbers.

Stud.—One of a series of slender wood structural members used as supporting elements in walls and partitions.

Straight grain.—(*See* Grain.)

Summerwood.—The portion of the annual growth ring that is formed after the springwood formation has ceased. It is usually denser and stronger mechanically than springwood.

Surfaced lumber.—(*See* Lumber.)

Symmetrical construction.—Plywood panels in which the plies on one side of a center ply or core are essentially equal in thickness, grain direction, properties, and arrangement to those on the other side of the core.

Tangential.—Strictly, coincident with a tangent at the circumference of a tree or log, or parallel to such a tangent. In practice, however, it often means roughly coincident with a growth ring. A tangential section is a longitudinal section through a tree or limb perpendicular to a radius. Flat-grained lumber is sawed tangentially.

Tension set.—A conditon of wood in which a group of fibers, owing to restraint imposed by adjoining fibers or by an external mechanical agency, are fixed or set in a condition of tension as a result of a restraint on normal shrinkage during a drop in moisture content.

Tension wood.—An abnormal form of wood found in leaning trees of some hardwood species and characterized by the presence of gelatinous fibers and excessive longitudinal shrinkage. Tension wood fibers hold together tenaciously, so that sawed surfaces usually have projecting fibers and planed surfaces often are torn or have raised grain. Tension wood may cause warping.

Texture.—A term often used interchangeably with grain. Sometimes used to combine the concepts of density and degree of contrast between springwood and summerwood. In this handbook texture refers to the finer structure of the wood (*see* Grain) rather than the annual rings.

Thermoplastic glues and resins.—Glues and resins that are cured by cooling from the heated condition but soften when subsequently subjected to high temperatures.

Thermosetting glues and resins.—Glues and resins that are cured with heat but do not soften when subsequently subjected to high temperatures.

Timber, standing.—Timber still on the stump.

Timbers.—(*See* Lumber.)

Tracheid. The elongated cells that constitute the greater part of the structure of the softwoods (frequently referred to as fibers). Also present in some hardwoods.

Transverse.—Directions in wood at right angles to the wood fibers. Includes radial and tangential directions. A transverse section is a section through a tree or timber at right angles to the pith.

Truss.—An assembly of members, such as beams, bars, rods, and the like, so combined as to form a rigid framework that cannot be deformed by the application of exterior force without deformation of one or more members.

Twist.—A distortion caused by the turning or winding of the edges of a board so that the four corners of any face are no longer in the same plane.

Tyloses.—Masses of parenchyma cells appearing somewhat like froth in the pores of some hardwoods, notably the white oak and black locust. Tyloses are formed by the extension of the cell wall of the living cells surrounding vessels of hardwood or sometimes in a similar manner by the extension of the cell wall into resin-passage cavities in the case of softwoods.

Vapor barrier.—A material with a high resistance to vapor movement, such as asphalt-impregnated paper, that is used in combination with insulation to control condensation.

Veneer.—A thin layer or sheet of wood cut on a veneer machine.

> *Rotary-cut veneer.*—Veneer cut in a lathe which rotates a log or bolt, chucked in the center, against a knife.
>
> *Sawed veneer.*—Veneer produced by sawing.
>
> *Sliced veneer.*—Veneer that is sliced off a log, bolt, or flitch with a knife.

Vertically laminated.—(*See* Laminated wood.)

Vertical grain.—(*See* Grain.)

Vessels.—Wood cells of comparatively large diameter that have open ends and are set one above the other so as to form continuous tubes. The openings of the vessels on the surface of a piece of wood are usually referred to as pores.

Virgin growth.—The original growth of mature trees.

Wane.—Bark or lack of wood from any cause on edge or corner of a piece.

Warp.—Any variation from a true or plane surface. Warp includes bow, crook, cup, and twist, or any combination thereof.

Wavy-grained wood.—(*See* Grain.)

Weathering.—The mechanical or chemical disintegration and discoloration of the surface of wood that is caused by exposure to light, the action of dust and sand carried by winds, and the alternate shrinking and swelling of the surface fibers with the continual variation in moisture content brought by changes in the weather. Weathering does not include decay.

Wet-bulb temperature.—The temperature indicated by the wet-bulb thermometer of a psychrometer (which see). Theoretically, the wet-bulb temperature is the temperature at which the atmosphere would become saturated by evaporation of water without loss or gain in total heat content of the air and vapor.

Wood flour.—Wood reduced to finely divided particles approximating those of cereal flours in size, appearance, and texture, and passing a 40–100 mesh screen.

Wood preservative.—(*See* Preservative.)

Wood substance.—The solid material of which wood is composed. It usually refers to the extractive-free solid substance of which the cell walls are composed, but this is not always true. There is no wide variation in chemical composition or specific gravity between the wood substance of various species, the characteristic differences of species being largely due to differences in infiltrated materials and variations in relative amounts of cell walls and cell cavities.

Workability.—The degree of ease and smoothness of cut obtainable with hand or machine tools.

Working stress.—(*See* Stress.)

Xylem.—The portion of the tree trunk, branches, and roots that lies between the pith and the cambium.

Yard brown stain.—(*See* Stain.)

Yard lumber.—(*See* Lumber.)

INDEX

www.ingramcontent.com/pod-product-compliance
Lightning Source LLC
Chambersburg PA
CBHW031804190326
41518CB00006B/197